SIDDHARTHA MUKHERJEE

Siddhartha Mukherjee is a cancer physician and researcher, a stem cell biologist and a cancer geneticist. He is the author of *The Laws of Medicine* and *The Emperor of All Maladies: A Biography of Cancer*, which won the 2011 Pulitzer Prize in general non-fiction and the Guardian First Book Award.

Mukherjee is an assistant professor of medicine at Columbia University. A Rhodes Scholar, he graduated from Stanford University, the University of Oxford, and Harvard Medical School. His laboratory has identified genes that regulate stem cell , and his team is internationally recognized for its discovery of skeletal stem cells and genetic alterations in blood cancers.

He has published work in *Nature, Cell, Neuron, The New England Journal of Medicine*, the *New* al other magazine and jour- with his family in New York City.

SIDDHARTHA MUKHERJEE

The Gene

An Intimate History

VINTAGE

1 3 5 7 9 10 8 6 4 2

Vintage
20 Vauxhall Bridge Road,
London SW1V 2SA

Vintage is part of the Penguin Random House group of companies
whose addresses can be found at global.penguinrandomhouse.com

Penguin
Random House
UK

First published in Vintage in 2017
First published in hardback by The Bodley Head in 2016
(First published in the United States by Scribner, an imprint of
Simon & Schuster Inc., in 2016)

Certain names and identifying characteristics have been changed.

penguin.co.uk/vintage

A CIP catalogue record for this book is available from the British Library

ISBN 9780099584575

Printed and bound by Clays Ltd, St Ives Plc

Penguin Random House is committed to a sustainable future
for our business, our readers and our planet. This book is made
from Forest Stewardship Council® certified paper.

MIX
Paper from
responsible sources
FSC
www.fsc.org   FSC® C018179

To Priyabala Mukherjee (1906–1985), who knew the perils;

to Carrie Buck (1906–1983), who experienced them.

An exact determination of the laws of heredity will probably work more change in man's outlook on the world, and in his power over nature, than any other advance in natural knowledge that can be foreseen.

—William Bateson

Human beings are ultimately nothing but carriers— passageways—for genes. They ride us into the ground like racehorses from generation to generation. Genes don't think about what constitutes good or evil. They don't care whether we are happy or unhappy. We're just means to an end for them. The only thing they think about is what is most efficient for them.

—Haruki Murakami, *1Q84*

Contents

THE
GENE

Prologue: Families

In the winter of 2012, I traveled from Delhi to Calcutta to visit my cousin Moni. My father accompanied me, as a guide and companion, but he was a sullen and brooding presence, lost in a private anguish that I could sense only dimly. My father is the youngest of five brothers, and Moni is his first-born nephew—the eldest brother's son. Since 2004, when he was forty, Moni has been confined to an institution for the mentally ill (a "lunatic home," as my father calls it), with a diagnosis of schizophrenia. He is kept densely medicated—awash in a sea of assorted antipsychotics and sedatives—and has an attendant watch, bathe, and feed him through the day.

My father has never accepted Moni's diagnosis. Over the years, he has waged a lonely countercampaign against the psychiatrists charged with his nephew's care, hoping to convince them that their diagnosis was a colossal error, or that Moni's broken psyche would somehow magically mend itself. My father has visited the institution in Calcutta twice—once without warning, hoping to see a transformed Moni, living a secretly normal life behind the barred gates.

But my father knew—and I knew—that there was more than just

avuncular love at stake for him in these visits. Moni is not the only member of my father's family with mental illness. Of my father's four brothers, two—not Moni's father, but two of Moni's uncles—suffered from various unravelings of the mind. Madness, it turns out, has been among the Mukherjees for at least two generations, and at least part of my father's reluctance to accept Moni's diagnosis lies in my father's grim recognition that some kernel of the illness may be buried, like toxic waste, in himself.

In 1946, Rajesh, my father's third-born brother, died prematurely in Calcutta. He was twenty-two years old. The story runs that he was stricken with pneumonia after spending two nights exercising in the winter rain—but the pneumonia was the culmination of another sickness. Rajesh had once been the most promising of the brothers—the nimblest, the supplest, the most charismatic, the most energetic, the most beloved and idolized by my father and his family.

My grandfather had died a decade earlier in 1936—he had been murdered following a dispute over mica mines—leaving my grandmother to raise five young boys. Although not the oldest, Rajesh had stepped rather effortlessly into his father's shoes. He was only twelve then, but he could have been twenty-two: his quick-fire intelligence was already being cooled by gravity, the brittle self-assuredness of adolescence already annealing into the self-confidence of adulthood.

But in the summer of '46, my father recalls, Rajesh had begun to behave oddly, as if a wire had been tripped in his brain. The most striking change in his personality was his volatility: good news triggered uncontained outbursts of joy, often extinguished only through increasingly acrobatic bouts of physical exercise, while bad news plunged him into inconsolable desolation. The emotions were normal in context; it was their extreme range that was abnormal. By the winter of that year, the sine curve of Rajesh's psyche had tightened in its frequency and gained in its amplitude. The fits of energy, tipping into rage and grandiosity, came often and more fiercely, and the sweeping undertow of grief that followed was just as strong. He ventured into the occult—organizing séances and planchette sessions at home, or meeting his friends to meditate at a crematorium at night. I don't know if he self-medicated—in the forties, the dens in Calcutta's Chinatown had ample supplies of opium from Burma and Afghani hashish to calm a young man's nerves—but my father recollects an altered brother: fearful at times, reckless at others, descending and ascending steep slopes of mood, irritable one morning and overjoyed the next (that word: *over-*

joyed. Used colloquially, it signals something innocent: an amplification of joy. But it also delineates a limit, a warning, an outer boundary of sobriety. Beyond *overjoy*, as we shall see, there is no *over-overjoy*; there is only madness and mania).

The week before the pneumonia, Rajesh had received news of a strikingly successful performance in his college exams and—elated—had vanished on a two-night excursion, supposedly "exercising" at a wrestling camp. When he returned, he was boiling up with a fever and hallucinating.

It was only years later, in medical school, that I realized that Rajesh was likely in the throes of an acute manic phase. His mental breakdown was the result of a near-textbook case of manic-depression—bipolar disorder.

①

Jagu—the fourth-born of my father's siblings—came to live with us in Delhi in 1975, when I was five years old. His mind was also crumbling. Tall and rail thin, with a slightly feral look in his eyes and a shock of matted, overgrown hair, he resembled a Bengali Jim Morrison. Unlike Rajesh, whose illness had surfaced in his twenties, Jagu had been troubled from childhood. Socially awkward, withdrawn to everyone except my grandmother, he was unable to hold a job or live by himself. By 1975, deeper cognitive problems had emerged: he had visions, phantasms, and voices in his head that told him what to do. He made up conspiracy theories by the dozens: a banana vendor who sold fruit outside our house was secretly recording Jagu's behavior. He often spoke to himself, with a particular obsession of reciting made-up train schedules ("Shimla to Howrah by Kalka mail, then transfer at Howrah to Shri Jagannath Express to Puri"). He was still capable of extraordinary bursts of tenderness—when I mistakenly smashed a beloved Venetian vase at home, he hid me in his bedclothes and informed my mother that he had "mounds of cash" stashed away that would buy "a thousand" vases in replacement. But this episode was symptomatic: even his love for me involved extending the fabric of his psychosis and confabulation.

Unlike Rajesh, who was never formally diagnosed, Jagu was. In the late 1970s, a physician saw him in Delhi and diagnosed him with schizophrenia. But no medicines were prescribed. Instead, Jagu continued to live at home, half-hidden away in my grandmother's room (as in many families in India, my grandmother lived with us). My grandmother—

besieged yet again, and now with doubled ferocity—assumed the role of public defender for Jagu. For nearly a decade, she and my father held a fragile truce between them, with Jagu living under her care, eating meals in her room and wearing clothes that she stitched for him. At night, when Jagu was particularly restless, consumed by his fears and fantasies, she put him to bed like a child, with her hand on his forehead. When she died in 1985, he vanished from our house and could not be persuaded to return. He lived with a religious sect in Delhi until his death in 1998.

①

Both my father and my grandmother believed that Jagu's and Rajesh's mental illnesses had been precipitated—even caused, perhaps—by the apocalypse of Partition, its political trauma sublimated into their psychic trauma. Partition, they knew, had split apart not just nations, but also minds; in Saadat Hasan Manto's "Toba Tek Singh"—arguably the best-known short story of Partition—the hero, a lunatic caught on the border between India and Pakistan, also inhabits a limbo between sanity and insanity. In Jagu's and Rajesh's case, my grandmother believed, the upheaval and uprooting from East Bengal to Calcutta had unmoored their minds, although in spectacularly opposite ways.

Rajesh had arrived in Calcutta in 1946, just as the city was itself losing sanity—its nerves fraying; its love depleted; its patience spent. A steady flow of men and women from East Bengal—those who had sensed the early political convulsions before their neighbors—had already started to fill the low-rises and tenements near Sealdah station. My grandmother was a part of this hardscrabble crowd: she had rented a three-room flat on Hayat Khan Lane, just a short walk from the station. The rent was fifty-five rupees a month—about a dollar in today's terms, but a colossal fortune for her family. The rooms, piled above each other like roughhousing siblings, faced a trash heap. But the flat, albeit minuscule, had windows and a shared roof from which the boys could watch a new city, and a new nation, being born. Riots were conceived easily on street corners; in August that year, a particularly ugly conflagration between Hindus and Muslims (later labeled the Great Calcutta Killing) resulted in the slaughtering of five thousand and left a hundred thousand evicted from their homes.

Rajesh had witnessed those rioting mobs in their tidal spate that summer. Hindus had dragged Muslims out of shops and offices in Lalbazar and gutted them alive on the streets, while Muslims had reciprocated, with equal and opposite ferocity, in the fish markets near Rajabazar and Harrison Road. Rajesh's mental breakdown had followed quickly on the heels of the riots. The city had stabilized and healed—but he had been left permanently scarred. Soon after the August massacres, he was hit by a volley of paranoid hallucinations. He grew increasingly fearful. The evening excursions to the gym became more frequent. Then came the manic convulsions, the spectral fevers, and the sudden cataclysm of his final illness.

If Rajesh's madness was the madness of arrival, then Jagu's madness, my grandmother was convinced, was the madness of departure. In his ancestral village of Dehergoti, near Barisal, Jagu's psyche had somehow been tethered to his friends and his family. Running wild in the paddy fields, or swimming in the puddles, he could appear as carefree and playful as any of the other kids—almost normal. In Calcutta, like a plant uprooted from its natural habitat, Jagu wilted and fell apart. He dropped out of college and parked himself permanently by one of the windows of the flat, looking blankly out at the world. His thoughts began to tangle, and his speech became incoherent. As Rajesh's mind was expanding to its brittle extreme, Jagu's contracted silently in his room. While Rajesh wandered the city at night, Jagu confined himself voluntarily at home.

Ⓓ

This strange taxonomy of mental illness (Rajesh as the town mouse and Jagu as the country mouse of psychic breakdown) was convenient while it lasted—but it shattered, finally, when Moni's mind also began to fail. Moni, of course, was not a "Partition child." He had never been uprooted; he had lived all his life in a secure home in Calcutta. Yet, uncannily, the trajectory of his psyche had begun to recapitulate Jagu's. Visions and voices had started to appear in his adolescence. The need for isolation, the grandiosity of the confabulations, the disorientation and confusion—these were all eerily reminiscent of his uncle's descent. In his teens, he had come to visit us in Delhi. We were supposed to go out to a film together, but he locked himself in our bathroom upstairs and refused to come out for nearly an hour, until my grandmother had ferreted him out. When she had found him inside, he was folded over in a corner, hiding.

In 2004, Moni was beaten up by a group of goons—allegedly for urinating in a public garden (he told me that an internal voice had commanded him, "Piss here; piss here"). A few weeks later, he committed a "crime" that was so comically egregious that it could only be a testament to his loss of sanity: he was caught flirting with one of the goon's sisters (again, he said that the voices had commanded him to act). His father tried, ineffectually, to intervene, but this time Moni was beaten up viciously, with a gashed lip and a wound in his forehead that precipitated a visit to the hospital.

The beating was meant to be cathartic (asked by the police, his tormentors later insisted that they had only meant to "drive the demons out of Moni")—but the pathological commanders in Moni's head only became bolder and more insistent. In the winter of that year, after yet another breakdown with hallucinations and hissing internal voices, he was institutionalized.

The confinement, as Moni told me, was partially voluntary: he was not seeking mental rehabilitation as much as a physical sanctuary. An assortment of antipsychotic medicines was prescribed, and he improved gradually—but never enough, apparently, to merit discharge. A few months later, with Moni still confined at the institution, his father died. His mother had already passed away years earlier, and his sister, his only other sibling, lived far away. Moni thus decided to remain in the institution, in part because he had nowhere else to go. Psychiatrists discourage the use of the archaic phrase *mental asylum*—but for Moni, the description had come to be chillingly accurate: this was the one place that offered him the shelter and safety that had been missing from his life. He was a bird that had voluntarily caged itself.

When my father and I visited him in 2012, I had not seen Moni for nearly two decades. Even so, I had expected to recognize him. But the person I met in the visiting room bore such little resemblance to my memory of my cousin that—had his attendant not confirmed the name—I could easily have been meeting a stranger. He had aged beyond his years. At forty-eight, he looked a decade older. The schizophrenia medicines had altered his body and he walked with the uncertainty and imbalance of a child. His speech, once effusive and rapid, was hesitant and fitful; the words emerged with a sudden, surprising force, as if he were spitting out strange pips of food that had been put into his mouth. He had little memory of my father, or me. When I mentioned my sister's name, he asked me if I had married her. Our conversation proceeded as if I were a newspaper reporter who had dropped out of the blue to interview him.

The most striking feature of his illness, though, was not the storm within his mind, but the lull in his eyes. The word *moni* means "gem" in Bengali, but in common usage it also refers to something ineffably beautiful: the shining pinpricks of light in each eye. But this, precisely, was what had gone missing in Moni. The twin points of light in his eyes had dulled and nearly vanished, as if someone had entered his eyes with a minute paintbrush and painted them gray.

℧

Throughout my childhood and adult life, Moni, Jagu, and Rajesh played an outsize role in my family's imagination. During a six-month flirtation with teenage angst, I stopped speaking to my parents, refused to turn in homework, and threw my old books in the trash. Anxious beyond words, my father dragged me glumly to see the doctor who had diagnosed Jagu. *Was his son now losing his mind?* As my grandmother's memory failed in the early eighties, she began to call me Rajeshwar—Rajesh—by mistake. She would correct herself at first, in a hot blush of embarrassment, but as she broke her final bonds with reality, she seemed to make the mistake almost willingly, as if she had discovered the illicit pleasure of that fantasy. When I met Sarah, now my wife, for the fourth or fifth time, I told her about the splintered minds of my cousin and two uncles. It was only fair to a future partner that I should come with a letter of warning.

By then, heredity, illness, normalcy, family, and identity had become recurrent themes of conversation in my family. Like most Bengalis, my parents had elevated repression and denial to a high art form, but even so, questions about this particular history were unavoidable. Moni; Rajesh; Jagu: three lives consumed by variants of mental illness. It was hard not to imagine that a hereditary component lurked behind this family history. Had Moni inherited a gene, or a set of genes, that had made him susceptible—the same genes that had affected our uncles? Had others been affected with different variants of mental illness? My father had had at least two psychotic fugues in his life—both precipitated by the consumption of *bhang* (mashed-up cannabis buds, melted in ghee, and churned into a frothing drink for religious festivals). Were these related to the same scar of history?

℗

In 2009, Swedish researchers published an enormous international study, involving thousands of families and tens of thousands of men and women. By analyzing families that possessed intergenerational histories of mental illness, the study found striking evidence that bipolar disorder and schizophrenia shared a strong genetic link. Some of the families described in the study possessed a crisscrossing history of mental illness achingly similar to my own: one sibling affected with schizophrenia, another with bipolar disorder, and a nephew or niece who also had schizophrenia. In 2012, several further studies corroborated these initial findings, strengthening the links between these variants of mental illness and family histories and deepening questions about their etiology, epidemiology, triggers, and instigators.

I read two of these studies on a winter morning on the subway in New York, a few months after returning from Calcutta. Across the aisle, a man in a gray fur hat was pinning down his son to put a gray fur hat on him. At Fifty-Ninth Street, a mother wheeled in a stroller with twins emitting, it seemed to my ears, identically pitched screams.

The study provided a strange interior solace—answering some of the questions that had so haunted my father and grandmother. But it also provoked a volley of new questions: If Moni's illness was genetic, then why had his father and sister been spared? What "triggers" had unveiled these predispositions? How much of Jagu's or Moni's illnesses arose from "nature" (i.e., genes that predisposed to mental illness) versus "nurture" (environmental triggers such as upheaval, discord, and trauma)? Might my father carry the susceptibility? Was I a carrier as well? What if I could know the precise nature of this genetic flaw? Would I test myself, or my two daughters? Would I inform them of the results? What if only one of them turned out to carry that mark?

℗

While my family's history of mental illness was cutting through my consciousness like a red line, my scientific work as a cancer biologist was also converging on the normalcy and abnormalcy of genes. Cancer, perhaps, is an ultimate perversion of genetics—a genome that becomes pathologically obsessed with replicating itself. The genome-as-self-replicating-machine co-opts the physiology of a cell, resulting in a shape-shifting illness that, despite significant advances, still defies our ability to treat or cure it.

But to study cancer, I realized, is to also study its obverse. What is the code of normalcy before it becomes corrupted by cancer's coda? What does the normal genome *do*? How does it maintain the constancy that makes us discernibly similar, and the variation that makes us discernibly different? How, for that matter, is constancy versus variation, or normalcy versus abnormalcy, defined or written into the genome?

And what if we learned to change our genetic code intentionally? If such technologies were available, who would control them, and who would ensure their safety? Who would be the masters, and who the victims, of this technology? How would the acquisition and control of this knowledge—and its inevitable invasion of our private and public lives—alter the way we imagine our societies, our children, and ourselves?

⟊

This book is the story of the birth, growth, and future of one of the most powerful and dangerous ideas in the history of science: the "gene," the fundamental unit of heredity, and the basic unit of all biological information.

I use that last adjective—*dangerous*—with full cognizance. Three profoundly destabilizing scientific ideas ricochet through the twentieth century, trisecting it into three unequal parts: the atom, the byte, the gene. Each is foreshadowed by an earlier century, but dazzles into full prominence in the twentieth. Each begins its life as a rather abstract scientific concept, but grows to invade multiple human discourses—thereby transforming culture, society, politics, and language. But the most crucial parallel between the three ideas, by far, is conceptual: each represents the irreducible unit—the building block, the basic organizational unit—of a larger whole: the atom, of matter; the byte (or "bit"), of digitized information; the gene, of heredity and biological information.*

Why does this property—being the least divisible unit of a larger form—imbue these particular ideas with such potency and force? The simple answer is that matter, information, and biology are inherently hierarchically organized: understanding that smallest part is crucial to understanding the whole. When the poet Wallace Stevens writes, "In the sum of the parts, there are only the parts," he is referring to the deep structural mystery that runs through language: you can only decipher the meaning of a sentence by deciphering every individual word—yet a sentence carries more meaning than any of the individual words. And so it is with genes. An organism is much more than its genes, of course, but to understand an organism, you must first understand its genes. When the Dutch biologist Hugo de Vries encountered the concept of the gene in the 1890s, he quickly intuited that the idea would reorganize our understanding of the natural world. "The whole organic world is the result of innumerable different combinations and permutations of relatively few factors. . . . Just as physics and chemistry go back to molecules and atoms, the biological sciences have to penetrate these units [genes] in order to explain . . . the phenomena of the living world."

The atom, the byte, and the gene provide fundamentally new scientific and technological understandings of their respective systems. You cannot explain the behavior of matter—why gold gleams; why hydrogen combusts with oxygen—without invoking the atomic nature of matter. Nor can you understand the complexities of computing—the nature of algorithms, or the storage or corruption of data—without comprehending the structural anatomy of digitized information. "Alchemy could not become chemistry until its fundamental units were discovered," a

---

* By *byte* I am referring to a rather complex idea—not only to the familiar byte of computer architecture, but also to a more general and mysterious notion that *all* complex information in the natural world can be described or encoded as a summation of discrete parts, containing no more than an "on" and "off" state. A more thorough description of this idea, and its impact on natural sciences and philosophy, might be found in *Information: A History, a Theory, a Flood* by James Gleick. This theory was most evocatively proposed by the physicist John Wheeler in the 1990s: "Every particle, every field of force, even the space-time continuum itself—derives its function, its meaning, its very existence entirely . . . from answers to yes-or-no questions, binary choices, bits . . . ; in short, that all things physical are information-theoretic in origin." The byte or bit is a man-made invention, but the theory of digitized information that underlies it is a beautiful natural law.

nineteenth-century scientist wrote. By the same token, as I argue in this book, it is impossible to understand organismal and cellular biology or evolution—or human pathology, behavior, temperament, illness, race, and identity or fate—without first reckoning with the concept of the gene.

There is a second issue at stake here. Understanding atomic science was a necessary precursor to manipulating matter (and, via the manipulation of matter, to the invention of the atomic bomb). Our understanding of genes has allowed us to manipulate organisms with unparalleled dexterity and power. The actual nature of the genetic code, it turns out, is astoundingly simple: there's just one molecule that carries our hereditary information and just one code. "That the fundamental aspects of heredity should have turned out to be so extraordinarily simple supports us in the hope that nature may, after all, be entirely approachable," Thomas Morgan, the influential geneticist, wrote. "Her much-advertised inscrutability has once more been found to be an illusion."

Our understanding of genes has reached such a level of sophistication and depth that we are no longer studying and altering genes in test tubes, but in their native context in human cells. Genes reside on chromosomes—long, filamentous structures buried within cells that contain tens of thousands of genes linked together in chains.* Humans have forty-six such chromosomes in total—twenty-three from one parent and twenty-three from another. The entire set of genetic instructions carried by an organism is termed a *genome* (think of the genome as the encyclopedia of all genes, with footnotes, annotations, instructions, and references). The human genome contains about between twenty-one and twenty-three thousand genes that provide the master instructions to build, repair, and maintain humans. Over the last two decades, genetic technologies have advanced so rapidly that we can decipher how several of these genes operate in time and space to enable these complex functions. And we can, on occasion, deliberately alter some of these genes to change their functions, thereby resulting in altered human states, altered physiologies, and changed beings.

This transition—from explanation to manipulation—is precisely what makes the field of genetics resonate far beyond the realms of science. It is one thing to try to understand how genes influence human identity or

---

* In certain bacteria, chromosomes can be circular.

sexuality or temperament. It is quite another thing to imagine altering identity or sexuality or behavior by altering genes. The former thought might preoccupy professors in departments of psychology, and their colleagues in the neighboring departments of neuroscience. The latter thought, inflected with both promise and peril, should concern us all.

Ⓘ

As I write this, organisms endowed with genomes are learning to change the heritable features of organisms endowed with genomes. I mean the following: in just the last four years—between 2012 and 2016—we have invented technologies that allow us to change human genomes intentionally and permanently (although the safety and fidelity of these "genomic engineering" technologies still need to be carefully evaluated). At the same time, the capacity to predict the future fate of an individual from his or her genome has advanced dramatically (although the true predictive capacities of these technologies still remain unknown). We can now "read" human genomes, and we can "write" human genomes in a manner inconceivable just three or four years ago.

It hardly requires an advanced degree in molecular biology, philosophy, or history to note that the convergence of these two events is like a headlong sprint into an abyss. Once we can understand the nature of fate encoded by individual genomes (even if we can predict this in likelihoods rather than in certainties) and once we acquire the technology to intentionally change these likelihoods (even if these technologies are inefficient and cumbersome) our future is fundamentally changed. George Orwell once wrote that whenever a critic uses the word *human*, he usually renders it meaningless. I doubt that I am overstating the case here: our capacity to understand and manipulate human genomes alters our conception of what it means to be "human."

The atom provides an organizing principle for modern physics—and it tantalizes us with the prospect of controlling matter and energy. The gene provides an organizing principle for modern biology—and it tantalizes us with the prospect of controlling our bodies and fates. Embedded in the history of the gene is "the quest for eternal youth, the Faustian myth of abrupt reversal of fortune, and our own century's flirtation with the perfectibility of man." Embedded, equally, is the desire to decipher our manual of instructions. *That* is what is at the center of this story.

①

This book is organized both chronologically and thematically. The overall arc is historical. We begin in Mendel's pea-flower garden, in an obscure Moravian monastery in 1864, where the "gene" is discovered and then quickly forgotten (the word *gene* only appears decades later). The story intersects with Darwin's theory of evolution. The gene entrances English and American reformers, who hope to manipulate human genetics to accelerate human evolution and emancipation. That idea escalates to its macabre zenith in Nazi Germany in the 1940s, where human eugenics is used to justify grotesque experiments, culminating in confinement, sterilization, euthanasia, and mass murder.

A chain of post–World War II discoveries launches a revolution in biology. DNA is identified as the source of genetic information. The "action" of a gene is described in mechanistic terms: *genes encode chemical messages to build proteins that ultimately enable form and function.* James Watson, Francis Crick, Maurice Wilkins, and Rosalind Franklin solve the three-dimensional structure of DNA, producing the iconic image of the double helix. The three-letter genetic code is deciphered.

Two technologies transform genetics in the 1970s: gene sequencing and gene cloning—the "reading" and "writing" of genes (the phrase *gene cloning* encompasses the gamut of techniques used to extract genes from organisms, manipulate them in test tubes, create gene hybrids, and produce millions of copies of such hybrids in living cells.) In the 1980s, human geneticists begin to use these techniques to map and identify genes linked to diseases, such as Huntington's disease and cystic fibrosis. The identification of these disease-linked genes augurs a new era of genetic management, enabling parents to screen fetuses, and potentially abort them if they carry deleterious mutations (any person who has tested their unborn child for Down syndrome, cystic fibrosis, or Tay-Sachs disease, or has been tested herself for, say, *BRCA1* or *BRCA2* has already entered this era of genetic diagnosis, management, and optimization. This is not a story of our distant future; it is already embedded in our present).

Multiple genetic mutations are identified in human cancers, leading to a deeper genetic understanding of that disease. These efforts reach their crescendo in the Human Genome Project, an international project to map and sequence the entire human genome. The draft sequence of the human genome is published in 2001. The genome project, in turn, inspires attempts to understand human variation and "normal" behavior in terms of genes.

The gene, meanwhile, invades discourses concerning race, racial discrimination, and "racial intelligence," and provides startling answers to some of the most potent questions coursing through our political and cultural realms. It reorganizes our understanding of sexuality, identity, and choice, thus piercing the center of some of the most urgent questions coursing through our personal realms.*

There are stories within each of these stories, but this book is also a very personal story—an intimate history. The weight of heredity is not an abstraction for me. Rajesh and Jagu are dead. Moni is confined to a mental institution in Calcutta. But their lives and deaths have had a greater impact on my thinking as a scientist, scholar, historian, physician, son, and father than I could possibly have envisioned. Scarcely a day passes in my adult life when I do not think about inheritance and family.

Most important, I owe a debt to my grandmother. She did not—she could not—outlive the grief of her inheritance, but she embraced and defended the most fragile of her children from the will of the strong. She weathered the buffets of history with resilience—but she weathered the buffets of heredity with something more than resilience: a grace that we, as her descendants, can only hope to emulate. It is to her that this book is dedicated.

---

* Some topics, such as genetically modified organisms (GMOs), the future of gene patents, the use of genes for drug discovery or biosynthesis, and the creation of new genetic species merit books in their own right, and lie outside the purview of this volume.

PART ONE

THE "MISSING SCIENCE OF HEREDITY"

The Discovery and Rediscovery of Genes
(1865–1935)

①

⟨①⟩

*This missing science of heredity, this unworked mine of
knowledge on the borderland of biology and anthropology,
which for all practical purposes is as unworked now as
it was in the days of Plato, is, in simple truth, ten times
more important to humanity than all the chemistry and
physics, all the technical and industrial science that ever
has been or ever will be discovered.*

—Herbert G. Wells, *Mankind in the Making*

JACK: *Yes, but you said yourself that a severe chill was
not hereditary.*
ALGERNON: *It usen't to be, I know—but I daresay it is
now. Science is always making wonderful improvements
in things.*

—Oscar Wilde, *The Importance of Being Earnest*

⟨①⟩

The Walled Garden

*The students of heredity, especially, understand all of
their subject except their subject. They were, I suppose,
bred and born in that brier-patch, and have really
explored it without coming to the end of it. That is, they
have studied everything but the question of what they
are studying.*
—G. K. Chesterton, *Eugenics and Other Evils*

Ask the plants of the earth, and they will teach you.
—Job 12:8

The monastery was originally a nunnery. The monks of Saint Augustine's
Order had once lived—as they often liked to grouse—in more lavish circumstances in the ample rooms of a large stone abbey on the top of a hill
in the heart of the medieval city of Brno (Brno in Czech, Brünn in German). The city had grown around them over four centuries, cascading
down the slopes and then sprawling out over the flat landscape of farms
and meadowlands below. But the friars had fallen out of favor with Emperor Joseph II in 1783. The midtown real estate was far too valuable
to house them, the emperor had decreed bluntly—and the monks were
packed off to a crumbling structure at the bottom of the hill in Old Brno,
the ignominy of the relocation compounded by the fact that they had
been assigned to live in quarters originally designed for women. The halls
had the vague animal smell of damp mortar, and the grounds were overgrown with grass, bramble, and weeds. The only perk of this fourteenth-
century building—as cold as a meathouse and as bare as a prison—was a
rectangular garden with shade trees, stone steps, and a long alley, where
the monks could walk and think in isolation.

The friars made the best of the new accommodations. A library was restored on the second floor. A study room was connected to it and out-fitted with pine reading desks, a few lamps, and a growing collection of nearly ten thousand books, including the latest works of natural history, geology, and astronomy (the Augustinians, fortunately, saw no conflict between religion and most science; indeed, they embraced science as yet another testament of the workings of the divine order in the world). A wine cellar was carved out below, and a modest refectory vaulted above it. One-room cells, with the most rudimentary wooden furniture, housed the inhabitants on the second floor.

In October 1843, a young man from Silesia, the son of two peasants, joined the abbey. He was a short man with a serious face, myopic, and tending toward portliness. He professed little interest in the spiritual life—but was intellectually curious, good with his hands, and a natural gardener. The monastery provided him with a home, and a place to read and learn. He was ordained on August 6, 1847. His given name was Johann, but the friars changed it to Gregor Johann Mendel.

For the young priest in training, life at the monastery soon settled into a predictable routine. In 1845, as part of his monastic education, Mendel attended classes in theology, history, and natural sciences at Brno's Theological College. The tumult of 1848—the bloody populist revolutions that swept fiercely through France, Denmark, Germany, and Austria and over-turned social, political, and religious orders—largely passed him by, like distant thunder. Nothing about Mendel's early years suggested even the faintest glimmer of the revolutionary scientist who would later emerge. He was disciplined, plodding, deferential—a man of habits among men in habits. His only challenge to authority, it seemed, was his occasional refusal to wear the scholar's cap to class. Admonished by his superiors, he politely complied.

In the summer of 1848, Mendel began work as a parish priest in Brno. He was, by all accounts, terrible at the job. "Seized by an unconquerable timidity," as the abbot described it, Mendel was tongue-tied in Czech (the language of most parishioners), uninspiring as a priest, and too neurotic to bear the emotional brunt of the work among the poor. Later that year, he schemed a perfect way out: he applied for a job to teach mathematics, natural sciences, and elementary Greek at the Znaim High School. With a helpful nudge from the abbey, Mendel was selected—although there was a catch. Knowing that he had never been trained as a teacher, the school asked Mendel to sit for the formal examination in the natural sciences for high school teachers.

In the late spring of 1850, an eager Mendel took the written version of the exam in Brno. He failed—with a particularly abysmal performance in geology ("arid, obscure and hazy," one reviewer complained of Mendel's writing on the subject). On July 20, in the midst of an enervating heat wave in Austria, he traveled from Brno to Vienna to take the oral part of the exam. On August 16, he appeared before his examiners to be tested in the natural sciences. This time, his performance was even worse—in biology. Asked to describe and classify mammals, he scribbled down an incomplete and absurd system of taxonomy—omitting categories, inventing others, lumping kangaroos with beavers, and pigs with elephants. "The candidate seems to know nothing about technical terminology, naming all animals in colloquial German, and avoiding systematic nomenclature," one of the examiners wrote. Mendel failed again.

In August, Mendel returned to Brno with his exam results. The verdict from the examiners had been clear: if Mendel was to be allowed to teach, he needed additional education in the natural sciences—more advanced training than the monastery library, or its walled garden, could provide. Mendel applied to the University of Vienna to pursue a degree in the natural sciences. The abbey intervened with letters and pleas; Mendel was accepted.

In the winter of 1851, Mendel boarded the train to enroll in his classes at the university. It was here that Mendel's problems with biology—and biology's problems with Mendel—would begin.

☾

The night train from Brno to Vienna slices through a spectacularly bleak landscape in the winter—the farmlands and vineyards buried in frost, the canals hardened into ice-blue venules, the occasional farmhouse blanketed in the locked darkness of Central Europe. The river Thaya crosses the land, half-frozen and sluggish; the islands of the Danube come into view. It is a distance of only ninety miles—a journey of about four hours in Mendel's time. But the morning of his arrival, it was as if Mendel had woken up in a new cosmos.

In Vienna, science was crackling, electric—alive. At the university, just a few miles from his back-alley boardinghouse on Invalidenstrasse,

Mendel began to experience the intellectual baptism that he had so ardently sought in Brno. Physics was taught by Christian Doppler, the redoubtable Austrian scientist who would become Mendel's mentor, teacher, and idol. In 1842, Doppler, a gaunt, acerbic thirty-nine-year-old, had used mathematical reasoning to argue that the pitch of sound (or the color of light) was not fixed, but depended on the location and velocity of the observer. Sound from a source speeding toward a listener would become compressed and register at a higher pitch, while sound speeding away would be heard with a drop in its pitch. Skeptics had scoffed: How could the same light, emitted from the same lamp, be registered as different colors by different viewers? But in 1845, Doppler had loaded a train with a band of trumpet players and asked them to hold a note as the train sped forward. As the audience on the platform listened in disbelief, a higher note came from the train as it approached, and a lower note emanated as it sped away.

Sound and light, Doppler argued, behaved according to universal and natural laws—even if these were deeply counterintuitive to ordinary viewers or listeners. Indeed, if you looked carefully, all the chaotic and complex phenomena of the world were the result of highly organized natural laws. Occasionally, our intuitions and perceptions might allow us to grasp these natural laws. But more commonly, a profoundly artificial experiment—loading trumpeters on a speeding train—might be necessary to understand and demonstrate these laws.

Doppler's demonstrations and experiments captivated Mendel as much as they frustrated him. Biology, his main subject, seemed to be a wild, overgrown garden of a discipline, lacking any systematic organizing principles. Superficially, there seemed to be a profusion of order—or rather a profusion of Orders. The reigning discipline in biology was taxonomy, an elaborate attempt to classify and subclassify all living things into distinct categories: Kingdoms, Phylae, Classes, Orders, Families, Genera, and Species. But these categories, originally devised by the Swedish botanist Carl Linnaeus in the mid-1700s, were purely descriptive, not mechanistic. The system described how to categorize living things on the earth, but did not ascribe an underlying logic to its organization. Why, a biologist might ask, were living things categorized in *this* manner? What maintained its constancy or fidelity: What kept elephants from morphing into pigs, or kangaroos into beavers? What was the mechanism of heredity? Why, or how, did like beget like?

①

The question of "likeness" had preoccupied scientists and philosophers for centuries. Pythagoras, the Greek scholar—half scientist, half mystic—who lived in Croton around 530 BC, proposed one of the earliest and most widely accepted theories to explain the similarity between parents and their children. The core of Pythagoras's theory was that hereditary information ("likeness") was principally carried in male semen. Semen collected these instructions by coursing through a man's body and absorbing mystical vapors from each of the individual parts (the eyes contributed their color, the skin its texture, the bones their length, and so forth). Over a man's life, his semen grew into a mobile library of every part of the body—a condensed distillate of the self.

This self-information—seminal, in the most literal sense—was transmitted into a female body during intercourse. Once inside the womb, semen matured into a fetus via nourishment from the mother. In reproduction (as in any form of production) men's work and women's work were clearly partitioned, Pythagoras argued. The father provided the essential information to create a fetus. The mother's womb provided nutrition so that this data could be transformed into a child. The theory was eventually called spermism, highlighting the central role of the sperm in determining all the features of a fetus.

In 458 BC, a few decades after Pythagoras's death, the playwright Aeschylus used this odd logic to provide one of history's most extraordinary legal defenses of matricide. The central theme of Aeschylus's *Eumenides* is the trial of Orestes, the prince of Argos, for the murder of his mother, Clytemnestra. In most cultures, matricide was perceived as an ultimate act of moral perversion. In *Eumenides*, Apollo, chosen to represent Orestes in his murder trial, mounts a strikingly original argument: he reasons that Orestes's mother is no more than a stranger to him. A pregnant woman is just a glorified human incubator, Apollo argues, an intravenous bag dripping nutrients through the umbilical cord into her child. The true forebear of all humans is the father, whose sperm carries "likeness." "Not the true parent is the woman's womb that bears the child," Apollo tells a sympathetic council of jurors. "She doth but nurse the seed, new-sown. The male is parent. She for him—as stranger for a stranger—just hoards the germ of life."

The evident asymmetry of this theory of inheritance—the male supplying all the "nature" and the female providing the initial "nurture" in her womb—didn't seem to bother Pythagoras's followers; indeed, they may have found it rather pleasing. Pythagoreans were obsessed with the mystical geometry of triangles. Pythagoras had learned the triangle theorem—that the length of the third side of a right-angled triangle can be deduced mathematically from the length of the other two sides—from Indian or Babylonian geometers. But the theorem became inextricably attached to his name (henceforth called the Pythagorean theorem), and his students offered it as proof that such secret mathematical patterns— "harmonies"—were lurking everywhere in nature. Straining to see the world through triangle-shaped lenses, Pythagoreans argued that in heredity too a triangular harmony was at work. The mother and the father were two independent sides and the child was the third—the biological hypotenuse to the parents' two lines. And just as a triangle's third side could arithmetically be derived from the two other sides using a strict mathematical formula, so was a child derived from the parents' individual contributions: nature from father and nurture from mother.

A century after Pythagoras's death, Plato, writing in 380 BC, was captivated by this metaphor. In one of the most intriguing passages in *The Republic*—borrowed, in part, from Pythagoras—Plato argued that if children were the arithmetic derivatives of their parents, then, at least in principle, the formula could be hacked: perfect children could be derived from perfect combinations of parents breeding at perfectly calibrated times. A "theorem" of heredity existed; it was merely waiting to be known. By unlocking the theorem and then enforcing its prescriptive combinations, any society could guarantee the production of the fittest children—unleashing a sort of numerological eugenics: "For when your guardians are ignorant of the law of births, and unite bride and bridegroom out of season, the children will not be goodly or fortunate," Plato concluded. The guardians of his republic, its elite ruling class, having deciphered the "law of births," would ensure that only such harmonious "fortunate" unions would occur in the future. A political utopia would develop as a consequence of genetic utopia.

①

It took a mind as precise and analytical as Aristotle's to systematically dismantle Pythagoras's theory of heredity. Aristotle was not a particularly ardent champion of women, but he nevertheless believed in using evidence as the basis of theory building. He set about dissecting the merits and problems of "spermism" using experimental data from the biological world. The result, a compact treatise titled *Generation of Animals*, would serve as a foundational text for human genetics just as Plato's *Republic* was a founding text for political philosophy.

Aristotle rejected the notion that heredity was carried exclusively in male semen or sperm. He noted, astutely, that children can inherit features from their mothers and grandmothers (just as they inherit features from their fathers and grandfathers), and that these features can even skip generations, disappearing for one generation and reappearing in the next. "And from deformed [parents] deformed [offspring] comes to be," he wrote, "just as lame come to be from lame and blind from blind, and in general they resemble often the features that are against nature, and have inborn signs such as growths and scars. Some of such features have even been transmitted through three [generations]: for instance, someone who had a mark on his arm and his son was born without it, but his grandson had black in the same place, but in a blurred way. . . . In Sicily a woman committed adultery with a man from Ethiopia; the daughter did not become an Ethiopian, but her [grand]daughter did." A grandson could be born with his grandmother's nose or her skin color, without that feature being visible in either parent—a phenomenon virtually impossible to explain in terms of Pythagoras's scheme of purely patrilineal heredity.

Aristotle challenged Pythagoras's "traveling library" notion that semen collected hereditary information by coursing through the body and obtaining secret "instructions" from each individual part. "Men generate before they yet have certain characters, such as a beard or grey hair," Aristotle wrote perceptively—but they pass on those features to their children. Occasionally, the feature transmitted through heredity was not even corporeal: a manner of walking, say, or a way of staring into space, or even a state of mind. Aristotle argued that such traits—not material to start with—could not materialize into semen. And finally, and perhaps more obviously, he attacked Pythagoras's scheme with the most self-evident of arguments: it could not possibly account for female anatomy. How could a father's sperm "absorb" the instructions to produce his daughter's "generative parts," Aristotle asked, when none of these parts was to be found anywhere in the father's body? Pythagoras's theory could explain every aspect of genesis except the most crucial one: genitals.

Aristotle offered an alternative theory that was strikingly radical for its time: perhaps females, like males, contribute actual material to the fetus—a form of female semen. And perhaps the fetus is formed by the *mutual* contributions of male and female parts. Grasping for analogies, Aristotle called the male contribution a "principle of movement." "Movement," here, was not literally motion, but instruction, or information—*code*, to use a modern formulation. The actual material exchanged during intercourse was merely a stand-in for a more obscure and mysterious exchange. Matter, in fact, didn't really matter; what passed from man to woman was not matter, but *message*. Like an architectural plan for a building, or like a carpenter's handiwork to a piece of wood, male semen carried the instructions to build a child. "[Just as] no material part comes from the carpenter to the wood in which he works," Aristotle wrote, "but the shape and the form are imparted from him to the material by means of the motion he sets up. . . . In like manner, Nature uses the semen as a tool."

Female semen, in contrast, contributed the physical raw material for the fetus—wood for the carpenter, or mortar for the building: the stuff and the stuffing of life. Aristotle argued that the actual material provided by females was menstrual blood. Male semen sculpted menstrual blood into the shape of a child (the claim might sound outlandish today, but here too Aristotle's meticulous logic was at work. Since the disappearance of menstrual blood is coincident with conception, Aristotle assumed that the fetus must be made from it).

Aristotle was wrong in his partitioning of male and female contributions into "material" and "message," but abstractly, he had captured one of the essential truths about the nature of heredity. The transmission of heredity, as Aristotle perceived it, was essentially the transmission of information. Information was then used to build an organism from scratch: message *became* material. And when an organism matured, it generated male or female semen again—transforming material back to message. In fact, rather than Pythagoras's triangle, there was a circle, or a cycle, at work: form begat information, and then information begat form. Centuries later, the biologist Max Delbrück would joke that Aristotle should have been given the Nobel Prize posthumously—for the discovery of DNA.

☉

But if heredity was transmitted as information, then how was that information encoded? The word *code* comes from the Latin *caudex*, the wooden pith of a tree on which scribes carved their writing. What, then, was the caudex of heredity? What was being transcribed, and how? How was the material packaged and transported from one body to the next? Who encrypted the code, and who translated it, to create a child?

The most inventive solution to these questions was the simplest: it dispensed of code altogether. Sperm, this theory argued, *already* contained a minihuman—a tiny fetus, fully formed, shrunken and curled into a minuscule package and waiting to be progressively inflated into a baby. Variations of this theory appear in medieval myths and folklore. In the 1520s, the Swiss-German alchemist Paracelsus used the minihuman-in-sperm theory to suggest that human sperm, heated with horse dung and buried in mud for the forty weeks of normal conception, would eventually grow into a human, although with some monstrous characteristics. The conception of a normal child was merely the transfer of this minihuman—the homunculus—from the father's sperm into the mother's womb. In the womb, the minihuman was expanded to the size of the fetus. There was no code; there was only miniaturization.

The peculiar charm of this idea—called *preformation*—was that it was infinitely recursive. Since the homunculus had to mature and produce its own children, it had to have preformed mini-homunculi lodged inside it—tiny humans encased inside humans, like an infinite series of Russian dolls, a great chain of beings that stretched all the way backward from the present to the first man, to Adam, and forward into the future. For medieval Christians, the existence of such a chain of humans provided a most powerful and original understanding of original sin. Since all future humans were encased within all humans, each of us had to have been physically present inside Adam's body—"floating . . . in our First Parent's loins," as one theologian described—during his crucial moment of sin. Sinfulness, therefore, was embedded within us thousands of years before we were born—from Adam's loins directly to his line. All of us bore its taint—not because our distant ancestor had been tempted in that distant garden, but because each of us, lodged in Adam's body, had actually tasted the fruit.

The second charm of preformation was that it dispensed of the problem of de-encryption. Even if early biologists could fathom encryption—the conversion of a human body into some sort of code (by osmosis, à la Pythagoras)—the reverse act, deciphering that code *back* into a human being, completely boggled the mind. How could something as complex as a human form emerge out of the union of sperm and egg? The homunculus dispensed of this conceptual problem. If a child came already preformed, then its formation was merely an act of expansion—a biological version of a blowup doll. No key or cipher was required for the deciphering. The genesis of a human being was just a matter of adding water.

The theory was so seductive—so artfully vivid—that even the invention of the microscope was unable to deal the expected fatal blow to the homunculus. In 1694, Nicolaas Hartsoeker, the Dutch physicist and microscopist, conjured a picture of such a minibeing, its enlarged head twisted in fetal position and curled into the head of a sperm. In 1699, another Dutch microscopist claimed to have found homuncular creatures floating abundantly in human sperm. As with any anthropomorphic fantasy—finding human faces on the moon, say—the theory was only magnified by the lenses of imagination: pictures of homunculi proliferated in the seventeenth century, with the sperm's tail reconceived into a filament of human hair, or its cellular head visualized as a tiny human skull. By the end of the seventeenth century, preformation was considered the most logical and consistent explanation for human and animal heredity. Men came from small men, as large trees came from small cuttings. "In nature there is no generation," the Dutch scientist Jan Swammerdam wrote in 1669, "but only propagation."

①

But not everyone could be convinced that miniature humans were infinitely encased inside humans. The principal challenge to preformation was the idea that something had to happen during embryogenesis that led to the formation of entirely *new* parts in the embryo. Humans did not come pre-shrunk and premade, awaiting only expansion. They had to be generated from scratch, using specific instructions locked inside the sperm and egg. Limbs, torsos, brains, eyes, faces—even temperaments or propensities that were inherited—had to be created anew each time an embryo unfurled into a human fetus. Genesis happened . . . well—by genesis.

By what impetus, or instruction, was the embryo, and the final organism, generated from sperm and egg? In 1768, the Berlin embryologist Caspar Wolff tried to finesse an answer by concocting a guiding principle—*vis essentialis corporis*, as he called it—that progressively shepherded the maturation of a fertilized egg into a human form. Like Aristotle, Wolff imagined that the embryo contained some sort of encrypted information—*code*—that was not merely a miniature version of a human, but instructions to make a human from scratch. But aside from inventing a Latinate name for a vague principle, Wolff could provide no further specifics. The instructions, he argued obliquely, were blended together in the fertilized egg. The *vis essentialis* then came along, like an invisible hand, and molded the formation of this mass into a human form.

℗

While biologists, philosophers, Christian scholars, and embryologists fought their way through vicious debates between preformation and the "invisible hand" throughout much of the eighteenth century, a casual observer may have been forgiven for feeling rather unimpressed by it all. This was, after all, stale news. "The opposing views of today were in existence centuries ago," a nineteenth-century biologist complained, rightfully. Indeed, preformation was largely a restatement of Pythagoras's theory—that sperm carried all the information to make a new human. And the "invisible hand" was, in turn, merely a gilded variant of Aristotle's idea—that heredity was carried in the form of messages to create materials (it was the "hand" that carried the instructions to mold an embryo).

In time, both the theories would be spectacularly vindicated, and spectacularly demolished. Both Aristotle and Pythagoras were partially right and partially wrong. But in the early 1800s, it seemed as if the entire field of heredity and embryogenesis had reached a conceptual impasse. The world's greatest biological thinkers, having pored over the problem of heredity, had scarcely advanced the field beyond the cryptic musings of two men who had lived on two Greek islands two thousand years earlier.

"The Mystery of Mysteries"

. . . They mean to tell us all was rolling blind
Till accidentally it hit on mind
In an albino monkey in a jungle,
And even then it had to grope and bungle,
Till Darwin came to earth upon a year . . .
——Robert Frost, "Accidentally on Purpose"

In the winter of 1831, when Mendel was still a schoolboy in Silesia, a young clergyman-in-training, Charles Darwin, boarded a ten-gun brig-sloop, the HMS *Beagle*, at Plymouth Sound, on the southwestern shore of England. Darwin was then twenty-two years old, the son and grand-son of prominent physicians. He had the square, handsome face of his father, the porcelain complexion of his mother, and the dense overhang of eyebrows that ran in the Darwin family over generations. He had tried, unsuccessfully, to study medicine at Edinburgh—but, horrified by the "screams of a strapped-down child amid the blood and sawdust of the . . . operating theater," had fled medicine to study theology at Christ's College in Cambridge. But Darwin's interest ranged far beyond theology. Holed up in a room above a tobacconist's shop on Sidney Street, he had occu-pied himself by collecting beetles, studying botany and geology, learning geometry and physics, and arguing hotly about God, divine intervention, and the creation of animals. More than theology or philosophy, Darwin was drawn to natural history—the study of the natural world using sys-tematic scientific principles. He apprenticed with another clergyman, John Henslow, the botanist and geologist who had created and curated the Cambridge Botanic Garden, the vast outdoor museum of natural history where Darwin first learned to collect, identify, and classify plant and animal specimens.

Two books particularly ignited Darwin's imagination during his student years. The first, *Natural Theology*, published in 1802 by William Paley, the former vicar of Dalston, made an argument that would resonate deeply with Darwin. Suppose, Paley wrote, a man walking across a heath happens upon a watch lying on the ground. He picks up the instrument and opens it to find an exquisite system of cogs and wheels turning inside, resulting in a mechanical device that is capable of telling time. Would it not be logical to assume that such a device could only have been manufactured by a watchmaker? The same logic had to apply to the natural world, Paley reasoned. The exquisite construction of organisms and human organs—"the pivot upon which the head turns, the ligament within the socket of the hip joint"—could point to only one fact: that all organisms were created by a supremely proficient designer, a divine watchmaker: God.

The second book, *A Preliminary Discourse on the Study of Natural Philosophy*, published in 1830 by the astronomer Sir John Herschel, suggested a radically different view. At first glance, the natural world seems incredibly complex, Herschel acknowledged. But science can reduce seemingly complex phenomena into causes and effects: motion is the result of a force impinging on an object; heat involves the transference of energy; sound is produced by the vibration of air. Herschel had little doubt that chemical, and, ultimately, biological phenomena, would also be attributed to such cause-and-effect mechanisms.

Herschel was particularly interested in the creation of biological organisms—and his methodical mind broke the problem down to its two basic components. The first was the problem of the creation of life from nonlife—genesis *ex nihilo*. Here, he could not bring himself to challenge the doctrine of the divine creation. "To ascend to the origin of things, and speculate on creation, is not the business of the natural philosopher," he wrote. Organs and organisms might behave according to the laws of physics and chemistry—but the genesis of life itself could never be understood through these laws. It was as if God had given Adam a nice little laboratory in Eden, but then forbidden him from peering over the walls of the garden.

But the second problem, Herschel thought, was more tractable: Once life had been created, what process generated the observed diversity of the natural world? How, for instance, did a new species of animal arise from another species? Anthropologists, studying language, had demonstrated that new languages arose from old languages through the transformation of words. Sanskrit and Latin words could be traced back to mutations and variations in an ancient Indo-European language, and English and Flemish had arisen from a common root. Geologists had proposed that the current shape of the earth—its rocks, chasms, and mountains—had been created by the transmutation of previous elements. "Battered relics of past ages," Herschel wrote, "contain . . . indelible records capable of intelligible interpretation." It was an illuminating insight: a scientist could understand the present and the future by examining the "battered relics" of the past. Herschel did not have the correct mechanism for the origin of species, but he posed the correct question. He called this the "mystery of mysteries."

☉

Natural history, the subject that gripped Darwin at Cambridge, was not particularly poised to solve Herschel's "mystery of mysteries." To the fiercely inquisitive Greeks, the study of living beings had been intimately linked to the question of the origin of the natural world. But medieval Christians were quick to realize that this line of inquiry could only lead to unsavory theories. "Nature" was God's creation—and to be safely consistent with Christian doctrine, natural historians had to tell the story of nature in terms of Genesis.

A descriptive view of nature—i.e., the identification, naming, and classification of plants and animals—was perfectly acceptable: in describing nature's wonders, you were, in effect, celebrating the immense diversity of living beings created by an omnipotent God. But a *mechanistic* view of nature threatened to cast doubt on the very basis of the doctrine of creation: to ask why and when animals were created—by what mechanism or force—was to challenge the myth of divine creation and edge dangerously close to heresy. Perhaps unsurprisingly, by the late eighteenth century, the discipline of natural history was dominated by so-called parson-naturalists—vicars, parsons, abbots, deacons, and monks who cultivated their gardens and collected plant and animal specimens to service the wonders of divine Creation, but generally veered away from questioning its fundamental assumptions. The church provided a safe haven for these scientists—but it also effectively neutered their curiosity. The injunctions against the wrong kinds of investigation were so sharp that the parson-naturalists did not even *question* the myths of creation; it was the perfect separation of church and mental state. The result was a peculiar distortion of the field. Even as taxonomy—the classification of plant and animal species—flourished, inquiries into the origin of living beings were relegated to the forbidden sidelines. Natural history devolved into the study of nature without history.

It was this static view of nature that Darwin found troubling. A natural historian should be able to describe the state of the natural world in terms of causes and effects, Darwin reasoned—just as a physicist might describe the motion of a ball in the air. The essence of Darwin's disruptive genius was his ability to think about nature not as fact—but as process, as progression, as history. It was a quality that he shared with Mendel. Both obsessive observers of the natural world, Darwin and Mendel made their crucial leaps by asking variants of the same question: How does "nature" come into being? Mendel's question was microscopic: How does a single organism transmit information to its offspring over a single generation? Darwin's question was macroscopic: How do organisms *transmute* information about their features over a thousand generations? In time, both visions would converge, giving rise to the most important synthesis in modern biology, and the most powerful understanding of human heredity.

①

In August 1831, two months after his graduation from Cambridge, Darwin received a letter from his mentor, John Henslow. An exploratory "survey" of South America had been commissioned, and the expedition required the service of a "gentleman scientist" who could assist in collecting specimens. Although more gentleman than scientist (having never published a major scientific paper), Darwin thought himself a natural fit. He was to travel on the *Beagle*—not as a "finished Naturalist," but as a scientist-in-training "amply qualified for collecting, observing and noting any thing worthy to be noted in Natural History."

The *Beagle* lifted anchor on December 27, 1831, with seventy-three sailors on board, clearing a gale and tacking southward toward Tenerife. By early January, Darwin was heading toward Cape Verde. The ship was smaller than he had expected, and the wind more treacherous. The sea churned constantly beneath him. He was lonely, nauseated, and dehydrated, surviving on a diet of dry raisins and bread. That month, he began writing notes in his journal. Slung on a hammock bed above the salt-starched survey maps, he pored over the few books that he had brought with him on the voyage—Milton's *Paradise Lost* (which seemed all too apposite to his condition), and Charles Lyell's *Principles of Geology*, published between 1830 and 1833.

Lyell's work in particular left an impression on him. Lyell had argued (radically, for his time) that complex geological formations, such as boulders and mountains, had been created over vast stretches of time, not by the hand of God but by slow natural processes such as erosion, sedimentation, and deposition. Rather than a colossal biblical Flood, Lyell argued, there had been millions of floods; God had shaped the earth not through singular cataclysms but through a million paper cuts. For Darwin, Lyell's central idea—of the slow heave of natural forces shaping and reshaping the earth, sculpting nature—would prove to be a potent intellectual spur. In February 1832, still "squeamish and uncomfortable," Darwin crossed over to the southern hemisphere. The winds changed direction, and the currents altered their flow, and a new world floated out to meet him.

①

Darwin, as his mentors had predicted, proved to be an excellent collector and observer of specimens. As the *Beagle* hopscotched its way down the eastern coast of South America, passing through Montevideo, Bahía Blanca, and Port Desire, he rifled through the bays, rain forests, and cliffs, hauling aboard a vast assortment of skeletons, plants, pelts, rocks, and shells—"cargoes of apparent rubbish," the captain complained. The land yielded not just a cargo of living specimens, but ancient fossils as well; laying them out on long lines along the deck, it was as if Darwin had created his own museum of comparative anatomy. In September 1832, exploring the gray cliffs and low-lying clay bays near Punta Alta, he discovered an astonishing natural cemetery, with fossilized bones of enormous extinct mammals splayed out before him. He pried out the jaw of one fossil from the rock, like a mad dentist, then returned the next week to extract a huge skull from the quartz. The skull belonged to a megatherium, a mammoth version of a sloth.

That month, Darwin found more bones strewn among the pebbles and rocks. In November, he paid eighteen pence to a Uruguayan farmer for a piece of a colossal skull of yet another extinct mammal—the rhino-like *Toxodon*, with giant squirrel teeth—that had once roamed the plains. "I have been wonderfully lucky," he wrote. "Some of the mammals were gigantic, and many of them are quite new." He collected fragments from a pig-size guinea pig, armor plates from a tanklike armadillo, more elephantine bones from elephantine sloths, and crated and shipped them to England.

The *Beagle* rounded the sharp jaw-bend of Tierra del Fuego and climbed the western coast of South America. In 1835, the ship left Lima, on the coast of Peru, and headed toward a lonely spray of charred volcanic islands west of Ecuador—the Galápagos. The archipelago was "black, dismal-looking heaps . . . of broken lava, forming a shore fit for pandemonium," the captain wrote. It was a Garden of Eden of a hellish sort: isolated, untouched, parched, and rocky—turds of congealed lava overrun by "hideous iguanas," tortoises, and birds. The ship wandered from island to island—there were about eighteen in all—and Darwin ventured ashore, scrambling through the pumice, collecting birds, plants, and lizards. The crew survived on a steady diet of tortoise meat, with every island yielding a seemingly unique variety of tortoise. Over five weeks, Darwin collected carcasses of finches, mockingbirds, blackbirds, grosbeaks, wrens, albatrosses, iguanas, and an array of sea and land plants. The captain grimaced and shook his head.

On October 20, Darwin returned to sea, headed toward Tahiti. Back in his room aboard the *Beagle*, he began to systematically analyze the corpses of the birds that he had collected. The mockingbirds, in particular, surprised him. There were two or three varieties, but each subtype was markedly distinct, and each was endemic to one particular island. Offhandedly, he scribbled one of the most important scientific sentences that he would ever write: "Each variety is constant in its own Island." Was the same pattern true of other animals—of the tortoises, say? Did each island have a unique tortoise type? He tried, belatedly, to establish the same pattern for the turtles—but it was too late. He and the crew had eaten the specimens for lunch.

①

When Darwin returned to England after five years at sea, he was already a minor celebrity among natural historians. His vast fossil loot from South America was being unpacked, preserved, cataloged, and organized; whole museums could be built around it. The taxidermist and bird painter John Gould had taken over the classification of the birds. Lyell himself displayed Darwin's specimens during his presidential address to the Geological Society. Richard Owen, the paleontologist who hovered over England's natural historians like a patrician falcon, descended from the Royal College of Surgeons, to verify and catalog Darwin's fossil skeletons.

But while Owen, Gould, and Lyell named and classified the South American treasures, Darwin turned his mind to other problems. He was not a splitter, but a lumper, a seeker of deeper anatomy. Taxonomy and nomenclature were, for him, merely means to an end. His instinctive genius lay in unearthing *patterns*—systems of organization—that lay behind the specimens; not in Kingdoms and Orders, but in kingdoms of order that ran through the biological world. The same question that would frustrate Mendel in his teaching examination in Vienna—why on earth were living things organized in *this* manner?—became Darwin's preoccupation in 1836.

Two facts stood out that year. First, as Owen and Lyell pored through the fossils, they found an underlying pattern in the specimens. They were typically skeletons of colossal, extinct versions of animals that were still in existence at the very same locations where the fossils had been discovered. Giant-plated armadillos once roamed in the very valley where small armadillos were now moving through the brush. Gargantuan sloths had foraged where smaller sloths now resided. The huge femoral bones that Darwin had extracted from the soil belonged to a vast, elephant-size llama; its smaller current version was unique to South America.

The second bizarre fact came from Gould. In the early spring of 1837, Gould told Darwin that the assorted varieties of wrens, warblers, blackbirds, and "Gross-beaks" that Darwin had sent him were not assorted or various at all. Darwin had misclassified them: they were all finches—an astonishing thirteen species. Their beaks, claws, and plumage were so distinct that only a trained eye could have discerned the unity lurking beneath. The thin-throated, wrenlike warbler and the ham-necked, pincer-beaked blackbirds were anatomical cousins—variants of the same species. The warbler likely fed on fruit and insects (hence that flutelike beak). The spanner-beaked finch was a seed-cracking ground forager (hence its nutcracker-like bill). And the mockingbirds that were endemic to each island were also three distinct species. Finches and finches everywhere. It was as if each site had produced its own variant—a bar-coded bird for each island.

How could Darwin reconcile these two facts? Already, the bare outline of an idea was coalescing in his mind—a notion so simple, and yet so deeply radical, that no biologist had dared to explore it fully: *What if all the finches had arisen from a common ancestral finch?* What if the small armadillos of today had arisen from a giant ancestral armadillo? Lyell had argued that the current landscape of the earth was the consequence of natural forces that had accumulated over millions of years. In 1796, the French physicist Pierre-Simon Laplace had proposed that even the current solar system had arisen from the gradual cooling and condensation of matter over millions of years (when Napoléon had asked Laplace why God was so conspicuously missing from his theory, Laplace had replied with epic cheekiness: "Sire, I had no need for *that* hypothesis"). What if the current forms of animals were also the consequence of natural forces that had accumulated over millennia?

☩

In July 1837, in the stifling heat of his study on Marlborough Street, Darwin began scribbling in a new notebook (the so-called B notebook), firing off ideas about how animals could change over time. The notes were cryptic, spontaneous, and raw. On one page, he drew a diagram that would return to haunt his thoughts: rather than all species radiating out from the central hub of divine creation, perhaps they arose like branches of a "tree," or like rivulets from a river, with an ancestral stem that divided and subdivided into smaller and smaller branches toward dozens of modern descendants. Like languages, like landscapes, like the slowly cooling cosmos, perhaps the animals and plants had *descended* from earlier forms through a process of gradual, continuous change.

It was, Darwin knew, an explicitly profane diagram. The Christian concept of speciation placed God firmly at the epicenter; all animals created by Him sprayed outward from the moment of creation. In Darwin's drawing, there was no center. The thirteen finches were not created by some divine whim, but by "natural descent"—cascading downward and outward from an original ancestral finch. The modern llama arose similarly, by descending from a giant ancestral beast. As an afterthought, he added, "I think," above the page, as if to signal his last point of departure from the mainlands of biological and theological thought.

But—with God shoved aside—what was the driving force behind the origin of species? What impetus drove the descent of, say, thirteen variants of finches down the fierce rivulets of speciation? In the spring of 1838, as Darwin tore into a new journal—the maroon C notebook—he had more thoughts on the nature of this driving force.

The first part of the answer had been sitting under his nose since his childhood in the farmlands of Shrewsbury and Hereford; Darwin had merely traveled eight thousand miles around the globe to rediscover it. The phenomenon was called variation—animals occasionally produced offspring with features different from the parental type. Farmers had been using this phenomenon for millennia—breeding and interbreeding animals to produce natural variants, and selecting these variants over multiple generations. In England, farm breeders had refined the creation of novel breeds and variants to a highly sophisticated science. The shorthorn bulls of Hereford bore little resemblance to the longhorns of Craven. A curious naturalist traveling from the Galápagos to England—a Darwin in reverse—might have been astonished to find that each region had its own species of cow. But as Darwin, or any bull breeder, could tell you, the breeds had not arisen by accident. They had been deliberately created by humans—by the selective breeding of variants from the same ancestral cow.

The deft combination of variation and artificial selection, Darwin knew, could produce extraordinary results. Pigeons could be made to look like roosters and peacocks, and dogs made short-haired, long-haired, pied, piebald, bowlegged, hairless, crop-tailed, vicious, mild-mannered, diffident, guarded, belligerent. But the force that had molded the selection of cows, dogs, and pigeons was the human hand. What hand, Darwin asked, had guided the creation of such different varieties of finches on those distant volcanic islands or made small armadillos out of giant precursors on the plains of South America?

Darwin knew that he was now gliding along the dangerous edge of the known world, tacking south of heresy. He could easily have ascribed the invisible hand to God. But the answer that came to him in October 1838, in a book by another cleric, the Reverend Thomas Malthus, had nothing to do with divinity.

①

Thomas Malthus had been a curate at the Okewood Chapel in Surrey by daytime, but he was a closet economist by night. His true passion was the study of populations and growth. In 1798, writing under a pseudonym, Malthus had published an incendiary paper—*An Essay on the Principle of Population*—in which he had argued that the human population was in constant struggle with its limited resource pool. As the population expanded, Malthus reasoned, its resource pool would be depleted, and competition between individuals would grow severe. A population's inherent inclination to expand would be severely counterbalanced by the limitations of resources; its natural wont met by natural want. And then potent apocalyptic forces—"sickly seasons, epidemics, pestilence and plague [would] advance in terrific array, and sweep off their thousands and tens of thousands"—leveling the "population with the food of the world." Those that survived this "natural selection" would restart the grim cycle again—Sisyphus moving from one famine to the next.

In Malthus's paper, Darwin immediately saw a solution to his quandary. This struggle for survival was the shaping hand. *Death* was nature's culler, its grim shaper. "It at once struck me," he wrote, "that under these circumstances [of natural selection], favourable variations would tend to be preserved and unfavourable ones to be destroyed. The results of this would be the formation of a new species."*

Darwin now had the skeletal sketch of his master theory. When animals reproduce, they produce variants that differ from the parents.† Individuals within a species are constantly competing for scarce resources. When these resources form a critical bottleneck—during a famine, for instance—a variant better adapted for an environment is "naturally selected." The best adapted—the "fittest"—survive (the phrase *survival of the fittest* was borrowed from the Malthusian economist Herbert Spencer). These survivors then reproduce to make more of their kind, thereby driving evolutionary change within a species.

Darwin could almost *see* the process unfolding on the salty bays of

---

* Darwin missed a crucial step here. Variation and natural selection offer cogent explanations of the mechanism by which evolution might occur *within* a species, but they do not explain the formation of species per se. For a new species to arise, organisms must no longer be able to reproduce viably with each other. This typically occurs when animals are isolated from each other by a physical barrier or another permanent form of isolation, ultimately leading to reproductive incompatibility. We will return to this idea in subsequent pages.

† Darwin was unsure how these variants were generated, another fact to which we will return in subsequent pages.

Punta Alta or on the islands of the Galápagos, as if an eons-long film were running on fast-forward, a millennium compressed to a minute. Flocks of finches fed on fruit until their population exploded. A bleak season came upon the island—a rotting monsoon or a parched summer—and fruit supplies dwindled drastically. Somewhere in the vast flock, a variant was born with a grotesque beak capable of cracking seeds. As famine raged through the finch world, this gross-beaked variant survived by feeding on hard seeds. It reproduced, and a new species of finch began to appear. The freak became the norm. As new Malthusian limits were imposed— diseases, famines, parasites—new breeds gained a stronghold, and the population shifted again. Freaks became norms, and norms became extinct. Monster by monster, evolution advanced.

⚬

By the winter of 1839, Darwin had assembled the essential outlines of his theory. Over the next few years, he tinkered and fussed obsessively with his ideas—arranging and rearranging "ugly facts" like his fossil specimens, but he never got around to publishing the theory. In 1844, he distilled the crucial parts of his thesis into a 255-page essay and mailed it to his friends to read privately. But he did not bother committing the essay to print. He concentrated, instead, on studying barnacles, writing papers on geology, dissecting sea animals, and tending to his family. His daughter Annie—the eldest, and his favorite—contracted an infection and died, leaving Darwin numb with grief. A brutal, internecine war broke out in the Crimean Peninsula. Men were hauled off to the battlefront and Europe plunged into a depression. It was as if Malthus and the struggle for survival had come alive in the real world.

In the summer of 1855, more than a decade and a half after Darwin had first read Malthus's essay and crystallized his ideas about speciation, a young naturalist, Alfred Russel Wallace, published a paper in the *Annals and Magazine of Natural History* that skirted dangerously close to Darwin's yet-unpublished theory. Wallace and Darwin had emerged from vastly different social and ideological backgrounds. Unlike Darwin—gentleman biologist, and soon to be England's most lauded natural historian—Wallace had been born to a middle-class family in Monmouthshire. He too had read Malthus's paper on populations—not in an armchair in his study, but on the hard-back benches of the free library at Leicester (Malthus's book was widely circulated in intellectual circles in Great Britain). Like Darwin, Wallace had also embarked on a seafaring journey—to Brazil—to collect specimens and fossils and had emerged transformed.

In 1854, having lost the little money that he possessed, and all the specimens that he had collected, in a shipping disaster, an even more deeply impoverished Wallace moved from the Amazon basin to another series of scattered volcanic islands—the Malay Archipelago—on the edge of southeastern Asia. There, like Darwin, he observed astonishing differences between closely related species that had been separated by channels of water. By the winter of 1857, Wallace had begun to formulate a general theory about the mechanism driving variation in these islands. That spring, lying in bed with a hallucinatory fever, he stumbled upon the last missing piece of his theory. He recalled Malthus's paper. "The answer was clearly . . . [that] the best fitted [variants] live. . . . In this way every part of an animal's organization could be modified exactly as required." Even the language of his thoughts—variation, mutation, survival, and selection—bore striking similarities to Darwin's. Separated by oceans and continents, buffeted by very different intellectual winds, the two men had sailed to the same port.

In June 1858, Wallace sent Darwin a tentative draft of his paper outlining his general theory of evolution by natural selection. Stunned by the similarities between Wallace's theory and his own, a panicked Darwin dashed his own manuscript off to his old friend Lyell. Cannily, Lyell advised Darwin to have both papers presented simultaneously at the meeting of the Linnean Society that summer so that both Darwin and Wallace could simultaneously be credited for their discoveries. On July 1, 1858, Darwin's and Wallace's papers were read back to back and discussed publicly in London. The audience was not particularly enthusiastic about either study. The next May, the president of the society remarked parenthetically that the past year had not yielded any particularly noteworthy discoveries.

①

Darwin now rushed to finish the monumental opus that he had originally intended to publish with all his findings. In 1859, he contacted the publisher John Murray hesitantly: "I heartily hope that my Book may be sufficiently successful that you may not repent of having undertaken it." On November 24, 1859, on a wintry Thursday morning, Charles Darwin's book *On the Origin of Species by Means of Natural Selection* appeared in bookstores in England, priced at fifteen shillings a copy. Twelve hundred and fifty copies had been printed. As Darwin noted, stunned, "All copies were sold [on the] first day."

A torrent of ecstatic reviews appeared almost immediately. Even the earliest readers of *Origin* were aware of the book's far-reaching implications. "The conclusions announced by Mr. Darwin are such as, if established, would cause a complete revolution in the fundamental doctrines of natural history," one reviewer wrote. "We imply that his work [is] one of the most important that for a long time past have been given to public."

Darwin had also fueled his critics. Perhaps wisely, he had been deliberately cagey about the implications of his theory for human evolution: the only line in *Origin* regarding human descent—"light will be thrown on the origin of man and his history"—might well have been the scientific understatement of the century. But Richard Owen, the fossil taxonomist—Darwin's frenemy—was quick to discern the philosophical implications of Darwin's theory. If the descent of species occurred as Darwin suggested, he reasoned, then the implication for human evolution was obvious. "Man might be a transmuted ape"—an idea so deeply repulsive that Owen could not even bear to contemplate it. Darwin had advanced the boldest new theory in biology, Owen wrote, without adequate experimental proof to support it; rather than fruit, he had provided "intellectual husks." Owen complained (quoting Darwin himself): "One's imagination must fill up very wide blanks."

The "Very Wide Blank"

Now, I wonder if Mr. Darwin ever took the trouble to think how long it would take to exhaust any given original stock of . . . gemmules . . . It seems to me if he had given it a casual thought, he surely would never have dreamt of "pangenesis."

—Alexander Wilford Hall, 1880

It is a testament to Darwin's scientific audacity that he was not particularly bothered by the prospect of human descent from apelike ancestors. It is also a testament to his scientific integrity that what *did* bother him, with far fiercer urgency, was the integrity of the internal logic of his own theory. One particularly "wide blank" had to be filled: heredity.

A theory of heredity, Darwin realized, was not peripheral to a theory of evolution; it was of pivotal importance. For a variant of gross-beaked finch to appear on a Galápagos island by natural selection, two seemingly contradictory facts had to be simultaneously true. First, a short-beaked "normal" finch must be able to occasionally produce a gross-beaked variant—a monster or freak (Darwin called these *sports*—an evocative word, suggesting the infinite caprice of the natural world. The crucial driver of evolution, Darwin understood, was not nature's sense of purpose, but her sense of humor). And second, once born, that gross-beaked finch must be able to *transmit* the same trait to its offspring, thereby fixing the variation for generations to come. If either factor failed—if reproduction failed to produce variants or if heredity failed to transmit the variations—then nature would be mired in a ditch, the cogwheels of evolution jammed. For Darwin's theory to work, heredity had to possess constancy *and* inconstancy, stability *and* mutation.

℗

Darwin wondered incessantly about a mechanism of heredity that could achieve these counterbalanced properties. In Darwin's time, the most commonly accepted mechanism of heredity was a theory advanced by the eighteenth-century French biologist Jean-Baptiste Lamarck. In Lamarck's view, hereditary traits were passed from parents to offspring in the same manner that a message, or story, might be passed—i.e., by instruction. Lamarck believed that animals adapted to their environments by strengthening or weakening certain traits—"with a power proportional to the length of time it has been so used." A finch forced to feed on hard seeds adapted by "strengthening" its beak. Over time, the finch's beak would harden and become pincer shaped. This adapted feature would then be transmitted to the finch's offspring by instruction, and their beaks would harden as well, having been *pre*-adapted to the harder seeds by their parents. By similar logic, antelopes that foraged on tall trees found that they had to extend their necks to reach the high foliage. By "use and disuse," as Lamarck put it, their necks would stretch and lengthen, and these antelopes would produce long-necked offspring—thereby giving rise to giraffes (note the similarities between Lamarck's theory—of the body giving "instructions" to sperm—and Pythagoras's conception of human heredity, with sperm collecting messages from all organs).

The immediate appeal of Lamarck's idea was that it offered a reassuring story of progress: all animals were progressively adapting to their environments, and thus progressively slouching along an evolutionary ladder toward perfection. Evolution and adaptation were bundled together into one continuous mechanism: adaptation *was* evolution. The scheme was not just intuitive, it was also conveniently divine—or close enough for a biologist's work. Although initially created by God, animals still had a chance to perfect their forms in the changing natural world. The Divine Chain of Being still stood. If anything, it stood even more upright: at the end of the long chain of adaptive evolution was the well-adjusted, best-erected, most perfected mammal of them all: humans.

Darwin had obviously split ways with Lamarck's evolutionary ideas. Giraffes hadn't arisen from straining antelopes needing skeletal neck-braces. They had emerged—loosely speaking—because an ancestral antelope had produced a long-necked variant that had been progressively selected by a natural force, such as a famine. But Darwin kept returning to the mechanism of heredity: What had made the long-necked antelope emerge in the first place?

Darwin tried to envision a theory of heredity that would be compatible with evolution. But here his crucial intellectual shortcoming came to the fore: he was not a particularly gifted experimentalist. Mendel, as we shall see, was an instinctual gardener—a breeder of plants, a counter of seeds, an isolator of traits; Darwin was a garden digger—a classifier of plants, an organizer of specimens, a taxonomist. Mendel's gift was experimentation—the manipulation of organisms, cross-fertilization of carefully selected sub-breeds, the testing of hypotheses. Darwin's gift was natural history—the reconstruction of history by observing nature. Mendel, the monk, was an isolator; Darwin, who had once aspired to be a parson, a synthesizer.

But observing nature, it turned out, was very different from experimenting with nature. Nothing about the natural world, at first glance, suggests the existence of a gene; indeed, you have to perform rather bizarre experimental contortions to uncover the idea of discrete particles of inheritance. Unable to arrive at a theory of heredity via experimental means, Darwin was forced to conjure one up from purely theoretical grounds. He struggled with the concept for nearly two years, driving himself to the brink of a mental breakdown, before he thought he had stumbled on an adequate theory. Darwin imagined that the cells of all organisms produce minute particles containing hereditary information—*gemmules*, he called them. These gemmules circulate in the parent's body. When an animal or plant reaches its reproductive age, the information in the gemmules is transmitted to germ cells (sperm and egg). Thus, the information about a body's "state" is transmitted from parents to offspring during conception. As with Pythagoras, in Darwin's model, every organism carried information to build organs and structures in miniaturized form—except in Darwin's case, the information was decentralized. An organism was built by parliamentary ballot. Gemmules secreted by the hand carried the instructions to manufacture a new hand; gemmules dispersed by the ear transmitted the code to build a new ear.

How were these gemmular instructions from a father and a mother applied to a developing fetus? Here, Darwin reverted to an old idea: the instructions from the male and female simply met in the embryo and blended together like paints or colors. This notion—blending inheritance—was already familiar to most biologists: it was a restatement of Aristotle's theory of mixing between male and female characters. Darwin had, it seemed, achieved yet another marvelous synthesis between opposing poles of biology. He had melded the Pythagorean homunculus (gemmules) with the Aristotelian notion of message and mixture (blending) into a new theory of heredity.

Darwin dubbed his theory pangenesis—"genesis from everything" (since all organs contributed gemmules). In 1867, nearly a decade after the publication of *Origin*, he began to complete a new manuscript, *The Variation of Animals and Plants Under Domestication*, in which he would fully explicate this view of inheritance. "It is a rash and crude hypothesis," Darwin confessed, "but it has been a considerable relief to my mind." He wrote to his friend Asa Gray, "Pangenesis will be called a mad dream, but at the bottom of my own mind, I think it contains a great truth."

①

Darwin's "considerable relief" could not have been long-lived; he would soon be awoken from his "mad dream." That summer, while *Variation* was being compiled into its book form, a review of his earlier book, *Origin*, appeared in the *North British Review*. Buried in the text of that review was the most powerful argument against pangenesis that Darwin would encounter in his lifetime.

The author of the review was an unlikely critic of Darwin's work: a mathematician-engineer and inventor from Edinburgh named Fleeming Jenkin, who had rarely written about biology. Brilliant and abrasive, Jenkin had diverse interests in linguistics, electronics, mechanics, arithmetic, physics, chemistry, and economics. He read widely and profusely—Dickens, Dumas, Austen, Eliot, Newton, Malthus, Lamarck. Having chanced upon Darwin's book, Jenkin read it thoroughly, worked swiftly through the implications, and immediately found a fatal flaw in the argument.

Jenkin's central problem with Darwin was this: if hereditary traits kept "blending" with each other in every generation, then what would keep any variation from being diluted out immediately by interbreeding? "The [variant] will be swamped by the numbers," Jenkin wrote, "and after a few generations its peculiarity will be obliterated." As an example—colored deeply by the casual racism of his era—Jenkin concocted a story: "Suppose a white man to have been wrecked on an island inhabited by negroes. . . . Our shipwrecked hero would probably become king; he would kill a great many blacks in the struggle for existence; he would have a great many wives and children."

But if genes blended with each other, then Jenkin's "white man" was fundamentally doomed—at least in a genetic sense. His children—

from black wives—would presumably inherit half his genetic essence. His grandchildren would inherit a quarter; his great-grandchildren, an eighth; his great-great-grandchildren, one-sixteenth, and so forth—until his genetic essence had been diluted, in just a few generations, into complete oblivion. Even if "white genes" *were* the most superior—the "fittest," to use Darwin's terminology—nothing would protect them from the inevitable decay caused by blending. In the end, the lone white king of the island would vanish from its genetic history—even though he had fathered more children than any other man of his generation, and even though his genes were best suited for survival.

The particular details of Jenkin's story were ugly—perhaps deliberately so—but its conceptual point was clear. If heredity had no means of *maintaining* variance—of "fixing" the altered trait—then all alterations in characters would eventually vanish into colorless oblivion by virtue of blending. Freaks would always remain freaks—unless they could guarantee the passage of their traits to the next generation. Prospero could safely afford to create a single Caliban on an isolated island and let him roam at large. Blending inheritance would function as his natural genetic prison: even if he mated—precisely *when* he mated—his hereditary features would instantly vanish into a sea of normalcy. Blending was the same as infinite dilution, and no evolutionary information could be maintained in the face of such dilution. When a painter begins to paint, dipping the brush occasionally to dilute the pigment, the water might initially turn blue, or yellow. But as more and more paints are diluted into the water, it inevitably turns to murky gray. Add more colored paint, and the water remains just as intolerably gray. If the same principle applied to animals and inheritance, then what force could possibly conserve any distinguishing feature of any variant organism? Why, Jenkin might ask, weren't all Darwin's finches gradually turning gray?*

℗

Darwin was deeply struck by Jenkin's reasoning. "Fleeming Jenkins [*sic*] has given me much trouble," he wrote, "but has been of more use to me

---

* Geographic isolation might have solved some of the "grey finch" problem—by restricting interbreeding between particular variants. But this would still be unable to explain why all finches in a single island did not gradually collapse to have identical characteristics.

than any other Essay or Review." There was no denying Jenkin's inescapable logic: to salvage Darwin's theory of evolution, he needed a congruent theory of heredity.

But what features of heredity might solve Darwin's problem? For Darwinian evolution to work, the mechanism of inheritance had to possess an intrinsic capacity to conserve information without becoming diluted or dispersed. Blending would not work. There had to be *atoms* of information—discrete, insoluble, indelible particles—moving from parent to child.

Was there any proof of such constancy in inheritance? Had Darwin looked carefully through the books in his voluminous library, he might have found a reference to an obscure paper by a little-known botanist from Brno. Unassumingly entitled "Experiments in Plant Hybridization" and published in a scarcely read journal in 1866, the paper was written in dense German and packed with the kind of mathematical tables that Darwin particularly despised. Even so, Darwin came tantalizingly close to reading it: in the early 1870s, poring through a book on plant hybrids, he made extensive handwritten notes on pages 50, 51, 53, and 54—but mysteriously skipped page 52, where the Brno paper on pea hybrids was discussed in detail.

If Darwin had actually read it—particularly as he was writing *Variation* and formulating pangenesis—this study might have provided the final critical insight to understand his own theory of evolution. He would have been fascinated by its implications, moved by the tenderness of its labor, and struck by its strange explanatory power. Darwin's incisive intellect would quickly have grasped its implications for the understanding of evolution. He may also have been pleased to note that the paper had been authored by another cleric who, in another epic journey from theology to biology, had also drifted off the edge of a map—an Augustine monk named Gregor Johann Mendel.

"Flowers He Loved"

We want only to disclose the [nature of] matter and its force. Metaphysics is not our interest.
　　　　　　　　　—The manifesto of the Brünn Natural
　　　　　　　　　　Science Society, where Mendel's paper
　　　　　　　　　　was first read in 1865

The whole organic world is the result of innumerable different combinations and permutations of relatively few factors. . . . These factors are the units which the science of heredity has to investigate. Just as physics and chemistry go back to molecules and atoms, the biological sciences have to penetrate these units in order to explain . . . the phenomena of the living world.
　　　　　　　　　　　　　　　　—Hugo de Vries

As Darwin was beginning to write his opus on evolution in the spring of 1856, Gregor Mendel decided to return to Vienna to retake the teacher's exam that he had failed in 1850. He felt more confident this time. Mendel had spent two years studying physics, chemistry, geology, botany, and zoology at the university in Vienna. In 1853, he had returned to the monastery and started work as a substitute teacher at the Brno Modern School. The monks who ran the school were very particular about tests and qualifications, and it was time to try the certifying exam again. Mendel applied to take the test.

Unfortunately, this second attempt was also a disaster. Mendel was ill, most likely from anxiety. He arrived in Vienna with a sore head and a foul temper—and quarreled with the botany examiner on the first day of the three-day test. The topic of disagreement is unknown, but likely concerned species formation, variation, and heredity. Mendel did not finish the exam. He returned to Brno reconciled to his destiny as a substitute teacher. He never attempted to obtain certification again.

①

Late that summer, still bruising from his failed exam, Mendel planted a crop of peas. It wasn't his first crop. He had been breeding peas inside the glass hothouse for about three years. He had collected thirty-four strains from the neighboring farms and bred them to select the strains that bred "true"—that is, every pea plant produced exactly identical offspring, with the same flower color or the same seed texture.* These plants "remained constant without exception," he wrote. Like always begat like. He had collected the founding material for his experiment.

The true-bred pea plants, he noted, possessed distinct traits that were hereditary and variant. Bred to themselves, tall-stemmed plants generated only tall plants; short plants only dwarf ones. Some strains produced only smooth seeds, while others produced only angular, wrinkled seeds. The unripe pods were either green or vividly yellow, the ripe pods either loose or tight. He listed the seven such true-breeding traits:

1. the texture of the seed (smooth versus wrinkled)
2. the color of seeds (yellow versus green)
3. the color of the flower (white versus violet)
4. the position of the flower (at the tip of the plant versus the branches)
5. the color of the pea pod (green versus yellow)
6. the shape of the pea pod (smooth versus crumpled)
7. the height of the plant (tall versus short)

Every trait, Mendel noted, came in at least two different variants. They were like two alternative spellings of the same word, or two colors of the same jacket (Mendel experimented with only two variants of the same trait, although, in nature, there might be multiple ones, such as white-, purple-, mauve-, and yellow-flowering plants). Biologists would later term these variants *alleles*, from the Greek word *allos*—loosely referring to two different subtypes of the same general kind. Purple and white were

---

* Mendel was aided in his studies by a long interest in breeding amongst farmers in and around Brno. The abbot, Cyril Knapp, also took an interest in breeding experiments.

two alleles of the same trait: flower color. Long and short were two alleles of another characteristic—height.

The purebred plants were only a starting point for his experiment. To reveal the nature of heredity, Mendel knew that he needed to breed hybrids; only a "bastard" (a word commonly used by German botanists to describe experimental hybrids) could reveal the nature of purity. Contrary to later belief, he was acutely aware of the far-reaching implication of his study: his question was crucial to "the history of the evolution of organic forms," he wrote. In two years, astonishingly, Mendel had produced a set of reagents that would allow him to interrogate some of the most important features of heredity. Put simply, Mendel's question was this: If he crossed a tall plant with a short one, would there be a plant of intermediate size? Would the two alleles—shortness and tallness—blend?

The production of hybrids was tedious work. Peas typically self-fertilize. The anther and the stigma mature inside the flower's clasplike keel, and the pollen is dusted directly from a flower's anther to its own stigma. *Cross*-fertilization was another matter altogether. To make hybrids, Mendel had to first neuter each flower by snipping off the anthers—emasculating it—and then transfer the orange blush of pollen from one flower to another. He worked alone, stooping with a paintbrush and forceps to snip and dust the flowers. He hung his outdoor hat on a harp, so that every visit to the garden was marked by the sound of a single, crystalline note. This was his only music.

It's hard to know how much the other monks in the abbey knew about Mendel's experiments, or how much they cared. In the early 1850s, Mendel had tried a more audacious variation of this experiment, starting with white and gray field mice. He had bred mice in his room—mostly undercover—to try to produce mice hybrids. But the abbot, although generally tolerant of Mendel's whims, had intervened: a monk coaxing mice to mate to understand heredity was a little too risqué, even for the Augustinians. Mendel had switched to plants and moved the experiments to the hothouse outside. The abbot had acquiesced. He drew the line at mice, but didn't mind giving peas a chance.

℗

By the late summer of 1857, the first hybrid peas had bloomed in the abbey garden, in a riot of purple and white. Mendel noted the colors of the flowers, and when the vines had hung their pods, he slit open the shells to examine the seeds. He set up new hybrid crosses—tall with short; yellow with green; wrinkled with smooth. In yet another flash of inspiration, he crossed some hybrids to each other, making hybrids of hybrids. The experiments went on in this manner for eight years. The plantings had, by then, expanded from the hothouse to a plot of land by the abbey—a twenty-foot-by-hundred-foot rectangle of loam that bordered the refectory, visible from his room. When the wind blew the shades of his window open, it was as if the entire room turned into a giant microscope. Mendel's notebook was filled with tables and scribblings, with data from thousands of crosses. His thumbs were getting numb from the shelling.

"How small a thought it takes to fill someone's whole life," the philosopher Ludwig Wittgenstein wrote. Indeed, at first glance, Mendel's life seemed to be filled with the smallest thoughts. Sow, pollinate, bloom, pluck, shell, count, repeat. The process was excruciatingly dull—but small thoughts, Mendel knew, often bloomed into large principles. If the powerful scientific revolution that had swept through Europe in the eighteenth century had one legacy, it was this: the laws that ran through nature were uniform and pervasive. The force that drove Newton's apple from the branch to his head was the same force that guided planets along their celestial orbits. If heredity too had a universal natural law, then it was likely influencing the genesis of peas as much as the genesis of humans. Mendel's garden plot may have been small—but he did not confuse its size with that of his scientific ambition.

"The experiments progress slowly," Mendel wrote. "At first a certain amount of patience was needed, but I soon found that matters went better when I was conducting several experiments simultaneously." With multiple crosses in parallel, the production of data accelerated. Gradually, he began to discern patterns in the data—unanticipated constancies, conserved ratios, numerical rhythms. He had tapped, at last, into heredity's inner logic.

①

The first pattern was easy to perceive. In the first-generation hybrids, the individual heritable traits—tallness and shortness, or green and yellow seeds—did not blend at all. A tall plant crossed with a dwarf inevitably produced *only* tall plants. Round-seeded peas crossed with wrinkled seeds produced *only* round peas. All seven of the traits followed this pattern. "The hybrid character" was not intermediate but "resembled one of the parental forms," he wrote. Mendel termed these overriding traits *dominant*, while the traits that had disappeared were termed *recessive*.

Had Mendel stopped his experiments here, he would already have made a major contribution to a theory of heredity. The existence of dominant and recessive alleles for a trait contradicted nineteenth-century theories of blending inheritance: the hybrids that Mendel had generated did not possess intermediate features. Only one allele had asserted itself in the hybrid, forcing the other variant trait to vanish.

But where had the recessive trait disappeared? Had it been consumed or eliminated by the dominant allele? Mendel deepened his analysis with his second experiment. He bred short-tall hybrids with short-tall hybrids to produce third-generation progeny. Since tallness was dominant, all the parental plants in this experiment were tall to start; the recessive trait had disappeared. But when crossed with each other, Mendel found, they yielded an entirely unexpected result. In some of these third-generation crosses, shortness *reappeared*—perfectly intact—after having disappeared for a generation. The same pattern occurred with all seven of the other traits. White flowers vanished in the second generation, the hybrids, only to reemerge in some members of the third. A "hybrid" organism, Mendel realized, was actually a *composite*—with a visible, dominant allele and a latent, recessive allele (Mendel's word to describe these variants was *forms*; the word *allele* would be coined by geneticists in the 1900s).

By studying the mathematical relationships—the ratios—between the various kinds of progeny produced by each cross, Mendel could begin to construct a model to explain the inheritance of traits.* Every trait, in Mendel's model, was determined by an independent, indivisible particle of information. The particles came in two variants, or two alleles: short versus tall (for height) or white versus violet (for flower color) and so forth.

---

* Several statisticians have examined Mendel's original data and accused him of faking the data. Mendel's ratios and numbers were not just accurate; they were too perfect. It was as if he had encountered no statistical or natural error in his experiments—an impossible situation. In retrospect, it is unlikely that Mendel actively faked his studies. More likely, he constructed a hypothesis from his earliest experiments, then used the later experiments to validate his hypothesis: he stopped counting and tabulating the peas once they had conformed to the expected values and ratios. This method, albeit unconventional, was not unusual for his time, but it also reflected Mendel's scientific naïveté.

Every plant inherited one copy from each parent—one allele from father, via sperm, and one from mother, via the egg. When a hybrid was created, both traits existed intact—although only one asserted its existence.

Ⓢ

Between 1857 and 1864, Mendel shelled bushel upon bushel of peas, compulsively tabulating the results for each hybrid cross ("yellow seeds, green cotyledons, white flowers"). The results remained strikingly consistent. The small patch of land in the monastery garden produced an overwhelming volume of data to analyze—twenty-eight thousand plants, forty thousand flowers, and nearly four hundred thousand seeds. "It requires indeed some courage to undertake a labor of such far-reaching extent," Mendel would write later. But *courage* is the wrong word here. More than courage, something else is evident in that work—a quality that one can only describe as *tenderness*.

It is a word not typically used to describe science, or scientists. It shares roots, of course, with *tending*—a farmer's or gardener's activity—but also with *tension*, the stretching of a pea tendril to incline it toward sunlight or to train it on an arbor. Mendel was, first and foremost, a gardener. His genius was not fueled by deep knowledge of the conventions of biology (thankfully, he had failed that exam—twice). Rather, it was his instinctual knowledge of the garden, coupled with an incisive power of observation—the laborious cross-pollination of seedlings, the meticulous tabulation of the colors of cotyledons—that soon led him to findings that could not be explained by the traditional understanding of inheritance.

Heredity, Mendel's experiments implied, could only be explained by the passage of *discrete pieces of information from parents to offspring*. Sperm brought one copy of this information (an allele); the egg brought the other copy (a second allele); an organism thus inherited one allele from each parent. When that organism generated sperm or eggs, the alleles were split up again—one was passed to the sperm, and one to the egg, only to become combined in the next generation. One allele might "dominate" the other when both were present. When the dominant allele was present, the recessive allele seemed to disappear, but when a plant received two recessive alleles, the allele reiterated its character. Throughout, the information carried by an individual allele remained indivisible. The particles themselves remained intact.

Doppler's example returned to Mendel: there was music behind noise, laws behind seeming lawlessness, and only a profoundly artificial experiment—creating hybrids out of purebred strains carrying simple traits—could reveal these underlying patterns. Behind the epic variance of natural organisms—tall; short; wrinkled; smooth; green; yellow; brown—there were corpuscles of hereditary information, moving from one generation to the next. Each trait was *unitary*—distinct, separate, and indelible. Mendel did not give this unit of heredity a name, but he had discovered the most essential features of a gene.*

Ⅾ

On February 8, 1865, seven years after Darwin and Wallace had read their papers at the Linnean Society in London, Mendel presented his paper, in two parts, at a much less august forum: he spoke to a group of farmers, botanists, and biologists at the Natural Science Society in Brno (the second part of the paper was read on March 8, a month later). Few records exist of this moment in history. The room was small, and about forty people attended. The paper, with dozens of tables and arcane symbols to denote traits and variants, was challenging even for statisticians. For biologists, it may have seemed like absolute mumbo jumbo. Botanists generally studied morphology, not numerology. The counting of variants in seeds and flowers across tens of thousands of hybrid specimens must have mystified Mendel's contemporaries; the notion of mystical numerical "harmonies" lurking in nature had gone out of fashion with Pythagoras. Soon after Mendel was done, a professor of botany stood up to discuss Darwin's *Origin* and the theory of evolution. No one in the audience perceived a link between the two subjects being discussed. Even if Mendel was aware of a potential connection between his "units of heredity" and

---

* Did Mendel know that he was trying to uncover general laws that govern heredity? Or was he, as some historians claim, merely trying to understand the nature of hybridity in peas? The answer might be found in Mendel's papers. It cannot be debated that the existence of a "gene" was utterly unknown to Mendel. But in his own words, the experiments were performed "in order to discover the relations in which the hybrid forms stand towards . . . their progenitors", and to understand the "unity in the developmental plan of organic life". Indeed, Mendel even uses variations of the word "inherit" in his paper. To this reader, at least, it can hardly be argued that Mendel was unaware of the far-reaching implications of his study: he was trying to unlock the material basis and laws of heredity.

evolution—his prior notes had certainly indicated that he had sought such a link—he made no explicit comments on the topic.

Mendel's paper was published in the annual *Proceedings of the Brno Natural Science Society*. A man of few words, Mendel was even more concise in his writing: he had distilled nearly a decade's work into forty-four spectacularly dreary pages. Copies were sent to the Royal Society and the Linnean Society in England, and to the Smithsonian in Washington, among dozens of institutions. Mendel himself requested forty reprints, which he mailed, heavily annotated, to many scientists. It is likely that he sent one to Darwin, but there is no record of Darwin's having actually read it.

What followed, as one geneticist wrote, was "one of the strangest silences in the history of biology." The paper was cited only four times between 1866 and 1900—virtually disappearing from scientific literature. Between 1890 and 1900, even as questions and concerns about human heredity and its manipulation became central to policy makers in America and Europe, Mendel's name and his work were lost to the world. The study that founded modern biology was buried in the pages of an obscure journal of an obscure scientific society, read mostly by plant breeders in a declining Central European town.

①

On New Year's Eve in 1866, Mendel wrote to the Swiss plant physiologist Carl von Nägeli in Munich, enclosing a description of his experiments. Nägeli replied two months later—already signaling distance with his tardiness—sending a courteous but icy note. A botanist of some repute, Nägeli did not think much of Mendel or his work. Nägeli had an instinctual distrust of amateur scientists and scribbled a puzzlingly derogatory note next to the first letter: "only empirical . . . cannot be proved rational"—as if experimentally deduced laws were worse than those created de novo by human "reason."

Mendel pressed on, with further letters. Nägeli was the scientific colleague whose respect Mendel most sought—and his notes to him took an almost ardent, desperate turn. "I knew that the results I obtained were not easily compatible with our contemporary science," Mendel wrote, and "an isolated experiment might be doubly dangerous." Nägeli remained cautious and dismissive, often curt. The possibility that Mendel had deduced a fundamental natural rule—a dangerous law—by tabulating pea hybrids seemed absurd and far-fetched to Nägeli. If Mendel believed in the priesthood, then he should stick to it; Nägeli believed in the priesthood of science.

Nägeli was studying another plant—the yellow-flowering hawkweed—and he urged Mendel to try to reproduce his findings on hawkweed as well. It was a catastrophically wrong choice. Mendel had chosen peas after deep consideration: the plants reproduced sexually, produced clearly identifiable variant traits, and could be cross-pollinated with some care. Hawkweeds—unknown to Mendel and Nägeli—could reproduce asexually (i.e., without pollen and eggs). They were virtually impossible to cross-pollinate and rarely generated hybrids. Predictably, the results were a mess. Mendel tried to make sense of the hawkweed hybrids (which were not hybrids at all), but he couldn't decipher any of the patterns that he had observed in the peas. Between 1867 and 1871, he pushed himself even harder, growing thousands of hawkweeds in another patch of garden, emasculating the flowers with the same forceps, and dusting pollen with the same paintbrush. His letters to Nägeli grew increasingly despondent. Nägeli replied occasionally, but the letters were infrequent and patronizing. He could hardly be bothered with the progressively lunatic ramblings of a self-taught monk in Brno.

In November 1873, Mendel wrote his last letter to Nägeli. He had been unable to complete the experiments, he reported remorsefully. He had been promoted to the position of abbot of the monastery in Brno, and his administrative responsibilities were now making it impossible for him to continue any plant studies. "I feel truly unhappy that I have to neglect my plants . . . so completely," Mendel wrote. Science was pushed to the wayside. Taxes piled up at the monastery. New prelates had to be appointed. Bill by bill, and letter by letter, his scientific imagination was slowly choked by administrative work.

Mendel wrote only one monumental paper on pea hybrids. His health declined in the 1880s, and he gradually restricted his work—all except his beloved gardening. On January 6, 1884, Mendel died of kidney failure in Brno, his feet swollen with fluids. The local newspaper wrote an obituary, but made no mention of his experimental studies. Perhaps more fitting was a short note from one of the younger monks in the monastery: "Gentle, free-handed, and kindly . . . Flowers he loved."

"A Certain Mendel"

The origin of species is a natural phenomenon.
—Jean-Baptiste Lamarck

The origin of species is an object of inquiry.
—Charles Darwin

The origin of species is an object of experimental investigation.
—Hugo de Vries

In the summer of 1878, a thirty-year-old Dutch botanist named Hugo de Vries traveled to England to see Darwin. It was more of a pilgrimage than a scientific visit. Darwin was vacationing at his sister's estate in Dorking, but de Vries tracked him down and traveled out to meet him. Gaunt, intense, and excitable, with Rasputin's piercing eyes and a beard that rivaled Darwin's, de Vries already looked like a younger version of his idol. He also had Darwin's persistence. The meeting must have been exhausting, for it lasted only two hours, and Darwin had to excuse himself to take a break. But de Vries left England transformed. With no more than a brief conversation, Darwin had inserted a sluice into de Vries's darting mind, diverting it forever. Back in Amsterdam, de Vries abruptly terminated his prior work on the movement of tendrils in plants and threw himself into solving the mystery of heredity.

By the late 1800s, the problem of heredity had acquired a near-mystical aura of glamour, like a Fermat's Last Theorem for biologists. Like Fermat—the odd French mathematician who had tantalizingly scribbled that he'd found a "remarkable proof" of his theorem, but failed to write it down because the paper's "margin was too small"—Darwin had desultorily announced that he had found a solution to heredity, but had never published it. "In another work I shall discuss, if time and health permit, the variability of organic beings in a state of nature," Darwin had written in 1868.

Darwin understood the stakes implicit in that claim. A theory of heredity was crucial to the theory of evolution: without any means to generate variation, and fix it across generations, he knew, there would be no mechanism for an organism to evolve new characteristics. But a decade had passed, and Darwin had never published the promised book on the genesis of "variability in organic beings." Darwin died in 1882, just four years after de Vries's visit. A generation of young biologists was now rifling through Darwin's works to find clues to the theory that had gone missing.

De Vries also pored through Darwin's books, and he latched onto the theory of pangenesis—the idea that "particles of information" from the body were somehow collected and collated in sperm and eggs. But the notion of messages emanating from cells and assembling in sperm as a manual for building an organism seemed particularly far-fetched; it was as if the sperm were trying to write the Book of Man by collecting telegrams.

And experimental proof against pangenes and gemmules was mounting. In 1883, with rather grim determination, the German embryologist August Weismann had performed an experiment that directly attacked Darwin's gemmule theory of heredity. Weismann had surgically excised the tails of five generations of mice, then bred the mice to determine if the offspring would be born tailless. But the mice—with equal and obdurate consistency—had been born with tails perfectly intact, generation upon generation. If gemmules existed, then a mouse with a surgically excised tail should produce a mouse without a tail. In total, Weismann had serially removed the tails of 901 animals. And mice with absolutely normal tails—not even marginally shorter than the tail of the original mouse—had kept arising; it was impossible to wash "the hereditary taint" (or, at least, the "hereditary tail") away. Grisly as it was, the experiment nonetheless announced that Darwin and Lamarck could not be right.

Weismann had proposed a radical alternative: perhaps hereditary information was contained *exclusively* in sperm and egg cells, with no direct mechanism for an acquired characteristic to be transmitted into sperm or eggs. No matter how ardently the giraffe's ancestor stretched its neck, it could not convey that information into its genetic material. Weismann called this hereditary material *germplasm* and argued that it was the only method by which an organism could generate another organism. Indeed, all of evolution could be perceived as the vertical transfer of germplasm from one generation to the next: an egg was the only way for a chicken to transfer information to another chicken.

①

But what was the material nature of germplasm? de Vries wondered. Was it like paint: Could it be mixed and diluted? Or was the information in germplasm discrete and carried in packets—like an unbroken, unbreakable message? De Vries had not encountered Mendel's paper yet. But like Mendel, he began to scour the countryside around Amsterdam to collect strange plant variants—not just peas, but a vast herbarium of plants with twisted stems and forked leaves, with speckled flowers, hairy anthers, and bat-shaped seeds: a menagerie of monsters. When he bred these variants with their normal counterparts, he found, like Mendel, that the variant traits did not blend away, but were maintained in a discrete and independent form from one generation to the next. Each plant seemed to possess a collection of features—flower color, leaf shape, seed texture—and each of these features seemed to be encoded by an independent, discrete piece of information that moved from one generation to the next.

But de Vries still lacked Mendel's crucial insight—that bolt of mathematical reasoning that had so clearly illuminated Mendel's pea-hybrid experiments in 1865. From his own plant hybrids, de Vries could dimly tell that variant features, such as stem size, were encoded by indivisible particles of information. But how many particles were needed to encode one variant trait? One? One hundred? A thousand?

In the 1880s, still unaware of Mendel's work, de Vries edged toward a more quantitative description of his plant experiments. In a landmark paper written in 1897, entitled *Hereditary Monstrosities*, de Vries analyzed his data and inferred that each trait was governed by a *single* particle of information. Every hybrid inherited two such particles—one from the sperm and one from the egg. And these particles were passed along, intact, to the next generation through sperm and egg. Nothing was ever blended. No information was lost. He called these particles "pangenes." It was a name that protested its own origin: even though he had systematically demolished Darwin's theory of pangenesis, de Vries paid his mentor a final homage.

①

While de Vries was still knee-deep in the study of plant hybrids in the spring of 1900, a friend sent him a copy of an old paper drudged up from the friend's library. "I know that you are studying hybrids," the friend wrote, "so perhaps the enclosed reprint of the year 1865 by a certain Mendel . . . is still of some interest to you."

It is hard not to imagine de Vries, in his study in Amsterdam on a gray March morning, slitting open that reprint and running his eyes through the first paragraph. Reading the paper, he must have felt that inescapable chill of déjà vu running through his spine: the "certain Mendel" had certainly preempted de Vries by more than three decades. In Mendel's paper, de Vries discovered a solution to his question, a perfect corroboration of his experiments—and a challenge to his originality. It seemed that he too was being forced to relive the old saga of Darwin and Wallace: the scientific discovery that he had hoped to claim as his own had actually been made by someone else. In a fit of panic, de Vries rushed his paper on plant hybrids to print in March 1900, pointedly neglecting any mention of Mendel's prior work. Perhaps the world had forgotten "a certain Mendel" and his work on pea hybrids in Brno. "Modesty is a virtue," he would later write, "yet one gets further without it."

①

De Vries was not alone in rediscovering Mendel's notion of independent, indivisible hereditary instructions. That same year de Vries published his monumental study of plant variants, Carl Correns, a botanist in Tübingen, published a study on pea and maize hybrids that precisely recapitulated Mendel's results. Correns had, ironically, been Nägeli's student in Munich. But Nägeli—who considered Mendel an amateur crank—had neglected to tell Correns about the voluminous correspondence on pea hybrids that he had once received from "a certain Mendel."

In his experimental gardens in Munich and Tübingen, about four hundred miles from the abbey, Correns thus laboriously bred tall plants with short plants and made hybrid-hybrid crosses—with no knowledge that he was just methodically repeating Mendel's prior work. When Correns completed his experiments and was ready to assemble his paper for publication, he returned to the library to find references to his scientific predecessors. He thus stumbled on Mendel's earlier paper buried in the Brno journal.

And in Vienna—the very place where Mendel had failed his botany exam in 1856—another young botanist, Erich von Tschermak-Seysenegg, also rediscovered Mendel's laws. Von Tschermak had been a graduate student at Halle and in Ghent, where, working on pea hybrids, he had also observed hereditary traits moving independently and discretely, like particles, across generations of hybrids. The youngest of the three scientists, von Tschermak had received news of two other parallel studies that fully corroborated his results, then waded back into the scientific literature to discover Mendel. He too had felt that ascending chill of déjà vu as he read the opening salvos of Mendel's paper. "I too still believed that I had found something new," he would later write, with more than a tinge of envy and despondency.

Being rediscovered once is proof of a scientist's prescience. Being rediscovered thrice is an insult. That three papers in the short span of three months in 1900 independently converged on Mendel's work was a demonstration of the sustained myopia of biologists, who had ignored his work for nearly forty years. Even de Vries, who had so conspicuously forgotten to mention Mendel in his first study, was forced to acknowledge Mendel's contribution. In the spring of 1900, soon after de Vries had published his paper, Carl Correns suggested that de Vries had appropriated Mendel's work deliberately—committing something akin to scientific plagiarism ("by a strange coincidence," Correns wrote mincingly, de Vries had even incorporated "Mendel's vocabulary" in his paper). Eventually, de Vries caved in. In a subsequent version of his analysis of plant hybrids, he mentioned Mendel glowingly and acknowledged that he had merely "extended" Mendel's earlier work.

But de Vries also took his experiments further than Mendel. He may have been preempted in the discovery of heritable units—but as de Vries delved more deeply into heredity and evolution, he was struck by a thought that must also have perplexed Mendel: *How did variants arise in the first place?* What force made tall versus short peas, or purple flowers and white ones?

The answer, again, was in the garden. Wandering through the countryside in one of his collecting expeditions, de Vries stumbled on an enormous, invasive patch of primroses growing in the wild—a species named (ironically, as he would soon discover) after Lamarck: *Oenothera lamarck-*

iana. De Vries harvested and planted fifty thousand seeds from the patch. Over the next years, as the vigorous *Oenothera* multiplied, de Vries found that eight hundred new variants had spontaneously arisen—plants with gigantic leaves, with hairy stems, or with odd-shaped flowers. Nature had spontaneously thrown up rare freaks—precisely the mechanism that Darwin had proposed as evolution's first step. Darwin had called these variants "sports," implying a streak of capricious whimsy in the natural world. De Vries chose a more serious-sounding word. He called them *mutants*—from the Latin word for "change."*

De Vries quickly realized the importance of his observation: these mutants had to be the missing pieces in Darwin's puzzle. Indeed, if you coupled the genesis of spontaneous mutants—the giant-leaved *Oenothera*, say—with natural selection, then Darwin's relentless engine was automatically set in motion. Mutations created variants in nature: long-necked antelopes, short-beaked finches, and giant-leaved plants arose spontaneously in the vast tribes of normal specimens (contrary to Lamarck, these mutants were not generated purposefully, but by random chance). These variant qualities were hereditary—carried as discrete instructions in sperm and eggs. As animals struggled to survive, the best-adapted variants—the fittest mutations—were serially selected. Their children inherited these mutations and thus generated new species, thereby driving evolution. Natural selection was not operating on organisms but on their units of heredity. A chicken, de Vries realized, was merely an egg's way of making a better egg.

☾

It had taken two excruciatingly slow decades for Hugo de Vries to become a convert to Mendel's ideas of heredity. For William Bateson, the English biologist, the conversion took about an hour—the time spent on a speeding train between Cambridge and London in May 1900.† That evening, Bateson was traveling to the city to deliver a lecture on heredity at

---

* De Vries's "mutants" might actually have been the result of backcrosses, rather than spontaneously arising variants.

† The story of Bateson's "conversion" to Mendel's theory during a train ride has been disputed by some historians. The story appears frequently in his biography, but may have been embellished by some of Bateson's students for dramatic flair.

the Royal Horticultural Society. As the train trundled through the darkening fens, Bateson read a copy of de Vries's paper—and was instantly transmuted by Mendel's idea of discrete units of heredity. This was to be Bateson's fateful journey: by the time he reached the society's office on Vincent Square, his mind was spinning. "We are in the presence of a new principle of the highest importance," he told the lecture hall. "To what further conclusions it may lead us cannot yet be foretold." In August that year, Bateson wrote to his friend Francis Galton: "I am writing to ask you to look up the paper of Mendl [*sic*] [which] seems to me one of the most remarkable investigations yet made on heredity and it is extraordinary that it should have got forgotten."

Bateson made it his personal mission to ensure that Mendel, once forgotten, would never again be ignored. First, he independently confirmed Mendel's work on plant hybrids in Cambridge. Bateson met de Vries in London and was impressed by his experimental rigor and his scientific vitality (although not by his continental habits. De Vries refused to bathe before dinner, Bateson complained: "His linen is foul. I daresay he puts on a new shirt once a week"). Doubly convinced by Mendel's experimental data, and by his own evidence, Bateson set about proselytizing. Nicknamed "Mendel's bulldog"—an animal that he resembled both in countenance and temperament—Bateson traveled to Germany, France, Italy, and the United States, giving talks on heredity that emphasized Mendel's discovery. Bateson knew that he was witnessing, or, rather, midwifing, the birth of a profound revolution in biology. Deciphering the laws of heredity, he wrote, would transform "man's outlook on the world, and his power over nature" more "than any other advance in natural knowledge that can be foreseen."

In Cambridge, a group of young students gathered around Bateson to study the new science of heredity. Bateson knew that he needed a name for the discipline that was being born around him. *Pangenetics* seemed an obvious choice—extending de Vries's use of the word *pangene* to denote the units of heredity. But *pangenetics* was overloaded with all the baggage of Darwin's mistaken theory of hereditary instructions. "No single word in common use quite gives this meaning [yet] such a word is badly wanted," Bateson wrote.

In 1905, still struggling for an alternative, Bateson coined a word of his own. *Genetics*, he called it: the study of heredity and variation—the word ultimately derived from the Greek *genno*, "to give birth."

Bateson was acutely aware of the potential social and political impact of the newborn science. "What will happen when . . . enlightenment actually comes to pass and the facts of heredity are . . . commonly known?" he wrote, with striking prescience, in 1905. "One thing is certain: mankind will begin to interfere; perhaps not in England, but in some country more ready to break with the past and eager for 'national efficiency.' . . . Ignorance of the remoter consequences of interference has never long postponed such experiments."

More than any scientist before him, Bateson also grasped the idea that the discontinuous nature of genetic information carried vast implications for the future of human genetics. *If genes were, indeed, independent particles of information, then it should be possible to select, purify, and manipulate these particles independently from one another.* Genes for "desirable" attributes might be selected or augmented, while undesirable genes might be eliminated from the gene pool. In principle, a scientist should be able to change the "composition of individuals," and of nations, and leave a permanent mark on human identity.

"When power is discovered, man always turns to it," Bateson wrote darkly. "The science of heredity will soon provide power on a stupendous scale; and in some country, at some time not, perhaps, far distant, that power will be applied to control the composition of a nation. Whether the institution of such control will ultimately be good or bad for that nation, or for humanity at large, is a separate question." He had preempted the century of the gene.

Eugenics

Improved environment and education may better the generation already born. Improved blood will better every generation to come.
—Herbert Walter, *Genetics*

Most Eugenists are Euphemists. I mean merely that short words startle them, while long words soothe them. And they are utterly incapable of translating the one into the other. . . . Say to them "The . . . citizen should . . . make sure that the burden of longevity in the previous generations does not become disproportionate and intolerable, especially to the females"; say this to them and they sway slightly to and fro. . . . Say to them "Murder your mother," and they sit up quite suddenly.
—G. K. Chesterton, *Eugenics and Other Evils*

In 1883, one year after Charles Darwin's death, Darwin's cousin Francis Galton published a provocative book—*Inquiries into Human Faculty and Its Development*—in which he laid out a strategic plan for the improvement of the human race. Galton's idea was simple: he would mimic the mechanism of natural selection. If nature could achieve such remarkable effects on animal populations through survival and selection, Galton imagined accelerating the process of refining humans via human intervention. The selective breeding of the strongest, smartest, "fittest" humans—*unnatural* selection—Galton imagined, could achieve over just a few decades what nature had been attempting for eons.

Galton needed a word for this strategy. "We greatly want a brief word to express the science of improving stock," he wrote, "to give the more suitable races or strains of blood a better chance of prevailing speedily over the less suitable." For Galton, the word *eugenics* was an opportune fit—"at least a neater word . . . than *viriculture*, which I once ventured to use." It combined the Greek prefix *eu*—"good"—with *genesis*: "good in stock, hereditarily endowed with noble qualities." Galton—who never blanched from the recognition of his own genius—was deeply satisfied with his coinage: "Believing, as I do, that human eugenics will become recognised before long as a study of the highest practical importance, it seems to me that no time ought to be lost in . . . compiling personal and family histories."

①

Galton was born in the winter of 1822—the same year as Gregor Mendel—and thirteen years after his cousin Charles Darwin. Slung between the two giants of modern biology, he was inevitably haunted by an acute sense of scientific inadequacy. For Galton, the inadequacy may have felt particularly galling because he too had been meant to become a giant. His father was a wealthy banker in Birmingham; his mother was the daughter of Erasmus Darwin, the polymath poet and doctor, who was also Charles Darwin's grandfather. A child prodigy, Galton learned to read at two, was fluent in Greek and Latin by five, and solved quadratic equations by eight. Like Darwin, he collected beetles, but he lacked his cousin's plodding, taxonomic mind and soon gave up his collection for more ambitious pursuits. He tried studying medicine, but then switched to mathematics at Cambridge. In 1843, he attempted an honors exam in mathematics, but suffered a nervous breakdown and returned home to recuperate.

In the summer of 1844, while Charles Darwin was writing his first essay on evolution, Galton left England to travel to Egypt and Sudan—the first of many trips he would take to Africa. But while Darwin's encounters with the "natives" of South America in the 1830s had strengthened his belief in the common ancestry of humans, Galton only saw difference: "I saw enough of savage races to give me material to think about all the rest of my life."

In 1859, Galton read Darwin's *Origin of Species*. Rather, he "devoured" the book: it struck him like a jolt of electricity, both paralyzing and galvanizing him. He simmered with envy, pride, and admiration. He had been "initiated into an entirely new province of knowledge," he wrote glowingly to Darwin.

65

The "province of knowledge" that Galton felt particularly inclined to explore was heredity. Like Fleeming Jenkin, Galton quickly realized that his cousin had got the principle right, but not the mechanism: the nature of inheritance was crucial to the understanding of Darwin's theory. Heredity was the yin to evolution's yang. The two theories had to be congenitally linked—each bolstering and completing the other. If "cousin Darwin" had solved half the puzzle, then "cousin Galton" was destined to crack the other.

In the mid-1860s, Galton began to study heredity. Darwin's "gemmule" theory—that hereditary instructions were thrown adrift by all cells and then floated in the blood, like a million messages in bottles—suggested that blood transfusions might transmit gemmules and thereby alter heredity. Galton tried transfusing rabbits with the blood of other rabbits to transmit the gemmules. He even tried working with plants—peas, of all things—to understand the basis of hereditary instructions. But he was an abysmal experimentalist; he lacked Mendel's instinctive touch. The rabbits died of shock, and the vines withered in his garden. Frustrated, Galton switched to the study of humans. Model organisms had failed to reveal the mechanism of heredity. The measurement of variance and heredity in humans, Galton reasoned, should unlock the secret. The decision bore the hallmarks of his overarching ambition: a top-down approach, beginning with the most complex and variant traits conceivable—intelligence, temperament, physical prowess, height. It was a decision that would launch him into a full-fledged battle with the science of genetics.

Galton was not the first to attempt to model human heredity by measuring variation in humans. In the 1830s and 1840s, the Belgian scientist Adolphe Quetelet—an astronomer-turned-biologist—had begun to systematically measure human features and analyze them using statistical methods. Quetelet's approach was rigorous and comprehensive. "Man is born, grows up and dies according to certain laws that have never been studied," Quetelet wrote. He tabulated the chest breadth and height of 5,738 soldiers to demonstrate that chest size and height were distributed along smooth, continuous, bell-shaped curves. Indeed, wherever Quetelet looked, he found a recurrent pattern: human features—even behaviors—were distributed in bell-shaped curves.

Galton was inspired by Quetelet's measurements and ventured deeper into the measurement of human variance. Were complex features such as intelligence, intellectual accomplishment, or beauty, say, variant in the same manner? Galton knew that no ordinary measuring devices existed for any of these characteristics. But where he lacked devices, he invented them ("Whenever you can, [you should] count," he wrote). As a surrogate for intelligence, he obtained the examination marks for the mathematical honors exam at Cambridge—ironically, the very test that he had failed—and demonstrated that, to the best approximation, even examination abilities followed this bell-curve distribution. He walked through England and Scotland tabulating "beauty"—secretly ranking the women he met as "attractive," "indifferent," or "repellent" using pinpricks on a card hidden in his pocket. It seemed no human attribute could escape Galton's sifting, evaluating, counting, tabulating eye: "Keenness of Sight and Hearing; Colour Sense; Judgment of Eye; Breathing Power; Reaction Time; Strength and Pull of Squeeze; Force of Blow; Span of Arms; Height . . . Weight."

Galton now turned from measurement to mechanism. Were these variations in humans inherited? And in what manner? Again, he veered away from simple organisms, hoping to jump straight into humans. Wasn't his own exalted pedigree—Erasmus as grandfather, Darwin as cousin—proof that genius ran in families? To marshal further evidence, Galton began to reconstruct pedigrees of eminent men. He found, for instance, that among 605 notable men who lived between 1453 and 1853, there were 102 familial relationships: one in six of all accomplished men were apparently related. If an accomplished man had a son, Galton estimated, chances were one in twelve that the son would be eminent. In contrast, only one in three thousand "randomly" selected men could achieve distinction. Eminence, Galton argued, was inherited. Lords produced lords—not because peerage was hereditary, but because intelligence was.

Galton considered the obvious possibility that eminent men might produce eminent sons because the son "will be placed in a more favorable position for advancement." Galton coined the memorable phrase *nature versus nurture* to discriminate hereditary and environmental influences. But his anxieties about class and status were so deep that he could not bear the thought that his own "intelligence" might merely be the by-product of privilege and opportunity. Genius had to be encrypted in genes. He had barricaded the most fragile of his convictions—that purely hereditary influences could explain such patterns of accomplishment—from any scientific challenge.

Galton published much of this data in an ambitious, rambling, often incoherent book—*Hereditary Genius*. It was poorly received. Darwin read the study, but he was not particularly convinced, damning his cousin with faint praise: "You have made a convert of an opponent in one sense, for I have always maintained that, excepting fools, men did not differ much in intellect, only in zeal and hard work." Galton swallowed his pride and did not attempt another genealogical study.

①

Galton must have realized the inherent limits of his pedigree project, for he soon abandoned it for a more powerful empirical approach. In the mid-1880s, he began to mail out "surveys" to men and women, asking them to examine their family records, tabulate the data, and mail him detailed measurements on the height, weight, eye color, intelligence, and artistic abilities of parents, grandparents, and children (Galton's family fortune—his most tangible inheritance—came in handy here; he offered a substantial fee to anyone who returned a satisfactory survey). Armed with real numbers, Galton could now find the elusive "law of heredity" that he had hunted so ardently for decades.

Much of what he found was relatively intuitive—albeit with a twist. Tall parents tended to have tall children, he discovered—but *on average*. The children of tall men and women were certainly taller than the mean height of the population, but they too varied in a bell-shaped curve, with some taller and some shorter than their parents.* If a general rule of inheritance lurked behind the data, it was that human features were distributed in continuous curves, and continuous variations reproduced continuous variations.

But did a law—an underlying pattern—govern the genesis of variants? In the late 1880s, Galton boldly synthesized all his observations into his most mature hypothesis on heredity. He proposed that every feature in a human—height, weight, intelligence, beauty—was a composite function generated by a conserved pattern of ancestral inheritance. The parents of a child provided, on average, half the content of that feature; the grand-

---

* Indeed, the *mean* height of the sons of exceptionally tall fathers tended to be slightly lower than the father's height—and closer to the population's average—as if an invisible force were always dragging extreme features toward the center. This discovery—called regression to the mean—would have a powerful effect on the science of measurement and the concept of variance. It would be Galton's most important contribution to statistics.

parents, a quarter; the great-grandparents, an eighth—and so forth, all the way back to the most distant ancestor. The sum of all contributions could be described by the series—½ + ¼ + ⅛ . . .—all of which conveniently added to 1. Galton called this the Ancestral Law of Heredity. It was a sort of mathematical homunculus—an idea borrowed from Pythagoras and Plato—but dressed up with fractions and denominators into a modern-sounding law.

Galton knew that the crowning achievement of the law would be its ability to accurately predict a real pattern of inheritance. In 1897, he found his ideal test case. Capitalizing on yet another English pedigree obsession—of dogs—Galton discovered an invaluable manuscript: the *Basset Hound Club Rules*, a compendium published by Sir Everett Millais in 1896, which documented the coat colors of basset hounds across multiple generations. To his great relief, Galton found that his law could accurately predict the coat colors of every generation. He had finally solved the code of heredity.

The solution, however satisfying, was short-lived. Between 1901 and 1905, Galton locked horns with his most formidable adversary—William Bateson, the Cambridge geneticist who was the most ardent champion of Mendel's theory. Dogged and imperious, with a handlebar mustache that seemed to bend his smile into a perpetual scowl, Bateson was un-moved by equations. The basset-hound data, Bateson argued, was either aberrant or inaccurate. Beautiful laws were often killed by ugly facts—and despite how lovely Galton's infinite series looked, Bateson's own experiments pointed decidedly toward one fact: that hereditary instructions were carried by individual units of information, not by halved and quartered messages from ghostly ancestors. Mendel, despite his odd scientific lineage, and de Vries, despite his dubious personal hygiene, were right. A child *was* an ancestral composite, but a supremely simple one: one-half from the mother, one-half from the father. Each parent contributed a set of instructions, which were decoded to create a child.

Galton defended his theory against Bateson's attack. Two prominent biologists—Walter Weldon and Arthur Darbishire—and the eminent mathematician Karl Pearson joined the effort to defend the "ancestral law," and the debate soured quickly into an all-out war. Weldon, once Bateson's teacher at Cambridge, turned into his most vigorous opponent. He labeled Bateson's experiments "utterly inadequate" and refused to be-lieve de Vries's studies. Pearson, meanwhile, founded a scientific journal,

Biometrika (its name drawn from Galton's notion of biological measurement), which he turned into a mouthpiece for Galton's theory.

In 1902, Darbishire launched a fresh volley of experiments on mice, hoping to disprove Mendel's hypothesis once and for all. He bred mice by the thousands, hoping to prove Galton right. But as Darbishire analyzed his own first-generation hybrids, and the hybrid-hybrid crosses, the pattern was clear: the data could only be explained by Mendelian inheritance, with indivisible traits moving vertically across the generations. Darbishire resisted at first, but he could no longer deny the data; he ultimately conceded the point.

In the spring of 1905, Weldon lugged copies of Bateson's and Darbishire's data to his vacation in Rome, where he sat, stewing with anger, trying, like a "mere clerk," to rework the data to fit Galtonian theory. He returned to England that summer, hoping to overturn the studies with his analysis, but was struck by pneumonia and died suddenly at home. He was only forty-six years old. Bateson wrote a moving obituary to his old friend and teacher. "To Weldon I owe the chief awakening of my life," he recalled, "but this is the personal, private obligation of my own soul."

℗

Bateson's "awakening" was not private in the least. Between 1900 and 1910, as evidence for Mendel's "units of heredity" mounted, biologists were confronted by the impact of the new theory. The implications were deep. Aristotle had recast heredity as the flow of information—a river of code moving from egg to the embryo. Centuries later, Mendel had stumbled on the essential structure of that information, the alphabet of the code. If Aristotle had described a current of information moving across generations, then Mendel had found its currency.

But perhaps an even greater principle was at stake, Bateson realized. The flow of biological information was not restricted to heredity. It was coursing through all of biology. The transmission of hereditary traits was just one instance of information flow—but if you looked deeply, squinting your conceptual lenses, it was easy to imagine information moving pervasively through the entire living world. The unfurling of an embryo; the reach of a plant toward sunlight; the ritual dance of bees—every biological activity required the decoding of coded instructions. Might Mendel, then, have also stumbled on the essential structure of these instructions?

Were units of information guiding each of these processes? "Each of us who now looks at his own patch of work sees Mendel's clues running through it," Bateson proposed. "We have only touched the edge of that new country which is stretching out before us. . . . The experimental study of heredity . . . is second to no branch of science in the magnitude of the results it offers."

The "new country" demanded a new language: Mendel's "units of heredity" had to be christened. The word *atom*, used in the modern sense, first entered scientific vocabulary in John Dalton's paper in 1808. In the summer of 1909, almost exactly a century later, the botanist Wilhelm Johannsen coined a distinct word to denote a unit of heredity. At first, he considered using de Vries's word, *pangene*, with its homage to Darwin. But Darwin, in all truth, had misconceived the notion, and *pangene* would always carry the memory of that misconception. Johannsen shortened the word to *gene*. (Bateson wanted to call it *gen*, hoping to avoid errors in pronunciation—but it was too late. Johannsen's coinage, and the continental habit of mangling English, were here to stay.)

As with Dalton and the atom, neither Bateson nor Johannsen had any understanding of what a gene *was*. They could not fathom its material form, its physical or chemical structure, its location within the body or inside the cell, or even its mechanism of action. The word was created to mark a function; it was an abstraction. A gene was defined by what a gene *does*: it was a carrier of hereditary information. "Language is not only our servant," Johannsen wrote, "[but] it may also be our master. It is desirable to create new terminology in all cases where new and revised conceptions are being developed. Therefore, I have proposed the word 'gene.' The 'gene' is nothing but a very applicable little word. It may be useful as an expression for the 'unit factors' . . . demonstrated by modern Mendelian researchers." "The word 'gene' is completely free of any hypothesis," Johannsen remarked. "It expresses only the evident fact that . . . many characteristics of the organism are specified . . . in unique, separate and thereby independent ways."

But in science, a word *is* a hypothesis. In natural language, a word is used to convey an idea. But in scientific language, a word conveys more than an idea—a mechanism, a consequence, a prediction. A scientific noun can launch a thousand questions—and the idea of the "gene" did exactly that. What was the chemical and physical nature of the gene? How was the set of genetic instructions, the *genotype*, translated into the actual physical manifestations, the *phenotype*, of an organism? How were genes transmitted? Where did they reside? How were they regulated? If genes were discrete particles specifying one trait, then how could this property be reconciled with the occurrence of human characteristics, say, height or skin color, in continuous curves? How does the gene permit genesis?

"The science of genetics is so new that it is impossible to say . . . what its boundaries may be," a botanist wrote in 1914. "In research, as in all business of exploration, the stirring time comes when a fresh region is unlocked by the discovery of a new key."

①

Cloistered in his sprawling town house on Rutland Gate, Francis Galton was oddly unstirred by the "stirring times." As biologists rushed to embrace Mendel's laws and grapple with their consequences, Galton adopted a rather benign indifference to them. Whether hereditary units were divisible or indivisible did not particularly bother him; what concerned him was whether heredity was *actionable* or *inactionable*: whether human inheritance could be manipulated for human benefit.

"All around [Galton]," the historian Daniel Kevles wrote, "the technology of the industrial revolution confirmed man's mastery of nature." Galton had been unable to discover genes, but he would not miss out on the creation of genetic technologies. Galton had already coined a name for this effort—*eugenics*, the betterment of the human race via artificial selection of genetic traits and directed breeding of human carriers. Eugenics was merely an applied form of genetics for Galton, just as agriculture was an applied form of botany. "What nature does blindly, slowly and ruthlessly, man may do providently, quickly, and kindly. As it lies within his power, so it becomes his duty to work in that direction," Galton wrote. He had originally proposed the concept in *Hereditary Genius* as early as 1869—thirty years before the rediscovery of Mendel—but left the idea unexplored, concentrating, instead, on the mechanism of heredity. But as Galton's hypothesis about "ancestral inheritance" had been dismantled, piece by piece, by Bateson and de Vries, Galton had taken a sharp turn from a descriptive impulse to a prescriptive one. He may have misunderstood the biological basis of human heredity—but at least he understood what to *do* about it. "This is not a question for the microscope," one of his protégés wrote—a sly barb directed at Bateson, Morgan, and de Vries. "It involves a study of . . . forces which bring greatness to the social group."

In the spring of 1904, Galton presented his argument for eugenics at a public lecture at the London School of Economics. It was a typical Bloomsbury evening. Coiffed and resplendent, the city's perfumed elite blew into the auditorium to hear Galton: George Bernard Shaw and H. G. Wells; Alice Drysdale-Vickery, the social reformer; Lady Welby, the philosopher of language; the sociologist Benjamin Kidd; the psychiatrist Henry Maudsley. Pearson, Weldon, and Bateson arrived late and sat apart, still seething with mutual distrust.

Galton's remarks lasted ten minutes. Eugenics, he proposed, had to be "introduced into the national consciousness, like a new religion." Its founding tenets were borrowed from Darwin—but they grafted the logic of natural selection onto human societies. "All creatures would agree that it was better to be healthy than sick, vigorous than weak, well-fitted than ill-fitted for their part in life; in short, that it was better to be good rather than bad specimens of their kind, whatever that kind might be. So with men."

The purpose of eugenics was to accelerate the selection of the well-fitted over the ill-fitted, and the healthy over the sick. To achieve this, Galton proposed to selectively breed the strong. Marriage, he argued, could easily be subverted for this purpose—but only if enough social pressure could be applied: "if unsuitable marriages from the eugenic point of view were banned socially . . . very few would be made." As Galton imagined it, a record of the best traits in the best families could be maintained by society—generating a human studbook, of sorts. Men and women would be selected from this "golden book"—as he called it—and bred to produce the best offspring, in a manner akin to basset hounds and horses.

⊕

Galton's remarks were brief—but the crowd had already grown restless. Henry Maudsley, the psychiatrist, launched the first attack, questioning Galton's assumptions about heredity. Maudsley had studied mental illness among families and concluded that the patterns of inheritance were vastly more complex than the ones Galton had proposed. Normal fathers produced schizophrenic sons. Ordinary families generated extraordinary children. The child of a barely known glove maker from the Midlands—

"born of parents not distinguished from their neighbors"—could grow up to be the most prominent writer of the English language. "He had five brothers," Maudsley noted, yet, while one boy, William, "rose to the extraordinary eminence that he did, none of his brothers distinguished themselves in any way." The list of "defective" geniuses went on and on: Newton was a sickly, fragile child; John Calvin was severely asthmatic; Darwin suffered crippling bouts of diarrhea and near-catatonic depression. Herbert Spencer—the philosopher who had coined the phrase *survival of the fittest*—had spent much of his life bedridden with various illnesses, struggling with his own fitness for survival.

But where Maudsley proposed caution, others urged speed. H. G. Wells, the novelist, was no stranger to eugenics. In his book *The Time Machine*, published in 1895, Wells had imagined a future race of humans that, having selected innocence and virtue as desirable traits, had inbred to the point of effeteness—degenerating into an etiolated, childlike race devoid of any curiosity or passion. Wells agreed with Galton's impulses to manipulate heredity as a means to create a "fitter society." But selective inbreeding via marriage, Wells argued, might paradoxically produce weaker and duller generations. The only solution was to consider the macabre alternative—the selective elimination of the weak. "It is in the sterilization of failure, and not in the selection of successes for breeding, that the possibility of an improvement of the human stock lies."

Bateson spoke in the end, sounding the darkest, and most scientifically sound, note of the meeting. Galton had proposed using physical and mental traits—human *phenotype*—to select the best specimens for breeding. But the real information, Bateson argued, was not contained in the features, but in the combination of genes that determined them—i.e., in the *genotype*. The physical and mental characteristics that had so entranced Galton—height, weight, beauty, intelligence—were merely the outer shadows of genetic characteristics lurking underneath. The real power of eugenics lay in the manipulation of genes—not in the selection of features. Galton may have derided the "microscope" of experimental geneticists, but the tool was far more powerful than Galton had presumed, for it could penetrate the outer shell of heredity into the mechanism itself. Heredity, Bateson warned, would soon be shown to "follow a precise law of remarkable simplicity." If the eugenicist learned these laws and then figured out how to hack them—à la Plato—he would acquire unprecedented power; by manipulating genes, he could manipulate the future.

Galton's talk might not have generated the effusive endorsement that he had expected—he later groused that his audience was "living forty years ago"—but he had obviously touched a raw nerve. Like many members of the Victorian elite, Galton and his friends were chilled by the fear of race degeneration (Galton's own encounter with the "savage races," symptomatic of Britain's encounter with colonial natives throughout the seventeenth and eighteenth centuries, had also convinced him that the racial purity of whites had to be maintained and protected against the forces of miscegenation). The Second Reform Act of 1867 had given working-class men in Britain the right to vote. By 1906, even the best-guarded political bastions had been stormed—twenty-nine seats in Parliament had fallen to the Labour Party—sending spasms of anxiety through English high society. The political empowerment of the working class, Galton believed, would just provoke their genetic empowerment: they would produce bushels of children, dominate the gene pool, and drag the nation toward profound mediocrity. The *homme moyen* would degenerate. The "mean man" would become even meaner.

"A pleasant sort o' soft woman may go on breeding you stupid lads [till] the world was turned topsy-turvy," George Eliot had written in *The Mill on the Floss* in 1860. For Galton, the continuous reproduction of softheaded women and men posed a grave genetic threat to the nation. Thomas Hobbes had worried about a state of nature that was "poor, nasty, brutish and short"; Galton was concerned about a future state overrun by genetic inferiors: poor, nasty, British—and short. The brooding masses, he worried, were also the *breeding* masses and, left to themselves, would inevitably produce a vast, unwashed inferior breed (he called this process *kakogenics*—"from bad genes").

Indeed, Wells had only articulated what many in Galton's inner circle felt deeply but had not dared to utter—that eugenics would only work if the selective breeding of the strong (so-called positive eugenics) was augmented with selective sterilization of the weak—negative eugenics. In 1911, Havelock Ellis, Galton's colleague, twisted the image of Mendel, the solitary gardener, to service his enthusiasm for sterilization: "In the great garden of life it is not otherwise than in our public gardens. We repress the license of those who, to gratify their own childish or perverted desires, would pluck up the shrubs or trample on the flowers, but in so doing we achieve freedom and joy for all. . . . We seek to cultivate the sense of order, to encourage sympathy and foresight, to pull up racial weeds by the roots. . . . In these matters, indeed, the gardener in his garden is our symbol and our guide."

①

In the last years of his life, Galton wrestled with the idea of negative eugenics. He never made complete peace with it. The "sterilization of failures"—the weeding and culling of the human genetic garden—haunted him with its many implicit moral hazards. But in the end, his desire to build eugenics into a "national religion" outweighed his qualms about negative eugenics. In 1909, he founded a journal, the *Eugenics Review*, which endorsed not just selective breeding but selective sterilization. In 1911, he produced a strange novel, entitled *Kantsaywhere*, about a future utopia in which roughly half the population was marked as "unfit" and severely restricted in its ability to reproduce. He left a copy of the novel with his niece. She found it so embarrassing that she burned large parts of it.

On July 24, 1912, one year after Galton's death, the first International Conference on Eugenics opened at the Cecil Hotel in London. The location was symbolic. With nearly eight hundred rooms and a vast, monolithic façade overlooking the Thames, the Cecil was Europe's largest, if not grandest, hotel—a site typically reserved for diplomatic or national events. Luminaries from twelve countries and diverse disciplines descended on the hotel to attend the conference: Winston Churchill; Lord Balfour; the lord mayor of London; the chief justice; Alexander Graham Bell; Charles Eliot, the president of Harvard University; August Weismann, the embryologist. Darwin's son Leonard Darwin presided over the meeting; Karl Pearson worked closely with Darwin on the program. Visitors—having walked through the domed, marble-hemmed entrance lobby, where a framed picture of Galton's pedigree was prominently displayed—were treated to talks on genetic manipulations to increase the average height of children, on the inheritance of epilepsy, on the mating patterns of alcoholics, and on the genetic nature of criminality.

Two presentations, among all, stood out in their particularly chilling fervor. The first was an enthusiastic and precise exhibit by the Germans endorsing "race hygiene"—a grim premonition of times to come. Alfred Ploetz, a physician, scientist, and ardent proponent of the race-hygiene theory, gave an impassioned talk about launching a racial-cleansing effort in Germany. The second presentation—even larger in its scope and ambition—was presented by the American contingent. If eugenics was becoming a cottage industry in Germany, it was already a full-fledged national operation in America. The father of the American movement was the patrician Harvard-trained zoologist Charles Davenport, who had founded a eugenics-focused research center and laboratory—the Eugenics Record Office—in 1910. Davenport's 1911 book, *Heredity in Relation to Eugenics*, was the movement's bible; it was also widely assigned as a textbook of genetics in colleges across the nation.

Davenport did not attend the 1912 meeting, but his protégé Bleecker Van Wagenen, the young president of the American Breeders' Association, gave a rousing presentation. Unlike the Europeans, still mired in theory and speculation, Van Wagenen's talk was all Yankee practicality. He spoke glowingly about the operational efforts to eliminate "defective strains" in America. Confinement centers—"colonies"—for the genetically unfit were already planned. Committees had already been formed to consider the sterilization of unfit men and women—epileptics, criminals, deaf-mutes, the feebleminded, those with eye defects, bone deformities, dwarfism, schizophrenia, manic depression, or insanity.

"Nearly ten percent of the total population . . . are of inferior blood," Van Wagenen suggested, and "they are totally unfitted to become the parents of useful citizens. . . . In eight of the states of the Union, there are laws authorizing or requiring sterilization." In "Pennsylvania, Kansas, Idaho, Virginia . . . there have been sterilized a considerable number of individuals. . . . Many thousands of sterilization operations have been performed by surgeons in both private and institutional practice. As a rule, these operations have been for purely pathological reasons, and it has been found difficult to obtain authentic records of the more remote effects of these operations."

"We endeavor to keep track of those who are discharged and receive reports from time to time," the general superintendent for the California State hospital concluded cheerfully in 1912. "We have found no ill effects."

"Three Generations of Imbeciles Is Enough"

If we enable the weak and the deformed to live and to propagate their kind, we face the prospect of a genetic twilight. But if we let them die or suffer when we can save or help them, we face the certainty of a moral twilight.
— Theodosius Grigorievich Dobzhansky,
Heredity and the Nature of Man

And from deformed [parents] deformed [offspring] come to be, just as lame come to be from lame and blind from blind, and in general they resemble often the features that are against nature, and have inborn signs such as growths and scars. Some of such features have even been transmitted through three [generations].
— Aristotle, *History of Animals*

In the spring of 1920, Emmett Adaline Buck—Emma for short—was brought to the Virginia State Colony for Epileptics and Feebleminded in Lynchburg, Virginia. Her husband, Frank Buck, a tin worker, had either bolted from home or died in an accident, leaving Emma to care for a young daughter, Carrie Buck.

Emma and Carrie lived in squalor, depending on charity, food donations, and makeshift work to support a meager lifestyle. Emma was rumored to have sex for money, to have contracted syphilis, and to drink her wages on weekends. In March that year, she was caught on the streets in town, booked, either for vagrancy or prostitution, and brought before a municipal judge. A cursory mental examination, performed on April 1, 1920, by two doctors, classified her as "feebleminded." Buck was packed off to the colony in Lynchburg.

"Feeblemindedness," in 1924, came in three distinct flavors: idiot, moron, and imbecile. Of these, an idiot was the easiest to classify—the US Bureau of the Census defined the term as a "mentally defective person with a mental age of not more than 35 months"—but imbecile and moron were more porous categories. On paper, the terms referred to less severe forms of cognitive disability, but in practice, the words were revolving semantic doors that swung inward all too easily to admit a diverse group of men and women, some with no mental illness at all—prostitutes, orphans, depressives, vagrants, petty criminals, schizophrenics, dyslexics, feminists, rebellious adolescents—anyone, in short, whose behavior, desires, choices, or appearance fell outside the accepted norm.

Feebleminded women were sent to the Virginia State Colony for confinement to ensure that they would not continue breeding and thereby contaminate the population with further morons or idiots. The word *colony* gave its purpose away: the place was never meant to be a hospital or an asylum. Rather, from its inception, it was designed to be a containment zone. Sprawling over two hundred acres in the windward shadow of the Blue Ridge Mountains, about a mile from the muddy banks of the James River, the colony had its own postal office, powerhouse, coal room, and a spur rail-track for off-loading cargo. There was no public transportation into or out of the colony. It was the Hotel California of mental illness: patients who checked in rarely ever left.

When Emma Buck arrived, she was cleaned and bathed, her clothes thrown away, and her genitals douched with mercury to disinfect them. A repeat intelligence test performed by a psychiatrist confirmed the initial diagnosis of a "Low Grade Moron." She was admitted to the colony. She would spend the rest of her lifetime in its confines.

℗

Before her mother had been carted off to Lynchburg in 1920, Carrie Buck had led an impoverished but still-normal childhood. A school report from 1918, when she was twelve, noted that she was "very good" in "deportment and lessons." Gangly, boyish, rambunctious—tall for her age, all elbows and knees, with a fringe of dark bangs, and an open smile—she liked to write notes to boys in school and fish for frogs and brookies in the local ponds. But with Emma gone, her life began to fall apart. Carrie was placed in foster care. She was raped by her foster parents' nephew and soon discovered that she was pregnant.

Stepping in quickly to nip the embarrassment, Carrie's foster parents brought her before the same municipal judge that had sent her mother, Emma, to Lynchburg. The plan was to cast Carrie as an imbecile as well: she was reported to be devolving into a strange dimwit, given to "hallucinations and outbreaks of temper," impulsive, psychotic, and sexually promiscuous. Predictably, the judge—a friend of Carrie's foster parents—confirmed the diagnosis of "feeblemindedness": like mother, like daughter. On January 23, 1924, less than four years after Emma's appearance in court, Carrie too was assigned to the colony.

On March 28, 1924, awaiting her transfer to Lynchburg, Carrie gave birth to a daughter, Vivian Elaine. By state order, the daughter was also placed in foster care. On June 4, 1924, Carrie arrived at the Virginia State Colony. "There is no evidence of psychosis—she reads and writes and keeps herself in tidy condition," her report read. Her practical knowledge and skills were found to be normal. Nonetheless, despite all the evidence to the contrary, she was classified as a "Moron, Middle Grade" and confined.

①

In August 1924, a few months after she arrived in Lynchburg, Carrie Buck was asked to appear before the Board of the Colony at the request of Dr. Albert Priddy.

A small-town doctor originally from Keysville, Virginia, Albert Priddy had been the colony's superintendent since 1910. Unbeknownst to Carrie and Emma Buck, he was in the midst of a furious political campaign. Priddy's pet project was "eugenic sterilizations" of the feebleminded. Endowed with extraordinary, Kurtz-like powers over his colony, Priddy was convinced that the imprisonment of "mentally defectives" in colonies was a temporary solution to the propagation of their "bad heredity." Once released, the imbeciles would start breeding again, contaminating and befouling the gene pool. Sterilization would be a more definitive strategy, a superior solution.

What Priddy needed was a blanket legal order that would authorize him to sterilize a woman on explicitly eugenic grounds; one such test case would set the standard for a thousand. When he broached the topic, he found that legal and political leaders were largely sympathetic to his ideas. On March 29, 1924, with Priddy's help, the Virginia Senate authorized eugenic sterilization within the state as long as the person to be sterilized had been screened by the "Boards of Mental-health institutions." On September 10, again urged by Priddy, the Board of the Virginia State Colony reviewed Buck's case during a routine meeting. Carrie Buck was asked a single question during the inquisition: "Do you care to say anything about having the operations performed on you?" She spoke only two sentences: "No, sir, I have not. It is up to my people." Her "people," whoever they were, did not rise to Buck's defense. The board approved Priddy's request to have Buck sterilized.

But Priddy was concerned that his attempts to achieve eugenic sterilizations would still be challenged by state and federal courts. At Priddy's instigation, Buck's case was next presented to the Virginia court. If the courts affirmed the act, Priddy believed, he would have complete authority to continue his eugenic efforts at the colony and even extend them to other colonies. The case—*Buck v. Priddy*—was filed in the Circuit Court of Amherst County in October 1924.

On November 17, 1925, Carrie Buck appeared for her trial at the courthouse in Lynchburg. She found that Priddy had arranged nearly a dozen witnesses. The first, a district nurse from Charlottesville, testified that Emma and Carrie were impulsive, "irresponsible mentally, and . . . feebleminded." Asked to provide examples of Carrie's troublesome behavior, she said Carrie had been found "writing notes to boys." Four other women then testified about Emma and Carrie. But Priddy's most important witness was yet to come. Unbeknownst to Carrie and Emma, Priddy had sent a social worker from the Red Cross to examine Carrie's eight-month-old child, Vivian, who was living with foster parents. If Vivian could also be shown to be feebleminded, Priddy reasoned, his case would be closed. With three generations—Emma, Carrie, and Vivian—affected by imbecility, it would be hard to argue against the heredity of their mental capacity.

The testimony did not go quite as smoothly as Priddy had planned. The social worker—veering sharply off script—began by admitting biases in her judgment:

"Perhaps my knowledge of the mother may prejudice me."

"Have you any impression about the child?" the prosecutor asked.

The social worker was hesitant again. "It is difficult to judge the probabilities of a child as young as that, but it seems to me not quite a normal baby. . . ."

"You would not judge the child as a normal baby?"

"There is a look about it that is not quite normal, but just what it is, I can't tell."

For a while, it seemed as if the future of eugenic sterilizations in America depended on the foggy impressions of a nurse who had been handed a cranky baby without toys.

The trial took five hours, including a break for lunch. The deliberation was brief, the decision clinical. The court affirmed Priddy's decision to sterilize Carrie Buck. "The act complies with the requirements of due process of law," the decision read. "It is not a penal statute. It cannot be said, as contended, that the act divides a natural class of persons into two."

Buck's lawyers appealed the decision. The case climbed to the Virginia Supreme Court, where Priddy's request to sterilize Buck was affirmed again. In the early spring of 1927, the trial reached the US Supreme Court. Priddy had died, but his successor, John Bell, the new superintendent of the colony, was the appointed defendant.

℗

Buck v. Bell was argued before the Supreme Court in the spring of 1927. Right from the onset, the case was clearly neither about Buck nor Bell. It was a charged time; the entire nation was frothing with anguish about its history and inheritance. The Roaring Twenties stood at the tail end of a historic surge of immigration to the United States. Between 1890 and 1924, nearly 10 million immigrants—Jewish, Italian, Irish, and Polish workers—streamed into New York, San Francisco, and Chicago, packing the streets and tenements and inundating the markets with foreign tongues, rituals, and foods (by 1927, new immigrants comprised more than 40 percent of the populations of New York and Chicago). And as much as class anxiety had driven the eugenic efforts of England in the 1890s, "race anxiety" drove the eugenic efforts of Americans in the 1920s.*

---

* Undoubtedly, the historical legacy of slavery was also an important factor driving American eugenics. White eugenicists in America had long convulsed with the fear that African slaves, with their inferior genes, would intermarry with whites and thereby contaminate the gene pool—but laws to prevent interracial marriages, promulgated during the 1860s, had calmed most of these fears. White *immigrants*, in contrast, were not so easy to identify and separate, thus amplifying the anxieties of ethnic contamination and miscegenation in the 1920s.

Galton may have despised the great unwashed masses, but they were, indisputably, great and unwashed *English* masses. In America, in contrast, the great unwashed masses were increasingly foreign—and their genes, like their accents, were identifiably alien.

Eugenicists such as Priddy had long worried that the flooding of America by immigrants would precipitate "race suicide." The right people were being overrun by the wrong people, they argued, and the right genes corrupted by the wrong ones. If genes were fundamentally indivisible—as Mendel had shown—then a genetic blight, once spread, could never be erased ("A cross between [any race] and a Jew is a Jew," Madison Grant wrote). The only way of "cutting off the defective germplasm," as one eugenicist described it, was to excise the organ that produced germplasm—i.e., to perform compulsory sterilizations of genetic unfits such as Carrie Buck. To protect the nation against "the menace of race deterioration," radical social surgery would need to be deployed. "The Eugenic ravens are croaking for reform [in England]," Bateson wrote with obvious distaste in 1926. The American ravens croaked even louder.

Counterpoised against the myth of "race suicide" and "race deterioration" was the equal and opposite myth of racial and genetic purity. Among the most popular novels of the early twenties, devoured by millions of Americans, was Edgar Rice Burroughs's *Tarzan of the Apes*, a bodice-ripping saga involving an English aristocrat who, orphaned as an infant and raised by apes in Africa, retains not just his parents' complexion, bearing, and physique, but their moral rectitude, Anglo-Saxon values, and even the instinctual use of proper dinnerware. Tarzan—"his straight and perfect figure, muscled as the best of the ancient Roman gladiators must have been muscled"—exemplified the ultimate victory of nature over nurture. If a white man raised by jungle apes could retain the integrity of a white man in a flannel suit, then surely racial purity could be maintained in any circumstance.

Against this backdrop, the US Supreme Court took scarcely any time to reach its decision on *Buck v. Bell*. On May 2, 1927, a few weeks before Carrie Buck's twenty-first birthday, the Supreme Court handed down its verdict. Writing the 8–1 majority opinion, Oliver Wendell Holmes Jr. reasoned, "It is better for all the world, if instead of waiting to execute degenerate offspring for crime, or to let them starve for their imbecility, society can prevent those who are manifestly unfit from continuing their kind. The principle that sustains compulsory vaccination is broad enough to cover cutting the Fallopian tubes."

Holmes—the son of a physician, a humanist, a scholar of history, a man widely celebrated for his skepticism of social dogmas, and soon to be one of the nation's most vocal advocates of judicial and political moderation—was evidently tired of the Bucks and their babies. "Three generations of imbeciles is enough," he wrote.

①

Carrie Buck was sterilized by tubal ligation on October 19, 1927. That morning, around nine o'clock, she was moved to the state colony's infirmary. At ten o'clock, narcotized on morphine and atropine, she lay down on a gurney in a surgical room. A nurse administered anesthesia, and Buck drifted into sleep. Two doctors and two nurses were in attendance—an unusual turnout for such a routine procedure, but this was a special case. John Bell, the superintendent, opened her abdomen with an incision in the midline. He removed a section of both fallopian tubes, tied the ends of the tubes, and sutured them shut. The wounds were cauterized with carbolic acid and sterilized with alcohol. There were no surgical complications.

The chain of heredity had been broken. "The first case operated on under the sterilization law" had gone just as planned, and the patient was discharged in excellent health, Bell wrote. Buck recovered in her room uneventfully.

①

Six decades and two years, no more than a passing glance of time, separate Mendel's initial experiments on peas and the court-mandated sterilization of Carrie Buck. Yet in this brief flash of six decades, the gene had transformed from an abstract concept in a botanical experiment to a powerful instrument of social control. As *Buck v. Bell* was being argued in the Supreme Court in 1927, the rhetoric of genetics and eugenics penetrated social, political, and personal discourses in the United States. In 1927, the state of Indiana passed a revised version of an earlier law to sterilize "confirmed criminals, idiots, imbeciles and rapists." Other states followed with even more draconian legal measures to sterilize and confine men and women judged to be genetically inferior.

While state-sponsored sterilization programs expanded throughout the nation, a grassroots movement to personalize genetic selection was also gaining popularity. In the 1920s, millions of Americans thronged to agricultural fairs where, alongside tooth-brushing demonstrations, popcorn machines, and hayrides, the public encountered Better Babies Contests, in which children, often as young as one or two years old, were proudly displayed on tables and pedestals, like dogs or cattle, as physicians, psychiatrists, dentists, and nurses in white coats examined their eyes and teeth, prodded their skin, and measured heights, weights, skull sizes, and temperaments to select the healthiest and fittest variants. The "fittest" babies were then paraded through the fairs. Their pictures were featured prominently on posters, newspapers, and magazines—generating passive support for a national eugenics movement. Davenport, the Harvard-trained zoologist famous for establishing the Eugenics Record Office, created a standardized evaluation form to judge the fittest babies. Davenport instructed his judges to examine the parents before judging the children: "You should score 50% for heredity before you begin to examine a baby." "A prize winner at two may be an epileptic at ten." These fairs often contained "Mendel booths," where the principles of genetics and the laws of inheritance were demonstrated using puppets.

In 1927, a film called *Are You Fit to Marry?*, by Harry Haiselden, another eugenics-obsessed doctor, played to packed audiences across the United States. The revival of an earlier film titled *The Black Stork*, the plot involved a physician, played by Haiselden himself, who refuses to perform lifesaving operations on disabled infants in an effort to "cleanse" the nation of defective children. The film ends with a woman who has a nightmare of bearing a mentally defective child. She awakens and decides that she and her fiancé must get tested before their marriage to ensure their genetic compatibility (by the late 1920s, premarital genetic-fitness tests, with assessments of family histories of mental retardation, epilepsy, deafness, skeletal diseases, dwarfism, and blindness, were being widely advertised to the American public). Ambitiously, Haiselden meant to market his film as a "date night" movie: it had love, romance, suspense, and humor—with some retail infanticide thrown in on the side.

As the front of the American eugenics movement advanced from imprisonment to sterilization to outright murder, European eugenicists watched the escalation with a mix of eagerness and envy. By 1936, less than a decade after *Buck v. Bell*, a vastly more virulent form of "genetic cleansing" would engulf that continent like a violent contagion, morphing the language of genes and inheritance into its most potent and macabre form.

PART TWO

"IN THE SUM OF THE PARTS, THERE ARE ONLY THE PARTS"

Deciphering the Mechanism of Inheritance
(1930–1970)

①

○

It was when I said
"Words are not forms of a single word.
In the sum of the parts, there are only the parts.
The world must be measured by eye."

—Wallace Stevens, "On the Road Home"

○

"Abhed"

*Genio y hechura, hasta sepultura. (Natures and features
last until the grave.)*

—Spanish saying

I am the family face:
Flesh perishes, I live on,
Projecting trait and trace
Through time to times anon,
And leaping from place to place
Over oblivion.

—Thomas Hardy, "Heredity"

The day before our visit with Moni, my father and I took a walk in Calcutta. We started near Sealdah station, where my grandmother had stepped off the train from Barisal in 1946, with five boys and four steel trunks in tow. From the edge of the station, we retraced their path, walking along Prafulla Chandra Road, past the bustling wet market, with open-air stalls of fish and vegetables on the left, and the stagnating pond of water hyacinths on the right, then turned left again, heading toward the city.

The road narrowed sharply and the crowd thickened. On both sides of the street, the larger apartments divided into tenements, as if driven by some furious biological process—one room splitting into two, two becoming four, and four, eight. The streets reticulated and the sky vanished. There was the clank of cooking, and the mineral smell of coal smoke. At a pharmacist's shop, we turned into the inlet of Hayat Khan Lane and walked toward the house that my father and his family had occupied. The rubbish heap was still there, breeding its multigenerational population of feral dogs. The front door of the house opened into a small courtyard. A woman was in the kitchen downstairs, about to behead a coconut with a scythe.

"Are you Bibhuti's daughter?" my father asked in Bengali, out of the blue. Bibhuti Mukhopadhyay had owned the house and rented it to my grandmother. He was no longer alive, but my father recalled two children—a son and a daughter.

The woman looked at my father warily. He had already stepped past the threshold and climbed onto the raised veranda, a few feet from the kitchen. "Does Bibhuti's family still live here?" The questions were launched without any formal introduction. I noted a deliberate change in his accent—the softened hiss of the consonants in his words, the dental *chh* of West Bengali softening into the sibilant *ss* of the East. In Calcutta, I knew, every accent is a surgical probe. Bengalis send out their vowels and consonants like survey drones—to test the identities of their listeners, to sniff out their sympathies, to confirm their allegiances.

"No, I'm his brother's daughter-in-law," the woman said. "We have lived here since Bibhuti's son died."

It is difficult to describe what happened next—except to say that it is a moment that occurs uniquely in the histories of refugees. A tiny bolt of understanding passed between them. The woman recognized my father—not the actual man, whom she had never met, but the *form* of the man: a boy returning home. In Calcutta—in Berlin, Peshawar, Delhi, Dhaka—men like this seem to turn up every day, appearing out of nowhere off the streets and walking unannounced into houses, stepping casually over thresholds into their past.

Her manner warmed visibly. "Were you the family that lived here once? Weren't there many brothers?" She asked all this matter-of-factly, as if this visit had been long overdue.

Her son, about twelve years old, peeked out from the window upstairs with a textbook in his hand. I knew that window. Jagu had parked himself there for days on end, staring into the courtyard.

"It's all right," she said to her son, motioning with her hands. He fled inside. She turned to my father. "Go upstairs if you'd like. Look around, but leave the shoes on the stairwell."

I removed my sneakers, and the ground felt instantly intimate on my soles, as if I had always lived here.

①

My father walked around the house with me. It was smaller than I had expected—as places reconstructed from borrowed memories inevitably are—but also duller and dustier. Memories sharpen the past; it is reality that decays. We climbed a narrow gullet of stairs to a small pair of rooms. The four younger brothers, Rajesh, Nakul, Jagu, and my father, had shared one of the rooms. The eldest boy, Ratan—Moni's father—and my grandmother had shared the adjacent room, but as Jagu's mind had involuted into madness, she had moved Ratan out with his brothers and taken Jagu in. Jagu would never again leave her room.

We climbed up to the balcony on the roof. The sky dilated at last. Dusk was falling so quickly that it seemed you could almost sense the curvature of the earth arching away from the sun. My father looked out toward the lights of the station. A train whistled in the distance like a desolate bird. He knew I was writing about heredity.

"Genes," he said, frowning.

"Is there a Bengali word?" I asked.

He searched his inner lexicon. There was no word—but perhaps he could find a substitute.

"*Abhed*," he offered. I had never heard him use the term. It means "indivisible" or "impenetrable," but it is also used loosely to denote "identity." I marveled at the choice; it was an echo chamber of a word. Mendel or Bateson might have relished its many resonances: indivisible; impenetrable; inseparable; identity.

I asked my father what he thought about Moni, Rajesh, and Jagu.

"*Abheder dosh*," he said.

A flaw in identity; a genetic illness; a blemish that cannot be separated from the self—the same phrase served all meanings. He had made peace with its indivisibility.

☉

For all the talk in the late 1920s about the links between genes and identity, the gene itself appeared to possess little identity of its own. If a scientist had been asked what a gene was made of, how it accomplished its function, or where it resided within the cell, there would be few satisfactory answers. Even as genetics was being used to justify sweeping changes in law and society, the gene itself had remained a doggedly abstract entity, a ghost lurking in the biological machine.

This black box of genetics was pried open, almost accidentally, by an unlikely scientist working on an unlikely organism. In 1907, when William Bateson visited the United States to give talks on Mendel's discovery, he stopped in New York to meet Thomas Hunt Morgan, the cell biologist. Bateson was not particularly impressed. "Morgan is a blockhead," he wrote to his wife. "He is in a continuous whirl—very active and inclined to be noisy."

Noisy, active, obsessive, eccentric—with a dervishlike mind that spiraled from one scientific question to the next—Thomas Morgan was a professor of zoology at Columbia University. His main interest was embryology. At first, Morgan was not even interested in whether units of heredity existed or how or where they were stored. The principal question he cared about concerned development: How does an organism emerge from a single cell?

Morgan had resisted Mendel's theory of heredity at first—arguing that it was unlikely that complex embryological information could be stored in discrete units in the cell (hence Bateson's "blockhead" comment). Eventually, however, Morgan had become convinced by Bateson's evidence; it was hard to argue against "Mendel's bulldog," who came armed with charts of data. Yet, even as he had come to accept the existence of genes, Morgan had remained perplexed about their material form. Cell biologists look; geneticists count; biochemists clean, the scientist Arthur Kornberg once said. Indeed, armed with microscopes, cell biologists had become accustomed to a cellular world in which visible structures performed identifiable functions within cells. But thus far, the gene had been "visible" only in a statistical sense. Morgan wanted to uncover the physical basis of heredity. "We are interested in heredity not primarily as a *mathematical* formulation," he wrote, "but rather as a problem concerning the cell, the egg and the sperm."

But where might genes be found within cells? Intuitively, biologists had long guessed that the best place to visualize a gene was the embryo. In the 1890s, a German embryologist working with sea urchins in Naples, Theodor Boveri, had proposed that genes resided in *chromosomes*, threadlike filaments that stained blue with aniline, and lived, coiled like springs, in the nucleus of cells (the word *chromosome* was coined by Boveri's colleague Wilhelm von Waldeyer-Hartz).

Boveri's hypothesis was corroborated by work performed by two other scientists. Walter Sutton, a grasshopper-collecting farm boy from the prairies of Kansas, had grown into a grasshopper-collecting scientist in New York. In the summer of 1902, working on grasshopper sperm and egg cells—which have particularly gigantic chromosomes—Sutton also postulated that genes were physically carried on chromosomes. And Boveri's own student, a biologist named Nettie Stevens, had become interested in the determination of sex. In 1905, using cells from the common mealworm, Stevens demonstrated that "maleness" in worms was determined by a unique factor—the Y chromosome—that was only present in male embryos, but never in female ones (under a microscope, the Y chromosome looks like any other chromosome—a squiggle of DNA that stains brightly blue—except that it is shorter and stubbier compared to the X chromosome). Having pinpointed the location of gender-carrying genes to a single chromosome, Stevens proposed that all genes might be carried on chromosomes.

☉

Thomas Morgan admired the work of Boveri, Sutton, and Stevens. But he still yearned for a more tangible description of the gene. Boveri had identified the chromosome as the physical residence for genes, but the deeper architecture of genes and chromosomes still remained unclear. How were genes organized on chromosomes? Were they strung along chromosomal filaments—like pearls on a string? Did every gene have a unique chromosomal "address"? Did genes overlap? Was one gene physically or chemically linked to another?

Morgan approached these questions by studying yet another model organism—fruit flies. He began to breed flies sometime around 1905 (some of Morgan's colleagues would later claim that his first stock came from a flock of flies above a pile of overripe fruit in a grocery store in Woods Hole, Massachusetts. Others suggested that he got his first flies from a colleague in New York). A year later, he was breeding maggots by the thousands, in milk bottles filled with rotting fruit in a third-floor laboratory at Columbia University.* Bunches of overripe bananas hung from sticks. The smell of fermented fruit was overpowering, and a haze of escaped flies lifted off the tables like a buzzing veil every time Morgan moved. The

---

* Some of the work was also performed at Woods Hole, where Morgan would move his lab every summer.

students called his laboratory the Fly Room. It was about the same size and shape as Mendel's garden—and in time it would become an equally iconic site in the history of genetics.

Like Mendel, Morgan began by identifying heritable traits—visible variants that he could track over generations. He had visited Hugo de Vries's garden in Amsterdam in the early 1900s and become particularly interested in de Vries's plant mutants. Did fruit flies have mutations as well? By scoring thousands of flies under the microscope, he began to catalog dozens of mutant flies. A rare white-eyed fly appeared spontaneously among the typically red-eyed flies. Other mutant flies had forked bristles; sable-colored bodies; curved legs; bent, batlike wings; disjointed abdomens; deformed eyes—a Halloween's parade of oddballs.

A flock of students joined him in New York, each one odd in his own right: a tightly wound, precise Midwesterner named Alfred Sturtevant; Calvin Bridges, a brilliant, grandiose young man given to fantasies about free love and promiscuity; and paranoid, obsessive Hermann Muller, who jostled daily for Morgan's attention. Morgan openly favored Bridges; it was Bridges, as an undergraduate student assigned to wash bottles, who had spotted, among hundreds of vermilion-eyed flies, the white-eyed mutant that would become the basis for many of Morgan's crucial experiments. Morgan admired Sturtevant for his discipline and his work ethic. Muller was favored the least: Morgan found him shifty, laconic, and disengaged from the other members of the lab. Eventually, all three students would quarrel fiercely, unleashing a cycle of envy and destructiveness that would blaze through the discipline of genetics. But for now, in a fragile peace dominated by the buzz of flies, they immersed themselves in experiments on genes and chromosomes. By breeding normal flies with mutants—mating white-eyed males with red-eyed females, say—Morgan and his students could track the inheritance of traits across multiple generations. The mutants, again, would prove crucial to these experiments: only the outliers could illuminate the nature of normal heredity.

①

To understand the significance of Morgan's discovery, we need to return to Mendel. In Mendel's experiments, every gene had behaved like an independent entity—a free agent. Flower color, for instance, had no link with seed texture or stem height. Each characteristic was inherited independently, and all combinations of traits were possible. The result of each cross was thus a perfect genetic roulette: if you crossed a tall plant with purple flowers with a short plant with white flowers, you would eventually produce all sorts of mixes—tall plants with white flowers and short plants with purple flowers and so forth.

But Morgan's fruit fly genes did not always behave independently. Between 1910 and 1912, Morgan and his students crossed thousands of fruit fly mutants with each other to create tens of thousands of flies. The result of each cross was meticulously recorded: white-eyed, sable-colored, bristled, short-winged. When Morgan examined these crosses, tabulated across dozens of notebooks, he found a surprising pattern: some genes acted as if they were "linked" to each other. The gene responsible for creating white eyes (called *white eyed*), for instance, was inescapably linked to the X chromosome: no matter how Morgan crossed his flies, the white-eyed trait tracked with that chromosome. Similarly, the gene for sable color was linked with the gene that specified the shape of a wing.

For Morgan, this genetic linkage could only mean one thing: genes had to be *physically* linked to each other. In flies, the gene for sable color was never (or rarely) inherited independently from the gene for miniature wings because they were both carried on the same chromosome. If two beads are on the same string, then they are always tied together, no matter how one attempts to mix and match strings. For two genes on the same chromosome, the same principle applied: there was no simple way to separate the forked-bristle gene from the coat-color gene. The inseparability of features had a material basis: the chromosome was a "string" along which certain genes were permanently strung.

⬡

Morgan had discovered an important modification to Mendel's laws. Genes did not travel separately; instead, they moved in packs. Packets of information were themselves packaged—into chromosomes, and ultimately in cells. But the discovery had a more important consequence: conceptually, Morgan had not just linked genes; he had linked two disciplines—cell biology and genetics. The gene was not a "purely theoretical unit." It was a material *thing* that lived in a particular location, and a particular form, within a cell. "Now that we locate them [genes] on chromosomes," Morgan reasoned, "are we justified in regarding them as material units; as chemical bodies of a higher order than molecules?"

①

The establishment of linkage between genes prompted a second, and third, discovery. Let us return to linkage: Morgan's experiments had established that genes that were physically linked to each other on the same chromosome were inherited together. If the gene that produces blue eyes (call it *B*) is linked to a gene that produces blond hair (*Bl*), then children with blond hair will inevitably tend to inherit blue eyes (the example is hypothetical, but the principle that it illustrates is true).

But there was an exception to linkage: occasionally, very occasionally, a gene could *unlink* itself from its partner genes and swap places from the paternal chromosome to the maternal chromosome, resulting in a fleetingly rare blue-eyed, *dark-haired* child, or, conversely, a dark-eyed, blond-haired child. Morgan called this phenomenon "crossing over." In time, as we shall see, the crossing over of genes would launch a revolution in biology, establishing the principle that genetic information could be mixed, matched, and swapped—not just between sister chromosomes, but between organisms and across species.

①

The final discovery prompted by Morgan's work was also the result of a methodical study of "crossing over." Some genes were so tightly linked that they never crossed over. These genes, Morgan's students hypothesized, were physically closest to each other on the chromosome. Other genes, although linked, were more prone to splitting apart. These genes had to be positioned farther apart on the chromosome. Genes that had no linkage whatsoever had to be present on entirely different chromosomes. The tightness of genetic linkage, in short, was a surrogate for the physical proximity of genes on chromosomes: by measuring how often two features—blond-hairedness and blue-eyedness—were linked or unlinked, you could measure the distance between their genes on the chromosome.

On a winter evening in 1911, Sturtevant, then a twenty-year-old undergraduate student in Morgan's lab, brought the available experimental data on the linkage of *Drosophila* (fruit fly) genes to his room and—neglecting his mathematics homework—spent the night constructing the first map of genes in flies. If A was tightly linked to B, and very loosely linked to C, Sturtevant reasoned, then the three genes must be positioned on the chromosome in that order and with proportional distance from each other:

$$A \, . \, B \, ... \, C \, .$$

If an allele that created notched wings (N) tended to be co-inherited with an allele that made short bristles (SB), then the two genes, N and SB, must be on the same chromosome, while the unlinked gene for eye color must be on a different chromosome. By the end of the evening, Sturtevant had sketched the first linear genetic map of half a dozen genes along a *Drosophila* chromosome.

Sturtevant's rudimentary genetic map would foreshadow the vast and elaborate efforts to map genes along the human genome in the 1990s. By using linkage to establish the relative positions of genes on chromosomes, Sturtevant would also lay the groundwork for the future cloning of genes tied to complex familial diseases, such as breast cancer, schizophrenia, and Alzheimer's disease. In about twelve hours, in an undergraduate dorm room in New York, he had poured the foundation for the Human Genome Project.

℗

Between 1905 and 1925, the Fly Room at Columbia was the epicenter of genetics, a catalytic chamber for the new science. Ideas ricocheted off ideas, like atoms splitting atoms. The chain reaction of discoveries—linkage, crossing over, the linearity of genetic maps, the distance between genes—burst forth with such ferocity that it seemed, at times, that genetics was not born but zippered into existence. Over the next decades, a spray of Nobel Prizes would be showered on the occupants of the room: Morgan, his students, his student's students, and even *their* students would all win the prize for their discoveries.

But beyond linkage and gene maps, even Morgan had a difficult time imagining or describing genes in a material form: What chemical could possibly carry information in "threads" and "maps"? It is a testament to the ability of scientists to accept abstractions as truths that fifty years after the publication of Mendel's paper—from 1865 to 1915—biologists knew genes only through the properties they produced: genes specified traits; genes could become mutated and thereby specify alternative traits; and genes tended to be chemically or physically linked to each other. Dimly, as if through a veil, geneticists were beginning to visualize patterns and themes: threads, strings, maps, crossings, broken and unbroken lines, chromosomes that carried information in a coded and compressed form. But no one had seen a gene in action or knew its material essence. The central quest of the study of heredity seemed like an object perceived only through its shadows, tantalizingly invisible to science.

①

If urchins, mealworms, and fruit flies seemed far removed from the world of humans—if the concrete relevance of Morgan's or Mendel's findings was ever in doubt—then the events of the violent spring of 1917 proved otherwise. In March that year, as Morgan was writing his papers on genetic linkage in his Fly Room in New York, a volley of brutal popular uprisings ricocheted through Russia, ultimately decapitating the czarist monarchy and culminating in the creation of the Bolshevik government.

At face value, the Russian Revolution had little to do with genes. The Great War had whipped a starving, weary population into a murderous frenzy of discontent. The czar was considered weak and ineffectual. The army was mutinous; the factory workers galled; inflation ran amok. By March 1917, Czar Nicholas II had been forced to abdicate the throne. But genes—and linkage—were certainly potent forces in this history. The czarina of Russia, Alexandra, was the granddaughter of Queen Victoria of England—and she carried the marks of that heritage: not just the carved obelisk of the nose, or the fragile enamel-like sheen of her skin, but also a gene that caused hemophilia B, a lethal bleeding disorder that had crisscrossed through Victoria's descendants.

Hemophilia is caused by a single mutation that disables a protein in the clotting of blood. In the absence of this protein, blood refuses to clot—and even a small nick or wound can accelerate into a lethal bleeding crisis. The name of the illness—from Greek *haimo* ("blood") and *philia* ("to like, or love")—is actually a wry comment on its tragedy: hemophiliacs like to bleed all too easily.

Hemophilia—like white eyes in fruit flies—is a sex-linked genetic illness. Females can be carriers and transmit the gene, but only males are typically afflicted by the disease. The mutation in the hemophilia gene, which affects the clotting of blood, had likely arisen spontaneously in Queen Victoria at birth. Her eighth child, Leopold, had inherited the gene and died of a brain hemorrhage at age thirty. The gene had also been passed from Victoria to her second daughter, Alice—and then from Alice to *her* daughter, Alexandra, the czarina of Russia.

In the summer of 1904, Alexandra—still an unsuspecting carrier of the gene—gave birth to Alexei, the czarevitch of Russia. Little is known about the medical history of his childhood, but his attendants must have noticed something amiss: that the young prince bruised all too easily, or that his nosebleeds were often unstoppable. While the precise nature of his ailment was kept secret, Alexei continued to be a pale, sickly boy. He bled frequently and spontaneously. A playful fall, or a nick in his skin—even a bumpy horse ride—could precipitate disaster.

As Alexei grew older, and the hemorrhages more life threatening, Alexandra began to rely on a Russian monk of legendary unctuousness, Grigory Rasputin, who promised to heal the czar-to-be. While Rasputin claimed that he kept Alexei alive using various herbs, salves, and strategically offered prayers, most Russians considered him an opportunistic fraud (he was rumored to be having an affair with the czarina). His continuous presence in the royal family and his growing influence on Alexandra were considered evidence of a crumbling monarchy gone utterly batty.

The economic, political, and social forces that unloosed themselves on the streets of Petrograd and launched the Russian Revolution were vastly more complex than Alexei's hemophilia or Rasputin's machinations. History cannot devolve into medical biography—but nor can it stand outside it. The Russian Revolution may not have been about genes, but it was very much about heredity. The disjunction between the prince's all-too-human genetic inheritance and his all-too-exalted political inheritance must have seemed particularly evident to the critics of the monarchy. The metaphorical potency of Alexei's illness was also undeniable—symptomatic of an empire gone sick, dependent on bandages and prayers, hemorrhaging at its core. The French had tired of a greedy queen who ate cake. The Russians were fed up with a sickly prince swallowing strange herbs to combat a mysterious illness.

Rasputin was poisoned, shot, slashed, bludgeoned, and drowned to death by his rivals on December 30, 1916. Even by the grim standards of Russian assassinations, the violence of this murder was a testimony to the visceral hatred that he had inspired in his enemies. In the early summer of 1918, the royal family was moved to Yekaterinburg and placed under house arrest. On the evening of July 17, 1918, a month shy of Alexei's fourteenth birthday, a firing squad instigated by the Bolsheviks burst into the czar's house and assassinated the whole family. Alexei was shot twice in the head. The bodies of the children were supposedly scattered and buried nearby, but Alexei's body was not found.

In 2007, an archaeologist exhumed two partially burned skeletons from a bonfire site near the house where Alexei had been murdered. One of the skeletons belonged to a thirteen-year-old boy. Genetic testing of the bones confirmed that the body was Alexei's. Had the full genetic sequence of the skeleton been analyzed, the investigators might have found the culprit gene for hemophilia B—the mutation that had crossed one continent and four generations and insinuated itself into a defining political moment of the twentieth century.

Truths and Reconciliations

All changed, changed utterly:
A terrible beauty is born.
—William Butler Yeats, *Easter, 1916*

The gene was born "outside" biology. By this, I mean the following: if you consider the major questions raging through the biological sciences in the late nineteenth century, heredity does not rank particularly high on that list. Scientists studying living organisms were far more preoccupied with other matters: embryology, cell biology, the origin of species, and evolution. How do cells function? How does an organism arise from an embryo? How do species originate? What generates the diversity of the natural world?

Yet, attempts to answer these questions had all become mired at precisely the same juncture. The missing link, in all cases, was *information*. Every cell, and every organism, needs information to carry out its physiological function—but where does that information come from? An embryo needs a message to become an adult organism—but what carries this message? Or how, for that matter, does one member of a species "know" that it is a member of that species and not another?

The ingenious property of the gene was that it offered a potential solution to all these problems in a single sweep. Information for a cell to carry out a metabolic function? It came from a cell's genes, of course. The message encrypted in an embryo? Again, it was all encoded in genes. When an organism reproduces, it transmits the instructions to build embryos, make cells function, enable metabolism, perform ritual mating dances, give wedding speeches, and produce future organisms of the same species—all in one grand, unified gesture. Heredity cannot be a peripheral question in biology; it must rank among its central questions. When we think of heredity in a colloquial sense, we think about the inheritance of unique or particular features across generations: a peculiar shape of a father's nose or the susceptibility to an unusual illness that runs through a family. But the real conundrum that heredity solves is much more general: What is the nature of instruction that allows an organism to build a nose—*any* nose—in the first place?

①

The delayed recognition of the gene as the answer to the central problem of biology had a strange consequence: genetics had to be reconciled with other major fields of biology as an afterthought. If the gene was the central currency of biological information, then major characteristics of the living world—not just heredity—should be explicable in terms of genes. First, genes had to explain the phenomenon of variation: How could discrete units of heredity explain that human eyes, say, do not have six discrete forms but seemingly 6 billion continuous variants? Second, genes had to explain evolution: How could the inheritance of such units explain that organisms have acquired vastly different forms and features over time? And third, genes had to explain development: How could individual units of instruction prescribe the code to create a mature organism out of an embryo?

We might describe these three reconciliations as attempts to explain nature's past, present, and future through the lens of the gene. Evolution describes nature's past: *How did living things arise?* Variation describes its present: *Why do they look like this now?* And embryogenesis attempts to capture the future: *How does a single cell create a living thing that will eventually acquire its particular form?*

In two transformative decades between 1920 and 1940, the first two of these questions—i.e., variation and evolution—would be solved by unique alliances between geneticists, anatomists, cell biologists, statisticians, and mathematicians. The third question—embryological development—would require a much more concerted effort to solve. Ironically, even though embryology had launched the discipline of modern genetics, the reconciliation between genes and genesis would be a vastly more engaging scientific problem.

①

In 1909, a young mathematician named Ronald Fisher entered Caius College in Cambridge. Born with a hereditary condition that caused a progressive loss of vision, Fisher had become nearly blind by his early teens. He had learned mathematics largely without paper or pen and thus acquired the ability to visualize problems in his mind's eye before writing equations on paper. Fisher excelled at math as a secondary school student, but his poor eyesight became a liability at Cambridge. Humiliated by his tutors, who were disappointed in his abilities to read and write mathematics, he switched to medicine, but failed his exams (like Darwin, like Mendel, and like Galton—the failure to achieve conventional milestones of success seems to be a running theme in this story). In 1914, as war broke out in Europe, he began working as a statistical analyst in the City of London.

By day, Fisher examined statistical information for insurance companies. By night, with the world almost fully extinguished to his vision, he turned to theoretical aspects of biology. The scientific problem that engrossed Fisher also involved reconciling biology's "mind" with its "eye." By 1910, the greatest minds in biology had accepted that discrete particles of information carried on chromosomes were the carriers of hereditary information. But everything *visible* about the biological world suggested near-perfect continuity: nineteenth-century biometricians such as Quetelet and Galton had demonstrated that human traits, such as height, weight, and even intelligence, were distributed in smooth, continuous, bell-shaped curves. Even the development of an organism—the most obviously inherited chain of information—seemed to progress through smooth, continuous stages, and not in discrete bursts. A caterpillar does not become a butterfly in stuttering steps. If you plot the beak sizes of finches, the points fit on a continuous curve. How could "particles of information"—pixels of heredity—give rise to the observed smoothness of the living world?

Fisher realized that the careful mathematical modeling of hereditary traits might resolve this rift. Mendel had discovered the discontinuous nature of genes, Fisher knew, because he had *chosen* highly discrete traits and crossed pure-breeding plants to begin with. But what if real-world traits, such as height or skin color, were the result of not a single gene, with just two states—"tall" and "short," "on" and "off"—but of multiple genes? What if there were five genes that governed height, say, or seven genes that controlled the shape of a nose?

The mathematics to model a trait controlled by five or seven genes, Fisher discovered, was not all that complex. With just three genes in question, there would be six alleles or gene variants in total—three from the mother and three from the father. Simple combinatorial mathematics yielded twenty-seven unique combinations of these six gene variants. And if each *combination* generated a unique effect on height, Fisher found, the result smoothened out.

If he started with five genes, the permutations were even greater in number, and the variations in height produced by these permutations seemed almost continuous. Add the effects of the environment—the impact of nutrition on height, or sunlight exposure on skin color—and Fisher could imagine even more unique combinations and effects, ultimately generating perfectly smooth curves. Consider seven pieces of transparent paper colored with the seven basic colors of the rainbow. By juxtaposing the pieces of paper against each other and overlapping one color with another, one can almost produce every shade of color. The "information" in the sheets of paper remains discrete. The colors do not actually blend with each other—but the result of their overlap creates a spectrum of colors that seems virtually continuous.

In 1918, Fisher published his analysis in a paper entitled "The Correlation between Relatives on the Supposition of Mendelian Inheritance." The title was rambling, but the message was succinct: if you mixed the effects of three to five variant genes on any trait, you could generate nearly perfect continuity in phenotype. "The exact amount of human variability," he wrote, could be explained by rather obvious extensions of Mendelian genetics. The individual effect of a gene, Fisher argued, was like a dot of a pointillist painting. If you zoomed in close enough, you might see the dots as individual, discrete. But what we observed and experienced in the natural world from afar was an aggregation of dots: pixels merging to form a seamless picture.

①

The second reconciliation—between genetics and evolution—required more than mathematical modeling; it hinged on experimental data. Darwin had reasoned that evolution works via natural selection—but for natural selection to work, there had to be something natural to select. A population of organisms in the wild must have enough natural variation such that winners and losers can be picked. A flock of finches on an island, for instance, needs to possess enough intrinsic diversity in beak sizes such that a season of drought might be able to select birds with the toughest or longest beaks. Take that diversity away—force all finches to have the same beak—and selection comes up empty-handed. All the birds go extinct in a fell swoop. Evolution grinds to a halt.

But what is the engine that generates natural variation in the wild? Hugo de Vries had proposed that *mutations* were responsible for variation: changes in genes created changes in forms that could be selected by natural forces. But de Vries's conjecture predated the molecular definition of the gene. Was there experimental proof that identifiable mutations in real genes were responsible for variation? Were mutations sudden and spontaneous, or were abundant natural genetic variations already present in wild populations? And what happened to genes upon natural selection?

In the 1930s, Theodosius Dobzhansky, a Ukrainian biologist who had emigrated to the United States, set out to describe the extent of genetic variation in wild populations. Dobzhansky had trained with Thomas Morgan in the Fly Room at Columbia. But to describe genes in the wild, he knew that he would have to go wild himself. Armed with nets, fly cages, and rotting fruit, he began to collect wild flies, first near the laboratory at Caltech, then on Mount San Jacinto and along the Sierra Nevada in California, and then in forests and mountains all over the United States. His colleagues, confined to their lab benches, thought that he had gone fully mad. He might as well have left for the Galápagos.

The decision to hunt for variation in wild flies proved critical. In a wild fly species named *Drosophila pseudoobscura,* for instance, Dobzhansky found multiple gene variants that influenced complex traits, such as life span, eye structure, bristle morphology, and wing size. The most striking examples of variation involved flies collected from the same region that possessed two radically different configurations of the same genes. Dobzhansky called these genetic variants "races." Using Morgan's technique of mapping genes by virtue of their placement along a chromosome, Dobzhansky made a map of three genes—A, B, and C. In some flies, the three genes were strung along the fifth chromosome in one configuration: A-B-C. In other flies, Dobzhansky found that configuration had been fully inverted to C-B-A. The distinction between the two "races" of flies by virtue of a single chromosomal inversion was the most dramatic example of genetic variation that any geneticist had ever seen in a natural population.

But there was more. In September 1943, Dobzhansky launched an attempt to demonstrate variation, selection, and evolution in a single experiment—to re-create the Galápagos in a carton. He inoculated two sealed, aerated cartons with a mixture of two fly strains—ABC and CBA—in a one-to-one ratio. One carton was exposed to a cold temperature. The other, inoculated with the same mixture of strains, was left at room temperature. The flies were fed, cleaned, and watered in that enclosed space for generation upon generation. The populations grew and fell. New larvae were born, matured into flies, and died in that carton. Lineages and families—kingdoms of flies—were established and extinguished. When Dobzhansky harvested the two cages after four months, he found that the populations had changed dramatically. In the "cold carton," the ABC strain had nearly doubled, while the CBA had dwindled. In the carton kept at room temperature, the two strains had acquired the opposite ratio.

He had captured all the critical ingredients of evolution. Starting with a population with natural variation in gene configurations, he had added a force of natural selection: temperature. The "fittest" organisms—those best adapted to low or high temperatures—had survived. As new flies had been born, selected, and bred, the gene frequencies had changed, resulting in populations with new genetic compositions.

①

To explain the intersection of genetics, natural selection, and evolution in formal terms, Dobzhansky resurrected two important words—*genotype* and *phenotype*. A genotype is an organism's genetic composition. It can refer to one gene, a configuration of genes, or even an entire genome. A phenotype, in contrast, refers to an organism's physical or biological attributes and characteristics—the color of an eye, the shape of a wing, or resistance to hot or cold temperatures.

Dobzhansky could now restate the essential truth of Mendel's discovery—*a gene determines a physical feature*—by generalizing that idea across multiple genes and multiple features:

a genotype determines a phenotype

But two important modifications to this rule were necessary to complete the scheme. First, Dobzhansky noted, genotypes were not the sole determinants of phenotypes. Obviously, the environment or the milieu that surrounds an organism contributes to its physical attributes. The shape of a boxer's nose is not just the consequence of his genetic heritage; it is determined by the nature of his chosen profession, and the number of physical assaults on its cartilage. If Dobzhansky had capriciously trimmed the wings of all the flies in one box, he would have affected their phenotypes—the shape of their wings—without ever touching their genes. In other words:

$$genotype + environment = phenotype$$

And second, some genes are activated by external triggers or by random chance. In flies, for instance, a gene that determines the size of a vestigial wing depends on temperature: you cannot predict the shape of the wing based on the fly's genes or on the environment alone; you need to combine the two pieces of information. For such genes, neither the genotype nor the environment is the sole predictor of outcome: it is the *intersection* of genes, environment, and chance.

In humans, a mutant *BRCA1* gene increases the risk for breast cancer—but not all women carrying the *BRCA1* mutation develop cancer. Such trigger-dependent or chance-dependent genes are described as having partial or incomplete "penetrance"—i.e., even if the gene is inherited, its capacity to *penetrate* into an actual attribute is not absolute. Or a gene may have variable "expressivity"—i.e., even if the gene is inherited, the extent of it becoming *actualized* into an attribute varies from one individual to another. One woman with the *BRCA1* mutation might develop an aggressive, metastatic variant of breast cancer at age thirty. Another woman with the same mutation might develop an indolent variant; and yet another might not develop breast cancer at all.

We still do not know what causes the difference of outcomes between these three women—but it is some combination of age, exposures, other genes, and bad luck. You cannot use just the genotype—*BRCA1* mutation—to predict the final outcome with certainty.

So the final modification might be read as:

$$genotype + environment + triggers + chance = phenotype$$

Succinct, yet magisterial, this formula captured the essence of the interactions between heredity, chance, environment, variation, and evolution in determining the form and fate of an organism. In the natural world, variations in genotype exist in wild populations. These variations intersect with different environments, triggers, and chance to determine the attributes of an organism (a fly with greater or lesser resistance to temperature). When a severe selection pressure is applied—a rise in temperature or a sharp restriction of nutrients—organisms with the "fittest" phenotype are selected. The selective survival of such a fly results in its ability to produce more larvae, which inherit part of the genotype of the parent fly, resulting in a fly that is more adapted to that selective pressure. The process of selection, notably, acts on a *physical or biological* attribute—and the underlying genes are selected passively as a result. A mis-shapen nose might be the result of a particularly bad day in the ring—i.e., it may have nothing to do with genes—but if a mating contest is judged only by the symmetry of noses, then the bearer of the wrong kind of nose will be eliminated. Even if that bearer possesses multiple other genes that are salubrious in the long run—a gene for tenacity or for withstanding excruciating pain—the entire gamut of these genes will be damned to extinction during the mating contest, all because of that damned nose.

Phenotype, in short, drags genotypes behind it, like a cart pulling a horse. It is the perennial conundrum of natural selection that it seeks one thing (fitness) and accidentally finds another (genes that produce fitness). Genes that produce fitness become gradually overrepresented in populations through the selection of phenotypes, thereby allowing organisms to become more and more adapted to their environments. There is no such thing as perfection, only the relentless, thirsty matching of an organism to its environment. *That* is the engine that drives evolution.

①

Dobzhansky's final flourish was to solve the "mystery of mysteries" that had preoccupied Darwin: the origin of species. The Galápagos-in-a-carton experiment had demonstrated how a population of interbreeding organisms—flies, say—evolves over time.* But if wild populations with variations in genotype keep interbreeding, Dobzhansky knew, a new species would never be formed: a species, after all, is fundamentally defined by its inability to interbreed with another.

For a new species to arise, then, some factor must arise that makes interbreeding impossible. Dobzhansky wondered if the missing factor was

---

* The first experiments on reproductive incompatibility and species formation were carried out before the selection experiments, but Dobzhansky and his students continued to work on both problems in the 1940s and 50s.

geographic isolation. Imagine a population of organisms with gene variants that are capable of interbreeding. The population is suddenly split into two by some sort of geographical rift. A flock of birds from one island is storm-blown to a distant island and cannot fly back to its island of origin. The two populations now evolve independently, à la Darwin—until particular gene variants are selected in the two sites that become biologically incompatible. Even if the new birds can return to their original island—on ships, say—they cannot breed with their long-lost cousins of cousins: the offspring produced by the two birds possess genetic incompatibilities—garbled messages—that do not allow them to survive or be fertile. Geographic isolation leads to genetic isolation, and to eventual reproductive isolation.

This mechanism of speciation was not just conjecture; Dobzhansky could demonstrate it experimentally. He mixed two flies from two "races" in the same cage. The flies mated, gave rise to progeny—but the larvae grew into infertile adults. Using linkage analysis, geneticists could even trace an actual configuration of genes that evolved to make the progeny infertile. This was the missing link in Darwin's logic: reproductive incompatibility, ultimately derived from genetic incompatibility, drove the origin of novel species.

By the late 1930s, Dobzhansky began to realize that his understanding of genes, variation, and natural selection had ramifications far beyond biology. The bloody revolution of 1917 that had swept through Russia attempted to erase all individual distinctions to prioritize a collective good. In contrast, a monstrous form of racism that was rising in Europe exaggerated and demonized individual distinctions. In both cases, Dobzhansky noted, the fundamental questions at stake were biological. What defines an individual? How does variation contribute to individuality? What is "good" for a species?

⓪

In the 1940s, Dobzhansky would attack these questions directly: he would eventually become one of the most strident scientific critics of Nazi eugenics, Soviet collectivization, and European racism. But his studies on wild populations, variation, and natural selection had already provided crucial insights to these questions.

First, it was evident that genetic variation was the norm, not the exception, in nature. American and European eugenicists insisted on artificial selection to promote human "good"—but in nature there was no single "good." Different populations had widely divergent genotypes, and these diverse genetic types coexisted and even overlapped in the wild. Nature was not as hungry to homogenize genetic variation as human eugenicists had presumed. Indeed, Dobzhansky recognized that natural variation was a vital reservoir for an organism—an asset that far outweighed its liabilities. Without this variation—without deep genetic diversity—an organism might ultimately lose its capacity to evolve.

Second, a mutation is just a variation by another name. In wild fly populations, Dobzhansky noted, no genotype was inherently superior: whether the ABC or CBA strain survived depended on the environment, and on gene-environment interactions. One man's "mutant" was another man's "genetic variant." A winter's night might choose one fly. A summer's day might choose quite another. Neither variant was morally or biologically superior; each was just more or less adapted to a particular environment.

And finally, the relationship between an organism's physical or mental attributes and heredity was much more complex than anticipated. Eugenicists such as Galton had hoped to select complex *phenotypes*—intelligence, height, beauty, and moral rectitude—as a biological shortcut to enrich genes for intelligence, height, beauty, and morality. But a phenotype was not determined by one gene in a one-to-one manner. Selecting phenotypes was going to be a flawed mechanism to guarantee genetic selection. If genes, environments, triggers, and chance were responsible for the ultimate characteristics of an organism, then eugenicists would be inherently thwarted in their capacity to enrich intelligence or beauty across generations without deconvoluting the relative effects of each of these contributions.

Each of Dobzhansky's insights was a powerful plea against the misuse of genetics and human eugenics. Genes, phenotypes, selection, and evolution were bound together by cords of relatively basic laws—but it was easy to imagine that these laws could be misunderstood and distorted. "Seek simplicity, but distrust it," Alfred North Whitehead, the mathematician and philosopher, once advised his students. Dobzhansky had sought simplicity—but he had also issued a strident moral warning against the oversimplification of the logic of genetics. Buried in textbooks and scientific papers, these insights would be ignored by powerful political forces that would soon embark on the most perverse forms of human genetic manipulations.

Transformation

If you prefer an "academic life" as a retreat from reality,
do not go into biology. This field is for a man or woman
who wishes to get even closer to life.

—Hermann Muller

We do deny that . . . geneticists will see genes under the
microscope. . . . The hereditary basis does not lie in some
special self-reproducing substance.

—Trofim Lysenko

The reconciliation between genetics and evolution was termed the Modern Synthesis or, grandly, the Grand Synthesis.* But even as geneticists celebrated the synthesis of heredity, evolution, and natural selection, the material nature of the gene remained an unsolved puzzle. Genes had been described as "particles of heredity," but that description carried no information about what that "particle" was in a chemical or physical sense. Morgan had visualized genes as "beads on a string," but even Morgan had no idea what his description meant in material form. What were the "beads" made of? And what was the nature of the "string"?

In part, the material composition of the gene had defied identification because biologists had never intercepted genes in their chemical form. Throughout the biological world, genes generally travel *vertically*—i.e., from parents to children, or from parent cells to daughter cells. The vertical transmission of mutations had allowed Mendel and Morgan to study

---

* Sewall Wright, J. B. S. Haldane and several other biologists also contributed to the Grand Synthesis. A fuller description of all the names and contributors lies outside the purview of this book.

the action of a gene by analyzing patterns of heredity (e.g., the movement of the white-eyed trait from parent flies to their offspring). But the problem with studying vertical transformation is that the gene never leaves the living organism or cell. When a cell divides, its genetic material divides within it and is partitioned to its daughters. Throughout the process, genes remain biologically visible, but chemically impenetrable—shuttered within the black box of the cell.

Rarely, though, genetic material can cross from one organism to another—not between parent and child, but between two unrelated strangers. This horizontal exchange of genes is called *transformation*. Even the word signals our astonishment: humans are accustomed to transmitting genetic information only through reproduction—but during transformation, one organism seems to metamorphose into another, like Daphne growing twigs (or rather, the movement of genes *transforms* the attributes of one organism into the attributes of another; in the genetic version of the fantasy, twig-growing genes must somehow enter Daphne's genome and enable the ability to extrude bark, wood, xylem, and phloem out of human skin).

Transformation almost never occurs in mammals. But bacteria, which live on the rough edges of the biological world, can exchange genes horizontally (to fathom the strangeness of the event, imagine two friends, one blue eyed and one brown eyed, who go out for an evening stroll—and return with altered eye colors, having casually exchanged genes). The moment of genetic exchange is particularly strange and wonderful. Caught in transit between two organisms, a gene exists momentarily as a pure chemical. A chemist seeking to understand the gene has no more opportune moment to capture the chemical nature of the gene.

①

Transformation was discovered by an English bacteriologist named Frederick Griffith. In the early 1920s, Griffith, a medical officer at the British Ministry of Health, began to investigate a bacterium named *Streptococcus pneumoniae* or pneumococcus. The Spanish flu of 1918 had raged through the continent, killing nearly 20 million men and women worldwide and ranking among the deadliest natural disasters in history. Victims of the flu often developed a secondary pneumonia caused by pneumococcus—an illness so rapid and fatal that doctors had termed it the "captain of the men of death." Pneumococcal pneumonia after influenza infection—the epidemic within the epidemic—was of such concern that the ministry had deployed teams of scientists to study the bacterium and develop a vaccine against it.

Griffith approached the problem by focusing on the microbe: Why was pneumococcus so fatal to animals? Following work performed in Germany by others, he discovered that the bacterium came in two strains. A "smooth" strain possessed a slippery, sugary coat on the cell surface and could escape the immune system with newtlike deftness. The "rough" strain, which lacked this sugary coat, was more susceptible to immune attack. A mouse injected with the smooth strain thus died rapidly of pneumonia. In contrast, mice inoculated with the rough strain mounted an immune response and survived.

Griffith performed an experiment that, unwittingly, launched the molecular biology revolution. First, he killed the virulent, smooth bacteria with heat, then injected the heat-killed bacteria into mice. As expected, the bacterial remnants had no effect on the mice: they were dead and unable to cause an infection. But when he mixed the dead material from the virulent strain with live bacteria of the nonvirulent strain, the mice died rapidly. Griffith autopsied the mice and found that the rough bacteria had changed: they had *acquired* the smooth coat—the virulence-determining factor—merely by contact with the debris from the dead bacteria. The harmless bacteria had somehow "transformed" into the virulent form.

How could heat-killed bacterial debris—no more than a lukewarm soup of microbial chemicals—have transmitted a genetic trait to a live bacterium by mere contact? Griffith was unsure. At first, he wondered whether the live bacteria had ingested the dead bacteria and thus changed their coats, like a voodoo ritual in which eating the heart of a brave man transmits courage or vitality to another. But once transformed, the bacteria maintained their new coats for several generations—long after any food source would have been exhausted.

The simplest explanation, then, was that genetic information had passed between the two strains in a chemical form. During "transformation," the gene that governed virulence—producing the smooth coat versus the rough coat—had somehow slipped out of the bacteria into the chemical soup, then out of that soup into live bacteria and become incorporated into the genome of the live bacterium. Genes could, in other words, be transmitted between two organisms without any form of reproduction. They were autonomous units—*material* units—that carried information. Messages were not whispered between cells via ethereal pangenes or gemmules. Hereditary messages were transmitted through a molecule, that molecule could exist in a chemical form outside a cell, and it was capable of carrying information from cell to cell, from organism to organism, and from parents to children.

Had Griffith publicized this startling result, he would have set all of biology ablaze. In the 1920s, scientists were just beginning to understand living systems in chemical terms. Biology was becoming chemistry. The cell was a beaker of chemicals, biochemists argued, a pouch of compounds bound by a membrane that were reacting to produce a phenomenon called "life." Griffith's identification of a chemical capable of carrying hereditary instructions between organisms—the "gene molecule"—would have sparked a thousand speculations and restructured the chemical theory of life.

But Griffith, an unassuming, painfully shy scientist—"this tiny man who . . . barely spoke above a whisper"—could hardly be expected to broadcast the broader relevance or appeal of his results. "Englishmen do everything on principle," George Bernard Shaw once noted—and the principle that Griffith lived by was utter modesty. He lived alone, in a nondescript apartment near his lab in London, and in a spare, white modernist cottage that he had built for himself in Brighton. Genes might have moved between organisms, but Griffith could not be forced to travel from his lab to his own lectures. To trick him into giving scientific talks, his friends would stuff him into a taxicab and pay a one-way fare to the destination.

In January 1928, after hesitating for months ("God is in no hurry, so why should I be?"), Griffith published his data in the *Journal of Hygiene*—a scientific journal whose sheer obscurity might have impressed even Mendel. Writing in an abjectly apologetic tone, Griffith seemed genuinely sorry that he had shaken genetics by its roots. His study discussed transformation as a curiosity of microbial biology, but never explicitly mentioned the discovery of a potential chemical basis of heredity. The most important conclusion of the most important biochemical paper of the decade was buried, like a polite cough, under a mound of dense text.

①

Although Frederick Griffith's experiment was the most definitive demonstration that the gene was a chemical, other scientists were also circling the idea. In 1920, Hermann Muller, the former student of Thomas Morgan's, moved from New York to Texas to continue studying fly genetics. Like Morgan, Muller hoped to use mutants to understand heredity. But naturally arising mutants—the bread and butter of fruit fly geneticists—were far too rare. The white-eyed or sable-bodied flies that Morgan and his students had discovered in New York had been fished out laboriously by hunting through massive flocks of insects over thirty years. Tired of mutant hunting, Muller wondered if he could accelerate the production of mutants—perhaps by exposing flies to heat or light or higher bursts of energy.

In theory, this sounded simple; in practice, it was tricky. When Muller first tried exposing flies to X-rays, he killed them all. Frustrated, he lowered the dose—and found that he had now sterilized them. Rather than mutants, he had created vast flocks of dead, and then infertile, flies. In the winter of 1926, acting on a whim, he exposed a cohort of flies to an even lower dose of radiation. He mated the x-rayed males with females and watched the maggots emerge in the milk bottles.

Even a cursory look confirmed a striking result: the newly born flies had accumulated mutations—dozens of them, perhaps hundreds. It was late at night, and the only person to receive the breaking news was a lone botanist working on the floor below. Each time Muller found a new mutant, he shouted down from the window, "I got another." It had taken nearly three decades for Morgan and his students to collect about fifty fly mutants in New York. As the botanist noted, with some chagrin, Muller had discovered nearly half that number in a single night.

Muller was catapulted into international fame by his discovery. The effect of radiation on the mutation rate in flies had two immediate implications. First, genes had to be made of matter. Radiation, after all, is merely energy. Frederick Griffith had made genes move between organisms. Muller had altered genes using energy. A gene, whatever it was, was capable of motion, transmission, and of energy-induced change—properties generally associated with chemical matter.

But more than the material nature of the gene, it was the sheer *malleability* of the genome—that X-rays could make such Silly Putty of genes—that stunned scientists. Even Darwin, among the strongest original proponents of the fundamental mutability of nature, would have found this rate of mutation surprising. In Darwin's scheme, the rate of change of an organism was generally fixed, while the rate of natural selection could be amplified to accelerate evolution or dampened to decelerate it. Muller's experiments demonstrated that heredity could be manipulated quite easily: the mutation rate was itself quite mutable. "There is no permanent *status quo* in nature," Muller later wrote. "All is a process of adjustment and readjustment, or else eventual failure." By altering mutation rates and selecting variants in conjunction, Muller imagined he could possibly push the evolutionary cycle into hyperdrive, even creating entirely new species and subspecies in his laboratory—acting like the lord of his flies.

Muller also realized that his experiment had broad implications for human eugenics. If fly genes could be altered with such modest doses of radiation, then could the alteration of human genes be far behind? If genetic alterations could be "induced artificially," he wrote, then heredity could no longer be considered the unique privilege of an "unreachable god playing pranks on us."

Like many scientists and social scientists of his era, Muller had been captivated by eugenics since the 1920s. As an undergraduate, he had formed a Biological Society at Columbia University to explore and support "positive eugenics." But by the late twenties, as he had witnessed the menacing rise of eugenics in the United States, he had begun to reconsider his enthusiasm. The Eugenics Record Office, with its preoccupation with racial purification, and its drive to eliminate immigrants, "deviants," and "defectives," struck him as frankly sinister. Its prophets—Davenport, Priddy, and Bell—were weird, pseudoscientific creeps.

As Muller thought about the future of eugenics and the possibility of altering human genomes, he wondered whether Galton and his collaborators had made a fundamental conceptual error. Like Galton and Pearson, Muller sympathized with the desire to use genetics to alleviate suffering. But unlike Galton, Muller began to realize that positive eugenics was achievable only in a society that had *already* achieved radical equality. Eugenics could not be the prelude to equality. Instead, equality had to be the precondition for eugenics. Without equality, eugenics would inevitably falter on the false premise that social ills, such as vagrancy, pauperism, deviance, alcoholism, and feeblemindedness were *genetic* ills—while, in fact, they merely reflected inequality. Women such as Carrie Buck weren't genetic imbeciles; they were poor, illiterate, unhealthy, and powerless—

victims of their social lot, not of the genetic lottery. The Galtonians had been convinced that eugenics would ultimately generate radical equality—transforming the weak into the powerful. Muller turned that reasoning on its head. Without equality, he argued, eugenics would degenerate into yet another mechanism by which the powerful could control the weak.

①

While Hermann Muller's scientific work was ascending to its zenith in Texas, his personal life was falling apart. His marriage faltered and failed. His rivalry with Bridges and Sturtevant, his former lab partners from Columbia University, reached a brittle end point, and his relationship with Morgan, never warm, devolved into icy hostility.

Muller was also hounded for his political proclivities. In New York, he had joined several socialist groups, edited newspapers, recruited students, and befriended the novelist and social activist Theodore Dreiser. In Texas, the rising star of genetics began to edit an underground socialist newspaper, *The Spark* (after Lenin's *Iskra*), which promoted civil rights for African-Americans, voting rights for women, the education of immigrants, and collective insurance for workers—hardly radical agendas by contemporary standards, but enough to inflame his colleagues and irk the administration. The FBI launched an investigation into his activities. Newspapers referred to him as a subversive, a commie, a Red nut, a Soviet sympathizer, a freak.

Isolated, embittered, increasingly paranoid and depressed, Muller disappeared from his lab one morning and could not be found in his classroom. A search party of graduate students found him hours later, wandering in the woods in the outskirts of Austin. He was walking in a daze, his clothes wrinkled from the drizzle of rain, his face splattered with mud, his shins scratched. He had swallowed a roll of barbiturates in an attempt to commit suicide, but had slept them off by a tree. The next morning, he returned sheepishly to his class.

The suicide attempt was unsuccessful, but it was symptomatic of his malaise. Muller was sick of America—its dirty science, ugly politics, and selfish society. He wanted to escape to a place where he could meld science and socialism more easily. Radical genetic interventions could only be imagined in radically egalitarian societies. In Berlin, he knew, an ambitious liberal democracy with socialist leanings was shedding the husk of its past and guiding the birth of a new republic in the thirties. It was the "newest city" of the world, Twain had written—a place where scientists, writers, philosophers, and intellectuals were gathering in cafés and salons to forge a free and futuristic society. If the full potential of the modern science of genetics was to be unleashed, Muller thought, it would be in Berlin.

In the winter of 1932, Muller packed his bags, shipped off several hundred strains of flies, ten thousand glass tubes, a thousand glass bottles, one microscope, two bicycles, and a '32 Ford—and left for the Kaiser Wilhelm Institute in Berlin. He had no inkling that his adopted city would, indeed, witness the unleashing of the new science of genetics, but in its most grisly form in history.

Lebensunwertes Leben
(Lives Unworthy of Living)

He who is bodily and mentally not sound and deserving
may not perpetuate this misfortune in the bodies of his
children. The völkische *[people's] state has to perform*
the most gigantic rearing-task here. One day, however,
it will appear as a deed greater than the most victorious
wars of our present bourgeois era.

—Hitler's order for the Aktion T4

He wanted to be God . . . to create a new race.
—Auschwitz prisoner on Josef Mengele's goals

A hereditarily ill person costs 50,000 reichsmarks on
average up to the age of sixty.

—Warning to high school students in
a Nazi-era German biology textbook

Nazism, the biologist Fritz Lenz once said, is nothing more than "applied biology."*

In the spring of 1933, as Hermann Muller began his work at the Kaiser Wilhelm Institute in Berlin, he watched Nazi "applied biology" swing into action. In January that year, Adolf Hitler, the Führer of the National Socialist German Workers' Party, was appointed the chancellor of Germany. In March, the German parliament endorsed the Enabling Act,

---

* The quote has also been attributed to Rudolf Hess, Hitler's deputy.

granting Hitler unprecedented power to enact laws without parliamentary involvement. Jubilant Nazi paramilitary troops marched through the streets of Berlin with firelit torches, hailing their victory.

"Applied biology," as the Nazis understood it, was really applied genetics. Its purpose was to enable *Rassenhygiene*—"racial hygiene." The Nazis were not the first to use the term: Alfred Ploetz, the German physician and biologist, had coined the phrase as early as 1895 (recall his sinister, impassioned speech at the International Conference on Eugenics in London in 1912). "Racial hygiene," as Ploetz described it, was the genetic cleansing of the race, just as personal hygiene was the physical cleaning of the self. And just as personal hygiene routinely purged debris and excrement from the body, racial hygiene eliminated genetic detritus, thereby resulting in the creation of a healthier and purer race.* In 1914, Ploetz's colleague Heinrich Poll, the geneticist, wrote: "Just as the organism ruthlessly sacrifices degenerate cells, just as the surgeon ruthlessly removes a diseased organ, both, in order to save the whole: so higher organic entities, such as the kinship group or the state, should not shy away in excessive anxiety from intervening in personal liberty to prevent the bearers of diseased hereditary traits from continuing to spread harmful genes throughout the generations."

Ploetz and Poll looked to British and American eugenicists such as Galton, Priddy, and Davenport as pioneers of this new "science." The Virginia State Colony for Epileptics and Feebleminded was an ideal experiment in genetic cleansing, they noted. By the early 1920s, as women like Carrie Buck were being identified and carted off to eugenic camps in America, German eugenicists were expanding their own efforts to create a state-sponsored program to confine, sterilize, or eradicate "genetically defective" men and women. Several professorships of "race biology" and racial hygiene were established at German universities, and racial science was routinely taught at medical school. The academic hub of "race science" was the Kaiser Wilhelm Institute for Anthropology, Human Heredity and Eugenics—a mere stone's throw away from Muller's new lab in Berlin.

℗

Hitler, imprisoned for leading the Beer Hall Putsch, the failed coup attempt to seize power in Munich, read about Ploetz and race science while

---

* Ploetz would join the Nazis in the 1930s.

jailed in the 1920s and was immediately transfixed. Like Ploetz, he believed that defective genes were slow-poisoning the nation and obstructing the rebirth of a strong, healthy state. When the Nazis seized power in the thirties, Hitler saw an opportunity to put these ideas into action. He did so immediately: in 1933, less than five months after the passage of the Enabling Act, the Nazis enacted the Law for the Prevention of Genetically Diseased Offspring—commonly known as the Sterilization Law. The outlines of the law were explicitly borrowed from the American eugenics program—if amplified for effect. "Anyone suffering from a hereditary disease can be sterilized by a surgical operation," the law mandated. An initial list of "hereditary diseases" was drawn up, including mental deficiency, schizophrenia, epilepsy, depression, blindness, deafness, and serious deformities. To sterilize a man or woman, a state-sponsored application was to be made to the Eugenics Court. "Once the Court has decided on sterilization," the law continued, "the operation must be carried out even against the will of the person to be sterilized. . . . Where other measures are insufficient, direct force may be used."

To drum up public support for the law, legal injunctions were bolstered by insidious propaganda—a formula that the Nazis would eventually bring to monstrous perfection. Films such as *Das Erbe* ("The Inheritance," 1935) and *Erbkrank* ("Hereditary Disease," 1936), created by the Office of Racial Policy, played to full houses in theaters around the country to showcase the ills of "defectives" and "unfits." In *Erbkrank*, a mentally ill woman in the throes of a breakdown fiddles repetitively with her hands and hair; a deformed child lies wasted in bed; a woman with shortened limbs walks on all fours like a pack animal. Counterposed against the grim footage of *Erbkrank* or *Das Erbe* were cinematic odes to the perfect Aryan body: in Leni Riefenstahl's *Olympia*, a film intended to celebrate German athletes, glistening young men with muscular bodies demonstrated calisthenics as showpieces of genetic perfection. The audience gawked at the "defectives" with repulsion—and at the superhuman athletes with envy and ambition.

While the state-run agitprop machine churned to generate passive consent for eugenic sterilizations, the Nazis ensured that the legal engines were also thrumming to extend the boundaries of racial cleansing. In November 1933, a new law allowed the state to sterilize "dangerous criminals" (including political dissidents, writers, and journalists) by force. In October 1935, the Nuremberg Laws for the Protection of the Hereditary Health of the German People sought to contain genetic mixing by barring Jews from marrying people of German blood or having sexual relations with anyone of Aryan descent. There was, perhaps, no more bizarre illustration of the conflation between cleansing and racial cleansing than a law that barred Jews from employing "German maids" in their houses.

The vast sterilization and containment programs required the creation of an equally vast administrative apparatus. By 1934, nearly five thousand adults were being sterilized every month, and two hundred Hereditary Health Courts (or Genetic Courts) had to work full-time to adjudicate appeals against sterilization. Across the Atlantic, American eugenicists applauded the effort, often lamenting their own inability to achieve such effective measures. Lothrop Stoddard, another protégé of Charles Davenport's, visited one such court in the late thirties and wrote admiringly of its surgical efficacy. On trial during Stoddard's visit was a manic-depressive woman, a girl with deaf-muteness, a mentally retarded girl, and an "ape-like man" who had married a Jewess and was apparently also a homosexual—a complete trifecta of crimes. From Stoddard's notes, it remains unclear how the hereditary nature of any of these symptoms was established. Nonetheless, all the subjects were swiftly approved for sterilization.

①

The slip from sterilization to outright murder came virtually unannounced and unnoticed. As early as 1935, Hitler had privately mused about ramping up his gene-cleansing efforts from sterilization to euthanasia—what quicker way to purify the gene pool than to *exterminate* the defectives?—but had been concerned about the public reaction. By the late 1930s, though, the glacial equanimity of the German public response to the sterilization program made the Nazis bolder. Opportunity presented itself in 1939. In the summer of that year, Richard and Lina Kretschmar petitioned Hitler to allow them to euthanize their child, Gerhard. Eleven months old, Gerhard had been born blind and with deformed limbs. The parents—ardent Nazis—hoped to service their nation by eliminating their child from the nation's genetic heritage.

Sensing his chance, Hitler approved the killing of Gerhard Kretschmar and then moved quickly to expand the program to other children. Working with Karl Brandt, his personal physician, Hitler launched the Scientific Registry of Serious Hereditary and Congenital Illnesses to administer a much larger, nationwide euthanasia program to eradicate genetic "defectives." To justify the exterminations, the Nazis had already begun to describe the victims using the euphemism *lebensunwertes Leben*—lives unworthy of living. The eerie phrase conveyed an escalation of the logic of eugenics: it was not enough to sterilize genetic defectives to cleanse the future state; it was necessary to exterminate them to cleanse the current state. This would be a genetic final solution.

The killing began with "defective" children under three years of age, but by September 1939 had smoothly expanded to adolescents. Juvenile delinquents were slipped onto the list next. Jewish children were disproportionately targeted—forcibly examined by state doctors, labeled "genetically sick," and exterminated, often on the most minor pretexts. By October 1939, the program was expanded to include adults. A richly appointed villa—No. 4 Tiergartenstrasse in Berlin—was designated the official headquarters of the euthanasia program. The program would eventually be called Aktion T4, after that street address.

Extermination centers were established around the nation. Particularly active among them was Hadamar, a castlelike hospital on a hill, and the Brandenburg State Welfare Institute, a brick building resembling a garrison, with rows of windows along its side. In the basements of these buildings, rooms were refitted into airtight chambers where victims were gassed to death with carbon monoxide. The aura of science and medical research was meticulously maintained, often dramatized to achieve an even greater effect on public imagination. Victims of euthanasia were brought to the extermination centers in buses with screened windows, often accompanied by SS officers in white coats. In rooms adjoining the gas chambers, makeshift concrete beds, surrounded by deep channels to collect fluids, were created, where doctors could dissect the corpses after euthanasia so as to preserve their tissues and brains for future genetic studies. Lives "unworthy of living" were apparently of extreme worth for the advancement of science.

To reassure families that their parents or children had been appropriately treated and triaged, patients were often moved to makeshift holding facilities first, then secretly relocated to Hadamar or Brandenburg for the extermination. After euthanasia, thousands of fraudulent death certificates were issued, citing diverse causes of death—some of them markedly absurd. Mary Rau's mother, who suffered from psychotic depression, was exterminated in 1939. Her family was told that she had died as a consequence of "warts on her lip." By 1941, Aktion T4 had exterminated nearly a quarter of a million men, women, and children. The Sterilization Law had achieved about four hundred thousand compulsory sterilizations between 1933 and 1943.

①

Hannah Arendt, the influential cultural critic who documented the perverse excesses of Nazism, would later write about the "banality of evil" that permeated German culture during the Nazi era. But equally pervasive, it seemed, was the credulity of evil. That "Jewishness" or "Gypsyness" was carried on chromosomes, transmitted through heredity, and thereby subject to genetic cleansing required a rather extraordinary contortion of belief—but the suspension of skepticism was the defining credo of the culture. Indeed, an entire cadre of "scientists"—geneticists, medical researchers, psychologists, anthropologists, and linguists—gleefully regurgitated academic studies to reinforce the scientific logic of the eugenics program. In a rambling treatise entitled *The Racial Biology of Jews*, Otmar von Verschuer, a professor at the Kaiser Wilhelm Institute in Berlin, argued, for instance, that neurosis and hysteria were intrinsic genetic features of Jews. Noting that the suicide rate among Jews had increased by sevenfold between 1849 and 1907, Verschuer concluded, astonishingly, that the underlying cause was not the systematic persecution of Jews in Europe but their neurotic overreaction to it: "only persons with psychopathic and neurotic tendencies will react in such a manner to such a change in their external condition." In 1936, the University of Munich, an institution richly endowed by Hitler, awarded a PhD to a young medical researcher for his thesis concerning the "racial morphology" of the human jaw—an attempt to prove that the anatomy of the jaw was racially determined and genetically inherited. The newly minted "human geneticist," Josef Mengele, would soon rise to become the most epically perverse of Nazi researchers, whose experiments on prisoners would earn him the title Angel of Death.

In the end, the Nazi program to cleanse the "genetically sick" was just a prelude to a much larger devastation to come. Horrific as it was, the extermination of the deaf, blind, mute, lame, disabled, and feebleminded would be numerically eclipsed by the epic horrors ahead—the extermination of 6 million Jews in camps and gas chambers during the Holocaust;

of two hundred thousand Gypsies; of several million Soviet and Polish citizens; and unknown numbers of homosexuals, intellectuals, writers, artists, and political dissidents. But it is impossible to separate this apprenticeship in savagery from its fully mature incarnation; it was in this kindergarten of eugenic barbarism that the Nazis learned the alphabets of their trade. The word *genocide* shares its root with *gene*—and for good reason: the Nazis used the vocabulary of genes and genetics to launch, justify, and sustain their agenda. The language of genetic discrimination was easily parlayed into the language of racial extermination. The dehumanization of the mentally ill and physically disabled ("they cannot think or act like us") was a warm-up act to the dehumanization of Jews ("they do not think or act like us"). Never before in history, and never with such insidiousness, had genes been so effortlessly conflated with identity, identity with defectiveness, and defectiveness with extermination. Martin Neimöller, the German theologian, summarized the slippery march of evil in his often-quoted statement:

> *First they came for the Socialists, and I did not speak out—*
> *Because I was not a Socialist.*
> *Then they came for the Trade Unionists, and I did not speak out—*
> *Because I was not a Trade Unionist.*
> *Then they came for the Jews, and I did not speak out—*
> *Because I was not a Jew.*
> *Then they came for me—and there was no one left to speak out for me.*

☽

As the Nazis were learning to twist the language of heredity to prop up a state-sponsored program of sterilization and extermination in the 1930s, another powerful European state was also contorting the logic of heredity and genes to justify its political agenda—although in precisely the opposite manner. The Nazis had embraced genetics as a tool for racial cleansing. In the Soviet Union in the 1930s, left-wing scientists and intellectuals proposed that nothing about heredity was inherent at all. In nature, everything—*everyone*—was changeable. Genes were a mirage invented by the bourgeoisie to emphasize the fixity of individual differences, whereas, in fact, nothing about features, identities, choices, or destinies was indelible. If the state needed cleansing, it would not be achieved through genetic selection, but through the reeducation of all individuals and the erasure of former selves. Brains—not genes—had to be washed clean.

As with the Nazis, the Soviet doctrine was also bolstered and reinforced by ersatz science. In 1928, an austere, stone-faced agricultural researcher named Trofim Lysenko—he "gives one the feeling of a toothache," one journalist wrote—claimed that he had found a way to "shatter" and reorient hereditary influences in animals and plants. In experiments performed on remote Siberian farms, Lysenko had supposedly exposed wheat strains to severe bouts of cold and drought and thereby caused the strains to acquire a hereditary resistance to adversity (Lysenko's claims would later be found to be either frankly fraudulent or based on experiments of the poorest scientific quality). By treating wheat strains with such "shock therapy," Lysenko argued that he could make the plants flower more vigorously in the spring and yield higher bounties of grain through the summer.

"Shock therapy" was obviously at odds with genetics. The exposure of wheat to cold or drought could no more produce permanent, heritable changes in its genes than the serial dismemberment of mice's tails could create a tailless mouse strain, or the stretching of an antelope's neck could produce a giraffe. To instill such a change in his plants, Lysenko would have had to mutate cold-resistance genes (à la Morgan or Muller), use natural or artificial selection to isolate mutant strains (à la Darwin), and crossbreed mutant strains with each other to fix the mutation (à la Mendel and de Vries). But Lysenko convinced himself and his Soviet bosses that he had "retrained" the crops through exposure and conditioning alone and thereby altered their inherent characteristics. He dismissed the notion of genes altogether. The gene, he argued, had been "invented by geneticists" to support a "rotting, moribund bourgeoisie" science. "The hereditary basis does not lie in some special self-reproducing substance." It was a hoary restatement of Lamarck's idea—of adaptation morphing directly into hereditary change—decades after geneticists had pointed out the conceptual errors of Lamarckism.

Lysenko's theory was immediately embraced by the Soviet political apparatus. It promised a new method to vastly increase agricultural production in a land teetering on the edge of famine: by "reeducating" wheat and rice, crops could be grown under any conditions, including the severest winters and the driest summers. Perhaps just as important, Stalin and his compatriots found the prospect of "shattering" and "retraining" genes via shock therapy satisfying ideologically. While Lysenko was retraining plants to relieve them of their dependencies on soil and climate, Soviet party workers were also reeducating political dissidents to relieve them of their ingrained dependence on false consciousness and material goods. The Nazis—believing in absolute genetic immutability ("a Jew is a Jew")—had resorted to eugenics to change the structure of their population. The Soviets—believing in absolute genetic reprogrammability ("anyone is everyone")—could eradicate all distinctions and thus achieve a radical collective good.

In 1940, Lysenko deposed his critics, assumed the directorship of the Institute of Genetics of the Soviet Union, and set up his own totalitarian fiefdom over Soviet biology. Any form of scientific dissent to his theories—especially any belief in Mendelian genetics or Darwinian evolution—was outlawed in the Soviet Union. Scientists were sent to gulags to "retrain" them in Lysenko's ideas (as with wheat, the exposure of dissident professors to "shock therapy" might convince them to change their minds). In August 1940, Nicolai Vavilov, a renowned Mendelian geneticist, was captured and sent to the notorious Saratov jail for propagating his "bourgeoisie" views on biology (Vavilov had dared to argue that genes were not so easily malleable). While Vavilov and other geneticists languished in prison, Lysenko's supporters launched a vigorous campaign to discredit genetics as a science. In January 1943, exhausted and malnourished, Vavilov was moved to a prison hospital. "I am nothing but dung now," he described himself to his captors, and died a few weeks later.

Nazism and Lysenkoism were based on dramatically opposed conceptions of heredity—but the parallels between the two movements are striking. Although Nazi doctrine was unsurpassed in its virulence, both Nazism and Lysenkoism shared a common thread: in both cases, a theory of heredity was used to construct a notion of human identity that, in turn, was contorted to serve a political agenda. The two theories of heredity may have been spectacularly opposite—the Nazis were as obsessed with the fixity of identity as the Soviets were with its complete pliability—but the language of genes and inheritance was central to statehood and progress: it is as difficult to imagine Nazism without a belief in the indelibility of inheritance as it is to conceive of a Soviet state without a belief in its perfect erasure. Unsurprisingly, in both cases, science was deliberately distorted to support state-sponsored mechanisms of "cleansing." By appropriating the language of genes and inheritance, entire systems of power and statehood were justified and reinforced. By the mid-twentieth century, the gene—or the denial of its existence—had already emerged as a potent political and cultural tool. It had become one of the most dangerous ideas in history.

①

Junk science props up totalitarian regimes. And totalitarian regimes produce junk science. Did the Nazi geneticists make any real contributions to the science of genetics?

Amid the voluminous chaff, two contributions stand out. The first was methodological: Nazi scientists advanced the "twin study"—although, characteristically, they soon morphed it into a ghastly form. Twin studies had originated in Francis Galton's work in the 1890s. Having coined the phrase *nature versus nurture*, Galton had wondered how a scientist might discern the influence of one over the other. How could one determine if any particular feature—height or intelligence, say—was the product of nature or nurture? How could one unbraid heredity and environment?

Galton proposed piggybacking on a natural experiment. Since twins share identical genetic material, he reasoned, any substantial similarities between them could be attributed to genes, while any differences were the consequence of environment. By studying twins, and comparing and contrasting similarities and differences, a geneticist could determine the precise contributions of nature versus nurture to important traits.

Galton was on the right track—except for a crucial flaw: he had not distinguished between identical twins, who are truly genetically identical, and fraternal twins, who are merely genetic siblings (identical twins are derived from the splitting of a single fertilized egg, thereby resulting in twins with identical genomes, while fraternal twins are derived from the simultaneous fertilization of two eggs by two sperm, thereby resulting in twins with nonidentical genomes). Early twin studies were thus confounded by this confusion, leading to inconclusive results. In 1924, Hermann Werner Siemens, the German eugenicist and Nazi sympathizer, proposed a twin study that advanced Galton's proposal by meticulously separating identical twins from fraternal twins.*

---

* Curtis Merriman, an American psychologist, and Walter Jablonski, a German ophthalmologist, also performed similar twin studies in the 1920s.

A dermatologist by training, Siemens was a student of Ploetz's and a vociferous early proponent of racial hygiene. Like Ploetz, Siemens realized that genetic cleansing could be justified only if scientists could first establish heredity: you could justify sterilizing a blind man only if you could establish that his blindness was inherited. For traits such as hemophilia, this was straightforward: one hardly needed twin studies to establish heredity. But for more complex traits, such as intelligence or mental illness, the establishment of heredity was vastly more complex. To deconvolute the effects of heredity and environment, Siemens suggested comparing fraternal twins to identical twins. The key test of heredity would be *concordance*. The term *concordance* refers to the fraction of twins who possess a trait in common. If twins share eye color 100 percent of the time, then the concordance is 1. If they share it 50 percent of the time, then the concordance is 0.5. Concordance is a convenient measure for whether genes influence a trait. If identical twins possess a strong concordance for schizophrenia, say, while fraternal twins—born and bred in an identical environment—show little concordance, then the roots of that illness can be firmly attributed to genetics.

For Nazi geneticists, these early studies provided the fuel for more drastic experiments. The most vigorous proponent of such experiments was Josef Mengele—the anthropologist-turned-physician-turned-SS-officer who, sheathed in a white coat, haunted the concentration camps at Auschwitz and Birkenau. Morbidly interested in genetics and medical research, Mengele rose to become physician in chief at Auschwitz, where he unleashed a series of monstrous experiments on twins. Between 1943 and 1945, more than a thousand twins were subjected to Mengele's experiments.* Egged on by his mentor, Otmar von Verschuer from Berlin, Mengele sought out twins for his studies by trawling through the ranks of incoming camp prisoners and shouting a phrase that would become etched into the memories of the camp dwellers: *Zwillinge heraus* ("Twins out") or *Zwillinge heraustreten* ("Twins step out").

Yanked off the ramps, the twins were marked by special tattoos, housed in separate blocks, and systematically victimized by Mengele and his assistants (ironically, as experimental subjects, twins were also more likely to survive the camp than nontwin children, who were more casually ex-

---

* The exact number is hard to place. See Gerald L. Posner and John Ware, *Mengele: The Complete Story*, for the breadth of Mengele's twin experiments.

terminated). Mengele obsessively measured their body parts to compare genetic influences on growth. "There isn't a piece of body that wasn't measured and compared," one twin recalled. "We were always sitting together—always nude." Other twins were murdered by gassing and their bodies dissected to compare the sizes of internal organs. Yet others were killed by the injection of chloroform into the heart. Some were subjected to unmatched blood transfusions, limb amputations, or operations without anesthesia. Twins were infected with typhus to determine genetic variations in the responses to bacterial infections. In a particularly horrific example, a pair of twins—one with a hunched back—were sewn together surgically to determine if a shared spine would correct the disability. The surgical site turned gangrenous, and both twins died shortly after.

Despite the ersatz patina of science, Mengele's work was of the poorest scientific quality. Having subjected hundreds of victims to experiments, he produced no more than a scratched, poorly annotated notebook with no noteworthy results. One researcher, examining the disjointed notes at the Auschwitz museum, concluded, "No scientist could take [them] seriously." Indeed, whatever early advances in twin studies were achieved in Germany, Mengele's experiments putrefied twin research so effectively, pickling the entire field in such hatred, that it would take decades for the world to take it seriously.

①

The second contribution of the Nazis to genetics was never intended as a contribution. By the mid-1930s, as Hitler ascended to power in Germany, droves of scientists sensed the rising menace of the Nazi political agenda and left the country. Germany had dominated science in the early twentieth century: it had been the crucible of atomic physics, quantum mechanics, nuclear chemistry, physiology, and biochemistry. Of the one hundred Nobel Prizes awarded in physics, chemistry, and medicine between 1901 and 1932, thirty-three were awarded to German scientists (the British received eighteen; the Americans only six). When Hermann Muller arrived in Berlin in 1932, the city was home to the world's preeminent scientific minds. Einstein was writing equations on the chalkboards of the Kaiser Wilhelm Institute of Physics. Otto Hahn, the chemist, was breaking apart atoms to understand their constituent subatomic particles. Hans Krebs, the biochemist, was breaking open cells to identify their constituent chemical components.

But the ascent of Nazism sent an immediate chill through the German scientific establishment. In April 1933, Jewish professors were abruptly evicted from their positions in state-funded universities. Sensing imminent danger, thousands of Jewish scientists migrated to foreign countries. Einstein left for a conference in 1933 and wisely declined to return. Krebs fled that same year, as did the biochemist Ernest Chain and physiologist Wilhelm Feldberg. Max Perutz, the physicist, moved to Cambridge University in 1937. For some non-Jews, such as Erwin Schrödinger and nuclear chemist Max Delbrück, the situation was morally untenable. Many resigned out of disgust and moved to foreign countries. Hermann Muller—disappointed by another false utopia—left Berlin for the Soviet Union, on yet another quest to unite science and socialism. (Lest we misconstrue the response of scientists to Nazi ascendency, let it be known that many German scientists maintained a deadly silence in response to Nazism. "Hitler may have ruined the long term prospects of German science," George Orwell wrote in 1945, but there was no dearth of "gifted [German] men to do necessary research on such things as synthetic oil, jet planes, rocket projectiles and the atomic bomb.")

Germany's loss was genetics' gain. The exodus from Germany allowed scientists to travel not just between nations, but also between disciplines. Finding themselves in new countries, they also found an opportunity to turn their attention to novel problems. Atomic physicists were particularly interested in biology; it was the unexplored frontier of scientific inquiry. Having reduced matter into its fundamental units, they sought to reduce life to similar material units. The ethos of atomic physics—the relentless drive to find irreducible particles, universal mechanisms, and systematic explanations—would soon permeate biology and drive the discipline toward new methods and new questions. The reverberations of this ethos would be felt for decades to come: as physicists and chemists drifted toward biology, they attempted to understand living beings in chemical and physical terms—through molecules, forces, structures, actions, and reactions. In time, these émigrés to the new continent would redraw its maps.

Genes drew the most attention. What were genes made of, and how did they function? Morgan's work had pinpointed their location on chromosomes, where they were supposedly strung like beads on a wire. Griffith's and Muller's experiments had pointed to a material substance, a chemical that could move between organisms and was quite easily altered by X-rays.

Biologists might have blanched at trying to describe the "gene molecule" on purely hypothetical grounds—but what physicist could resist taking a ramble in weird, risky territory? In 1943, speaking in Dublin, the quantum theorist Erwin Schrödinger audaciously attempted to describe the molecular nature of the gene based on purely theoretical principles (a lecture later published as the book *What Is Life?*). The gene, Schrödinger posited, had to be made of a peculiar kind of chemical; it had to be a molecule of contradictions. It had to possess chemical regularity—otherwise, routine processes such a copying and transmission would not work—but it also had to be capable of extraordinary *irregularity*—or else, the enormous diversity of inheritance could not be explained. The molecule had to be able to carry vast amounts of information, yet be compact enough to be packaged into cells.

Schrödinger imagined a chemical with multiple chemical bonds stretching out along the length of the "chromosome fiber." Perhaps the sequence of bonds encoded the code script—a "variety of contents compressed into [some] miniature code." Perhaps the *order* of beads on the string carried the secret code of life.

Similarity and difference; order and diversity; message and matter. Schrödinger was trying to conjure up a chemical that would capture the divergent, contradictory qualities of heredity—a molecule to satisfy Aristotle. In his mind's eye, it was almost as if he had seen DNA.

"That Stupid Molecule"

Never underestimate the power of . . . stupidity.
—Robert Heinlein

Oswald Avery was fifty-five in 1933 when he heard of Frederick Griffith's transformation experiment. His appearance made him seem even older than his years. Frail, small, bespectacled, balding, with a birdlike voice and limbs that hung like twigs in winter, Avery was a professor at the Rockefeller University in New York, where he had spent a lifetime studying bacteria—particularly pneumococcus. He was sure that Griffith had made some terrible mistake in his experiment. How could chemical debris carry genetic information from one cell to another?

Like musicians, like mathematicians—like elite athletes—scientists peak early and dwindle fast. It isn't creativity that fades, but stamina: science is an endurance sport. To produce that single illuminating experiment, a thousand nonilluminating experiments have to be sent into the trash; it is battle between nature and nerve. Avery had established himself as a competent microbiologist, but had never imagined venturing into the new world of genes and chromosomes. "The Fess"—as his students affectionately called him (short for "professor")—was a good scientist but unlikely to become a revolutionary one. Griffith's experiment may have stuffed genetics into a one-way taxicab and sent it scuttling toward a strange future—but Avery was reluctant to climb on that bandwagon.

①

If the Fess was a reluctant geneticist, then DNA was a reluctant "gene molecule." Griffith's experiment had generated widespread speculations about the molecular identity of the gene. By the early 1940s, biochemists had broken cells apart to reveal their chemical constituents and identified various molecules in living systems—but the molecule that carried the code of heredity was still unknown.

Chromatin—the biological structure where genes resided—was known to be made of two types of chemicals: proteins and nucleic acids. No one knew or understood the chemical structure of chromatin, but of the two "intimately mixed" components, proteins were vastly more familiar to biologists, vastly more versatile, and vastly more likely to be gene carriers. Proteins were known to carry out the bulk of functions in the cell. Cells depend on chemical reactions to live: during respiration, for instance, sugar combines chemically with oxygen to make carbon dioxide and energy. None of these reactions occurs spontaneously (if they did, our bodies would be constantly ablaze with the smell of flambéed sugar). Proteins coax and control these fundamental chemical reactions in the cell—speeding some and slowing others, pacing the reactions just enough to be compatible with living. Life may be chemistry, but it's a special circumstance of chemistry. Organisms exist not because of reactions that are possible, but because of reactions that are *barely* possible. Too much reactivity and we would spontaneously combust. Too little, and we would turn cold and die. Proteins enable these barely possible reactions, allowing us to live on the edges of chemical entropy—skating perilously, but never falling in.

Proteins also form the structural components of the cell: filaments of hair, nails, cartilage, or the matrices that trap and tether cells. Twisted into yet other shapes, they also form receptors, hormones, and signaling molecules, allowing cells to communicate with one another. Nearly every cellular function—metabolism, respiration, cell division, self-defense, waste disposal, secretion, signaling, growth, even cellular death—requires proteins. They are the workhorses of the biochemical world.

Nucleic acids, in contrast, were the dark horses of the biochemical world. In 1869—four years after Mendel had read his paper to the Brno Society—a Swiss biochemist, Friedrich Miescher, had discovered this new class of molecules in cells. Like most of his biochemist colleagues, Miescher was also trying to classify the molecular components of cells by breaking cells apart and separating the chemicals that were released. Of the various components, he was particularly intrigued by one kind of chemical. He had precipitated it in dense, swirling strands out of white blood cells that he had wrung out of human pus in surgical dressings. He had found the same white swirl of a chemical in salmon sperm. He called the molecule *nuclein* because it was concentrated in a cell's nucleus. Since the chemical was acidic, its name was later modified to *nucleic acids*—but the cellular function of nuclein had remained mysterious.

By the early 1920s, biochemists had acquired a deeper understanding of the structure of nucleic acids. The chemical came in two forms—DNA and RNA, molecular cousins. Both were long chains made of four components, called bases, strung together along a stringlike chain or backbone. The four bases protruded out from the backbone, like leaves emerging out of the tendril of ivy. In DNA, the four "leaves" (or bases) were adenine, guanine, cytosine, and thymine—abbreviated A, G, C, and T. In RNA, the thymine was switched into uracil—hence A, C, G, and U.* Beyond these rudimentary details, nothing was known about the structure or function of DNA and RNA.

To the biochemist Phoebus Levene, one of Avery's colleagues at Rockefeller University, the comically plain chemical composition of DNA—four bases strung along a chain—suggested an extremely "unsophisticated" structure. DNA must be a long, monotonous polymer, Levene reasoned. In Levene's mind, the four bases were repeated in a defined order: AGCT-AGCT-AGCT-AGCT and so forth ad nauseam. Repetitive, rhythmic, regular, austere, this was a conveyer belt of a chemical, the nylon of the biochemical world. Max Delbruck, the scientist, called it a "stupid molecule."

Even a cursory look at Levene's proposed structure for DNA disqualified it as a carrier of genetic information. Stupid molecules could not carry clever messages. Monotonous to the extreme, DNA seemed to be quite the opposite of Schrödinger's imagined chemical—not just a stupid molecule but worse: a boring one. In contrast, proteins—diverse, chatty, versatile, capable of assuming Zelig-like shapes and performing Zelig-like functions—were infinitely more attractive as gene carriers. If chromatin, as Morgan had suggested, was a string of beads, then proteins had to be the active component—the beads—while DNA was likely the string. The nucleic acid in a chromosome, as one biochemist put it, was merely the "structure-determining, supporting substance"—a glorified molecular scaffold for genes. Proteins carried the real stuff of heredity. DNA was the stuffing.

---

* The "backbone" or spine of DNA and RNA is made of a chain of sugars and phosphates strung together. In RNA, the sugar is ribose—hence Ribo-Nucleic Acid (RNA). In DNA, the sugar is a slightly different chemical: deoxyribose—hence Deoxyribo-Nucleic Acid (DNA).

Ⓘ

In the spring of 1940, Avery confirmed the key result of Griffith's experiment. He separated the crude bacterial debris from the virulent smooth strain, mixed it with the live bacteria of the nonvirulent rough strain, and injected the mix into mice. Smooth-coated, virulent bacteria emerged faithfully—and killed the mice. The "transforming principle" had worked. Like Griffith, Avery observed that the smooth-coated bacteria, once transformed, retained their virulence generation upon generation. In short, genetic information must have been transmitted between two organisms in a purely chemical form, allowing that transition from the rough-coated to the smooth-coated variant.

But what chemical? Avery fiddled with the experiment as only a microbiologist could, growing the bacteria in various cultures, adding beef-heart broth, removing contaminant sugars, and growing the colonies on plates. Two assistants, Colin MacLeod and Maclyn McCarty, joined his laboratory to help with the experiments. The early technical fussing was crucial; by early August, the three had achieved the transformation reaction in a flask and distilled the "transforming principle" into a highly concentrated form. By October 1940, they began to sift through the concentrated bacterial detritus, painstakingly separating each chemical component, and testing each fraction for its capacity to transmit genetic information.

First, they removed all the remaining fragments of the bacterial coat from the debris. The transforming activity remained intact. They dissolved the lipids in alcohol—but there was no change in transformation. They stripped away the proteins by dissolving the material in chloroform. The transforming principle was untouched. They digested the proteins with various enzymes; the activity remained unaltered. They heated the material to sixty-five degrees—hot enough to warp most proteins—then added acids to curdle the proteins, and the transmission of genes was still unaltered. The experiments were meticulous, exhaustive, and definitive. Whatever its chemical constituents, the transforming principle was not composed of sugars, lipids, or proteins.

What was it, then? It could be frozen and thawed. Alcohol precipitated it. It settled out of solution in a white "fibrous substance . . . that wraps itself about a glass rod like a thread on a spool." Had Avery placed the fibrous spool on his tongue, he might have tasted the faint sourness of the acid, followed by the aftertaste of sugar and the metallic note of salt—like the taste of the "primordial sea," as one writer described it. An enzyme that digested RNA had no effect. The only way to eradicate transformation was to digest the material with an enzyme that, of all things, degraded DNA.

DNA? Was DNA the carrier of genetic information? Could the "stupid molecule" be the carrier of the most complex information in biology? Avery, MacLeod, and McCarty unleashed a volley of experiments, testing the transforming principle using UV light, chemical analysis, electrophoresis. In every case, the answer was clear: the transforming material was indubitably DNA. "Who could have guessed it?" Avery wrote hesitantly to his brother in 1943. "If we are right—and of course that's not yet proven—then nucleic acids are not merely structurally important but functionally active substances . . . that induce <u>predictable and hereditary</u> changes in cells [the underlined words are Avery's]."

Avery wanted to be doubly sure before he published any results: "It is hazardous to go off half-cocked, and embarrassing to have to retract it later." But he fully understood the consequences of his landmark experiment: "The problem bristles with implications. . . . This is something that has long been the dream of geneticists." As one researcher would later describe it, Avery had discovered "the material substance of the gene"—the "cloth from which genes were cut."

☉

Oswald Avery's paper on DNA was published in 1944—the very year that the Nazi exterminations ascended to their horrific crescendo in Germany. Each month, trains disgorged thousands of deported Jews into the camps. The numbers swelled: in 1944 alone, nearly 500,000 men, women, and children were transported to Auschwitz. Satellite camps were added, and new gas chambers and crematoria were constructed. Mass graves overflowed with the dead. That year, an estimated 450,000 were gassed to death. By 1945, 900,000 Jews, 74,000 Poles, 21,000 Gypsies (Roma), and 15,000 political prisoners had been killed.

At the start of 1945, as the soldiers of the Soviet Red Army approached Auschwitz and Birkenau through the frozen landscape, the Nazis attempted to evacuate nearly sixty thousand prisoners from the camps and their satellites. Exhausted, cold, and severely malnourished, many of these prisoners died during the evacuation. On the morning of January 27, 1945, Soviet troops entered the camps and liberated the remaining seven thousand prisoners—a minuscule remnant of the number killed and buried in the camp. By then the language of eugenics and genetics had long become subsidiary to the more malevolent language of racial hatred. The pretext of genetic cleansing had largely been subsumed by its progression into ethnic cleansing. Even so, the mark of Nazi genetics remained, like an indelible scar. Among the bewildered, emaciated prisoners to walk out of the camp that morning were one family of dwarfs and several twins—the few remaining survivors of Mengele's genetic experiments.

①

This, perhaps, was the final contribution of Nazism to genetics: it placed the ultimate stamp of shame on eugenics. The horrors of Nazi eugenics inspired a cautionary tale, prompting a global reexamination of the ambitions that had spurred the effort. Around the world, eugenic programs came to a shamefaced halt. The Eugenics Record Office in America had lost much of its funding in 1939 and shrank drastically after 1945. Many of its most ardent supporters, having developed a convenient collective amnesia about their roles in encouraging the German eugenicists, renounced the movement altogether.

"Important Biological Objects
Come in Pairs"

One could not be a successful scientist without realizing that, in contrast to the popular conception supported by newspapers and the mothers of scientists, a goodly number of scientists are not only narrow-minded and dull, but also just stupid.

—James Watson

It is the molecule that has the glamour, not the scientists.
—Francis Crick

Science [would be] ruined if—like sports—it were to put competition above everything else.

—Benoit Mandelbrot

Oswald Avery's experiment achieved another "transformation." DNA, once the underdog of all biological molecules, was thrust into the limelight. Although some scientists initially resisted the idea that genes were made of DNA, Avery's evidence was hard to shrug off (despite three nominations, however, Avery was still denied the Nobel Prize because Einar Hammarsten, the influential Swedish chemist, refused to believe that DNA could carry genetic information). As additional proof from other laboratories and experiments accumulated in the 1950s,* even the most

---

* Experiments carried out by Alfred Hershey and Martha Chase in 1952 and 1953 also confirmed that DNA was the carrier of genetic information.

hidebound skeptics had to convert into believers. The allegiances shifted: the handmaiden of chromatin was suddenly its queen.

Among the early converts to the religion of DNA was a young physicist from New Zealand, Maurice Wilkins. The son of a country doctor, Wilkins had studied physics at Cambridge in the 1930s. The gritty frontier of New Zealand—far away and upside down—had already produced a force that had turned twentieth-century physics on its head: Ernest Rutherford, another young man who had traveled to Cambridge on scholarship in 1895, and torn through atomic physics like a neutron beam on the loose. In a blaze of unrivaled experimental frenzy, Rutherford had deduced the properties of radioactivity, built a convincing conceptual model of the atom, shredded the atom into its constituent subatomic pieces, and launched the new frontier of subatomic physics. In 1919, Rutherford had become the first scientist to achieve the medieval fantasy of chemical transmutation: by bombarding nitrogen with radioactivity, he had converted it into oxygen. Even elements, Rutherford had proved, were not particularly elemental. The atom—the fundamental unit of matter—was actually made of even more fundamental units of matter: electrons, protons, and neutrons.

Wilkins had followed in Rutherford's wake, studying atomic physics and radiation. He had moved to Berkeley in the 1940s, briefly joining scientists to separate and purify isotopes for the Manhattan Project. But on returning to England, Wilkins—following the trend among many physicists—had edged away from physics toward biology. He had read Schrödinger's *What Is Life?* and become instantly entranced. The gene—the fundamental unit of heredity—must also be made of subunits, he reasoned, and the structure of DNA should illuminate these subunits. Here was a chance for a physicist to solve the most seductive mystery of biology. In 1946, Wilkins was appointed assistant director of the new Biophysics Unit at King's College in London.

℗

Biophysics. Even that odd word, the mishmash of two disciplines, was a sign of new times. The nineteenth-century realization that the living cell was no more than a bag of interconnected chemical reactions had launched a powerful discipline fusing biology and chemistry—biochemistry. "Life . . . is a chemical incident," Paul Ehrlich, the chemist, had once said, and biochemists, true to form, had begun to break open cells and characterize the constituent "living chemicals" into classes and functions. Sugars provided energy. Fats stored it. Proteins enabled chemical reactions, speeding and controlling the pace of biochemical processes, thereby acting as the switchboards of the biological world.

But *how* did proteins make physiological reactions possible? Hemoglobin, the oxygen carrier in blood, for instance, performs one of the simplest and yet most vital reactions in physiology. When exposed to high levels of oxygen, hemoglobin binds oxygen. Relocated to a site with low oxygen levels, it willingly releases the bound oxygen. This property allows hemoglobin to shuttle oxygen from the lung to the heart and the brain. But what feature of hemoglobin allows it to act as such an effective molecular shuttle?

The answer lies in the structure of the molecule. Hemoglobin A, the most intensively studied version of the molecule, is shaped like a four-leaf clover. Two of its "leaves" are formed by a protein called alpha-globin; the other two are created by a related protein, beta-globin.* Each of these leaves clasps, at its center, an iron-containing chemical named *heme* that can bind oxygen—a reaction distantly akin to a controlled form of rusting. Once all the oxygen molecules have been loaded onto heme, the four leaves of hemoglobin tighten around the oxygen like a saddle clasp. When unloading oxygen, the same saddle-clasp mechanism loosens. The unbinding of one molecule of oxygen coordinately relaxes all the other clasps, like the crucial pin-piece pulled out from a child's puzzle. The four leaves of the clover now twist open, and hemoglobin yields its cargo of oxygen. The controlled binding and unbinding of iron and oxygen—the cyclical rusting and unrusting of blood—allows effective oxygen delivery into tissues. Hemoglobin allows blood to carry seventyfold more oxygen than what could be dissolved in liquid blood alone. The body plans of vertebrates depend on this property: if hemoglobin's capacity to deliver oxygen to distant sites was disrupted, our bodies would be forced to be small and cold. We might wake up and find ourselves transformed into insects.

It is the *form* of hemoglobin, then, that permits its function. The physical structure of the molecule enables its chemical nature, the chemical

---

* Hemoglobin has multiple variants, including some that are specific to the fetus. This discussion applies to the most common, and best-studied, variant, which exists abundantly in blood.

nature enables its physiological function, and its physiology ultimately permits its biological activity. The complex workings of living beings can be perceived in terms of these layers: physics enabling chemistry, and chemistry enabling physiology. To Schrödinger's *"What is life?"* a biochemist might answer, "If not chemicals." And what are chemicals—a biophysicist might add—if not molecules of matter?

This description of physiology—as the exquisite matching of form and function, down to the molecular level—dates back to Aristotle. For Aristotle, living organisms were nothing more than exquisite assemblages of machines. Medieval biology had departed from that tradition, conjuring up "vital" forces and mystical fluids that were somehow unique to life—a last-minute deus ex machina to explain the mysterious workings of living organisms (and justify the existence of the *deus*). But biophysicists were intent on restoring a rigidly mechanistic description to biology. Living physiology should be explicable in terms of physics, biophysicists argued—forces, motions, actions, motors, engines, levers, pulleys, clasps. The laws that drove Newton's apples to the ground should also apply to the growth of the apple tree. Invoking special vital forces or inventing mystical fluids to explain life was unnecessary. Biology was physics. Machina *en* deus.

⌽

Wilkins's pet project at King's was solving the three-dimensional structure of DNA. If DNA was truly the gene carrier, he reasoned, then its structure should illuminate the nature of the gene. Just as the terrifying economy of evolution had stretched the length of the giraffe's neck and perfected the four-armed saddle clasp of hemoglobin, that same economy should have generated a DNA molecule whose form was exquisitely matched to its function. The gene molecule had to somehow *look* like a gene molecule.

To decipher the structure of DNA, Wilkins had decided to corral a set of biophysical techniques invented in nearby Cambridge—crystallography and X-ray diffraction. To understand the basic outline of this technique, imagine trying to deduce the shape of a minute three-dimensional object—a cube, say. You cannot "see" this cube nor feel its edges—but it shares the one property that all physical objects must possess: it generates shadows. Imagine that you can shine light at the cube from various angles and record the shadows that are formed. Placed directly in front of the light, a cube casts a square shadow. Illuminated obliquely, it forms a diamond. Move the light source again, and the shadow is a trapezoid. The process is almost absurdly laborious—like sculpting a face out of a million silhouettes—but it works: piece by piece, a set of two-dimensional images can be transmuted into a three-dimensional form.

X-ray diffraction arises out of analogous principles—the "shadows" are actually the scatters of X-rays generated by a crystal—except to illuminate molecules and generate scatters in the molecular world, one needs the most powerful source of light: X-rays. And there's a subtler problem: molecules generally refuse to sit still for their portraits. In liquid or gas form, molecules whiz dizzily in space, moving randomly, like particles of dust. Shine light on a million moving cubes and you only get a hazy, moving shadow, a molecular version of television static. The only solution to the problem is ingenious: transform a molecule from a solution to a *crystal*—and its atoms are instantly locked into position. Now the shadows become regular; the lattices generate ordered and readable silhouettes. By shining X-rays at a crystal, a physicist can decipher its structure in three-dimensional space. At Caltech, two physical chemists, Linus Pauling and Robert Corey, had used this technique to solve the structures of several protein fragments—a feat that would win Pauling the Nobel Prize in 1954.

This, precisely, is what Wilkins hoped to do with DNA. Shining X-rays on DNA did not require much novelty or expertise. Wilkins found an X-ray diffraction machine in the chemistry department and housed it— "in solitary splendor"—in a lead-lined room under the embankment wing, just below the level of the neighboring river Thames. He had all the crucial material for his experiment. Now his main challenge was to make DNA sit still.

①

Wilkins was plowing methodically through his work in the early 1950s when he was interrupted by an unwelcome force. In the winter of 1950, the head of the Biophysics Unit, J. T. Randall, recruited an additional young scientist to work on crystallography. Randall was patrician, a small, genteel, cricket-loving dandy who nonetheless ran his unit with Napoleonic authority. The new recruit, Rosalind Franklin, had just finished studying coal crystals in Paris. In January 1951, she came up to London to visit Randall.

Wilkins was away on vacation with his fiancée—a decision he would later regret. It is not clear how much Randall had anticipated future collisions when he suggested a project to Franklin. "Wilkins has already found that fibers of [DNA] give remarkably good diagrams," he told her. Perhaps Franklin would consider studying the diffraction patterns of these fibers and deduce a structure? He had offered her DNA.

When Wilkins returned from vacation, he expected Franklin to join him as his junior assistant; DNA had, after all, always been *his* project. But Franklin had no intention of assisting anyone. A dark-haired, dark-eyed daughter of a prominent English banker, with a gaze that bored through her listeners like X-rays, Franklin was a rare specimen in the lab—an independent female scientist in a world dominated by men. With a "dogmatic, pushy father," as Wilkins would later write, Franklin grew up in a household where "her brothers and father resented R.F.'s greater intelligence." She had little desire to work as anyone's assistant—let alone for Maurice Wilkins, whose mild manner she disliked, whose values, she opined, were hopelessly "middle-class," and whose project—deciphering DNA—was on a direct collision course with hers. It was, as one friend of Franklin's would later put it, "hate at first sight."

Wilkins and Franklin worked cordially at first, meeting for occasional coffee at the Strand Palace Hotel, but the relationship soon froze into frank, glacial hostility. Intellectual familiarity bred a slow, glowering contempt; in a few months, they were barely on speaking terms. (She "barks often, doesn't succeed in biting me," Wilkins later wrote.) One morning, out with separate groups of friends, they found themselves punting on the Cam River. As Franklin charged down the river toward Wilkins, the boats came close enough to collide. "Now she's trying to drown me," he exclaimed in mock horror. There was nervous laughter—the kind when a joke cuts too close to the truth.

What she was trying to drown, really, was noise. The chink of beer mugs in pubs infested by men; the casual bonhomie of men discussing science in their male-only common room at King's. Franklin found most of her male colleagues "positively repulsive." It was not just sexism—but the *innuendo* of sexism that was exhausting: the energy spent parsing perceived slights or deciphering unintended puns. She would rather work on other codes—of nature, of crystals, of invisible structures. Unusually for his time, Randall was not averse to hiring women scientists; there were several women working with Franklin at King's. And female trailblazers had come before her: severe, passionate Marie Curie, with her chapped palms and char-black dresses, who had distilled radium out of a cauldron of black sludge and won not one Nobel Prize but two; and matronly, ethereal Dorothy Hodgkin at Oxford, who would later win her own Nobel for solving the crystal structure of penicillin (an "affable looking housewife," as one newspaper described her). Yet Franklin fit neither model: she was neither affable housewife nor cauldron-stirrer in a boiled wool robe, neither Madonna nor witch.

The noise that bothered Franklin most was the fuzzy static in the DNA pictures. Wilkins had obtained some highly purified DNA from a Swiss lab and stretched it into thin, uniform fibers. By stringing the fiber along a gap in a stretch of wire—a bent paper clip worked marvelously—he hoped to diffract X-rays and obtain images. But the material had proved difficult to photograph; it generated scattered, fuzzy dots on film. What made a purified molecule so difficult to image? she wondered. Soon, she stumbled on the answer. In its pure state, DNA came in two forms. In the presence of water, the molecule was in one configuration, and as it dried out, it switched to another. As the experimental chamber lost its humidity, the DNA molecules relaxed and tensed—exhaling, inhaling, exhaling, like life itself. The switch between the two forms was partly responsible for the noise that Wilkins had been struggling to minimize.

Franklin adjusted the humidity of the chamber using an ingenious apparatus that bubbled hydrogen through salt water. As she increased the wetness of DNA in the chamber, the fibers seemed to relax permanently. She had tamed them at last. Within weeks, she was taking pictures of DNA of a quality and clarity that had never before been seen. J. D. Bernal, the crystallographer, would later call them the "most beautiful X ray photographs of any substance ever taken."

①

In the spring of 1951, Maurice Wilkins gave a scientific talk at the Zoological Station in Naples—at the laboratory where Boveri and Morgan had once worked on urchins. The weather was just beginning to warm up, although the sea might still send a blast of chill through the corridors of the city. In the audience that morning—"shirttails flying, knees in the air, socks down around his ankles . . . cocking his head like a rooster"— was a biologist Wilkins had never heard of, an excitable, voluble young man named James Watson. Wilkins's talk on the structure of DNA was dry and academic. One of his last slides, presented with little enthusiasm, was an early X-ray diffraction picture of DNA. The photograph flickered onto the screen at the end of a long talk, and Wilkins showed little, if any, excitement about the fuzzy image. The pattern was still a muddle— Wilkins was still hampered by the quality of his sample and the dryness of his chamber—but Watson was instantly gripped by it. The general conclusion was unmistakable: in principle, DNA could be crystallized into a form amenable to X-ray diffraction. "Before Maurice's talk, I had worried about the possibility that the gene might be fantastically irregular," Watson would later write. The image, however, quickly convinced Watson otherwise: "Suddenly, I was excited about chemistry." He tried to talk to Wilkins about the image, but "Maurice was English, and [didn't] talk to strangers." Watson slunk away.

Watson knew "nothing about the X-ray diffraction technique," but he had an unfailing intuition about the importance of certain biological problems. Trained as an ornithologist at the University of Chicago, he had assiduously "avoid[ed] taking any chemistry or physics courses which looked of even medium difficulty." But a kind of homing instinct had led him to DNA. He too had read Schrödinger's *What Is Life?* and been captivated. He had been working on the chemistry of nucleic acids in Copenhagen—"a complete flop," as he would later describe it—but Wilkins's photograph entranced him. "The fact that I was unable to interpret it did not bother me. It was certainly better to imagine myself becoming famous than maturing into a stifled academic who had never risked a thought."

Impetuously, Watson returned to Copenhagen and asked to be transferred to Max Perutz's lab at Cambridge (Perutz, the Austrian biophysicist, had fled Nazi Germany and moved to England during the exodus of the 1930s). Perutz was working on molecular structures, and it was the closest that Watson could get to Wilkins's image, whose haunting, prophetic shadows he could not get out of his brain. Watson had decided that he was going to solve the structure of DNA—"the Rosetta stone for unraveling the true secret of life." He would later say, "As a geneticist, it was the only problem worth solving." He was all of twenty-three years old.

①

Watson had moved to Cambridge for the love of a photograph. The very first day that he landed in Cambridge, he fell in love again—with a man

146

named Francis Crick, another student in Perutz's lab. It was not an erotic love, but a love of shared madness, of conversations that were electric and boundless, of ambitions that ran beyond realities.* "A youthful arrogance, a ruthlessness, and an impatience with sloppy thinking came naturally to both of us," Crick would later write.

Crick was thirty-five—a full twelve years older than Watson, and still without a PhD (in part because he had worked for the Admiralty during the war years). He was not conventionally "academic," and he certainly was not "stifled." A former physics student with an expansive personality and a booming voice that often sent his coworkers running for cover and a bottle of aspirin, he too had read Schrödinger's *What Is Life?*—that "small book that had started a revolution"—and become transfixed by biology.

Englishmen hate many things, but no one is despised more than the man who sits by you on the morning train and solves your crossword puzzle. Crick's intelligence was as free ranging and as audacious as his voice; he thought nothing of invading the problems of others and suggesting solutions. To make things worse, he was usually right. In the late 1940s, switching from physics to graduate work in biology, he had taught himself much of the mathematical theory of crystallography—that swirl of nested equations that made it possible to transmute silhouettes into three-dimensional structures. Like most of his colleagues in Perutz's lab, Crick focused his initial studies on the structures of proteins. But unlike many others, he had been intrigued by DNA from the start. Like Watson, and like Wilkins and Franklin, he was also instinctively drawn to the structure of a molecule capable of carrying hereditary information.

The two of them—Watson and Crick—talked so volubly, like children let loose in a playroom, that they were assigned a room to themselves, a yellow brick chamber with wooden rafters where they were left to their own devices and dreams, their "mad pursuit[s]." They were complementary strands, interlocked by irreverence, zaniness, and fiery brilliance. They despised authority but craved its affirmation. They found the scien-

---

* In 1951, long before James Watson would become a household name around the world, the novelist Doris Lessing took a three-hour walk with the young Watson, whom she knew through a friend of a friend. During the entire walk, across the heaths and fens near Cambridge, Lessing did all the talking; Watson said not one word. At the end of the walk, "exhausted, wanting only to escape," Lessing at last heard the sound of human speech from her companion: "The trouble is, you see, that there is only one other person in the world that I can talk to."

tific establishment ridiculous and plodding, yet they knew how to insinu-
ate themselves into it. They imagined themselves quintessential outsiders,
yet felt most comfortable sitting in the inner quadrangles of Cambridge
colleges. They were self-appointed jesters in a court of fools.

The one scientist they did revere, if begrudgingly, was Linus Pauling—
the larger-than-life Caltech chemist who had recently announced that
he had solved an important conundrum in the structure of proteins.
Proteins are made of chains of amino acids. The chains fold in three-
dimensional space to form substructures, which then fold into larger
structures (imagine a chain that first coils into a spring and then a spring
that further jumbles into a spherical or globular shape). Working with
crystals, Pauling had found that proteins frequently folded into an archetypal substructure—a single helix coiled like a spring. Pauling had
revealed his model at a meeting at Caltech with the dramatic flair of a
sorcerer pulling a molecular bunny out of a hat: the model had been hidden behind a curtain until the end of the talk, and then—presto!—it had
been revealed to a stunned, applauding audience. Rumor had it that Pauling had now turned his attention from proteins to the structure of DNA.
Five thousand miles away, in Cambridge, Watson and Crick could almost
feel Pauling breathing down their necks.

Pauling's seminal paper on the protein helix was published in April
1951. Festooned with equations and numbers, it was intimidating to read,
even for experts. But to Crick, who knew the mathematical formulas as
intimately as anyone, Pauling had hidden his essential method behind
the smoke-and-mirrors algebra. Crick told Watson that Pauling's model
was, in fact, the "product of common sense, not the result of complicated
mathematical reasoning." The real magic was imagination. "Equations occasionally crept into his argument, but in most cases words would have
sufficed. . . . The alpha-helix had not been found by staring at X-ray pictures; the essential trick, instead, was to ask which atoms like to sit next
to each other. In place of pencil and paper, the main working tools were
a set of molecular models superficially resembling the toys of preschool
children."

Here, Watson and Crick took their most intuitive scientific leap. What
if the solution to the structure of DNA could be achieved by the same
"tricks" that Pauling had pulled? X-ray pictures would help, of course—
but trying to determine structures of biological molecules using experimental methods, Crick argued, was absurdly laborious—"like trying

to determine the structure of a piano by listening to the sound it made while being dropped down a flight of stairs." But what if the structure of DNA was so simple—so *elegant*—that it could be deduced by "common sense," by model building? What if a stick-and-stone assemblage could solve DNA?

☊

Fifty miles away, at King's College in London, Franklin had little interest in building models with toys. With her laserlike focus on experimental studies, she had been taking photograph after photograph of DNA—each with increasing clarity. The pictures would provide the answer, she reasoned; there was no need for guesswork. The experimental data would generate the models, not the other way around. Of the two forms of DNA—the "dry" crystalline form and a "wet" form—the wet form seemed to have a less convoluted structure. But when Wilkins proposed that they collaborate to solve the wet structure, she would have none of it. A collaboration, it seemed to her, was a thinly disguised capitulation. Randall was soon forced to intervene to formally separate them, like bickering children. Wilkins was to continue with the wet form, while Franklin was to concentrate on the dry form.

The separation hobbled both of them. Wilkins's DNA preparations were of poor quality and wouldn't generate good photographs. Franklin had pictures, but she found them difficult to interpret. ("How dare you interpret my data for me?" she once snapped at him.) Although they worked no more than a few hundred feet apart, the two of them might as well have inhabited two warring continents.

On November 21, 1951, Franklin gave a talk at King's. Watson was invited to the talk by Wilkins. The gray afternoon was fouled by the soupy London fog. The room was an old, damp lecture hall buried in the innards of the college; it resembled a dreary accountant's chamber in a Dickens novel. About fifteen people attended. Watson sat in the audience—"skinny and awkward . . . pop-eyed, and wrote down nothing."

Franklin spoke "in a quick nervous style [with] . . . not a trace of warmth or frivolity in her words," Watson would later write. "Momentarily I wondered how she would look if she took off her glasses and did something novel with her hair." There was something purposefully severe and offhanded in Franklin's manner of speaking; she delivered her lecture as if reading the Soviet evening news. Had anyone paid real attention to her subject—and not the styling of her hair—he might have noticed that she was circling a monumental conceptual advance, albeit with deliberate caginess. "Big helix with several chains,"* she had written in her notes, "phosphates on the outside." She had begun to glimpse the skeleton of an exquisite structure. But she gave only some cursory measurements, pointedly declined to specify any details about the structure, and then brought a witheringly dull academic seminar to its close.

The next morning, Watson excitedly brought news of Franklin's talk to Crick. They were boarding a train for Oxford to meet Dorothy Hodgkin, the grande dame of crystallography. Rosalind Franklin had said little in her talk except to provide a few preliminary measurements. But when Crick quizzed Watson about the precise numbers, Watson could provide only vague answers. He had not even bothered to scribble numbers on the back of a napkin. He had attended one of the most important seminars in his scientific life—and failed to take notes.

Still, Crick got enough of a sense of Franklin's preliminary thoughts to hurry back to Cambridge and begin building a model. They started the next morning, with lunch at the nearby Eagle Pub and some gooseberry pie. "Superficially, the X-ray data was compatible with two, three or four strands," they realized. The question was, how to put the strands together and make a model of an enigmatic molecule?

①

A single strand of DNA consists of a backbone of sugars and phosphates, and four bases—A, T, G, and C—attached to the backbone, like teeth jutting out from a zipper strand. To solve the structure of DNA, Watson and Crick had to first figure out how many zippers were in each DNA molecule, what part was at the center, and which part at the periphery. It looked like a relatively simple problem—but it was fiendishly difficult to build a

---

* In her initial studies on DNA, Franklin was not convinced that the X-ray patterns suggested a helix, most likely because she was working on the dry form of DNA. Indeed, at one point Franklin and her student had sent around a cheeky note announcing the "death of the helix." However, as her X-ray images improved, she gradually began to envision the helix with the phosphates on the outside, as indicated by her notes. Watson once told a journalist that Franklin's fault lay in her dispassionate approach to her own data: "She did not live DNA."

simple model. "Even though only about fifteen atoms were involved, they kept falling out of the awkward pincers set up to hold them."

By teatime, still tinkering with an awkward model set, Watson and Crick had come up with a seemingly satisfactory answer: three chains, twisted around each other, in a helical formation, with the sugar phosphate backbone compressed in the center. A triple helix. Phosphates on the inside. "A few of the atomic contacts were still too close for comfort," they admitted—but perhaps these would be fixed by additional fiddling. It wasn't a particularly elegant structure—but maybe that was asking too much. The next step, they realized, was to "check it with Rosy's quantitative measurements." And then, on a whim—in a misstep that they would later come to regret—they called Wilkins and Franklin to come and have a look.

Wilkins, Franklin, and her student, Ray Gosling, took the train down from King's the next morning to inspect the Watson and Crick model. The journey to Cambridge was loaded with expectations. Franklin was lost in her thoughts.

When the model was unveiled at last, it was an epic letdown. Wilkins found the model "disappointing"—but held his tongue. Franklin was not as diplomatic. One look at the model was enough to convince her that it was nonsense. It was worse than wrong; it was unbeautiful—an ugly, bulging, falling-apart catastrophe, a skyscraper after an earthquake. As Gosling recalled, "Rosalind let rip in her best pedagogical style: 'you're wrong for the following reasons' . . . which she proceeded to enumerate as she demolished their proposal." She may as well have kicked the model with her feet.

Crick had tried to stabilize the "wobbly unstable chains" by putting the phosphate backbone in the center. But phosphates are negatively charged. If they faced *inside* the chain, they would repel each other, forcing the molecule to fly apart in a nanosecond. To solve the problem of repulsion, Crick had inserted a positively charged magnesium ion at the center of the helix—like a last-minute dab of molecular glue to hold the structure together. But Franklin's measurements suggested that magnesium could not be at the center. Worse, the structure modeled by Watson and Crick was so tightly packed that it could not accommodate any significant number of water molecules. In their rush to build a model, they had even forgotten Franklin's *first* discovery: the remarkable "wetness" of DNA.

The viewing had turned into an inquisition. As Franklin picked the model apart, molecule by molecule, it was as if she were extracting bones from their bodies. Crick looked progressively deflated. "His mood," Watson recalled, "was no longer that of a confident master lecturing hapless colonial children." By now, Franklin was frankly exasperated at the "adolescent blather." The boys and their toys had turned out to be a monumental waste of her time. She caught the 3:40 train home.

Ⓘ

In Pasadena, meanwhile, Linus Pauling was also trying to solve the structure of DNA. Pauling's "assault on DNA," Watson knew, would be nothing short of formidable. He would come at it with a bang, deploying his deep understanding of chemistry, mathematics, and crystallography—but more important, his instinctual grasp of model building. Watson and Crick feared that they would wake up one morning, open the pages of an august scientific journal, and find the solved structure of DNA staring back at them. Pauling's name—not theirs—would be attached to the article.

In the first weeks of January 1953, that nightmare seemed to come true: Pauling and Robert Corey wrote a paper proposing a structure of DNA and sent a preliminary copy to Cambridge. It was a bombshell casually lobbed across the Atlantic. For a moment, it seemed to Watson that "all was lost." He rifled through paper like a madman until he had found the crucial figure. But as he stared at the proposed structure, Watson knew instantly "that something was not right." By coincidence, Pauling and Corey had also suggested a triple helix, with the bases A, C, G, and T pointed outside. The phosphate backbone twisted inside, like the central shaft of a spiral staircase, with its treads facing out. But Pauling's proposal did not have any magnesium to "glue" the phosphates together. Instead, he proposed that the structure would be held together by much weaker bonds. This magician's sleight of hand did not go unnoticed. Watson knew immediately that the structure would not work: it was energetically unstable. One colleague of Pauling's would later write, "If that were the structure of DNA, it would explode." Pauling had not produced a bang; he had created a molecular Big Bang.

"The blooper," as Watson described it, "was too unbelievable to keep secret for more than a few minutes." He dashed over to a chemist friend in the neighboring lab to show him Pauling's structure. The chemist concurred, "The giant [Pauling] had forgotten elementary college chemistry." Watson told Crick, and both took off for the Eagle, their favorite pub, where they celebrated Pauling's failure with shots of schadenfreude-infused whiskey.

℗

Late in January 1953, James Watson went to London to visit Wilkins. He stopped to see Franklin in her office. She was working at her bench, with dozens of photographs strewn around her, and a book full of notes and equations on her desk. They spoke stiffly, arguing about Pauling's paper. At one point, exasperated by Watson, Franklin moved quickly across the lab. Fearing "that in her hot anger, she might strike [him]," Watson retreated through the front door.

Wilkins, at least, was more welcoming. As the two commiserated about Franklin's radioactive temper, Wilkins opened up to Watson to a degree he never before had. What happened next is a twisted braid of mixed signals, distrust, miscommunication, and conjecture. Wilkins told Watson that Rosalind Franklin had taken a series of new photographs of the fully wet form of DNA over the summer—pictures so staggeringly crisp that the essential skeleton of the structure virtually jumped out of them.

On May 2, 1952, a Friday evening, she and Gosling had exposed a DNA fiber to X-rays overnight. The picture was technically perfect—although the camera had cocked a little off center. "V.Good. Wet Photo," she had written in her red notebook. At half past six the next evening—she worked on Saturday nights, of course, while the rest of the staff went to the pub—she set up the camera again with Gosling's help. On Tuesday afternoon, she exposed the photograph. It was even crisper than the previous one. It was the most perfect image that she had ever seen. She had labeled it "Photograph 51."

Wilkins walked over to the next room, pulled the crucial photograph out of a drawer, and showed it to Watson. Franklin was still in her office, smoldering with irritation. She had no knowledge that Wilkins had just revealed the most precious piece of her data to Watson.* ("Perhaps I should have asked Rosalind's permission and I didn't," a contrite Wilkins would later write. "Things were very difficult. . . . If there had been any-

---

* But was it *her* photograph? Wilkins later maintained that the photograph had been given to him by Gosling, Franklin's student—and therefore it was his to do with what he desired. Franklin was leaving King's College to take up a new job at Birkbeck College, and Wilkins thought that she was abandoning the DNA project.

thing like a normal situation here, I'd have asked her permission naturally, though if there had been anything like a normal situation, the whole question of permission wouldn't have come up. . . . I had this photograph, and there was a helix right on the picture, you couldn't miss it.")

Watson was immediately transfixed. "The instant I saw the picture my mouth fell open and my pulse started to race. The pattern was unbelievably simpler than those obtained previously. . . . The black cross could arise only from a helical structure. . . . After only a few minutes' calculations, the number of chains in the molecule could be fixed."

In the icy compartment of the train that sliced across the fens back to Cambridge that evening, Watson sketched what he remembered of the picture on the edge of a newspaper. He had come back the first time from London without notes. He wasn't going to repeat the same error. By the time he had returned to Cambridge and jumped over the back gate of the college, he was convinced that DNA had to be made of two intertwined, helical chains: "important biological objects come in pairs."

Ⅰ

The next morning, Watson and Crick raced down to the lab and started model building in earnest. Geneticists count; biochemists clean. Watson and Crick played. They worked methodically, diligently, and carefully—but left enough room for their key strength: lightness. If they were to win this race, it would be through whimsy and intuition; they would laugh their way to DNA. At first, they tried to salvage the essence of their first model, placing the phosphate backbone in the middle, and the bases projecting out to the sides. The model wobbled uneasily, with molecules jammed too close together for comfort. After coffee, Watson capitulated: perhaps the backbone was on *the outside*, and the bases—A, T, G, and C—faced in, apposed against each other. But solving one problem just created a bigger problem. With the bases facing outside, there had been no trouble fitting them: they had simply circled around the central backbone, like a spiral rosette. But with the bases turned inside, they had to be jammed and tucked against each other. The zipper's teeth had to intercalate. For A, T, G, and C to sit in the interior of the DNA double helix, they had to have some interaction, some relationship. But what did one base—A, say—have to do with another base?

One lone chemist had suggested, insistently, that the bases of DNA must have something to do with each other. In 1950, the Austrian-born biochemist Erwin Chargaff, working at Columbia University in New York, had found a peculiar pattern. Whenever Chargaff digested DNA and analyzed the base composition, he always found that the A and the T were present in nearly identical proportion, as were the G and the C. Something, mysteriously, had *paired* A to T and G to C, as if these chemicals were congenitally linked. But although Watson and Crick knew this rule, they had no idea how it might apply to the final structure of DNA.

A second problem arose with fitting the bases inside the helix: the precise measurement of the outer backbone became crucial. It was a packing problem, obviously constrained by the dimensions of the space. Once again, unbeknownst to Franklin, her data came to the rescue. In the winter of 1952, a visiting committee had been appointed to review the work being performed at King's College. Wilkins and Franklin had prepared a report on their most recent work on DNA and included many of their preliminary measurements. Max Perutz had been a member of the committee; he had obtained a copy of the report and handed it to Watson and Crick. The report was not explicitly marked "Confidential," but nor was it evident anywhere that it was to be made freely available to others, to Franklin's competitors, in particular.

Perutz's intentions, and his feigned naïveté about scientific competition, have remained mysterious (he would later write defensively, "I was inexperienced and casual in administrative matters and, since the report was not 'Confidential,' I saw no reason for withholding it."). The deed, nonetheless, was done: Franklin's report found its way to Watson's and Crick's hands. And with the sugar-phosphate backbone placed on the outside, and the general parameters of the measurements ascertained, the model builders could begin the most exacting phase of model building. At first, Watson tried to jam the two helices together, with the A on one strand matched with an A on the other—like bases paired with like. But the helix bulged and thinned inelegantly, like the Michelin Man in a wet suit. Watson tried to massage the model into shape, but it wouldn't fit. By the next morning, it had to be abandoned.

Sometime on the morning of February 28, 1953, Watson, still playing with cardboard cutouts in the shape of the bases, began to wonder if the interior of the helix contained mutually opposing bases that were *unlike* each other. What if A was paired with the T, and C with the G? "Suddenly I became aware that an adenine thymine pair $(A \to T)$ was identical in shape to a guanine cytosine pair (G→C) . . . no fudging was required to make the two types of base pairs identical in shape."

He realized that the base pairs could now easily be stacked atop each other, facing inward into the center of the helix. And the importance of Chargaff's rules became obvious in retrospect—A and T, and G and C, had to be present in identical amounts because they were always complementary: they were the two mutually opposing teeth in the zipper. The most important biological objects had to come in pairs. Watson could hardly wait for Crick to walk into the office. "Upon his arrival, Francis did not get halfway through the door before I let loose that the answer to everything was in our hands."

One look at the opposing bases convinced Crick. The precise details of the model still needed to be worked out—the A:T and G:C pairs still needed to be placed inside the skeleton of the helix—but the nature of the breakthrough was clear. The solution was so beautiful that it could not possibly be wrong. As Watson recalled, Crick "winged into the Eagle to tell everyone within the hearing distance that we had found the secret of life."

Like Pythagoras's triangle, like the cave paintings at Lascaux, like the Pyramids in Giza, like the image of a fragile blue planet seen from outer space, the double helix of DNA is an iconic image, etched permanently into human history and memory. I rarely reproduce biological diagrams in text—the mind's eye is usually richer in detail. But sometimes one must break rules for exceptions:

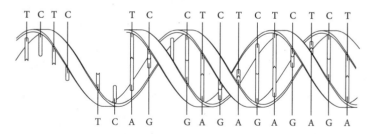

A schematic of the double-helical structure of DNA, showing a single helix (left) and its paired double helix (right). Note the complementarity of bases: A is paired with T, and G with C. The winding "backbone" of DNA is made of a chain of sugars and phosphates.

The helix contains two intertwined strands of DNA. It is "right-handed"—twisting upward as if driven by a right-handed screw. Across the molecule, it measures twenty-three angstroms—one-thousandth of one-thousandth of a millimeter. One million helices stacked side by side would fit in this letter: *o*. The biologist John Sulston wrote, "We see it as a rather stubby double helix, for they seldom show its other striking feature: it is immensely long and thin. In every cell in your body, you have two meters of the stuff; if we were to draw a scaled-up picture of it with the DNA as thick as sewing thread, that cell's worth would be about 200 kilometers long."

Each strand of DNA, recall, is a long sequence of "bases"—A, T, G, and C. The bases are linked together by the sugar-phosphate backbone. The backbone twists on the outside, forming a spiral. The bases face in, like treads in a circular staircase. The opposite strand contains the opposing bases: A matched with T and G matched with C. Thus, both strands contain the same information—except in a complementary sense: each is a "reflection," or echo, of the other (the more appropriate analogy is a yin-and-yang structure). Molecular forces between the A:T and G:C pairs lock the two strands together, as in a zipper. A double helix of DNA can thus be envisioned as a code written with four alphabets—ATGCCCTACGGGCCCATCG . . . — forever entwined with its mirror-image code.

"To see," the poet Paul Valéry once wrote, "is to forget the name of the things that one sees." To see DNA is to forget its name or its chemical formula. Like the simplest of human tools—hammer, scythe, bellows, ladder, scissors—the function of the molecule can be entirely comprehended from its structure. To "see" DNA is to immediately perceive its function as a repository of information. The most important molecule in biology needs no name to be understood.

Ⓓ

Watson and Crick built their complete model in the first week of March 1953. Watson ran down to the metal shop in the basement of the Cavendish labs to expedite the fabrication of the modeling parts. The hammering, soldering, and polishing took hours, while Crick paced impatiently upstairs. With the shiny metallic parts in hand, they began to build the model, adding part to part, as if building a house of cards. Every piece had to fit—and it had to match the known molecular measurements. Each time Crick frowned as he added another component, Watson's stomach took a turn—but in the end, the whole thing fit together, like a perfectly solved puzzle. The next day, they came back with a plumb line and a ruler to measure every distance between every component. Every measurement—every angle and width, all the spaces separating the molecules—was nearly perfect.

Maurice Wilkins came to take a look at the model the next morning. He needed but "a minute's look . . . to like it." "The model was standing high on a lab table," Wilkins later recalled. "[It] had a life of its own—rather like looking at a baby that had just been born. . . . The model seemed to speak for itself, saying—'I don't care what you think—I know I am right.'" He returned to London and confirmed that his most recent crystallographic data, as well as Franklin's, clearly supported a double helix. "I think you're a couple of old rogues, but you may well have something," Wilkins wrote from London on March 18, 1953. "I like the idea."

Franklin saw the model later that fortnight, and she too was quickly convinced. At first, Watson feared that her "sharp, stubborn mind, caught in her self-made . . . trap" would resist the model. But Franklin needed no further convincing. Her steel-trap mind knew a beautiful solution when it saw one. "The positioning of the backbone on the outside [and] the uniqueness of the A-T and G-C pairs was a fact that she saw no reason to argue about." The structure, as Watson described it, "was too pretty not to be true."

On April 25, 1953, Watson and Crick published their paper—"Molecular Structure of Nucleic Acids: A Structure for Deoxyribose Nucleic Acid"—in *Nature* magazine. Accompanying the article was another, by Gosling and Franklin, providing strong crystallographic evidence for the double-helical structure. A third article, from Wilkins, corroborated the evidence further with experimental data from DNA crystals.

In keeping with the grand tradition of counterposing the most significant discoveries in biology with supreme understatement—recall Mendel, Avery, and Griffith—Watson and Crick added a final line to their paper: "It has not escaped our notice that the specific pairing we have postulated immediately suggests a possible copying mechanism for the genetic material." The most important function of DNA—its capacity to transmit copies of information from cell to cell, and organism to organism—was buried in the structure. Message; movement; information; form; Darwin; Mendel; Morgan: all was writ into that precarious assemblage of molecules.

In 1962, Watson, Crick, and Wilkins won the Nobel Prize for their discovery. Franklin was not included in the prize. She had died in 1958, at the age of thirty-seven, from diffusely metastatic ovarian cancer—an illness ultimately linked to mutations in genes.

①

In London, where the river Thames arches away from the city near Belgravia, one might begin a walk at Vincent Square, the trapezoid-shaped park that abuts the office of the Royal Horticultural Society. It was here, in 1900, that William Bateson brought news of Mendel's paper to the scientific world, thereby launching the era of modern genetics. From the square, a brisk stroll northwest, past the southern edge of Buckingham Palace, brings us to the elegant town houses of Rutland Gate, where, in the 1900s, Francis Galton conjured up the theory of eugenics, hoping to manipulate genetic technologies to achieve human perfection.

About three miles due east, across the river, sits the former site of the Pathological Laboratories of the Ministry of Health, where, in the early 1920s, Frederick Griffith discovered the transformation reaction—the transfer of genetic material from one organism to another, the experiment that led to the identification of DNA as the "gene molecule." Cross the river to the north, and you arrive at the King's College labs, where Rosalind Franklin and Maurice Wilkins began their work on DNA crystals in the early 1950s. Veer southwest again, and the journey brings you to the Science Museum on Exhibition Road to encounter the "gene molecule" in person. The original Watson and Crick model of DNA, with its hammered metal plates and rickety rods twisting precariously around a steel laboratory stand, is housed behind a glass case. The model looks like a latticework corkscrew invented by a madman, or an impossibly fragile spiral staircase that might connect the human past to its future. Crick's handwritten scribbles—A, C, T, and G—still adorn the plates.

The revelation of the structure of DNA by Watson, Crick, Wilkins, and Franklin brought one journey of genes to its close, even as it threw open new directions of inquiry and discovery. "Once it had been known that DNA had a highly regular structure," Watson wrote in 1954, "the enigma of how the vast amount of genetic information needed to specify all the characteristics of a living organism could be stored in such a regular structure had to be solved." Old questions were replaced by new ones. What features of the double helix enabled it to bear the code of life? How did that code become transcribed and translated into actual form and function of an organism? Why, for that matter, were there two helices, and not one, or three or four? Why were the two strands complementary to each other—A matched with T, and G matched with C—like a molecular yin and yang? Why was *this* structure, of all structures, chosen as the central repository of all biological information? "It isn't that [DNA] looks so beautiful," Crick later remarked. "It is the *idea* of what it does."

Images crystallize ideas—and the image of a double-helical molecule that carried the instructions to build, run, repair, and reproduce humans crystallized the optimism and wonder of the 1950s. Encoded in that molecule were the loci of human perfectibility and vulnerability: once we learned to manipulate this chemical, we would rewrite our nature. Diseases would be cured, fates changed, futures reconfigured.

The Watson and Crick model of DNA marked the end of one conception of the gene—as a mysterious carrier of messages across generations—to another: as a chemical, or a molecule, capable of encoding, storing, and transferring information between organisms. If the keyword of early-twentieth-century genetics was *message*, then the keyword of late-twentieth-century genetics might be *code*. That genes carried messages had been abundantly clear for half a century. The question was, could humans decipher their code?

"That Damned, Elusive Pimpernel"

> *In the protein molecule, Nature has devised an instrument in which an underlying simplicity is used to express great subtlety and versatility; it is impossible to see molecular biology in proper perspective until this peculiar combination of virtues has been clearly grasped.*
> —Francis Crick

The word *code*, I wrote before, comes from *caudex*—the pith of the tree that was used to scratch out early manuscripts. There is something evocative in the idea that the material used to write code gave rise to the word itself: form became function. With DNA too, Watson and Crick realized, the form of the molecule had to be intrinsically linked to function. The genetic code had to be written into the material of DNA—just as intimately as scratches are etched into pith.

But what was genetic code? How did four bases in a molecular string of DNA—A, C, G, and T (or A,C,G,U in RNA)—determine the consistency of hair, the color of an eye, or the quality of the coat of a bacterium (or, for that matter, the propensity for mental illness or a deadly bleeding disease in a family)? How did Mendel's abstract "unit of heredity" become manifest as a physical trait?

⌾

In 1941, three years before Avery's landmark experiment, two scientists, George Beadle and Edward Tatum, working in a basement tunnel at Stanford University, discovered the missing link between genes and physical traits. Beadle—or "Beets," as his colleagues liked to call him—had been a student of Thomas Morgan's at Caltech. The red-eyed flies and the white-eyed mutants puzzled Beadle. A "gene for redness," Beets understood, is a

161

unit of hereditary information, and it is carried from a parent to its children in an indivisible form in DNA—in genes, in chromosomes. "Redness," the physical trait, in contrast, was the consequence of a chemical pigment in the eye. But how did a hereditary particle transmute into an eye pigment? What was the link between a "gene for redness" and "redness" itself—between information and its physical or anatomical form?

Fruit flies had transformed genetics by virtue of rare mutants. Precisely *because* they were rare, the mutants had acted like lamps in the darkness, allowing biologists to track "the action of a gene," as Morgan had described it, across generations. But the "action" of a gene—still a vague, mystical concept—intrigued Beadle. In the late 1930s, Beadle and Tatum reasoned that isolating the actual eye pigment of a fruit fly might solve the riddle of gene action. But the work stalled; the connection between genes and pigments was far too complicated to yield a workable hypothesis. In 1937, at Stanford University, Beadle and Tatum switched to an even simpler organism called *Neurospora crassa*, a bread mold originally found as a contaminant in a Paris bakery, to try to solve the gene-to-trait connection.

Bread molds are scrappy, fierce creatures. They can be grown in petri dishes layered with nutrient-rich broth—but, in fact, they do not need much to survive. By systematically depleting nearly all the nutrients from the broth, Beadle found that the mold strains could still grow on a minimal broth containing nothing more than a sugar and a vitamin called biotin. Evidently, the cells of the mold could build all the molecules needed for survival from basic chemicals—lipids from glucose, DNA and RNA from precursor chemicals, and complex carbohydrates out of simple sugars: wonder from Wonder Bread.

This capacity, Beadle understood, was due to the presence of enzymes within the cell—proteins that acted as master builders and could synthesize complex biological macromolecules out of basic precursor chemicals. For a bread mold to grow successfully in minimal media, then, it needed all its metabolic, molecule-building functions to be intact. If a mutation inactivated even one function, the mold would be unable to grow—unless the missing ingredient was supplied back into the broth. Beadle and Tatum could thus use this technique to track the missing metabolic function in every mutant: if a mutant needed the substance X, say, to grow in minimal media, then it must lack the enzyme to synthesize that substance, X, from scratch. This approach was intensely laborious—but patience was a virtue that Beadle possessed in abundance: he had once spent an entire afternoon teaching a graduate student how to marinate a steak, adding one spice at a time, over precisely timed intervals.

The "missing ingredient" experiment propelled Beadle and Tatum toward a new understanding of genes. Every mutant, they noted, was missing a single metabolic function, corresponding to the activity of a single protein enzyme. And genetic crosses revealed that every mutant was defective in only one gene.

But if a mutation disrupts the function of an enzyme, then the normal gene must specify the information to make the normal enzyme. A unit of heredity must *carry the code* to build a metabolic or cellular function specified by a protein. "A gene," Beadle wrote in 1945, "can be visualized as directing the final configuration of a protein molecule." *This* was the "action of the gene" that a generation of biologists had been trying to comprehend: *a gene "acts" by encoding information to build a protein, and the protein actualizes the form or function of the organism.* *

Or, in terms of information flow:

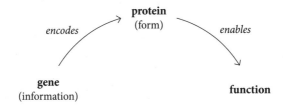

Beadle and Tatum shared a Nobel Prize in 1958 for their discovery, but the Beadle/Tatum experiment raised a crucial question that remained unanswered: How did a gene "encode" information to build a protein? A protein is created from twenty simple chemicals named *amino acids*—Methionine, Glycine, Leucine, and so forth—strung together in a chain. Unlike a chain of DNA, which exists primarily in the form of a double helix, a protein chain can twist and turn in space idiosyncratically, like a wire that has been sculpted into a unique shape. This shape-acquiring ability allows proteins to execute diverse functions in cells. They can exist

---

* This idea of the "gene" will be modified and extended in further pages. A gene is more than a set of protein-building instructions, but Beadle and Tatum's experiments provided a mechanistic basis for a gene's function.

as long, stretchable fibers in muscle (myosin). They can become globular in shape and enable chemical reactions—i.e., enzymes (DNA polymerase). They can bind colored chemicals and become pigments in the eye, or in a flower. Twisted into saddle clasps, they can act as transporters for other molecules (hemoglobin). They can specify how a nerve cell communicates with another nerve cell and thus become the arbiters of normal cognition and neural development.

But how could a sequence of DNA—ATGCCCC . . . etc.—carry instructions to build a protein? Watson had always suspected that DNA was first converted into an intermediate message. It was this "messenger molecule," as he called it, that carried the instructions to build a protein based on a gene's code. "For over a year," he wrote in 1953, "I had been telling Francis [Crick] that the genetic information in DNA chains must be first copied into that of complementary RNA molecules," and the RNA molecules must be used as "messages" to build proteins.

In 1954, the Russian-born physicist-turned-biologist George Gamow teamed with Watson to form a "club" of scientists to decipher the mechanism of protein synthesis. "Dear Pauling," Gamow wrote to Linus Pauling in 1954, with his characteristically liberal interpretation of grammar and spelling, "I am playing with complex organic molecules (what I never did before!) and geting [sic] some amusing results and would like your opinnion [sic] about it."

Gamow called it the RNA Tie Club. "The Club never met as a whole," Crick recalled: "It always had a rather ethereal existence." There were no formal conferences or rules or even basic principles of organization. Rather, the Tie Club was loosely clustered around informal conversations. Meetings happened by chance, or not at all. Letters proposing madcap, unpublished ideas, often accompanied by hand-scribbled figures, were circulated among the members; it was a blog before blogs. Watson got a tailor in Los Angeles to embroider green woolen ties with a golden strand of RNA, and Gamow sent a tie, and a pin, to each in the group of friends that he had handpicked as club members. He printed a letterhead and added his own motto: "Do or die, or don't try."

℗

In the mid-1950s, a pair of bacterial geneticists working in Paris, Jacques Monod and François Jacob, had also performed experiments that had dimly suggested that an intermediate molecule—a messenger—was required for the translation of DNA into proteins. Genes, they proposed, did not specify instructions for proteins directly. Rather, genetic information in DNA was first converted into a soft copy—a draft form—and it was this copy, not the DNA original, that was translated into a protein.

In April 1960, Francis Crick and Jacob met at Sydney Brenner's cramped apartment in Cambridge to discuss the identity of this mysterious intermediate. The son of a cobbler from South Africa, Brenner had come to England to study biology on a scholarship; like Watson and Crick, he too had become entranced by Watson's "religion of genes" and DNA. Over a barely digested lunch, the three scientists realized that this intermediate molecule had to shuttle from the cell's nucleus, where genes were stored, to the cytoplasm, where proteins were synthesized.

But what was the chemical identity of the "message" that was built from a gene? Was it a protein or a nucleic acid or some other kind of molecule? What was its relationship to the sequence of the gene? Although they still lacked concrete evidence, Brenner and Crick also suspected that it was RNA—DNA's molecular cousin. In 1959, Crick wrote a poem to the Tie Club, although he never sent it out:

> *What are the properties of Genetic RNA*
> *Is he in heaven, is he in hell?*
> *That damned, elusive Pimpernel.*

℗

In the early spring of 1960, Jacob flew to Caltech to work with Matthew Meselson to trap the "damned, elusive Pimpernel." Brenner arrived in early June, a few weeks later.

Proteins, Brenner and Jacob knew, are synthesized within a cell by a specialized cellular component called the *ribosome*. The surest means to purify the messenger intermediate was to halt protein synthesis abruptly—using a biochemical equivalent of a cold shower—and purify the shivering molecules associated with the ribosomes, thereby trapping the elusive Pimpernel.

The principle seemed obvious, but the actual experiment proved mysteriously daunting. At first, Brenner reported, all he could see in the experiment was the chemical equivalent of thick "California fog—wet, cold, silent." The fussy biochemical setup had taken weeks to perfect—except each time the ribosomes were caught, they crumbled and fell apart. Inside cells, ribosomes seemed to stay glued together with absolute equanimity. Why, then, did they degenerate outside cells, like fog slipping through fingers?

The answer appeared out of the fog—literally. Brenner and Jacob were sitting on the beach one morning when Brenner, ruminating on his basic biochemistry lessons, realized a profoundly simple fact: their solutions must be missing an essential chemical factor that kept ribosomes intact within cells. But what factor? It had to be something small, common, and ubiquitous—a tiny dab of molecular glue. He shot up from the sand, his hair flying, sand dribbling from his pockets, screaming, *"It's the magnesium. It's the magnesium."*

It was the magnesium. The addition of the ion was critical: with the solution supplemented with magnesium, the ribosome remained glued together, and Brenner and Jacob finally purified a minuscule amount of the messenger molecule out of bacterial cells. It was RNA, as expected—but RNA of a special kind.* The messenger was generated afresh when a gene was translated. Like DNA, these RNA molecules were built by stringing together four bases—A, G, C, and U (in the RNA copy of a gene, remember, the T found in DNA is substituted for U). Notably, Brenner and Jacob later discovered the messenger RNA was a *facsimile* of the DNA chain—a copy made from the original. The RNA copy of a gene then moved from the nucleus to the cytosol, where its message was decoded to build a protein. The messenger RNA was neither an inhabitant of heaven nor of hell—but a professional go-between. The generation of an RNA copy of a gene was termed *transcription*—referring to the rewriting of a word or sentence in a language close to the original. A gene's code (ATGGGCC . . .) was transcribed into an RNA code (AUGGGCC . . .).

The process was akin to a library of rare books that is accessed for translation. The master copy of information—i.e., the gene—was stored permanently in a deep repository or vault. When a "translation request" was generated by a cell, a photocopy of the original was summoned from the vault of the nucleus. This facsimile of a gene (i.e., RNA) was used as a working source for translation into a protein. The process allowed multi-

---

* A team led by James Watson and Walter Gilbert at Harvard also discovered the "RNA intermediate" in 1960. The Watson/Gilbert and Brenner/Jacob papers were published back to back in *Nature*.

ple copies of a gene to be in circulation at the same time, and for the RNA copies to be increased or decreased on demand—facts that would soon prove to be crucial to the understanding of a gene's activity and function.

①

But transcription solved only half the problem of protein synthesis. The other half remained: How was the RNA "message" decoded into a protein? To make an RNA copy of a gene, the cell used a rather simple transposition: every A,C,T, and G in a gene was copied to an A, C, U, and G in the messenger RNA (i.e., ACT CCT GGG→ACU CCU GGG). The only difference in code between the gene's original and the RNA copy was the substitution of the thymine to a uracil (T→U). *But once transposed into RNA, how was a gene's "message" decoded into a protein?*

To Watson and Crick, it was immediately clear that no single base— A, C, T, or G—could carry sufficient genetic message to build any part of a protein. There are twenty amino acids in all, and four letters could not specify twenty alternative states by themselves. The secret had to be in the combination of bases. "It seems likely," they wrote, "that the precise *sequence* of the bases is the code that carries the genetical information."

An analogy to natural language illustrates the point. The letters A, C, and T convey very little meaning by themselves, but can be combined in ways to produce substantially different messages. It is, once again, the *sequence* that carries the message: the words *act*, *tac*, and *cat*, for instance, arise from the same letters, yet signal vastly different meanings. The key to solving the actual genetic code was to map the elements of a sequence in an RNA chain to the sequence of a protein chain. It was like deciphering the Rosetta Stone of genetics: Which combination of letters (in RNA) specified which combination of letters (in a protein)? Or, conceptually:

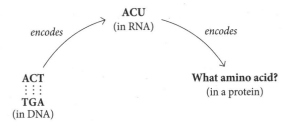

Through a series of ingenious experiments, Crick and Brenner realized that the genetic code had to occur in a "triplet" form—i.e., three bases of DNA (e.g., ACT) had to specify one amino acid in a protein.*

But which triplet specified which amino acid? By 1961, several laboratories around the world had joined the race to decipher the genetic code. At the National Institutes of Health in Bethesda, Marshall Nirenberg, Heinrich Matthaei, and Philip Leder used a biochemical approach to try to crack the cipher. An Indian-born chemist, Har Khorana, supplied crucial chemical reagents that made code breaking possible. And a Spanish biochemist in New York, Severo Ochoa, launched a parallel effort to map the triplet code to corresponding amino acids.

As with all code breaking, the work proceeded misstep by misstep. At first, one triplet seemed to overlap with another—making the prospect of a simple code impossible. Then, for a while, it seemed that some triplets did not work at all. But by 1965, all of these studies—and especially Nirenberg's—had successfully mapped every DNA triplet to a corresponding amino acid. ACT, for instance, specified the amino acid Threonine. CAT, in contrast, specified a different amino acid—Histidine. CGT specified Arginine. A particular sequence of DNA—ACT-GAC-CAC-GTG—was therefore used to build an RNA chain, and the RNA chain was translated into a chain of amino acids, ultimately leading to the construction of a protein. One triplet (ATG) was the code to start the building of a protein, and three triplets (TAA, TAG, TGA) represented codes to stop it. The basic alphabet of the genetic code was complete.

The flow of information could be visualized simply:

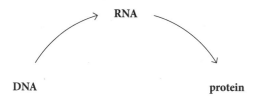

* This "triplet code" hypothesis was also supported by elementary mathematics. If a two-letter code was used—i.e., two bases in a sequence (AC or TC) encoded an amino acid in a protein—you could only achieve 16 combinations, obviously insufficient to specify all twenty amino acids. A triplet-based code had 64 combinations—enough for all twenty amino acids, with extra ones still left over to specify other coding functions, such as "stopping" or "starting" a protein chain. A quadruplet code would have 256 permutations—far more than needed to encode twenty amino acids. Nature was degenerate, but not *that* degenerate.

Or, at a conceptual level:

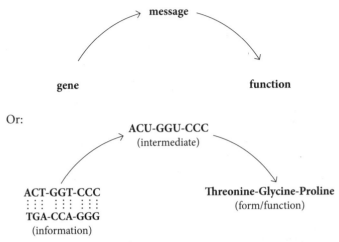

Or:

Francis Crick called this flow of information "the central dogma" of biological information. The word *dogma* was an odd choice (Crick later admitted that he never understood the linguistic implications of *dogma*, which implies a fixed, immutable belief)—but the *central* was an accurate description. Crick was referring to the striking universality of the flow of genetic information throughout biology.* From bacteria to elephants—from red-eyed flies to blue-blooded princes—biological information flowed through living systems in a systematic, archetypal manner: DNA provided instructions to build RNA. RNA provided instructions to build proteins. Proteins ultimately enabled structure and function—bringing genes to life.

Ⅎ

Perhaps no illness illustrates the nature of this information flow, and its penetrating effects on human physiology, as powerfully as sickle-cell anemia. As early as the sixth century BC, ayurvedic practitioners in India had recognized the general symptoms of anemia—the absence of adequate red cells in blood—by the characteristic pallor of the lips, skin,

---

* In Crick's original formulation, information was allowed to flow "backwards" from RNA to DNA. Watson, however, simplified the diagram to indicate information flow from DNA to RNA to protein that was later called the "central dogma".

and fingers. Termed *pandu roga* in Sanskrit, anemias were further sub-divided into categories. Some variants of the illness were known to be caused by nutritional deficiencies. Others were thought to be precipi-tated by episodes of blood loss. But sickle-cell anemia must have seemed the strangest—for it was hereditary, often appeared in fits and starts, and was accompanied by sudden, wrenching bouts of pain in the bones, joints, and chest. The Ga tribe in West Africa called the pain *chwech-weechwe* (body beating). The Ewe named it *nuiduidui* (body twisting)—onomatopoeic words whose very sounds seemed to capture the relentless nature of a pain that felt like corkscrews driven into the marrow.

In 1904, a single image captured under a microscope provided a unify-ing cause for all these seemingly disparate symptoms. That year, a young dentistry student named Walter Noel presented to his doctor in Chicago with an acute anemic crisis, accompanied by the characteristic chest and bone pain. Noel was from the Caribbean, of West African descent, and had suffered several such episodes over the prior years. Having ruled out a heart attack, the cardiologist, James Herrick, assigned the case rather casually to a medical resident named Ernest Irons. Acting on a whim, Irons decided to look at Noel's blood under the microscope.

Irons found a bewildering alteration. Normal red blood cells are shaped like flattened disks—a shape that allows them to be stacked atop each other, and thus move smoothly through networks of arteries and capil-laries and veins, bringing oxygen to the liver, heart, and brain. In Noel's blood, the cells had morphed, mysteriously, into shriveled, scythe-shaped crescents—"sickle cells," as Irons later described them.

But what made a red blood cell acquire a sickle shape? And why was the illness hereditary? The natural culprit was an abnormality in the gene for hemoglobin—the protein that carries oxygen and is present abun-dantly in red cells. In 1951, working with Harvey Itano at Caltech, Linus Pauling demonstrated that the variant of hemoglobin found in sickle cells was different from the hemoglobin in normal cells. Five years later, scien-tists in Cambridge pinpointed the difference between the protein chain of normal hemoglobin and "sickled" hemoglobin to a change in a single amino acid.*

But if the protein chain was altered by exactly *one amino acid*, then its

---

* The alteration of the single amino acid was discovered by Vernon Ingram, a former student of Max Perutz's.

gene had to be different by precisely *one triplet* ("one triplet encodes one amino acid"). Indeed, as predicted, when the gene encoding the hemoglobin B chain was later identified and sequenced in sickle-cell patients, there was a single change: one triplet in DNA—GAG—had changed to another—GTG. This resulted in the substitution of one amino acid for another: glutamate was switched to valine. That switch altered the folding of the hemoglobin chain: rather than twisting into its neatly articulated, clasplike structure, the mutant hemoglobin protein accumulated in stringlike clumps within red cells. These clumps grew so large, particularly in the absence of oxygen, that they tugged the membrane of the red cell until the normal disk was warped into a crescent-shaped, dysmorphic "sickle cell." Unable to glide smoothly through capillaries and veins, sickled red cells jammed into microscopic clots throughout the body, interrupting blood flow and precipitating the excruciating pain of a sickling crisis.

It was a Rube Goldberg disease. A change in the sequence of a gene caused the change in the sequence of a protein; that warped its shape; that shrank a cell; that clogged a vein; that jammed the flow; that racked the body (that genes built). Gene, protein, function, and fate were strung in a chain: one chemical alteration in one base pair in DNA was sufficient to "encode" a radical change in human fate.

Regulation, Replication, Recombination

Nécessité absolue trouver origine de cet emmerdement
[It is absolutely necessary to find the origin of this pain
in the ass].

—Jacques Monod

Just as the formation of a giant crystal can be seeded by the formal arrangement of a few critical atoms at its core, the birth of a great body of science can be nucleated by the interlocking of a few crucial concepts. Before Newton, generations of physicists had thought about phenomena such as *force, acceleration, mass,* and *velocity.* But Newton's genius involved defining these terms rigorously and linking them to each other via a nest of equations—thereby launching the science of mechanics.

By similar logic, the interlocking of just a few crucial concepts—

—relaunched the science of genetics. In time, as with Newtonian mechanics, the "central dogma" of genetics would be vastly refined, modified, and reformulated. But its effect on the nascent science was profound: it locked a system of thinking into place. In 1909, Johannsen, coining the word *gene,* had declared it "free of any hypothesis." By the early 1960s, however, the gene had vastly exceeded a "hypothesis." Genetics had found a means to describe the flow of information from organism to organism, and—within an organism—from encryption to form. A *mechanism* of heredity had emerged.

But how did this flow of biological information achieve the observed complexity of living systems? Take sickle-cell anemia as a case in point. Walter Noel had inherited two abnormal copies of the hemoglobin B gene. Every cell in his body carried the two abnormal copies (every cell in the body inherits the same genome). But *only* red blood cells were affected by the altered genes—not Noel's neurons or kidneys or liver cells or muscle cells. What enabled the selective "action" of hemoglobin in red blood cells? Why was there no hemoglobin in his eye or his skin—even though eye cells and skin cells and, indeed, every cell in the human body possessed identical copies of the same gene? How, as Thomas Morgan had put it, did "the properties implicit in genes become explicit in [different] cells?"

①

In 1940, an experiment on the simplest of organisms—a microscopic, capsule-shaped, gut-dwelling bacterium named *Escherichia coli*—provided the first crucial clue to this question. *E. coli* can survive by feeding on two very different kinds of sugars—glucose and lactose. Grown on either sugar alone, the bacterium begins to divide rapidly, doubling in number every twenty minutes or so. The curve of growth can be plotted as an exponential line—1, 2-, 4-, 8-, 16-fold growth—until the culture turns turbid, and the sugar source has been exhausted.

The relentless ogive of growth fascinated Jacques Monod, the French biologist. Monod had returned to Paris in 1937, having spent a year studying flies with Thomas Morgan at Caltech. Monod's visit to California had not been particularly fruitful—he had spent most of his time playing Bach with the local orchestra and learning Dixie and jazz—but Paris was utterly depressing, a city under siege. By the summer of 1940, Belgium and Poland had fallen to the Germans. In June 1940, France, having suffered devastating losses in battle, signed an armistice that allowed the German army to occupy much of Northern and Western France.

Paris was declared an "open city"—spared from bombs and ruin, but fully accessible to Nazi troops. The children were evacuated, the museums emptied of paintings, the storefronts shuttered. *"Paris will always be Paris,"* Maurice Chevalier sang, if pleadingly, in 1939—but the City of Lights was rarely illuminated. The streets were ghostly. The cafés were empty. At night, regular blackouts plunged it into an infernally bleak darkness.

In the fall of 1940, with red-and-black flags bearing swastikas hoisted on all government buildings, and German troops announcing nightly curfews on loudspeakers along the Champs-Élysées, Monod was working on *E. coli* in an overheated, underlit attic of the Sorbonne (he would secretly join the French resistance that year, although many of his colleagues would never know his political proclivities). That winter, with his lab now nearly frozen by the chill—he had to wait penitently until noon, listening to Nazi propaganda on the streets while waiting for some of the acetic acid to thaw—Monod repeated the bacterial growth experiment, but with a strategic twist. This time, he added both glucose *and* lactose— two different sugars—to the culture.

If sugar was sugar was sugar—if the metabolism of lactose was no different from that of glucose—then one might have expected bacteria fed on the glucose/lactose mix to exhibit the same smooth arc of growth. But Monod stumbled on a kink in his results—literally so. The bacteria grew exponentially at first, as expected, but then paused for a while before resuming growth again. When Monod investigated this pause, he discovered an unusual phenomenon. Rather than consuming both sugars equally, the *E. coli* cells had selectively consumed glucose first. Then the bacterial cells had stopped growing, as if reconsidering their diet, switched to lactose, and resumed growth again. Monod called this *diauxie*—"double growth."

That bend in the growth curve, small though it was, perplexed Monod. It bothered him, like a sand grain in the eye of his scientific instinct. Bacteria feeding on sugars should grow in smooth arcs. Why should a switch in sugar consumption cause a pause in growth? How might a bacterium even "know," or sense, that the sugar source had been switched? And why was one sugar consumed first, and only then the second, like a two-course bistro lunch?

By the late 1940s, Monod had discovered that the kink was the result of a metabolic readjustment. When bacteria switched from glucose to lactose consumption, they induced specific lactose-digesting enzymes. When they switched back to glucose, these enzymes disappeared and *glucose-digesting* enzymes reappeared. The induction of these enzymes during the switch—like changing cutlery between dinner courses (remove the fish knife; set the dessert fork)—took a few minutes, thereby resulting in the observed pause in growth.

To Monod, diauxie suggested that genes could be regulated by metabolic inputs. If enzymes—i.e., proteins—were being induced to appear and disappear in a cell, then *genes* must be being turned on and off, like molecular switches (enzymes, after all, are encoded by genes). In the early 1950s, Monod, joined by François Jacob in Paris, began to systematically explore the regulation of genes by *E. coli* by making mutants—the method used with such spectacular success by Morgan with fruit flies.*

As with flies, the bacterial mutants proved revealing. Monod and Jacob, working with Arthur Pardee, a microbial geneticist from America, discovered three cardinal principles that governed the regulation of genes. First, when a gene was turned on or off, the DNA master copy was always kept intact in a cell. *The real action was in RNA:* when a gene was turned on, it was induced to make more RNA messages and thereby produce more sugar-digesting enzymes. A cell's metabolic identity—i.e., whether it was consuming lactose or glucose—could be ascertained not by the sequence of its genes, which was always constant, but by the amount of RNA that a gene was producing. During lactose metabolism, the RNAs for lactose-digesting enzymes were abundant. During glucose metabolism, those messages were repressed, and the RNAs for glucose-digesting enzymes became abundant.

Second, *the production of RNA messages was coordinately regulated.* When the sugar source was switched to lactose, the bacteria turned on an entire module of genes—several lactose-metabolizing genes—to digest lactose. One of the genes in the module specified a "transporter protein" that allowed lactose to enter the bacterial cell. Another gene encoded an enzyme that was needed to break down lactose into parts. Yet another specified an enzyme to break those chemical parts into subparts. Surprisingly, all the genes dedicated to a particular metabolic pathway were physically present next to each other on the bacterial chromosome—like library books stacked by subject—and they were induced simultaneously

* Monod and Jacob knew each other distantly; both were close associates of the microbial geneticist André Lwoff. Jacob worked at the other end of the attic, experimenting with a virus that infected *E. coli*. Although their experimental strategies were superficially dissimilar, both were studying gene regulation. Monod and Jacob had compared notes and found, to their astonishment, that both were working on two aspects of the same general problem, and they had combined some parts of their work in the 1950s.

in cells. The metabolic alteration produced a profound genetic alteration in a cell. It wasn't just a cutlery switch; the whole dinner service was altered in a single swoop. A functional circuit of genes was switched on and off, as if operated by a common spool or a master switch. Monod called one such gene module an *operon*.*

The genesis of proteins was thus perfectly synchronized with the requirements of the environment: supply the correct sugar, and a set of sugar-metabolizing genes would be turned on together. The terrifying economy of evolution had again produced the most elegant solution to gene regulation. No gene, no message, and no protein labored in vain.

⟐

How did a lactose-sensing protein recognize and regulate only a lactose-digesting gene—and not the thousands of other genes in a cell? The third cardinal feature of gene regulation, Monod and Jacob discovered, was that *every gene had specific regulatory DNA sequences appended to it that acted like recognition tags*. Once a sugar-sensing protein had detected sugar in the environment, it would recognize one such tag and turn the target genes on or off. *That* was a gene's signal to make more RNA messages and thereby generate the relevant enzyme to digest the sugar.

A gene, in short, possessed not just information to encode a protein, but also information about when and where to make that protein. All that data was encrypted in DNA, typically appended to the front of every gene (although regulatory sequences can also be appended to the ends and middles of genes). The combination of regulatory sequences and the protein-encoding sequence defined a gene.

Once again, we might return to our analogy to an English sentence. When Morgan had discovered gene linkage in 1910, he had found no

---

* In 1957, Pardee, Monod, and Jacob discovered that the lactose operon was controlled by a single master switch—a protein eventually called the repressor. The repressor functioned like a molecular lock. When lactose was added to the growth medium, the repressor protein sensed the lactose, altered its molecular structure, and "unlocked" the lactose-digesting and lactose-transporting genes (i.e., allowed the genes to be activated), thereby enabling a cell to metabolize lactose. When another sugar, such as glucose, was present, the lock remained intact, and no lactose-digesting genes were allowed to be activated. In 1966, Walter Gilbert and Benno Muller-Hill isolated the repressor protein from bacterial cells—thereby proving Monod's operon hypothesis beyond doubt. Another repressor, from a virus, was isolated by Mark Ptashne and Nancy Hopkins in 1966.

seeming logic to why one gene was physically strung with another on a chromosome: the sable-colored and the white-eyed genes seemed to have no common functional connection, yet sat, cheek by jowl, on the same chromosome. In Jacob and Monod's model, in contrast, bacterial genes were strung together for a reason. Genes that operated on the same metabolic pathway were physically linked to each other: if you worked together, then you lived together in the genome. Specific sequences of DNA were appended to a gene that provided context for its activity—its "work." These sequences, meant to turn genes on and off, might be likened to punctuation marks and annotations—inverted quotes, a comma, a capitalized letter—in a sentence: they provide context, emphasis, and meaning, informing a reader what parts are to be read together, and when to pause for the next sentence:

> "This is the structure of your genome. It contains, among other things, independently regulated modules. Some words are gathered into sentences; others are separated by semicolons, commas, and dashes."

Pardee, Jacob, and Monod published their monumental study on the lactose operon in 1959, six years after the Watson and Crick paper on the structure of DNA. Called the Pa-Ja-Mo—or, colloquially, the Pajama—paper, after the initials of the three authors, the study was instantly a classic, with vast implications for biology. Genes, the Pajama paper argued, were not just passive blueprints. Even though every cell contains the same set of genes—an identical genome—the selective activation or repression of particular subsets of genes allows an individual cell to respond to its environments. The genome was an *active* blueprint—capable of deploying selected parts of its code at different times and in different circumstances.

Proteins act as regulatory sensors, or master switches, in this process—turning on and turning off genes, or even combinations of genes, in a coordinate manner. Like the master score of a bewitchingly complex symphonic work, the genome contains the instructions for the development and maintenance of organisms. But the genomic "score" is inert without proteins. Proteins actualize this information—by activating or repressing genes (some of these regulatory proteins are also called transcription factors). They *conduct* the genome, thereby playing out its music—activating the viola at the fourteenth minute, a crash of cymbals during the arpeggio, a roll of drums at the crescendo. Or conceptually:

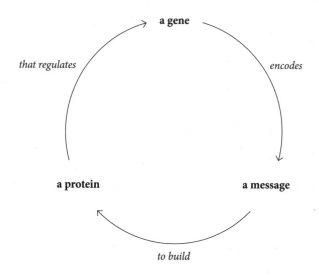

The Pa-Ja-Mo paper laid a central question of genetics to bed: How can an organism have a fixed set of genes, yet respond so acutely to changes in the environment? But it also suggested a solution to the central question in embryogenesis: How can thousands of cell types arise from an embryo out of the same set of genes? The *regulation* of genes—the selective turning on and off of certain genes in certain cells, and at certain times—must interpose a crucial layer of complexity on the unblinking nature of biological information.

It was through gene regulation, Monod argued, that cells could achieve their unique functions in time and space. "The genome contains not only a series of blue-prints [i.e., genes], but a co-ordinated *program* . . . and a means of controlling its execution," Monod and Jacob concluded. Walter Noel's red blood cells and liver cells contained the same genetic information—but gene regulation ensured that the hemoglobin protein was only present in red blood cells, and not in the liver. The caterpillar and the butterfly carry precisely the same genome—but gene regulation enables the metamorphosis of one into the other.

Embryogenesis could be reimagined as the gradual unfurling of gene regulation from a single-celled embryo. *This* was the "movement" that

Aristotle had so vividly imagined centuries before. In a famous story, a medieval cosmologist is asked what holds the earth up.

"Turtles," he says.

"And what holds up the turtles?" he is asked.

"More turtles."

"And those turtles?"

"You don't understand." The cosmologist stamps his foot. "It's turtles all the way."

To a geneticist, the development of an organism could be described as the sequential induction (or repression) of genes and genetic circuits. Genes specified proteins that switched on genes that specified proteins that switched on genes—and so forth, all the way to the very first embryological cell. It was genes, all the way.*

Ⅰ

Gene regulation—the turning on and off of genes by proteins—described the mechanism by which combinatorial complexity could be generated from the one hard copy of genetic information in a cell. But it could not explain the copying of genes themselves: How are genes replicated when a cell divides into two cells, or when a sperm or egg is generated?

To Watson and Crick, the double-helix model of DNA—with two complementary "yin-yang" strands counterposed against each other—instantly suggested a mechanism for replication. In the last sentence of the 1953 paper, they noted: "It has not escaped our notice that the specific pairing [of DNA] we have postulated immediately suggests a possible copying mechanism for the genetic material." Their model of DNA was not just a pretty picture; the structure predicted the most important features of the function. Watson and Crick proposed that each DNA strand was used to generate a copy of itself—thereby generating two double helices from the original double helix. During replication, the yin-yang strands of DNA were peeled apart. The yin was used as a template to create a yang, and the yang to make a yin—and this resulted in two yin-yang pairs (in 1958, Matthew Meselson and Frank Stahl proved this mechanism).

---

* Unlike cosmological turtles, this view is not absurd. In principle, the single-celled embryo *does* possess all the genetic information to specify a full organism. The question of how sequential genetic circuits can "actualize" the development of an organism is addressed in a subsequent chapter.

But a DNA double helix cannot autonomously make a copy of itself; otherwise, it might replicate without self-control. An enzyme was likely dedicated to copying DNA—a replicator protein. In 1957, the biochemist Arthur Kornberg set out to isolate the DNA-copying enzyme. If such an enzyme existed, Kornberg reasoned, the easiest place to find it would be in an organism that was dividing rapidly—*E. coli* during its furious phase of growth.

By 1958, Kornberg had distilled and redistilled the bacterial sludge into a nearly pure enzyme preparation ("A geneticist counts; a biochemist cleans," he once told me). He called it DNA polymerase (DNA is a polymer of A, C, G, and T, and this was the polymer-making enzyme). When he added the purified enzyme to DNA, supplied a source of energy and a reservoir of fresh nucleotide bases—A, T, G, and C—he could witness the formation of new strands of nucleic acid in a test tube: DNA made DNA in its own image.

"Five years ago," Kornberg wrote in 1960, "the synthesis of DNA was also regarded as a 'vital' process"—a mystical reaction that could not be reproduced in a test tube by the addition or subtraction of mere chemicals. "Tampering with the very genetic apparatus [of life] itself," this theory ran, "would surely produce nothing but disorder." But Kornberg's synthesis of DNA had created order out of disorder—a gene out of its chemical subunits. The unassailability of genes was no longer a barrier.

There is a recursion here that is worth noting: like all proteins, DNA polymerase, the enzyme that enables DNA to replicate, is itself the product of a gene.* Built into every genome, then, are the codes for proteins that will allow that genome to reproduce. This additional layer of complexity— that DNA encodes a protein that allows DNA to replicate—is important because it provides a critical node for regulation. DNA replication can be turned on and turned off by other signals and regulators, such as the age or the nutritional status of a cell, thus allowing cells to make DNA copies only when they are ready to divide. This scheme has a collateral rub: when the regulators themselves go rogue, nothing can stop a cell from replicating continuously. That, as we will soon learn, is the ultimate disease of malfunctioning genes—cancer.

---

* DNA replication requires many more proteins than just DNA polymerase to unfold the twisted double helix and to ensure that the genetic information is copied accurately. And there are multiple DNA polymerases, with slightly different functions, found in cells.

①

Genes make proteins that *regulate* genes. Genes make proteins that *replicate* genes. The third *R* of the physiology of genes is a word that lies outside common human vocabulary, but is essential to the survival of our species: *recombination*—the ability to generate new combinations of genes.

To understand recombination, we might, yet again, begin with Mendel and Darwin. A century of exploration of genetics illuminated how organisms transmit "likeness" to each other. Units of hereditary information, encoded in DNA and packaged on chromosomes, are transmitted through sperm and egg into an embryo, and from the embryo to every living cell in an organism's body. These units encode messages to build proteins—and the messages and proteins, in turn, enable the form and function of a living organism.

But while this description of the mechanism of heredity solved Mendel's question—how does like beget like?—it failed to solve Darwin's converse riddle: How does like beget *unlike*? For evolution to occur, an organism must be able to generate genetic variation—i.e., it must produce descendants that are genetically different from either parent. If genes typically transmit likeness, then how can they transmit "unlikeness"?

One mechanism of generating variation in nature is mutation—i.e., alterations in the sequence of DNA (an A switched to a T) that may change the structure of a protein and thereby alter its function. Mutations occur when DNA is damaged by chemicals or X-rays, or when the DNA replication enzyme makes a spontaneous error in copying genes. But a second mechanism of generating genetic diversity exists: genetic information can be swapped between chromosomes. DNA from the maternal chromosome can exchange positions with DNA from the paternal chromosome—potentially generating a gene hybrid of maternal and paternal genes. Recombination is also a form of "mutation"—except whole chunks of genetic material are swapped between chromosomes.[*]

The movement of genetic information from one chromosome to another occurs only under extremely special circumstances. The first occurs when sperm and eggs are generated for reproduction. Just before the spermiogenesis and oogenesis, the cell turns briefly into a playpen for genes. The paired maternal and paternal chromosomes hug each other

---

[*] The geneticist Barbara McClintock discovered genetic elements that can move around within the genome—so-called "jumping genes"; she would win the Nobel Prize in 1983.

and readily swap genetic information. The swapping of genetic information between paired chromosomes is crucial to the mixing and matching of hereditary information between parents. Morgan called this phenomenon *crossing over* (his students had used crossing over to map genes in flies). The more contemporary term is *recombination*—the ability to generate combinations of combinations of genes.

The second circumstance is more portentous. When DNA is damaged by a mutagen, such as X-rays, genetic information is obviously threatened. When such damage occurs, the gene can be recopied from the "twin" copy on the paired chromosome: part of the maternal copy may be redrafted from the paternal copy, again resulting in the creation of hybrid genes.

Once again, the pairing of bases is used to build the gene back. The yin fixes the yang, the image restores the original: with DNA, as with Dorian Gray, the prototype is constantly reinvigorated by its portrait. Proteins chaperone and coordinate the entire process—guiding the damaged strand to the intact gene, copying and correcting the lost information, and stitching the breaks together—ultimately resulting in the transfer of information from the undamaged strand to the damaged strand.

℗

Regulation. Replication. Recombination. Remarkably, the three *R*'s of gene physiology are acutely dependent on the molecular structure of DNA—on the Watson-Crick base pairing of the double helix.

Gene regulation works through the transcription of DNA into RNA—which depends on base pairing. When a strand of DNA is used to build the RNA message, it is the pairing of bases between DNA and RNA that allows a gene to generate its RNA copy. During replication, DNA is, once again, copied using its image as a guide. Each strand is used to generate a complementary version of itself, resulting in one double helix that splits into two double helices. And during the recombination of DNA, the strategy of interposing base against base is deployed yet again to restore damaged DNA. The damaged copy of a gene is reconstructed using the complementary strand, or the second copy of the gene, as its guide.*

---

* The fact that the genome also encodes genes to *repair* damage to the genome was discovered by several geneticists, including Evelyn Witkin and Steve Elledge. Witkin and Elledge, working independently, identified an entire cascade of proteins that sensed DNA damage, and activated a cellular response to repair or temporize the damage (if the damage was cata-

The double helix has solved all three of the major challenges of genetic physiology using ingenious variations on the same theme. Mirror-image chemicals are used to generate mirror-image chemicals, reflections used to reconstruct the original. Pairs used to maintain the fidelity and fixity of information. "Monet is but an eye," Cézanne once said of his friend, "but, God, what an eye." DNA, by that same logic, is but a chemical—but, God, what a chemical.

ⵔ

In biology, there is an old distinction between two camps of scientists—anatomists, and physiologists. Anatomists describe the nature of materials, structures, and body parts: they describe how things *are*. Physiologists concentrate, instead, on the mechanisms by which these structures and parts interact to enable the functions of living organisms; they concern themselves with how things *work*.

This distinction also marks a seminal transition in the story of the gene. Mendel, perhaps, was the original "anatomist" of the gene: in capturing the movement of information across generations of peas, he had described the essential structure of the gene as an indivisible corpuscle of information. Morgan and Sturtevant extended that anatomical strand in the 1920s, demonstrating that genes were material units, spread linearly along chromosomes. In the 1940s and 1950s, Avery, Watson, and Crick identified DNA as the gene molecule, and described its structure as a double helix—thereby bringing the anatomical conception of the gene to its natural culmination.

Between the late 1950s and the 1970s, however, it was the *physiology* of genes that dominated scientific inquiry. That genes could be regulated—i.e., turned "on" and "off" by particular cues—deepened the understanding of how genes function in time and space to specify the unique features of distinct cells. That genes could also be reproduced, recombined between chromosomes, and repaired by specific proteins, explained how cells and organisms manage to conserve, copy, and reshuffle genetic information across generations.

---

strophic, it would halt cell division). Mutations in these genes can lead to the accumulation of DNA damage—and thus, more mutations—ultimately leading to cancer. The fourth *R* of gene physiology, essential to both the survival and mutability of organisms, might be "repair."

For human biologists, each of these discoveries came with enormous payoffs. As genetics moved from a material to a mechanistic conception of genes—from what genes are to what they *do*—human biologists began to perceive long-sought connections between genes, human physiology, and pathology. A disease might arise not just from an alteration in the genetic code for a protein (e.g., hemoglobin in the case of sickle-cell disease), but as a consequence of gene regulation—the inability to turn the right gene "on" or "off" in the appropriate cell at the right time. Gene replication must explain how a multicellular organism emerges from a single cell—and errors in replication might elucidate how a spontaneous metabolic illness, or a devastating mental disease, might arise in a previously unaffected family. The similarities between genomes must explain the likeness between parents and their children, and mutations and recombination might explain their differences. Families must share not just social and cultural networks—but networks of active genes.

Just as nineteenth-century human anatomy and physiology laid the foundation for twentieth-century medicine, the anatomy and physiology of genes would lay the foundation for a powerful new biological science. In the decades to come, this revolutionary science would extend its domain from simple organisms to complex ones. Its conceptual vocabulary—*gene regulation, recombination, mutation, DNA repair*—would vault out of basic science journals into medical textbooks, and then permeate wider debates in society and culture (the word *race*, as we shall see, cannot be understood meaningfully without first understanding recombination and mutation). The new science would seek to explain how genes build, maintain, repair, and reproduce humans—and how variations in the anatomy and physiology of genes might contribute to the observed variations in human identity, fate, health, and disease.

From Genes to Genesis

In the beginning, there was simplicity.
—Richard Dawkins, *The Selfish Gene*

Am not I
A fly like thee?
Or art not thou
A man like me?
—William Blake, "The Fly"

While the molecular description of the gene clarified the mechanism of the transmission of heredity, it only deepened the puzzle that had preoccupied Thomas Morgan in the 1920s. For Morgan, the principal mystery of organismal biology was not the gene but genesis: How did "units of heredity" enable the formation of animals and maintain the functions of organs and organisms? ("Excuse my big yawn," he once told a student, "but I just came from my own lecture [on genetics].")

A gene, Morgan had noted, was an extraordinary solution to an extraordinary problem. Sexual reproduction demands the collapse of an organism into a single cell, but then requires that single cell to expand back into an organism. The gene, Morgan realized, solves one problem—the transmission of heredity—but creates another: the development of organisms. A single cell must be capable of carrying the entire set of instructions to build an organism from scratch—hence genes. But how do genes make a whole organism grow back out of a single cell?

①

It might seem intuitive for an embryologist to approach the problem of genesis forward—from the earliest events in the embryo to the development of a body plan of a full-fledged organism. But for necessary reasons, as we shall see, the understanding of organismal development emerged like a film run in reverse. The mechanism by which genes specify macroscopic anatomical features—limbs, organs, and structures—was the first to be deciphered. Then came the mechanism by which an organism determines where these structures are to be placed: front or back, left or right, above or below. The very earliest events in the specification of an embryo—the specification of the body axis, of front and back, and left versus right—were among the last to be understood.

The reason for this reversed order might be obvious. Mutations in genes that specified macroscopic structures, such as limbs and wings, were the easiest to spot and the first to be characterized. Mutations in genes that specified the basic elements of the body plan were more difficult to identify, since the mutations sharply decreased the survival of organisms. And mutants in the very first steps of embryogenesis were nearly impossible to capture alive since the embryos, with scrambled heads and tails, died instantly.

①

In the 1950s, Ed Lewis, a fruit fly geneticist at Caltech, began to reconstruct the formation of fruit fly embryos. Like an architectural historian obsessed with only one building, Lewis had been studying the construction of fruit flies for nearly two decades. Bean shaped and smaller than a speck of sand, the fruit fly embryo begins its life in a whir of activity. About ten hours after fertilization of the egg, the embryo divides into three broad segments, head, thorax, and abdomen, and each segment divides into further subcompartments. Each of these embryonic segments, Lewis knew, gives rise to a congruent segment found in the adult fly. One embryonic segment becomes the second section of the thorax and grows two wings. Three of the segments grow the fly's six legs. Yet other segments sprout bristles or grow antennae. As with humans, the basic plan for the adult body is furled into an embryo. The maturation of a fly is the serial unfolding of these segments, like the stretching of a live accordion.

But how does a fly embryo "know" to grow a leg out of the second thoracic segment or an antenna out of its head (and not vice versa)? Lewis studied mutants in which the organization of these segments was disrupted. The peculiar feature of the mutants, he discovered, was that the essential *plan* of macroscopic structures was often maintained—only the segment switched its position or identity in the body of the fly. In one mutant, for instance, an extra thoracic segment—fully intact and nearly functional—appeared in a fly, resulting in a four-winged insect (one set of wings from the normal thoracic segment, and a new set from the extra thoracic segment). It was as if the *build-a-thorax* gene had incorrectly been commanded in the wrong compartment—and had sanguinely launched its command. In another mutant, two legs sprouted out of the antenna in a fly's head—as if the *build-a-leg* command had mistakenly been launched in the head.

The building of organs and structures, Lewis concluded, is encoded by master-regulatory "effector" genes that work like autonomous units or subroutines. During the normal genesis of a fly (or any other organism), these effector genes kick into action at specified sites and at specified times and determine the identities of segments and organs. These master-regulatory genes work by turning other genes on and off; they can be likened to circuits in a microprocessor. Mutations in the genes thus result in malformed, ectopic segments and organs. Like the Queen of Hearts's bewildered servants in *Alice in Wonderland*, the genes scurry about to enact the instructions—*build a thorax, make a wing*—but in the wrong places or at the wrong times. If a master regulator shouts, *"ON with an antenna,"* then the antenna-building subroutine is turned on and an antenna is built—even if that structure happens to be growing out of the thorax or abdomen of a fly.

⌗

But who commands the commanders? Ed Lewis's discovery of master-regulatory genes that controlled the development of segments, organs, and structures solved the problem of the final stage of embryogenesis, but it raised a seemingly infinite recursive conundrum. If the embryo is built, segment upon segment and organ by organ, by genes that command the identity of each segment and organ, then how does a segment know its own identity in the first place? How, for instance, does a wing-making master gene "know" to build a wing in the second thoracic segment, and not, say, the first or third segment? If genetic modules are so autonomous, then why—to turn Morgan's riddle on its head—are legs *not* growing out of fly's heads, or humans *not* born with thumbs emerging from our noses?

To answer these questions, we need to turn the clock of embryological development backward. In 1979, one year after Lewis had published his paper on the genes that govern limb and wing development, two embryologists, Christiane Nüsslein-Volhard and Eric Wieschaus, working in Heidelberg, began to create fruit fly mutants to capture the very first steps that govern the formation of the embryo.

The mutants generated by Nüsslein-Volhard and Wieschaus were even more dramatic than the ones described by Lewis. In some mutants, whole segments of the embryo disappeared, or the thorax or abdominal compartments were drastically shortened—analogous to a human fetus born with no middle or with no hind segment. The genes altered in these mutants, Nüsslein-Volhard and Wieschaus reasoned, determine the basic architectural plan of the embryo. They are the mapmakers of the embryonic world. They divide the embryo into its basic subsegments. They then activate Lewis's commander genes to start building organs and body parts in some (and only those) compartments—an antenna on the head, a wing in the fourth segment of the thorax, and so forth. Nüsslein-Volhard and Wieschaus termed these *segmentation genes*.

But even the segmentation genes have to have *their* masters: How does the second segment of the fly thorax "know" to be a thoracic segment, and not an abdominal segment? Or how does a head know not to be a tail? Every segment of an embryo can be defined on an axis that stretches from head to tail. The head functions like an internal GPS system, and the position relative to the head and the tail gives each segment a unique "address" in the embryo. But how does an embryo develop its basic, original asymmetry—i.e., its "headness" versus "tailness"?

In the late 1980s, Nüsslein-Volhard and her students began to characterize a final flock of fly mutants in which asymmetrical organization of the embryo had been abrogated. These mutants—often headless or tailless—were arrested in development long before segmentation (and certainly long before the growth of structures and organs). In some, the embryonic head was malformed. In others, the front and back of the embryo could not be distinguished, resulting in strange mirror-image embryos (the most notorious of the mutants was called *bicoid*—literally "two-tailed"). The mutants clearly lacked some factor—a chemical—that determines the front versus the back of the fly. In 1986, in an astonishing experiment, Nüsslein-Volhard's students learned to prick a normal fly embryo with a minuscule needle, withdraw a droplet of liquid from its head, and transplant it into the headless mutants. Amazingly, the cellular surgery worked: the droplet of liquid from a normal head was sufficient to force an embryo to grow a head in the position of its tail.

In a volley of pathbreaking papers published between 1986 and 1990, Nüsslein-Volhard and her colleagues definitively identified several of the factors that provide the signal for "headness" and "tailness" in the embryo. We now know that about eight such chemicals—mostly proteins—are made by the fly during the development of the egg and deposited asymmetrically in the egg. These *maternal factors* are made and placed in the egg by the mother fly. The asymmetric deposition is only possible because the *egg itself* is placed asymmetrically in the mother fly's body—thereby enabling her to deposit some of these maternal factors on the head end of the egg, and others on the tail end.

The proteins create a gradient within the egg. Like sugar diffusing out of a cube in a cup of coffee, they are present at high concentration on one end of the egg, and low concentration on the other. The diffusion of a chemical through a matrix of protein can even create distinct, three-dimensional patterns—like a pool of syrup ribboning into oatmeal. Specific genes are activated at the high-concentration end versus at the low-concentration end, thereby allowing the head-tail axis to be defined, or other patterns to be formed.

The process is infinitely recursive—the ultimate chicken-and-egg story. Flies with heads and tails make eggs with heads and tails, which make embryos with heads and tails, which grow into flies with heads and tails, and so forth, ad infinitum. Or at a molecular level: Proteins in the early embryo are deposited preferentially at one end by the mother. They activate and silence genes, thereby defining the embryo's axis from head to tail. These genes, in turn, activate "mapmaker" genes that make segments and split the body into its broad domains. The mapmaker genes activate and silence genes that make organs and structures.* Finally, organ-formation

---

* This begs the question of how the first asymmetric organisms appeared in the natural world. We do not know, and perhaps we never will. Somewhere in evolutionary history, an organism evolved to separate the functions of one part of its body from another. Perhaps one end faced a rock, while the other faced the ocean. A lucky mutant was born with the miraculous ability to localize a protein to the mouth end, and not the foot end. Discriminating mouth from foot gave that mutant a selective advantage: each asymmetric part could be further specialized for its particular task, resulting in an organism more suited to its environment. Our heads and tails are the fortunate descendants of that evolutionary innovation.

and segment-identity genes activate and silence genetic subroutines that result in the creation of organs, structures, and parts.

The development of the human embryo is also likely achieved through three similar levels of organization. As with the fly, "maternal effect" genes organize the early embryo into its main axes—head versus tail, front versus back, and left versus right—using chemical gradients. Next, a series of genes analogous to the segmentation genes in the fly initiates the division of the embryo into its major structural parts—brain, spinal cord, skeleton, skin, guts, and so forth. Finally, organ-building genes authorize the construction of organs, parts, and structures—limbs, fingers, eyes, kidneys, liver, and lungs.

"Is it sin, which makes the worm a chrysalis, and the chrysalis a butterfly, and the butterfly dust?" the German theologian Max Müller asked in 1885. A century later, biology offered an answer. It wasn't sin; it was a fusillade of genes.

①

In Leo Lionni's classic children's book *Inch by Inch*, a tiny worm is saved by a robin because it promises to "measure things" using its inch-long body as a metric. The worm measures the robin's tail, the toucan's beak, the flamingo's neck, and the heron's legs; the world of birds thus gets its first comparative anatomist.

Geneticists too had learned the usefulness of small organisms to measure, compare, and understand much larger things. Mendel had shelled bushels of peas. Morgan had measured mutation rates in flies. The seven hundred suspenseful minutes between the birth of a fly embryo and the creation of its first segments—arguably the most intensively scrutinized block of time in the history of biology—had partly solved one of the most important problems in biology: How can genes be orchestrated to create an exquisitely complex organism out of a single cell?

It took an even smaller organism—a worm of less than an inch—to solve the remaining half of the puzzle: How do cells arising in an embryo "know" what to become? Fly embryologists had produced a broad outline of organismal development as the serial deployment of three phases—axis determination, segment formation, and organ building—each governed by a cascade of genes. But to understand embryological development at the deepest level, geneticists needed to understand how genes could govern the destinies of individual cells.

In the mid-1960s, in Cambridge, Sydney Brenner began to hunt for an organism that could help solve the puzzle of cell-fate determination. Minuscule as it was, even the fly—"compound eyes, jointed legs, and elaborate behavior patterns"—was much too big for Brenner. To understand how genes instruct the fates of cells, Brenner needed an organism so small and simple that *each cell* arising from the embryo could be counted and followed in time and space (as a point of comparison, humans have about 37 trillion cells. A cell-fate map of humans would outstrip the computing powers of the most powerful computers).

Brenner became a connoisseur of tiny organisms, a god of small things. He pored through nineteenth-century zoology textbooks to find an animal that would satisfy his requirements. In the end, he settled on a minuscule soil-dwelling worm called *Caenorhabditis elegans*—*C. elegans* for short. Zoologists had noted that the worm was *eutelic*: once it reached adulthood, every worm had a fixed number of cells. To Brenner, the constancy of that number was like a latchkey to a new cosmos: if every worm had exactly the same number of cells, then genes must be capable of carrying instructions to specify the fate of *every cell* in a worm's body. "We propose to identify every cell in the worm and trace lineages," he wrote to Perutz. "We shall also investigate the constancy of development and study its genetic control by looking for mutants."

The counting of cells began in earnest in the early 1970s. First, Brenner convinced John White, a researcher in his lab, to map the location of every cell in the worm's nervous system—but Brenner soon broadened the scope to track the lineage of every cell in the worm's body. John Sulston, a postdoctoral researcher, was conscripted to the cell-counting effort. In 1974, Brenner and Sulston were joined by a young biologist fresh from Harvard named Robert Horvitz.

It was exhausting, hallucination-inducing work, "like watching a bowl of hundreds of grapes" for hours at a time, Horvitz recalled, and then mapping each grape as it changed its position in time and space. Cell by cell, a comprehensive atlas of cellular fate fell into place. Adult worms come in two different types—hermaphrodites and males. Hermaphrodites had 959 cells. Males had 1,031. By the late 1970s, the lineage of each of those 959 adult cells had been traced back to the original cell. This too was a map—although a map unlike any other in the history of science: a map of fate. The experiments on cell lineage and identity could now begin.

①

Three features of the cellular map were striking. The first was its invariance. Each of the 959 cells in every worm arose in a precisely stereotypical manner. "You could look at the map and recapitulate the construction of an organism, cell by cell," Horvitz said. You could say, "In twelve hours, this cell will divide once, and in forty-eight hours, it will become a neuron, and sixty hours later, it will move to that part of the worm's nervous system and stay there for the rest of its life. And you'd be perfectly right: The cell would do exactly that. It would move exactly there, in *exactly* that time."

What determined the identity of each cell? By the late seventies, Horvitz and Sulston had created dozens of worm mutants in which normal cell lineages were disrupted. If flies bearing legs on their heads had been strange, then these worm mutants were part of an even stranger menagerie. In some mutants, for instance, the genes that make the worm's vulva, the organ that forms the exit of the uterus, failed to function. The eggs laid by the vulvaless worm could not leave their mother's womb, and the worm was thus literally swallowed alive by its own unborn progeny, like some monster from a Teutonic myth. The genes altered in these mutants controlled the identity of an individual vulva cell. Yet other genes controlled the time that a cell divided to form two cells, its movement to a particular position in the animal, or the final shape and size that a cell would assume.

"There is no history; there is only biography," Emerson once wrote. For the worm, certainly, history had collapsed into a cellular biography. Every cell knew what to "be" because genes told it what to "become" (and *where* and *when* to become). The anatomy of the worm was all genetic clockwork and nothing else: there was no chance, no mystery, no ambiguity—no fate. Cell by cell, an animal was assembled from genetic instructions. Genesis was *gene*-sis.

①

If the exquisite orchestration of the birth, position, shape, size, and identity of every cell by genes was remarkable, then the final series of worm mutants generated an even more remarkable revelation. By the early 1980s, Horvitz and Sulston began to discover that even the *death* of cells was governed by genes. Every adult hermaphrodite worm has 959 cells—

but if you counted the cells generated during worm development, a total of 1,090 cells were actually born. It was a small discrepancy, but it fascinated Horvitz endlessly: 131 extra cells had somehow disappeared. They had been produced during development—but then killed during the maturation of the worm. These cells were the castaways of development, the lost children of genesis. When Sulston and Horvitz used their lineage maps to track the deaths of the 131 lost cells, they found that only specific cells, produced at specific times, were killed. It was a selective purge: like everything else in the worm's development, nothing was left to chance. The death of these cells—or rather their planned, self-willed suicide—also seemed genetically "programmed."

Programmed *death*? Geneticists were just contending with the programmed *life* of worms. Was death too controlled by genes? In 1972, John Kerr, an Australian pathologist, had observed a similar pattern of cell death in normal tissues and in cancers. Until Kerr's observations, biologists had thought of death as a largely accidental process caused by trauma, injury, or infection—a phenomenon called *necrosis*—literally, "blackening." Necrosis was typically accompanied by the decomposition of tissues, leading to the formation of pus or gangrene. But in certain tissues, Kerr noted, dying cells seemed to activate specific structural changes in anticipation of death—as if turning on a "death subroutine." The dying cells did not elicit gangrene, wounds, or inflammation; they acquired a pearly, wilting translucence, like lilies in a vase before they die. If necrosis was blackening, then this was death by whiteout. Instinctively, Kerr surmised that the two forms of dying were fundamentally different. This "controlled cell deletion," he wrote, "is an active, *inherently programmed* phenomenon," controlled by "genes of death." Seeking a word to describe the process, he called it *apoptosis*, an evocative Greek word for the falling off of leaves from trees, or petals from a flower.

But what did these "genes of death" look like? Horvitz and Sulston made yet another series of mutants—except these were not altered in cell lineage, but in patterns of cellular death. In one mutant, the contents of the dying cells could not be adequately fragmented into pieces. In another mutant, dead cells were not removed from the worm's body, resulting in carcasses of cells littering its edges, like Naples on a trash strike. The genes altered in these mutants, Horvitz surmised, were the executioners, scavengers, cleaners, and cremators of the cellular world—the active participants in the killing.

The next set of mutants had even more dramatic distortions in patterns of death: the carcasses were not even formed. In one worm, all 131 dying cells remained alive. In another, specific cells were spared from death. Horvitz's students nicknamed the mutant worms *the undead* or *wombies,* for "worm zombies." The inactivated genes in these worms were the master regulators of the death cascade in cells. Horvitz named them *ced* genes—for *C. elegans* death.

Remarkably, several genes that regulate cell death would soon be implicated in human cancers. Human cells also possess genes that orchestrate their death via apoptosis. Many of these genes are ancient—and their structures and functions are similar to those of the death genes found in worms and flies. In 1985, the cancer biologist Stanley Korsmeyer discovered that a gene named *BCL2* is recurrently mutated in lymphomas.* *BCL2,* it turned out, was the human counterpart to one of Horvitz's death-regulating worm genes, called *ced9.* In worms, *ced9* prevents cell death by sequestering the cell-death-related executioner proteins (hence the "undead" cells in the worm mutants). In human cells, the activation of *BCL2* results in a cell in which the death cascade is blocked, creating a cell that is pathologically unable to die: cancer.

☉

But was the fate of every cell in the worm dictated by genes, and only genes? Horvitz and Sulston discovered occasional cells in the worm—rare pairs—that could choose one fate or another randomly, as if by coin flip. The fate of these cells was not determined by their genetic destiny, but by their proximity to other cells. Two worm biologists working in Colorado, David Hirsh and Judith Kimble, called this phenomenon *natural ambiguity.*

But even natural ambiguity was sharply constrained, Kimble found. The identity of an ambiguous cell was, in fact, regulated by signals from neighboring cells—but the neighboring cells were themselves genetically preprogrammed. The God of Worms had evidently left tiny loopholes of chance in the worm's design, but He still wouldn't throw dice.

A worm was thus constructed from two kinds of inputs—"intrinsic" inputs from genes, and "extrinsic" inputs from cell-cell interactions. Jokingly,

---

* The death-defying function of *BCL2* was also discovered by David Vaux and Suzanne Cory in Australia.

Brenner called it the "British model" versus the "American model." The British way, Brenner wrote, "is for cells to do their own thing and not to talk to their neighbors very much. Ancestry is what counts, and once a cell is born in a certain place it will stay there and develop according to rigid rules. The American way is quite the opposite. Ancestry does not count. . . . What counts is the interactions with its neighbors. It frequently exchanges information with its fellow cells and often has to move to accomplish its goals and find its proper place."

What if you forcibly introduced chance—fate—into the life of a worm? In 1978, Kimble moved to Cambridge and began to study the effects of sharp perturbations on cell fates. She used a laser to singe and kill single cells in a worm's body. The ablation of one cell could change the fate of a neighboring cell, she found, but under severe constraints. Cells that had already been genetically predetermined had almost no leeway in altering their destinies. In contrast, cells that were "naturally ambiguous" were more pliant—but even so, their capacity to alter their destiny was limited. Extrinsic cues could alter intrinsic determinants, but to a point. You could whisk the man in a gray flannel suit off the Piccadilly line and stuff him on the Brooklyn-bound F train. He would be transformed—but still emerge from the tunnels wanting beef pasties for lunch. Chance played a role in the microscopic world of worms, but it was severely constrained by genes. The gene was the lens through which chance was filtered and refracted.

⓪

The discoveries of gene cascades that governed the lives and deaths of flies and worms were revelations for embryologists—but their impact on genetics was just as powerful. In solving Morgan's puzzle—"How do genes specify a fly?"—embryologists had also solved a much deeper riddle: How can units of heredity generate the bewildering complexity of organisms?

The answer lies in organization and interaction. A single master-regulatory gene might encode a protein with rather limited function: an on-and-off switch for twelve other target genes, say. But suppose the activity of the switch depends on the *concentration* of the protein, and the protein can be layered in a gradient across the body of an organism, with a high concentration at one end and a low concentration at the other. This protein might flick on all twelve of its targets in one part of an organism, eight in another segment, and only three in yet another. Each combination of target genes (twelve, eight, and three) might then intersect with yet other protein gradients, and activate and repress yet other genes. Add the dimensions of time and space to this recipe—i.e., when and where a gene might be activated or repressed—and you can begin to construct intricate fantasias of form. By mixing and matching hierarchies, gradients, switches, and circuits of genes and proteins, an organism can create the observed complexity of its anatomy and physiology.

As one scientist described it, ". . . individual genes are not particularly clever—this one cares only about that molecule, that one only about some other molecule . . . But that simplicity is no barrier to building enormous complexity. If you can build an ant colony with just a few different kinds of simpleminded ants (workers, drones, and the like), think about what you can do with 30,000 cascading genes, deployed at will."

The geneticist Antoine Danchin once used the parable of the Delphic boat to describe the process by which individual genes could produce the observed complexity of the natural world. In the proverbial story, the oracle at Delphi is asked to consider a boat on a river whose planks have begun to rot. As the wood decays, each plank is replaced, one by one—and after a decade, no plank is left from the original boat. Yet, the owner is convinced that it is the same boat. How can the boat be the same boat—the riddle runs—if every physical element of the original has been replaced?

The answer is that the "boat" is not made of planks but of the *relationship* between planks. If you hammer a hundred strips of wood atop each other, you get a wall; if you nail them side to side, you get a deck; only a particular configuration of planks, held together in particular relationship, in a particular order, makes a boat.

Genes operate in the same manner. Individual genes specify individual functions, but the relationship among genes allows physiology. The genome is inert without these relationships. That humans and worms have about the same number of genes—around twenty thousand—and yet the fact that only one of these two organisms is capable of painting the ceiling of the Sistine Chapel suggests that the number of genes is largely unimportant to the physiological complexity of the organism. "It is not what you have," as a certain Brazilian samba instructor once told me, "it is what you *do* with it."

☊

Perhaps the most useful metaphor to explain the relationship between genes, forms, and functions is one proposed by the evolutionary biologist and writer Richard Dawkins. Some genes, Dawkins suggests, behave like actual blueprints. A blueprint, Dawkins continues, is an exact architectural or mechanical plan, with a one-to-one correspondence between every feature of that plan and the structure that it encodes. A door is scaled down precisely twenty times, or a mechanical screw is placed precisely seven inches from the axle. "Blueprint" genes, by that same logic, encode the instructions to "build" one structure (or protein). The factor VIII gene makes only one protein, which serves mainly one function: it enables blood to form clots. Mutations in factor VIII are akin to mistakes in a blueprint. Their effect, like a missing doorknob or forgotten widget, is perfectly predictable. The mutated factor VIII gene fails to enable normal blood clotting, and the resulting disorder—bleeding without provocation—is the direct consequence of the function of the protein.

The vast majority of genes, however, do not behave like blueprints. They do not specify the building of a single structure or part. Instead, they collaborate with cascades of other genes to enable a complex physiological function. These genes, Dawkins argues, are not like blueprints, but like recipes. In a recipe for a cake, for instance, it makes no sense to think that the sugar specifies the "top," and the flour specifies the "bottom"; there is usually no one-to-one correspondence between an individual component of a recipe and one structure. A recipe provides instructions about *process*.

A cake is a developmental consequence of sugar, butter, and flour meeting each other in the right proportion, at the right temperature, and the right time. Human physiology, by analogy, is the developmental consequence of certain genes intersecting with other genes in the right sequence, in the right space. A gene is one line in a recipe that specifies an organism. The human genome is the recipe that specifies a human.

①

By the early 1970s, as biologists began to decipher the mechanism by which genes were deployed to generate the astounding complexities of organisms, they also confronted the inevitable question of the intentional manipulation of genes in living beings. In April 1971, the US National Institutes of Health organized a conference to determine whether the introduction of deliberate genetic changes in organisms was conceivable in the near future. Provocatively titled *Prospects for Designed Genetic Change*, the meeting hoped to update the public on the possibility of gene manipulations in humans, and consider the social and political implications of such technologies.

No such method to manipulate genes (even in simple organisms) was available in 1971, the panelists noted—but its development, they felt confident, was only a matter of time. "This is not science fiction," one geneticist declared. "Science fiction is when you [. . .] can't do anything experimentally . . . it is now conceivable that not within 100 years, not within 25 years, but perhaps within the next five to ten years, certain inborn errors . . . will be treated or cured by the administration of a certain gene that is lacking—and we have a lot of work to do in order to prepare society for this kind of change."

If such technologies were invented, their implications would be immense: the recipe of human instruction might be rewritten. Genetic mutations are selected over millennia, one scientist observed at the meeting, but cultural mutations can be introduced and selected in just a few years. The capacity to introduce "designed genetic changes" in humans might bring genetic change to the speed of cultural change. Some human diseases might be eliminated, the histories of individuals and families changed forever; the technology would reshape our notions of heredity, identity, illness, and future. As Gordon Tomkins, the biologist from UCSF, noted: "So for the first time, large numbers of people are beginning to ask themselves: What are we doing?"

Ⓞ

A memory: It is 1978 or '79, and I am about eight or nine. My father has returned from a business trip. His bags are still in the car, and a glass of ice water is sweating on a tray on the dining room table. It is one of those blistering afternoons in Delhi when the ceiling fans seem to slosh heat around the room, making it feel even warmer. Two of our neighbors are waiting for him in the living room. The air seems tense with anxiety, although I cannot discern why.

My father enters the living room, and the men talk to him for a few minutes. It is, I sense, not a pleasant conversation. Their voices rise and their words sharpen, and I can make out the contours of most of the sentences, even through the concrete walls of the adjacent room, where I am supposed to be doing homework.

Jagu has borrowed money from both of them—not large sums, but enough to bring them to our house, demanding repayment. He has told one of the men that he needs the cash for medicines (he has never been prescribed any), and the other man that he needs it to buy a train ticket to Calcutta to visit his other brothers (no such trip has been planned; it would be impossible for Jagu to travel alone). "You should learn to control him," one of the men says accusingly.

My father listens silently, patiently—but I can feel the fiery meniscus of rage rising in him, coating his throat with bile. He walks to the steel closet, where we keep the household cash, and brings it to the men, making a point of not bothering to count the notes. He can spare a few extra rupees; they should keep the change.

By the time the men leave, I know that there will be a bruising altercation at home. With the instinctual certainty of wild animals that run uphill before tsunamis, our cook has left the kitchen to summon my grandmother. The tension between my father and Jagu has been building, thickening, for a while: Jagu's behavior at home has been particularly disruptive in the last few weeks—and this episode seems to have pushed my father beyond some edge. His face is hot with embarrassment. The fragile varnish of class and normalcy that he has struggled so hard to seal is being cracked open, and the secret life of his family is pouring out through the fissures. Now the neighbors know of Jagu's madness, of his confabulations. My father has been shamed in their eyes: he is cheap, mean, hard-hearted, foolish, unable to control his brother. Or worse: defiled by a mental illness that runs in his family.

He walks into Jagu's room and yanks him bodily off the bed. Jagu wails desolately, like a child who is being punished for a transgression that he does not understand. My father is livid, glowing with anger, dangerous. He shoves Jagu across the room. It is an inconceivable act of violence for him; he has never lifted a finger at home. My sister runs upstairs to hide. My mother is in the kitchen, crying. I watch the scene rise to its ugly crescendo from behind the living room curtains, as if watching a film in slow motion.

And then my grandmother emerges from her room, glowering like a she-wolf. She is screaming at my father, doubling-down on his violence. Her eyes are alight like coals, her tongue forked with fire. *Don't you dare touch him.*

"*Get out,*" she urges Jagu, who retreats quickly behind her.

I have never seen her more formidable. Her Bengali furls backward, like a fuse, toward its village origins. I can make out some words, thick with accent and idiom, sent out like airborne missiles: *womb, wash, taint.* When I piece the sentence together, its poison is remarkable: *If you hit him, I will wash my womb with water to clean your taint. I will wash my womb,* she says.

My father is also frothing with tears now. His head hangs heavily. He seems infinitely tired. *Wash it,* he says under his breath, pleadingly. *Wash it, clean it, wash it.*

PART THREE

"THE DREAMS OF GENETICISTS"

Sequencing and Cloning of Genes (1970–2001)

①

Progress in science depends on new techniques, new discoveries and new ideas, probably in that order.

—Sydney Brenner

If we are right ... it is possible to induce <u>predictable and hereditary</u> changes in cells. This is something that has long been the dream of geneticists.

—Oswald T. Avery

①

"Crossing Over"

What a piece of work is a man! How noble in reason,
how infinite in faculties, in form and moving how express and admirable,
in action how like an angel, in apprehension how like a god!
> —William Shakespeare, *Hamlet*, act 2, scene 2

In the winter of 1968, Paul Berg returned to Stanford after an eleven-month sabbatical at the Salk Institute in La Jolla, California. Berg was forty-one years old. Built powerfully, like an athlete, he had a manner of walking with his shoulders rolling in front of him. He had a remnant trace of his Brooklyn childhood in his habits—the way, for instance, he might raise his hand and begin his sentence with the word *look* when provoked by a scientific argument. He admired artists, especially painters, and especially the abstract expressionists: Pollock and Diebenkorn, Newman and Frankenthaler. He was entranced by their transmutation of old vocabularies into new ones, their ability to repurpose essential elements from the tool kit of abstraction—light, lines, forms—to create giant canvases pulsing with extraordinary life.

A biochemist by training, Berg had studied with Arthur Kornberg at Washington University in St. Louis and moved with Kornberg to set up the new department of biochemistry at Stanford. Berg had spent much of his academic life studying the synthesis of proteins—but the La Jolla sabbatical had given him a chance to think about new themes. Perched high on a mesa above the Pacific, often closed in by a dense wall of morning fog, the Salk was like an open-air monk's chamber. Working with Renato Dulbecco, the virologist, Berg had focused on studying animal viruses. He had spent his sabbatical thinking about genes, viruses, and the transmission of hereditary information.

One particular virus intrigued Berg: Simian virus 40, or SV40 for short—

"simian" because it infects monkey and human cells. In a conceptual sense, every virus is a professional gene carrier. Viruses have a simple structure: they are often no more than a set of genes wrapped inside a coat—a "piece of bad news wrapped in a protein coat," as Peter Medawar, the immunologist, had described them. When a virus enters a cell, it sheds its coat, and begins to use the cell as a factory to copy its genes, and manufacture new coats, resulting in millions of new viruses budding out of the cell. Viruses have thus distilled their life cycle to its bare essentials. They live to infect and reproduce; they infect and reproduce to live.

Even in a world of distilled essentials, SV40 is a virus distilled to the extreme essence. Its genome is no more than a scrap of DNA—six hundred thousand times shorter than the human genome, with merely seven genes to the human genome's 21,000. Unlike many viruses, Berg learned, SV40 could coexist quite peaceably with certain kinds of infected cells. Rather than producing millions of new virions after infection—and often killing the host cell as a result, as other viruses do—SV40 could insert its DNA into the host cell's chromosome, and then lapse into a reproductive lull, until activated by specific cues.

The compactness of the SV40 genome, and the efficiency with which it could be delivered into cells, made it an ideal vehicle to carry genes into human cells. Berg was gripped by the idea: if he could equip SV40 with a decoy "foreign" gene (foreign to the virus, at least), the viral genome would smuggle that gene into a human cell, thereby altering a cell's hereditary information—a feat that would open novel frontiers for genetics. But before he could envision modifying the human genome, Berg had to confront a technical challenge: he needed a method to insert a foreign gene into a viral genome. He would have to artificially engineer a genetic "chimera"—a hybrid between a virus's genes and a foreign gene.

℗

Unlike human genes that are strung along chromosomes, like beads on open-ended strings, SV40 genes are strung into a circle of DNA. The genome resembles a molecular necklace. When the virus infects the cell and inserts its genes into chromosomes, the necklace unclasps, becomes linearized, and attaches itself to the middle of a chromosome. To add a foreign gene into the SV40 genome, Berg would need to forcibly open the clasp, insert the gene into the open circle, and seal the ends to close it again. The viral genome would do all the rest: it would carry the gene into a human cell, and insert it into a human chromosome.*

Berg was not the only biologist thinking about unclasping and clasping viral DNA to insert foreign genes. In 1969, a graduate student working in a laboratory down the hall from Berg's lab at Stanford, Peter Lobban, had written a thesis for his third qualifying exam in which he had proposed performing a similar kind of genetic manipulation on a different virus. Lobban had come to Stanford from MIT, where he had been an undergraduate. He was an engineer by training—or, perhaps more accurately, an engineer by *feeling*. Genes, Lobban argued in his proposal, were no different from steel girders; they could also be retooled, altered, shaped to human specifications, and put to use. The secret was in finding the right tool kit for the right job. Working with his thesis adviser, Dale Kaiser, Lobban had even launched preliminary experiments using standard enzymes found in biochemistry to shuttle genes from one molecule of DNA to another.

In fact, the real secret, as Berg and Lobban had independently figured out, was to forget that SV40 was a virus at all, and treat its genome as if it were a chemical. Genes may have been "inaccessible" in 1971—but DNA was perfectly accessible. Avery, after all, had boiled it in solution as a naked chemical, and it had still transmitted information between bacteria. Kornberg had added enzymes to it and made it replicate in a test tube. To insert a gene into the SV40 genome, all that Berg needed was a series of reactions. He needed an enzyme to cut open the genome circle, and an enzyme to "paste" a piece of foreign DNA into the SV40 genome necklace. Perhaps the virus—or, rather, the information contained in the virus—would then spring to life again.

⓪

But where might a scientist find enzymes that would cut and paste DNA? The answer, as so often in the history of genetics, came from the bacterial world. Since the 1960s, microbiologists had been purifying enzymes from bacteria that could be used to manipulate DNA in test tubes. A bac-

---

* If a gene is added to an SV40 genome, it can no longer generate a virus because the DNA becomes too large to package it into the viral coat or shell. Despite this, the expanded SV40 genome, with its foreign gene, remains perfectly capable of inserting itself, and its payload gene, into an animal cell. It was this property of gene delivery that Berg hoped to use.

terial cell—any cell, for that matter—needs its own "tool kit" to maneuver its own DNA: each time a cell divides, repairs damaged genes, or flips its genes across chromosomes, it needs enzymes to copy genes or to fill in gaps created by damage.

The "pasting" of two fragments of DNA was part of this tool kit of reactions. Berg knew that even the most primitive organisms possess the capacity to stitch genes together. Strands of DNA, recall, can be split by damaging agents, such as X-rays. DNA damage occurs routinely in cells, and to repair the split strands, cells make specific enzymes to paste the broken pieces together. One of these enzymes, called "ligase" (from the Latin word *ligare*—"to tie together"), chemically stitches the two pieces of the broken backbone of DNA together, thus restoring the integrity of the double helix. Occasionally, the DNA-copying enzyme, "polymerase," might also be recruited to fill in the gap and repair a broken gene.

The cutting enzymes came from a more unusual source. Virtually all cells have ligases and polymerases to repair broken DNA, but there is little reason for most cells to have a DNA-cutting enzyme on the loose. But bacteria—organisms that live on the roughest edges of life, where resources are drastically limited, growth is fierce, and competition for survival is intense—possess such knifelike enzymes to defend themselves against viruses. They use DNA-cutting enzymes, like switchblades, to slice open the DNA of invaders, thereby rendering their hosts immune to attack. These proteins are called "restriction" enzymes because they restrict infections by certain viruses. Like molecular scissors, these enzymes recognize unique sequences in DNA and cut the double helix at very specific sites. The specificity is key: in the molecular world of DNA, a targeted gash at the jugular can be lethal. One microbe can paralyze an invading microbe by cutting its chain of information.

These enzymatic tools, borrowed from the microbial world, would form the basis of Berg's experiment. The crucial components to engineer genes, Berg knew, were frozen away in about five separate refrigerators in five laboratories. He just needed to walk to the labs, gather the enzymes, and string the reactions in a chain. Cut with one enzyme, paste with another—and any two fragments of DNA could be stitched together, allowing scientists to manipulate genes with extraordinary dexterity and skill.

Berg understood the implications of the technology that was being created. Genes could be combined to create new combinations, or combinations of combinations; they could be altered, mutated, and shuttled between organisms. A frog gene could be inserted into a viral genome and thus introduced into a human cell. A human gene could be shuttled into bacterial cells. If the technology was pushed to its extreme limits, genes would become infinitely malleable: you could create new mutations or erase them; you could even envision modifying heredity—washing its marks, cleaning it, changing it at will. To produce such genetic chimeras, Berg recalled, "none of the individual procedures, manipulations, and reagents used to construct this recombinant DNA was novel; the novelty lay in the specific way they were used in combination." The truly radical advance was the cutting and pasting of *ideas*—the reassortment and annealing of insights and techniques that already existed in the realm of genetics for nearly a decade.

①

In the winter of 1970, Berg and David Jackson, a postdoctoral researcher in Berg's lab, began their first attempts to cut and join two pieces of DNA. The experiments were tedious—"a biochemist's nightmare," as Berg described them. The DNA had to be purified, mixed with the enzymes, then repurified on ice-cold columns, and the process repeated, until each of the individual reactions could be perfected. The problem was that the cutting enzymes had not been optimized, and the yield was minuscule. Although preoccupied with his own construction of gene hybrids, Lobban continued to provide crucial technical insights to Jackson. He had found a method to add fragments to the ends of DNA to make two clasplike pieces that stuck together like latch and key, thereby vastly increasing the efficiency with which gene hybrids could be formed.

Despite the forbidding technical hurdles, Berg and Jackson managed to join the entire genome of SV40 to a piece of DNA from a bacterial virus called Lambda bacteriophage (or phage λ) and three genes from the bacterium *E. coli.*

This was no mean achievement. Although λ and SV40 are both "viruses," they are as different from each other, say, as a horse and a seahorse (SV40 infects primate cells, while phage λ only infects bacteria). And *E. coli* was an altogether different beast—a bacterium from the human intestine. The result was a strange chimera: genes from far branches of the evolutionary tree stitched together to form a single contiguous piece of DNA.

Berg called the hybrids "recombinant DNA". It was a cannily chosen phrase, harkening back to the natural phenomenon of "recombination," the genesis of hybrid genes during sexual reproduction. In nature, genetic information is frequently mixed and matched between chromosomes to generate diversity: DNA from the paternal chromosome swaps places with DNA from the maternal chromosome to generate "father:mother" gene hybrids—"crossing over," as Morgan had called the phenomenon. Berg's genetic hybrids, produced with the very tools that allowed genes to be cut, pasted, and repaired in their natural state in organisms, extended this principle beyond reproduction. Berg was also synthesizing gene hybrids, albeit with genetic material from different organisms, mixed and matched in test tubes. Recombination without reproduction: he was crossing over to a new cosmos of biology.

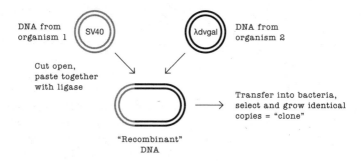

Figure adapted from Paul Berg's paper on "Recombinant" DNA.
By combining genes from any organisms, scientists could engineer genes at will,
foreshadowing human gene therapy and human genome engineering.

℗

That winter, a graduate student named Janet Mertz decided to join Berg's lab. Tenacious, unabashedly vocal about her opinions—"smart as all hell," as Berg described her—Mertz was an anomaly in the world of biochemists: the second woman to join Stanford's biochemistry department in nearly a decade. Like Lobban, Mertz had also come to Stanford from MIT, where she had majored in both engineering and biology. Mertz was intrigued by Jackson's experiments and was keen on the idea of synthesizing chimeras between genes of different organisms.

But what if she inverted Jackson's experimental goal? Jackson had inserted genetic material from a bacterium into the SV40 genome. What if she made genetic hybrids with SV40 genes inserted into the *E. coli* genome? Rather than viruses carrying bacterial genes, what might happen if Mertz created bacteria carrying viral genes?

The inversion of logic—or rather, the inversion of organisms—carried a crucial technical advantage. Like many bacteria, *E. coli* carry minuscule extra chromosomes, called mini-chromosomes or plasmids. As with the SV40 genome, plasmids also exist as circular necklaces of DNA, and they live and replicate within the bacteria. As bacterial cells divide and grow, the plasmids are also replicated. If Mertz could insert SV40 genes into an *E. coli* plasmid, she realized, she could use the bacteria as a "factory" for the new gene hybrids. As the bacteria grew and divided, the plasmid—and the foreign gene inside it—would be amplified manyfold. Copy upon copy of the modified chromosome, and its payload of foreign genes, would be created by the bacteria. There would ultimately be millions of exact replicas of a piece of DNA—"clones."

⬭

In June 1971, Mertz traveled from Stanford to Cold Spring Harbor in New York to attend a course on animal cells and viruses. As part of the course, students were expected to describe the research projects that they wished to pursue in the future. During her presentation, Mertz spoke about her plans to make genetic chimeras of SV40 and *E. coli* genes, and potentially propagate these hybrids in bacterial cells.

Graduate talks in summer courses typically don't generate much excitement. By the time Mertz was done with her slides, though, it was clear that this wasn't a typical graduate talk. There was silence at the end of Mertz's presentation—and then the students and instructors broke upon her with a tidal wave of questions: Had she contemplated the risks of generating such hybrids? What if the genetic hybrids that Berg and Mertz were about to generate were let loose on human populations? Had they considered the ethical aspects of making novel genetic elements?

Immediately after the session, Robert Pollack, a virologist and an instructor at the course, called Berg urgently. Pollack argued that the dangers implicit in "bridging evolutionary barriers that had existed since the last common ancestors between bacterium and people" were far too great to continue the experiment casually.

The issue was particularly thorny because SV40 was known to cause tumors in hamsters, and *E. coli* was known to live in the human intestine (current evidence suggests that SV40 is not likely to cause cancer in humans, but the risks were still unknown in the 1970s). What if Berg and Mertz ended up concocting the perfect storm of a genetic catastrophe—a human intestinal bacterium carrying a human cancer-causing gene? "You can stop splitting the atom; you can stop visiting the moon; you can stop using aerosol. . . . But you cannot recall a new form of life," Erwin Chargaff, the biochemist, wrote. "[The new genetic hybrids] will survive you and your children and your children's children. . . . The hybridization of Prometheus with Herostratus is bound to give evil results."

Berg spent weeks deliberating over the concerns raised by Pollack and Chargaff. "My first reaction was: this was absurd. I didn't really see any risk to it." The experiments were being carried out in a contained facility, with sterilized equipment; SV40 had never been implicated directly in human cancers. Indeed, many virologists had become infected with SV40, and no one had acquired any cancers. Frustrated with the constant public hysteria around the issue, Dulbecco had even offered to *drink* SV40 to prove that there was no link to human cancers.

But with his feet slung on the edge of a potential precipice, Berg could not afford to be cavalier. He wrote to several cancer biologists and microbiologists, asking them for independent opinions of the risk. Dulbecco was adamant about SV40, but could any scientist realistically estimate an unknown risk? In the end, Berg concluded that the biohazard was extremely minimal—but not zero. "In truth, I knew the risk was little," Berg said. "But I could not convince myself that there would be *no* risk. . . . I must have realized that I'd been wrong many, many times in predicting the outcomes of an experiment, and if I was wrong about the outcome of the risk, then the consequences were not something that I would want to live with." Until he had determined the precise nature of the risk, and made a plan for containment, Berg placed a self-imposed moratorium. For now, the DNA hybrids containing pieces of the SV40 genome would remain in a test tube. They would not be introduced into living organisms.

Mertz, meanwhile, had made another crucial discovery. The initial cutting and pasting of DNA, as envisaged by Berg and Jackson, required six tedious enzymatic steps. Mertz found a useful shortcut. Using a DNA-cutting enzyme—called EcoR1—obtained from Herb Boyer, a microbiologist in San Francisco, Mertz found that the pieces could be cut and pasted together in just two steps, rather than six.* "Janet really made the process vastly more efficient," Berg recalled. "Now, in just a few chemical reactions, we could generate new pieces of DNA. . . . She cut them, mixed them, added an enzyme that could join ends to ends, and then showed that she had gotten a product that shared the properties of both the starting materials." Mertz had created "recombinant DNA"—although, with the self-imposed moratorium in Berg's lab, she could not transfer the gene-hybrids into living bacterial cells.

Ⓘ

In November 1972, while Berg was weighing the risks of virus-bacteria hybrids, Herb Boyer, the San Francisco scientist who had supplied the DNA-cutting enzymes to Mertz, traveled to Hawaii for a meeting on microbiology. Born in a mining town in Pennsylvania in 1936, Boyer had discovered biology as a high school student and had grown up idealizing Watson and Crick (he had named his two Siamese cats after them). He had applied to medical school in the early sixties, but was rejected, unable to live down a D in metaphysics; instead, he had switched to studying microbiology as a graduate student.

Boyer had arrived in San Francisco in the summer of '66—with an afro, the requisite leather vest, and cutoff jeans—as an assistant professor at University of California, San Francisco (UCSF). Much of his work concerned the isolation of novel DNA-cutting enzymes, such as the one that he had sent to Berg's lab. Boyer had heard from Mertz about her DNA-cutting reaction, and the consequent simplification of the process of generating DNA hybrids.

Ⓘ

The conference in Hawaii was about bacterial genetics. Much of the excitement at the meeting involved the newly discovered plasmids in E. coli— the circular mini-chromosomes that replicated within bacteria, and could be transmitted between bacterial strains. After a long morning of presen-

---

* Mertz's discovery, made with Ron Davis, involved a fortuitous quality of enzymes such as EcoR1. If she cut the bacterial plasmid and the SV40 genome with EcoR1, she found, the ends came out naturally "sticky," like complementary pieces of Velcro, thereby making it easier to join them together into gene hybrids.

211

tations, Boyer fled to the beach for a respite, and spent the afternoon nursing a glass of rum and coconut juice.

Late that evening, Boyer ran into Stanley Cohen, a professor at Stanford. Boyer knew Cohen from his scientific papers, but they had never met in person. With a neatly trimmed, graying beard, owlish spectacles, and a cautious, deliberate manner of speaking, Cohen had the "physical persona of a Talmudic scholar," one scientist recalled—and a Talmudic knowledge of microbial genetics. Cohen worked on plasmids. He had also learned Frederick Griffith's "transformation" reaction—the technique needed to deliver DNA into bacterial cells.

Dinner had ended, but Cohen and Boyer were still hungry. With Stan Falkow, a fellow microbiologist, they strolled out of the hotel toward a quiet, dark street in a commercial strip near Waikiki beach. A New York–style deli, with bright flashing signs and neon-lit fixtures, loomed providentially out of the shadows of the volcanoes, and they found an open booth inside it. The waiter couldn't tell a *kishke* from a *knish*, but the menu offered corned beef and chopped liver. Over pastrami sandwiches, Boyer, Cohen, and Falkow talked about plasmids, gene chimeras, and bacterial genetics.

Both Boyer and Cohen knew about Berg and Mertz's successes at creating gene-hybrids in the lab. The discussion moved casually to Cohen's work. Cohen had isolated several plasmids from *E. coli*, including one that could be reliably purified out of the bacteria, and easily transmitted from one *E. coli* strain into another. Some of these plasmids carried genes to confer resistance to antibiotics—to tetracycline or penicillin, say.

But what if Cohen cut out an antibiotic-resistance gene from one plasmid and shuttled it to another plasmid? *Wouldn't a bacterium previously killed by the antibiotic now survive, thrive, and grow selectively, while the bacteria carrying the non-hybrid plasmids would die?*

The idea flashed out of shadows, like a neon sign on a darkening island. In Berg's and Jackson's initial experiments, there had been no simple method to identify the bacteria or viruses that had acquired the "foreign" gene (the hybrid plasmid had to be purified out of the biochemical gumbo using its size alone: A + B was larger than A or B). Cohen's plasmids, carrying antibiotic-resistance genes, in contrast, provided a powerful means to identify genetic recombinants. *Evolution* would be conscripted to help their experiment. Natural selection, deployed in a petri dish, would naturally select their hybrid plasmids. The transference of antibiotic resistance from one bacterium to another bacterium would confirm that the gene hybrid, or recombinant DNA, had been created.

But what of Berg and Jackson's technical hurdles? If the genetic chimeras were produced at a one-in-a-million frequency, then no selection method, however deft or powerful, would work: there would be no hybrids to select. On a whim, Boyer began to describe the DNA-cutting enzymes and Mertz's improved process to generate gene hybrids with greater efficiency. There was silence, as Cohen and Boyer tossed the idea around in their minds. The convergence was inevitable. Boyer had purified enzymes to create gene hybrids with vastly improved efficiency; Cohen had isolated plasmids that could be selected and propagated easily in bacteria. "The thought," Falkow recalls, was "too obvious to slip by unnoticed."

Cohen spoke in a slow, clear voice: "That means—"

Boyer cut him off mid-thought: "That's right . . . it should be possible. . . ."

"Sometimes in science, as in the rest of life," Falkow later wrote, "it is not necessary to finish the sentence or thought." The experiment was straightforward enough—so magnificently simple that it could be performed over the course of a single afternoon with standard reagents: "mix *Eco*R1-cut plasmid DNA molecules and rejoin them and there should be a proportion of recombinant plasmid molecules. Use antibiotic resistance to select the bacteria that had acquired the foreign gene, and you would select the hybrid DNA. Grow one such bacterial cell into its million descendants, and you would amplify the hybrid DNA a millionfold. You would clone recombinant DNA."

The experiment was not just innovative and efficient; it was also potentially safer. Unlike Berg and Mertz's experiment—involving virus-bacteria hybrids—Cohen and Boyer's chimeras were composed entirely of bacterial genes, which they considered far less hazardous. They could find no reason to halt the creation of these plasmids. Bacteria, after all, were capable of trading genetic material like gossip, with scarcely an afterthought; free trade in genes was a hallmark of the microbial world.

⊕

Through that winter, and into the early spring of 1973, Boyer and Cohen worked furiously to make their genetic hybrids. Plasmids and enzymes shuttled between UCSF and Stanford, up and down Highway 101, on a Volkswagen Beetle driven by a research assistant from Boyer's lab. By the end of the summer, Boyer and Cohen had successfully created their gene hybrids—two pieces of genetic material from two bacteria stitched together to form a single chimera. Boyer later recalled the moment of discovery with immense clarity: "I looked at the first gels and I remember tears coming into my eyes, it was so nice." Hereditary identities borrowed from two organisms had been shuffled around to form a new one; it was as close to metaphysics as one could get.

In February 1973, Boyer and Cohen were ready to propagate the first artificially produced genetic chimera in living cells. They cut two bacterial plasmids open with restriction enzymes and swapped the genetic material from one plasmid into another. The plasmid carrying the hybrid DNA was locked shut with ligase, and the resultant chimera introduced into bacterial cells using a modified version of the transformation reaction. The bacteria containing the gene hybrids were grown on petri dishes to form tiny translucent colonies, glistening like pearls on agar.

Sometime late one evening, Cohen inoculated a vat of sterile bacterial broth with a single colony of bacterial cells with the gene hybrids. Cells grew overnight in a shaking beaker. A hundred, a thousand, and then a million copies of the genetic chimera were replicated, each containing a mixture of genetic material from two completely different organisms. The birth of a new world was announced with no more noise than the mechanical *tick-tick-tick* of a bacterial incubator rocking through the night.

The New Music

Each generation needs a new music.

—Francis Crick

People now made music from everything.

—Richard Powers, *Orfeo*

While Berg, Boyer, and Cohen were mixing and matching gene fragments in test tubes at Stanford and UCSF, an equally seminal breakthrough in genetics was emerging from a laboratory in Cambridge, England. To understand the nature of this discovery, we must return to the formal language of genes. Genetics, like any language, is built out of basic structural elements—alphabet, vocabulary, syntax, and grammar. The "alphabet" of genes has only four letters: the four bases of DNA—A, C, G, and T. The "vocabulary" consists of the triplet code: three bases of DNA are read together to encode one amino acid in a protein; ACT encodes Threonine, CAT encodes Histidine, GGT encodes Glycine, and so forth. A protein is the "sentence" encoded by a gene, using alphabets strung together in a chain (ACT-CAT-GGT encodes Threonine-Histidine-Glycine). And the regulation of genes, as Monod and Jacob had discovered, creates a context for these words and sentences to generate meaning. The regulatory sequences appended to a gene—i.e., signals to turn a gene on or off at certain times and in certain cells—can be imagined as the internal grammar of the genome.

But the alphabet, grammar, and syntax of genetics exist exclusively within cells; humans are not native speakers. For a biologist to be able to read and write the language of genes, a novel set of tools had to be invented. To "write" is to mix and match words in unique permutations to generate new meanings. At Stanford, Berg, Cohen, and Boyer were beginning to write genes using gene cloning—generating words and sentences in DNA that had never existed in nature (a bacterial gene combined with a viral gene to form a new genetic element). But the "reading" of genes—the deciphering of the precise sequence of bases in a stretch of DNA—was still an enormous technical hurdle.

Ironically, the very features that enable a *cell* to read DNA are the features that make it incomprehensible to humans—to chemists, in particular. DNA, as Schrödinger had predicted, was a chemical built to defy chemists, a molecule of exquisite contradictions—monotonous and yet infinitely varied, repetitive to the extreme and yet idiosyncratic to the extreme. Chemists generally piece together the structure of a molecule by breaking the molecule down into smaller and smaller parts, like puzzle pieces, and then assembling the structure from the constituents. But DNA, broken into pieces, degenerates into a garble of four bases—A, C, G, and T. You cannot read a book by dissolving all its words into alphabets. With DNA, as with words, the *sequence* carries the meaning. Dissolve DNA into its constituent bases, and it turns into a primordial four-letter alphabet soup.

☉

How might a chemist determine the sequence of a gene? In Cambridge, England, in a hutlike laboratory buried half-underground near the fens, Frederick Sanger, the biochemist, had struggled with gene sequencing since the 1960s. Sanger had an obsessive interest in the chemical structures of complex biological molecules. In the early 1950s, Sanger had solved the sequence of a protein—insulin—using a variant of the conventional disintegration method. Insulin, first purified from dozens of pounds of ground-up dog pancreases in 1921 by a Toronto surgeon, Frederick Banting, and his medical student Charles Best, was the grand prize of protein purification—a hormone that, injected into diabetic children, could rapidly reverse their wasting, lethal, sugar-choking disease. By the late 1920s, the pharmaceutical company Eli Lilly was manufacturing grams of insulin out of vast vats of liquefied cow and pig pancreases.

Yet, despite several attempts, insulin remained doggedly resistant to molecular characterization. Sanger brought a chemist's methodological rigor to the problem: the solution—as any chemist knew—was always in dissolution. Every protein is made of a sequence of amino acids strung into a chain—Methionine-Histidine-Arginine-Lysine or Glycine-Histidine-Arginine-Lysine, and so forth. To identify the sequence of a protein, Sanger realized, he would have to run a sequence of degradation reactions. He would snap off one amino acid from the end of the chain, dissolve it in solvents, and characterize it chemically—Methionine. And he would repeat the process, snapping off the next amino acid: Histidine. The degradation and identification would be repeated again and again—Arginine . . . snap . . . Lysine . . . snap—until he reached the end of the protein. It was like *un*stringing a necklace, bead by bead—reversing the cycle used by a cell to build a protein. Piece by piece, the disintegration of insulin would reveal the structure of its chain. In 1958, Sanger won the Nobel Prize for this landmark discovery.

Between 1955 and 1962, Sanger used variations of this disintegration method to solve the sequences of several important proteins—but left the problem of DNA sequencing largely untouched. These were his "lean years," he wrote; he lived in the leeward shadow of his fame. He published rarely—immensely detailed papers on protein sequencing that others characterized as magisterial—but he counted none of these as major successes. In the summer of 1962, Sanger moved to another laboratory in Cambridge—the Medical Research Council (MRC) Building—where he was surrounded by new neighbors, among them Crick, Perutz, and Sydney Brenner, all immersed in the cult of DNA.

The transition of labs marked a seminal transition in Sanger's focus. Some scientists—Crick, Wilkins—were born into DNA. Others—Watson, Franklin, Brenner—had acquired it. Fred Sanger had DNA thrust upon him.

Ⓓ

In the mid-1960s, Sanger switched his focus from proteins to nucleic acids and began to consider DNA sequencing seriously. But the methods that had worked so marvelously for insulin—breaking, dissolving, breaking, dissolving—refused to work for DNA. Proteins are chemically structured such that amino acids can be serially snapped off the chain—but with DNA, no such tools existed. Sanger tried to reconfigure his degradation technique, but the experiments only produced chemical chaos. Cut into pieces and dissolved, DNA turned from genetic information to gobbledygook.

Inspiration came to Sanger unexpectedly in the winter of 1971—in the form of an inversion. He had spent decades learning to break molecules apart to solve their sequence. But what if he turned his own strategy up-side down and tried to *build* DNA, rather than break it down? To solve a gene sequence, Sanger reasoned, one must think like a gene. Cells build genes all the time: each time a cell divides, it makes a copy of every gene. If a biochemist could strap himself to the gene-copying enzyme (DNA polymerase), straddling its back as it made a copy of DNA and keeping tabs as the enzyme added base upon base—A, C, T, G, C, C, C, and so forth—the sequence of a gene would become known. It was like eaves-dropping on a copying machine: you could reconstruct the original from the copy. Once again, the mirror image would illuminate the original—Dorian Gray would be re-created, piece upon piece, from his reflection.

In 1971, Sanger began to devise a gene-sequencing technique using the copying reaction of DNA polymerase. (At Harvard, Walter Gilbert and Allan Maxam were also devising a system to sequence DNA, although using different reagents. Their method also worked, but was soon out-moded by Sanger's.) At first, Sanger's method was inefficient and prone to inexplicable failures. In part, the problem was that the copying reaction was too fast: polymerase raced along the strand of DNA, adding nucleo-tides at such a breakneck pace that Sanger could not catch the intermedi-ate steps. In 1975, Sanger made an ingenious modification: he spiked the copying reaction with a series of chemically altered bases—ever-so-slight variants of A, C, G, and T—that were still recognized by DNA polymerase, but jammed its copying ability. As polymerase stalled, Sanger could use the slowed-down reaction to map a gene by its jams—an A here, a T there, a G there, and so forth—for thousands of bases of DNA.

On February 24, 1977, Sanger used this technique to reveal the full se-quence of a virus—ΦX174—in a paper in *Nature*. Only 5,386 base pairs in length, phi was a tiny virus—its entire genome was smaller than some of the smallest human genes—but the publication announced a transfor-mative scientific advance. "The sequence identifies many of the features responsible for the production of the proteins of the nine known genes of the organism," he wrote. Sanger had learned to read the language of genes.

Φ

The new techniques of genetics—gene sequencing and gene cloning—immediately illuminated novel characteristics of genes and genomes. The first, and most surprising, discovery concerned a unique feature of the genes of animals and animal viruses. In 1977, two scientists working independently, Richard Roberts and Phillip Sharp, discovered that most animal proteins were not encoded in long, continuous stretches of DNA, but were actually split into modules. In bacteria, every gene is a continuous, uninterrupted stretch of DNA, starting with the first triplet code (ATG) and running contiguously to the final "stop" signal. Bacterial genes do not contain separate modules, and they are not split internally by spacers. But in animals, and in animal viruses, Roberts and Sharp found that a gene was typically split into parts and interrupted by long stretches of stuffer DNA.

As an analogy, consider the word *structure*. In bacteria, the gene is embedded in the genome in precisely that format, *structure*, with no breaks, stuffers, interpositions, or interruptions. In the human genome, in contrast, the word is interrupted by intermediate stretches of DNA: *s . . . tru . . . ct . . . ur . . . e.*

The long stretches of DNA marked by the ellipses (. . .) do not contain any protein-encoding information. When such an interrupted gene is used to generate a message—i.e., when DNA is used to build RNA—the stuffer fragments are excised from the RNA message, and the RNA is stitched together again with the intervening pieces removed: *s . . . tru . . . ct . . . ur . . . e* became simplified to *structure*. Roberts and Sharp later coined a phrase for the process: *gene splicing* or *RNA splicing* (since the RNA message of the gene was "spliced" to remove the stuffer fragments).

At first, this split structure of genes seemed puzzling: Why would an animal genome waste such long stretches of DNA splitting genes into bits and pieces, only to stitch them back into a continuous message? But the inner logic of split genes soon became evident: by splitting genes into modules, a cell could generate bewildering combinations of messages out of a single gene. The word *s . . . tru . . . c . . . t . . . ur . . . e* can be spliced to yield *cure* and *true* and so forth, thereby creating vast numbers of variant messages—called isoforms—out of a single gene. From *g . . . e . . . n . . . om . . . e* you can use splicing to generate *gene, gnome,* and *om*. And modular genes also had an evolutionary advantage: the individual modules from different genes could be mixed and matched to build entirely new kinds of genes (*c . . . om . . . e . . . t*). Wally Gilbert, the Harvard geneticist, created a new word for these modules; he called them *exons*. The in-between stuffer fragments were termed *introns*.

Introns are not the exception in human genes; they are the rule. Human introns are often enormous—spanning several hundreds of thousands of bases of DNA. And genes themselves are separated from each other by long stretches of intervening DNA, called intergenic DNA. Intergenic DNA and introns—spacers *between* genes and stuffers *within* genes—are thought to have sequences that allow genes to be regulated in context. To return to our analogy, these regions might be described as long ellipses scattered with occasional punctuation marks. The human genome can thus be visualized as:

This . . . . . . *is* . . . . *the* . . . . . . *(. . .)* . . . *s* . . . *truc* . . . *ture* . . . . . .
of . . . . . . *your* . . . . . . *gen* . . . *om* . . . *e;*

The words represent genes. The long ellipses between the words represent the stretches of intergenic DNA. The shorter ellipses within the words (*gen . . . ome . . . e*) are introns. The parentheses and semicolons—punctuation marks—are regions of DNA that regulate genes.

The twin technologies of gene sequencing and gene cloning also rescued genetics from an experimental jam. In the late 1960s, genetics had found itself caught in a deadlock. Every experimental science depends, crucially, on the capacity to perturb a system intentionally, and to measure the effects of that perturbation. But the only way to *alter* genes was by creating mutants—essentially a random process—and the only means to *read* the alteration was through changes in form and function. You could shower fruit flies with X-rays, as Muller had, to make wingless or eyeless flies, but you had no means to intentionally manipulate the genes that controlled eyes or wings, or to understand exactly how the wing or eye gene had been changed. "The gene," as one scientist described it, "was something inaccessible."

The inaccessibility of the gene had been particularly frustrating to the messiahs of the "new biology"—James Watson among them. In 1955, two years after his discovery of the structure of DNA, Watson had moved to the Department of Biology at Harvard and instantly raised the hackles of some of its most venerated professors. Biology, as Watson saw it, was a discipline splitting through its middle. On one side sat its old guard—natural historians, taxonomists, anatomists, and ecologists who were still preoccupied by the classifications of animals and by largely qualitative descriptions of organismal anatomy and physiology. The "new" biologists, in contrast, studied molecules and genes. The old school spoke of diversity and variation. The new school: of universal codes, common mechanisms, and "central dogmas."*

"Each generation needs a new music," Crick had said; Watson was frankly scornful of the old music. Natural history—a largely "descriptive" discipline, as Watson characterized it—would be replaced by a vigorous, muscular experimental science that he had helped create. The dinosaurs who studied dinosaurs would soon become extinct in their own right. Watson called the old biologists "stamp collectors"—mocking their preoccupation with the collection and classification of biological specimens.†

But even Watson had to admit that the inability to perform directed genetic interventions, or to read the exact nature of gene alterations, was a frustration for new biology. If genes could be sequenced and manipulated, then a vast experimental landscape would be thrown open. Until then, biologists would be stuck probing gene function with the only available tool—the genesis of random mutations in simple organisms. To Watson's insult, a natural historian might have hurled an equal and opposite injury. If old biologists were "stamp collectors," then the new molecular biologists were "mutant hunters."

Between 1970 and 1980, the mutant hunters transformed into gene manipulators and gene decoders. Consider this: In 1969, if a disease-linked gene was found in humans, scientists had no simple means to understand the nature of the mutation, no mechanism to compare the altered gene to normal form, and no obvious method to reconstruct the gene mutation in a different organism to study its function. By 1979, that same gene could be shuttled into bacteria, spliced into a viral vector, delivered into the genome of a mammalian cell, cloned, sequenced, and compared to the normal form.

In December 1980, in recognition of these seminal advancements in genetic technologies, the Nobel Prize in Chemistry was awarded jointly

---

* Notably, Darwin and Mendel had both bridged the gap between the old and the new biology. Darwin had started out as a natural historian—a fossil collector—but had then radically altered that discipline by seeking the *mechanism* behind natural history. Mendel, too, had started out as a botanist and a naturalist and radically swerved that discipline by seeking the mechanism that drove heredity and variation. Both Darwin and Mendel observed the natural world to seek deeper causes behind its organization.

† Watson borrowed this memorable phrase from Ernest Rutherford, who, in one of his characteristically brusque moments, had declared, "All science is either physics or stamp collecting."

to Fred Sanger, Walter Gilbert, and Paul Berg—the readers and writers of DNA. The "arsenal of chemical manipulations [of genes]," as one science journalist put it, was now fully stocked. "Genetic engineering," Peter Medawar, the biologist, wrote, "implies deliberate genetic change brought about by the manipulation of DNA, the vector of hereditary information. . . . Is it not a major truth of technology that anything which is in principle possible will be done . . . ? Land on the moon? Yes, assuredly. Abolish smallpox? A pleasure. Make up for deficiencies in the human genome? Mmmm, yes, though that's more difficult and will take longer. We aren't there yet, but we are certainly moving in the right direction."

ℚ

The technologies to manipulate, clone, and sequence genes may have been initially invented to shuttle genes between bacteria, viruses, and mammalian cells (à la Berg, Boyer, and Cohen) but the impact of these technologies reverberated broadly through organismal biology. Although the phrases *gene cloning* or *molecular cloning* were initially coined to refer to the production of identical copies of DNA (i.e., "clones") in bacteria or viruses, they would soon become shorthand for the entire gamut of techniques that allowed biologists to extract genes from organisms, manipulate these genes in test tubes, produce gene hybrids, and propagate the genes in living organisms (you could only clone genes, after all, by using a combination of all these techniques). "By learning to manipulate genes experimentally," Berg said, "you could learn to manipulate organisms experimentally. And by mixing and matching gene-manipulation and gene-sequencing tools, a scientist could interrogate not just genetics, but the whole universe of biology with a kind of experimental audacity that was unimaginable in the past."

Say an immunologist was trying to solve a fundamental riddle in immunology: the mechanism by which T cells recognize and kill foreign cells in the body. For decades, it had been known that T cells sense the presence of invading cells and virus-infected cells by virtue of a sensor found on the surface of the T cell. The sensor, called the *T cell receptor*, is a protein made uniquely by T cells. The receptor recognizes proteins on the surface of foreign cells and binds to them. The binding, in turn, triggers a signal to kill the invading cell, and thereby acts as a defense mechanism for an organism.

But what was the nature of the T cell receptor? Biochemists had approached the problem with their typical penchant for reduction: they had obtained vats upon vats of T cells, used soaps and detergents to dissolve the cell's components into a gray, cellular froth, then distilled the membranes and lipids away, and purified and repurified the material into smaller and smaller parts to hunt down the culprit protein. Yet the receptor protein, dissolved somewhere in that infernal soup, had remained elusive.

A gene cloner might take an alternative approach. Assume, for a moment, that the distinctive feature of the T cell receptor protein is that it is synthesized only in T cells, not in neurons, or ovaries, or liver cells. The *gene* for the receptor must exist in every human cell—human neurons, liver cells, and T cells have identical genomes, after all—but the *RNA* is made only in T cells. Could one compare the "RNA catalog" of two different cells, and thereby clone a functionally relevant gene from that catalog? The biochemist's approach pivots on concentration: find the protein by looking where it's most likely to be concentrated, and distill it out of the mix. The geneticist's approach, in contrast, pivots on *information*: find the gene by searching for differences in "databases" created by two closely related cells and multiply the gene in bacteria via cloning. The biochemist distills forms; the gene cloner amplifies information.

In 1970, David Baltimore and Howard Temin, two virologists, made a pivotal discovery that made such comparisons possible. Working independently, Baltimore and Temin discovered an enzyme found in retroviruses that could build DNA from an RNA template. They called the enzyme reverse transcriptase—"reverse" because it inverted the normal direction of information flow: from RNA *back* to DNA, or from a gene's message backward to a gene, thereby violating one version of the "central dogma" (that genetic information only moved from genes to messages, but never backward).

Using reverse transcriptase, every RNA in a cell could be used as a template to build its corresponding gene. A biologist could thus generate a catalog, or "library" of all "active" genes in a cell—akin to a library of books grouped by subject.* There would be a library of genes for T cells and another for red blood cells, a library for neurons in the retina, for insulin-secreting cells of the pancreas, and so forth. By comparing li-

* These libraries were conceived and created by Tom Maniatis in collaboration with Argiris Efstratiadis and Fotis Kafatos. Maniatis had been unable to work on gene cloning at Harvard because of concerns about the safety of recombinant DNA. He had moved to Cold Spring Harbor on Watson's invitation so that he could work on gene cloning in peace.

braries derived from two cells—a T cell and a pancreas cell, say—an immunologist could fish out genes that were active in one cell and not the other (e.g., insulin or the T cell receptor). Once identified, that gene could be amplified a millionfold in bacteria. The gene could be isolated and sequenced, its RNA and protein sequence determined, its regulatory regions identified; it could be mutated and inserted into a different cell to decipher the gene's structure and function. In 1984, this technique was deployed to clone the T cell receptor—a landmark achievement in immunology.

Biology, as one geneticist later recalled, was "liberated by cloning . . . and the field began to erupt with surprises." Mysterious, important, elusive genes sought for decades—genes for blood-clotting proteins, for regulators of growth, for antibodies and hormones, for transmitters between nerves, genes to control the replication of other genes, genes implicated in cancer, diabetes, depression, and heart disease—would soon be purified and cloned using gene "libraries" derived from cells as their source.

Every field of biology was transformed by gene-cloning and gene-sequencing technology. If experimental biology was the "new music," then the gene was its conductor, its orchestra, its assonant refrain, its principal instrument, its score.

Einsteins on the Beach

There is a tide in the affairs of men,
Which, taken at the flood, leads on to fortune;
Omitted, all the voyage of their life
Is bound in shallows and in miseries.
On such a full sea are we now afloat.
          —William Shakespeare,
          Julius Caesar, act 4, scene 3

I believe in the inalienable right of all adult scientists to
make absolute fools of themselves in private.
          —Sydney Brenner

In Erice, near the western coast of Sicily, a twelfth-century Norman fortress rises two thousand feet above the ground on a furl of rock. Viewed from afar, the fortress seems to have been created by some natural heave of the landscape, its stone flanks emerging from the rock face of the cliff as if through metamorphosis. The Erice Castle, or Venus Castle, as some call it, was built on the site of an ancient Roman temple. The older building was dismantled, stone by stone, and reassembled to form the walls, turrets, and towers of the castle. The shrine of the original temple has long vanished, but it was rumored to be dedicated to Venus. The Roman goddess of fertility, sex, and desire, Venus was conceived unnaturally from the spume spilled from Caelus's genitals into the sea.

In the summer of 1972, a few months after Paul Berg had created the first DNA chimeras at Stanford, he traveled to Erice to give a scientific seminar at a meeting. He arrived in Palermo late in the evening and took a two-hour taxi ride toward the coast. Night fell quickly. When he asked a stranger to give him directions to the town, the man gestured vaguely into the darkness where a flickering decimal point of light seemed suspended two thousand feet in the air.

The meeting began the next morning. The audience comprised about eighty young men and women from Europe, mostly graduate students in biology and a few professors. Berg gave an informal lecture—"a rap session," he called it—presenting his data on gene chimeras, recombinant DNA, and the production of the virus-bacteria hybrids.

The students were electrified. Berg was inundated with questions, as he had expected—but the direction of the conversation surprised him. At Janet Mertz's presentation at Cold Spring Harbor in 1971, the biggest concern had been safety: How could Berg or Mertz guarantee that their genetic chimeras would not unleash biological chaos on humans? In Sicily, in contrast, the conversation turned quickly to politics, culture, and ethics. What about the "spectre of genetic engineering in humans, behavior control?" Berg recalled. "What if we could cure genetic diseases?" the students asked. "[Or] program people's eye color? Intelligence? Height? . . . What would the implications be for humans and human societies?"

Who would ensure that genetic technologies would not be seized and perverted by powerful forces—as once before on that continent? Berg had obviously stoked an old fire. In America, the prospect of gene manipulation had principally raised the specter of future biological dangers. In Italy—not more than a few hundred miles from the sites of the former Nazi extermination camps—it was the moral hazards of genetics, more than the biohazards of genes, that haunted the conversation.

That evening, a German student gathered an impromptu group of his peers to continue the debate. They climbed the ramparts of the Venus Castle and looked out toward the darkening coast, with the lights of the city blinking below. Berg and the students stayed up late into the night for a second session, drinking beers and talking about natural and unnatural conceptions—"the beginning of a new era . . . [its] possible hazards, and the prospects of genetic engineering."

①

In January 1973, a few months after the Erice trip, Berg decided to organize a small conference in California to address the growing concerns about gene-manipulation technologies. The meeting was held at the Pacific Groves Conference Center at Asilomar, a sprawling, wind-buffeted complex of buildings on the edge of the ocean near Monterey Bay, about eighty miles from Stanford. Scientists from all disciplines—virologists, geneticists, biochemists, microbiologists—attended. "Asilomar I," as Berg would later call the meeting, generated enormous interest, but few recommendations. Much of the meeting focused on biosafety issues. The use of SV40 and other human viruses was hotly discussed. "Back then, we were still using our mouths to pipette viruses and chemicals," Berg told me. Berg's assistant Marianne Dieckmann once recalled a student who had accidentally flecked some liquid onto the tip of a cigarette (it was not unusual, for that matter, to have half-lit cigarettes, smoldering in ashtrays, strewn across the lab). The student had just shrugged and continued to smoke, with the droplet of virus disintegrating into ash.

The Asilomar conference produced an important book, *Biohazards in Biological Research*, but its larger conclusion was in the negative. As Berg described it, "What came out of it, frankly, was the recognition of how little we know."

Concerns about gene cloning were further inflamed in the summer of 1973 when Boyer and Cohen presented their experiments on bacterial gene hybrids at another conference. At Stanford, meanwhile, Berg was being flooded with requests from researchers around the world asking for gene recombination reagents. One researcher from Chicago proposed inserting genes of the highly pathogenic human herpes virus into bacterial cells, thereby creating a human intestinal bacterium loaded with a lethal toxin gene, ostensibly to study the toxicity of herpes virus genes. (Berg politely declined.) Antibiotic-resistance genes were routinely being swapped between bacteria. Genes were being shuffled between species and genera, leaping across a million years of evolutionary rift as if casually stepping over thin lines in sand. Noting the growing swirl of uncertainties, the National Academy of Sciences called on Berg to lead a study panel on gene recombination.

The panel—eight scientists, including Berg, Watson, David Baltimore, and Norton Zinder—met at MIT, in Boston, on a chilly spring afternoon in April 1973. They instantly got to work, brainstorming possible mechanisms to control and regulate gene cloning. Baltimore suggested the development of " 'safe' viruses, plasmids and bacteria, which would be crippled"—and thereby be unable to cause disease. But even that safety measure was not foolproof. Who would ensure that "crippled" viruses would remain permanently crippled? Viruses and bacteria were, after all, not passive, inert objects. Even within laboratory environments, they were living, evolving, moving targets. One mutation—and a previously disabled bacterium might spring to virulent life again.

The debate had gone on for several hours when Zinder proposed a plan that seemed almost reactionary: "Well, if we had any guts at all, we'd just tell people not to do these experiments." The proposal created a quiet stir around the table. It was far from an ideal solution—there was something obviously disingenuous about scientists telling scientists to restrict their scientific work—but it would at least act as a temporary stay order. "Unpleasant as it was, we thought it might just work," Berg recalled. The panel drafted a formal letter, pleading for a "moratorium" on certain kinds of recombinant DNA research. The letter weighed the risks and benefits of gene recombination technologies and suggested that certain experiments be deferred until the safety issues had been addressed. "Not every conceivable experiment was dangerous," Berg noted, but "some were clearly more hazardous than others." Three types of procedures involving recombinant DNA, in particular, needed to be sharply restricted: "Don't put toxin genes into *E. coli*. Don't put drug-resistant genes into *E. coli*, and don't put cancer genes into *E. coli*," Berg advised. With a moratorium in place, Berg and his colleagues argued, scientists could buy some time to consider the implications of their work. A second meeting was proposed for 1975, where the issues could be debated among a larger group of scientists.

In 1974, the "Berg letter" ran in *Nature*, *Science*, and *Proceedings of the National Academy of Sciences*. It drew instant attention around the globe. In Britain, a committee was formed to address the "potential benefits and potential hazards" of recombinant DNA and gene cloning. In France, reactions to the letter were published in *Le Monde*. That winter, François Jacob (of gene-regulation fame) was asked to review a grant application that proposed inserting a human muscle gene into a virus. Following Berg's footsteps, Jacob urged tabling such proposals until a national response to recombinant DNA technology had been drafted. At a meeting in Germany in 1974, many geneticists reiterated a similar caution. Sharp constraints on experiments with recombinant DNA research were essential until the risks had been delineated, and recommendations formalized.

The research, meanwhile, was steamrolling ahead, knocking down biological and evolutionary barriers as if they had been propped up on toothpicks. At Stanford, Boyer, Cohen, and their students grafted a gene for penicillin resistance from one bacterium onto another and thereby created drug-resistant *E. coli*. In principle, any gene could be transferred from one organism to the next. Audaciously, Boyer and Cohen projected forward: "It may be practical . . . to introduce genes specifying metabolic or synthetic functions [that are] indigenous to other biological classes, such as plants and animals." Species, Boyer declared jokingly, "are specious."

On New Year's Day 1974, a researcher working with Cohen at Stanford reported that he had inserted a frog gene into a bacterial cell. Another evolutionary border was casually crossed, another boundary transgressed. In biology, "being natural," as Oscar Wilde once put it, was turning out to be "simply a pose."

①

Asilomar II—one of the most unusual meetings in the history of science—was organized by Berg, Baltimore, and three other scientists for February 1975. Once again, geneticists returned to the windy beach dunes to discuss genes, recombination, and the shape of the future. It was an evocatively beautiful season. Monarch butterflies were migrating along the coast on their annual visit to the grasslands of Canada, and the redwoods and scrub pines were suddenly alit by a flotilla of red, orange, and black.

The human visitors arrived on February 24—but not just biologists. Cannily, Berg and Baltimore had asked lawyers, journalists, and writers to join the conference. If the future of gene manipulation was to be discussed, they wanted opinions not just from scientists, but from a much larger group of thinkers. The wood-decked pathways around the conference center allowed discursive conversations; walking on the decks or on the sand flats, biologists could trade notes on recombination, cloning, and gene manipulation. In contrast, the central hall—a stone-walled, cathedral-like space ablaze with sepulchral California light—was the epicenter of the conference, where the fiercest debates on gene cloning would soon erupt.

Berg spoke first. He summarized the data and outlined the scope of the problem. In the course of investigating methods to chemically alter DNA, biochemists had recently discovered a relatively facile technique to mix and match genetic information from different organisms. The technology, as Berg put it, was so "ridiculously simple" that even an amateur biologist could produce chimeric genes in a lab. These hybrid DNA molecules—recombinant DNA—could be propagated and expanded (i.e., cloned) in bacteria to generate millions of identical copies. Some of these molecules could be shuttled into mammalian cells. Recognizing the profound potential and risks of this technology, a preliminary meeting had suggested a temporary moratorium on experiments. The Asilomar II meeting had been convened to deliberate on the next steps. Eventually, this second meeting would so far overshadow the first in its influence and scope that it would be called simply the Asilomar Conference—or just Asilomar.

Tensions and tempers flared quickly on the first morning. The main issue was still the self-imposed moratorium: Should scientists be restricted in their experiments with recombinant DNA? Watson was against it. He wanted perfect freedom: let the scientists loose on the science, he urged. Baltimore and Brenner reiterated their plan to create "crippled" gene carriers to ensure safety. Others were deeply divided. The scientific opportunities were enormous, they argued, and a moratorium might paralyze progress. One microbiologist was particularly incensed by the severity of the proposed restrictions: "You *fucked* the plasmid group," he accused the committee. At one point, Berg threatened to sue Watson for failing to adequately acknowledge the nature of the risk of recombinant DNA. Brenner asked a journalist from the *Washington Post* to turn off his recorder during a particularly sensitive session on the risks of gene cloning; "I believe in the inalienable right of all adult scientists to make absolute fools of themselves in private," he said. He was promptly accused of "being a fascist."

The five members of the organizing committee—Berg, Baltimore, Brenner, Richard Roblin, and Maxine Singer, the biochemist—anxiously made rounds of the room, assessing the rising temperature. "Arguments went on and on," one journalist wrote. "Some people got sick of it all and went out to the beach to smoke marijuana." Berg sat in his room, glowering, worried that the conference would end with no conclusions at all.

Nothing had been formalized by the last evening of the conference, until the lawyers took the stage. The five attorneys asked to discuss the legal ramifications of cloning and laid out a grim vision of potential risks: if a single member of a laboratory was infected by a recombinant microbe, and that infection led to even the palest manifestations of a disease, they argued, the laboratory head, the lab, and the institution would be held legally liable. Whole universities would shut down. Labs would be closed indefinitely, their front doors picketed by activists and locked by hazmat men in astronaut suits. The NIH would be flooded with queries; all hell would break loose. The federal government would respond by proposing draconian regulations—not just on recombinant DNA, but on a larger swath of biological research. The result could be restrictions vastly more stringent than any rules that scientists might be willing to impose on themselves.

The lawyers' presentation, held strategically on the last day of Asilomar II, was the turning point for the entire meeting. Berg realized that the meeting should not—*could* not—end without formal recommendations. That evening, Baltimore, Berg, Singer, Brenner, and Roblin stayed up late in their cabana, eating Chinese takeout from paper cartons, scribbling on a blackboard, and drafting a plan for the future. At five thirty in the morning, disheveled and bleary-eyed, they emerged from the beach house smelling of coffee and typewriter ink, with a document in hand. The document began with the recognition of the strange parallel universe of biology that scientists had unwittingly entered with gene cloning. "The new techniques, which permit combination of genetic information from very different organisms, place us in an arena of biology with many unknowns. . . . It is this ignorance that has compelled us to conclude that it would be wise to exercise considerable caution in performing this research."

To mitigate the risks, the document proposed a four-level scheme to rank the biohazard potentials of various genetically altered organisms, with recommended containment facilities for each level (inserting a cancer-causing gene into a human virus, for instance, would merit the highest level of containment, while placing a frog gene into a bacterial cell might merit minimal containment). As Baltimore and Brenner had insisted, it proposed the development of crippled gene-carrying organisms and vectors to further contain them in laboratories. Finally, it urged continuous review of recombination and containment procedures, with the possibility of loosening or tightening restrictions in the near future.

When the meeting opened at eight thirty on the last morning, the five members of the committee worried that the proposal would be rejected. Surprisingly, it was near unanimously accepted.

①

In the aftermath of the Asilomar Conference, several historians of science have tried to grasp the scope of the meeting by seeking an analogous moment in scientific history. There is none. The closest one gets to a similar document, perhaps, is a two-page letter written in August 1939 by Albert Einstein and Leo Szilard to alert President Roosevelt to the alarming possibility of a powerful war weapon in the making. A "new and important source of energy" had been discovered, Einstein wrote, through which "vast amounts of power . . . might be generated." "This new phenomenon would also lead to the construction of bombs, and it is conceivable . . . that extremely powerful bombs of a new type may thus be constructed. A single bomb of this type, carried by boat and exploded in a port, might very well destroy the whole port." The Einstein-Szilard letter had generated an immediate response. Sensing the urgency, Roosevelt had appointed a scientific commission to investigate it. Within a few months, Roosevelt's commission would become the Advisory Committee on Uranium. By 1942, it would morph further into the Manhattan Project and ultimately culminate in the creation of the atomic bomb.

But Asilomar was different: here, scientists were alerting *themselves* to the perils of their own technology and seeking to regulate and constrain their own work. Historically, scientists had rarely sought to become self-regulators. As Alan Waterman, the head of the National Science Foundation, wrote in 1962, "Science, in its pure form, is not interested in where discoveries may lead. . . . Its disciples are interested only in discovering the truth."

But with recombinant DNA, Berg argued, scientists could no longer afford to focus merely on "discovering the truth." The truth was complex and inconvenient, and it required sophisticated assessment. Extraordinary technologies demand extraordinary caution, and political forces could hardly be trusted to assess the perils or the promise of gene cloning (nor, for that matter, had political forces been particularly wise about handling genetic technologies in the past—as the students had pointedly reminded Berg at Erice). In 1973, less than two years before Asilomar, Nixon, fed up with his scientific advisers, had vengefully scrapped the Office of Science and Technology, sending spasms of anxiety through the scientific community. Impulsive, authoritarian, and suspicious of science even at the best of times, the president might impose arbitrary control on scientists' autonomy at any time.

A crucial choice was at stake: scientists could relinquish the control of gene cloning to unpredictable regulators and find their work arbitrarily constrained—or they could become science regulators themselves. How were biologists to confront the risks and uncertainties of recombinant DNA? By using the methods that they knew best: gathering data, sifting evidence, evaluating risks, making decisions under uncertainty—and quarreling relentlessly. "The most important lesson of Asilomar," Berg said, "was to demonstrate that scientists were capable of self-governance." Those accustomed to the "unfettered pursuit of research" would have to learn to fetter themselves.

The second distinctive feature of Asilomar concerned the nature of communications between scientists and the public. The Einstein-Szilard letter had been deliberately shrouded in secrecy; Asilomar, in contrast, sought to air the concerns about gene cloning in the most public forum possible. As Berg put it, "The public's trust was undeniably increased by the fact that more than ten percent of the participants were from the news media. They were free to describe, comment on, and criticize the discussions and conclusions. . . . The deliberations, bickering, bitter accusations, wavering views, and the arrival at a consensus were widely chronicled by the reporters that attended."

A final feature of Asilomar deserves commentary—notably for its absence. While the biological risks of gene cloning were extensively discussed at the meeting, virtually no mention was made of the ethical and moral dimensions of the problem. What would happen once human genes were manipulated in human cells? What if we began to "write" new material into our own genes, and potentially our genomes? The conversation that Berg had started in Sicily was never rejuvenated.

Later, Berg reflected on this lacuna: "Did the organizers and participants of the Asilomar conference deliberately limit the scope of the concerns? . . . Others have been critical of the conference because it did not confront the potential misuse of the recombinant DNA technology or the ethical dilemmas that would arise from applying the technology to genetic screening and . . . gene therapy. It should not be forgotten that these possibilities were still far in the future. . . . In short, the agenda for the three-day meeting had to focus on an assessment of the [biohazard] risks. We accepted that the other issues would be dealt with as they became imminent and estimable." The absence of this discussion was noted by several participants, but it was never addressed during the meeting itself. It is a theme to which we will return.

℗

In the spring of 1993, I traveled to Asilomar with Berg and a group of researchers from Stanford. I was a student in Berg's lab then, and this was the annual retreat for the department. We left Stanford in a caravan of cars and vans, hugging the coast at Santa Cruz and then heading out toward the narrow cormorant neck of the Monterey Peninsula. Kornberg and Berg drove ahead. I was in a rental van driven by a graduate student and accompanied, improbably, by an opera-diva-turned-biochemist who worked on DNA replication and occasionally burst into strains of Puccini.

On the last day of our meeting, I took a walk through the scrub-pine groves with Marianne Dieckmann, Berg's long-term research assistant and collaborator. Dieckmann guided me through an unorthodox tour of Asilomar, pointing out the places where the fiercest mutinies and arguments had broken out. This was an expedition through a landscape of disagreements. Asilomar, she told me, was the most quarrelsome meeting that she had ever attended.

What did these quarrels achieve? I asked. Dieckmann paused, looking toward the sea. The tide had gone out, leaving the beach carved in the shadows of waves. She used her toe to draw a line on the wet sand. More than anything, Asilomar marked a transition, she said. The capacity to manipulate genes represented nothing short of a transformation in genetics. We had learned a new language. We needed to convince ourselves, and everyone else, that we were responsible enough to use it.

It is the impulse of science to try to understand nature, and the impulse of technology to try to manipulate it. Recombinant DNA had pushed genetics from the realm of science into the realm of technology. Genes were not abstractions anymore. They could be liberated from the genomes of organisms where they had been trapped for millennia, shuttled between species, amplified, purified, extended, shortened, altered, remixed, mutated, mixed, matched, cut, pasted, edited; they were infinitely malleable to human intervention. Genes were no longer just the subjects of study, but the instruments of study. There is an illuminated moment in the development of a child when she grasps the recursiveness of language: just as thoughts can be used to generate words, she realizes, words can be used to generate thoughts. Recombinant DNA had made the language of genetics recursive. Biologists had spent decades trying to interrogate the nature of the gene—but now it was the gene that could be used to interrogate biology. We had graduated, in short, from thinking *about* genes, to thinking *in* genes.

Asilomar, then, marked the crossing of these pivotal lines. It was a celebration, an appraisal, an assembly, a confrontation, a warning. It began with a speech and ended with a document. It was the graduation ceremony for the new genetics.

"Clone or Die"

If you know the question, you know half.
— Herb Boyer

Any sufficiently advanced technology is indistinguishable from magic.
— Arthur C. Clarke

Stan Cohen and Herb Boyer had also gone to Asilomar to debate the future of recombinant DNA. They found the conference irritating—even deflating. Boyer could not bear the infighting and the name-calling; he called the scientists "self-serving" and the meeting a "nightmare." Cohen refused to sign the Asilomar agreement (although as a grantee of the NIH, he would eventually have to comply with it).

Back in their own laboratories, they returned to an issue that they had neglected amid the commotion. In May 1974, Cohen's lab had published the "frog prince" experiment—the transfer of a frog gene into a bacterial cell. When asked by a colleague how he had identified the bacteria expressing the frog genes, Cohen had jokingly said that he had kissed the bacteria to check which ones would transform into a prince.

At first, the experiment had been an academic exercise; it had only turned biochemists' heads. (Joshua Lederberg, the Nobel Prize–winning biologist and Cohen's colleague at Stanford, was among the few who wrote, presciently, that the experiment "may completely change the pharmaceutical industry's approach to making biological elements, such as insulin and antibiotics.") But slowly, the media awoke to the potential impact of the study. In May, the *San Francisco Chronicle* ran a story on Cohen, focusing on the possibility that gene-modified bacteria might someday be used as biological "factories" for drugs or chemicals. Soon, articles on gene-cloning technologies had appeared in *Newsweek* and the *New York Times*. Cohen also received a quick baptism on the seamy side of scientific journalism. Having spent an afternoon talking patiently to a newspaper reporter about recombinant DNA and bacterial gene transfer, he awoke the next morning to the hysterical headline: "Man-made Bugs Ravage the Earth."

At Stanford University's patent office, Niels Reimers, a savvy former engineer, read about Cohen and Boyer's work through these news outlets and was intrigued by its potential. Reimers—less patent officer and more talent scout—was active and aggressive: rather than waiting for inventors to bring him inventions, he scoured the scientific literature on his own for possible leads. Reimers approached Boyer and Cohen, urging them to file a joint patent on their work on gene cloning (Stanford and UCSF, their respective institutions, would also be part of that patent). Both Cohen and Boyer were surprised. During their experiments, they had not even broached the idea that recombinant DNA techniques could be "patent-able," or that the technique could carry future commercial value. In the winter of 1974, still skeptical, but willing to humor Reimers, Cohen and Boyer filed a patent for recombinant DNA technology.

News of the gene-cloning patent filtered back to scientists. Kornberg and Berg were furious. Cohen and Boyer's claims "to commercial owner-ship of the techniques for cloning all possible DNAs, in all possible vec-tors, joined in all possible ways, in all possible organisms [is] dubious, presumptuous, and hubristic," Berg wrote. The patent would privatize the products of biological research that had been paid for with public money, they argued. Berg also worried that the recommendations of the Asilo-mar Conference could not be adequately policed and enforced in private companies. To Boyer and Cohen, however, all of this seemed much ado about nothing. Their "patent" on recombinant DNA was no more than a sheaf of paper making its way between legal offices—worth less, perhaps, than the ink that had been used to print it.

In the fall of 1975, with mounds of paperwork still moving through legal channels, Cohen and Boyer parted scientific ways. Their collabora-tion had been immensely productive—together they had published eleven landmark papers over five years—but their interests had begun to drift apart. Cohen became a consultant to a company called Cetus in Califor-nia. Boyer returned to his lab in San Francisco to concentrate on his ex-periments on bacterial gene transfer.

In the winter of 1975, a twenty-eight-year-old venture capitalist, Robert Swanson, called Herb Boyer out of the blue to suggest a meeting. A connoisseur of popular-science magazines and sci-fi films, Swanson had also heard about a new technology called "recombinant DNA." Swanson had an instinct for technology; even though he knew barely any biology, he had sensed that recombinant DNA represented a tectonic shift in thinking about genes and heredity. He had dug up a dog-eared handbook from the Asilomar meeting, made a list of important players working on gene-cloning techniques, and had started working down the list alphabetically. *Berg* came before *Boyer*—but Berg, who had no patience for opportunistic entrepreneurs making cold calls to his lab, turned Swanson down. Swanson swallowed his pride and kept going down the list. B . . . Boyer was next. Would Boyer consider a meeting? Immersed in experiments, Herb Boyer fielded Swanson's phone call distractedly one morning. He offered ten minutes of his time on a Friday afternoon.

Swanson came to see Boyer in January 1976. The lab was located in the grimy innards of the Medical Sciences Building at UCSF. Swanson wore a dark suit and tie. Boyer appeared amid mounds of half-rotting bacterial plates and incubators in jeans and his trademark leather vest. Boyer knew little about Swanson—only that he was a venture capitalist looking to form a company around recombinant DNA. Had Boyer investigated further, he might have discovered that nearly all of Swanson's prior investments in fledgling ventures had failed. Swanson was out of work, living in a rent-shared apartment in San Francisco, driving a broken Datsun, and eating cold-cut sandwiches for lunch and dinner.

The assigned ten minutes grew into a marathon meeting. They walked to a neighborhood bar, talking about recombinant DNA and the future of biology. Swanson proposed starting a company that would use gene-cloning techniques to make medicines. Boyer was fascinated. His own son had been diagnosed with a potential growth disorder, and Boyer had been gripped by the possibility of producing human growth hormone, a protein to treat such growth defects. He knew that he might be able to make growth hormone in his lab by using his own method of stitching genes and inserting them into bacterial cells, but it would be useless: no sane person would inject his or her child with bacterial broth grown in a test tube in a science lab. To make a medical product, Boyer needed to create a new kind of pharmaceutical company—one that would make medicines out of genes.

Three hours and three beers later, Swanson and Boyer had reached a tentative agreement. They would pitch in $500 each to cover legal fees to start such a company. Swanson wrote up a six-page plan. He approached his former employers, the venture firm Kleiner Perkins, for $500,000 in seed money. The firm took a quick look at the proposal and slashed that number fivefold to $100,000. ("This investment is highly speculative," Perkins later wrote apologetically to a California regulator, "but we are in the business of making highly speculative investments.")

Boyer and Swanson had nearly all the ingredients for a new company—except for a product and a name. The first potential product, at least, was obvious from the start: insulin. Despite many attempts to synthesize it using alternative methods, insulin was still being produced from mashed-up cow and pig innards, a pound of hormone from eight thousand pounds of pancreas—a near-medieval method that was inefficient, expensive, and outdated. If Boyer and Swanson could express insulin as a protein via gene manipulation in cells, it would be a landmark achievement for a new company. That left the issue of the name. Boyer rejected Swanson's suggestion of HerBob, which sounded like a hair salon on the Castro. In a flash of inspiration, Boyer suggested a condensation of Genetic Engineering Technology—Gen-en-tech.

①

Insulin: the Garbo of hormones. In 1869, a Berlin medical student, Paul Langerhans, had looked through a microscope at the pancreas, a fragile leaf of tissue tucked under the stomach, and discovered minute islands of distinct-looking cells studded across it. These cellular archipelagoes were later named the *islets of Langerhans*, but their function remained mysterious. Two decades later, two surgeons, Oskar Minkowski and Josef von Mering, had surgically removed the pancreas from a dog to identify the function of the organ. The dog was struck by an implacable thirst and began to urinate on the floor.

Mering and Minkowski were mystified: Why had removing an abdominal organ precipitated this odd syndrome? The clue emerged from a throwaway fact. A few days later, an assistant noted that the lab was buzzing with flies; they were swarming on the pools of dog urine that had now congealed and turned sticky, like treacle.* When Mering and Minkowski tested the urine and the dog's blood, both were overflowing with sugar. The dog had become severely diabetic. Some factor synthesized in the pancreas, they realized, must regulate blood sugar, and its dysfunction must cause diabetes. The sugar-regulating factor was later found to be a hormone, a protein secreted into the blood by those "islet cells" that Langerhans had identified. The hormone was called isletin, and then insulin—literally, "island protein."

The identification of insulin in pancreatic tissue led to a race to purify it—but it took two further decades to isolate the protein from animals. Ultimately, in 1921, Banting and Best extracted a few micrograms of the substance out of dozens of pounds of cow pancreases. Injected into diabetic children, the hormone rapidly restored proper blood sugar levels and stopped their thirst and urination. But the hormone was notoriously difficult to work with: insoluble, heat-labile, temperamental, unstable, mysterious—insular. In 1953, after three more decades, Fred Sanger deduced the amino acid sequence of insulin. The protein, Sanger found, was made of two chains, one larger and one smaller, cross-linked by chemical bonds. U-shaped, like a tiny molecular hand, with clasped fingers and an opposing thumb, the protein was poised to turn the knobs and dials that so potently regulated sugar metabolism in the body.

Boyer's plan for the synthesis of insulin was almost comically simple. He did not have the gene for human insulin at hand—no one did—but he would build it from scratch using DNA chemistry, nucleotide by nucleotide, triplet upon triplet—ATG, CCC, TCC, and so forth, all the way from the first triplet code to the last. He would make one gene for the A chain, and another gene for the B chain. He would insert both the genes in bacteria and trick them into synthesizing the human proteins. He would purify the two protein chains and then stitch them chemically to obtain the U-shaped molecule. It was a child's plan. He would build the most ardently sought molecule in clinical medicine block by block, out of an Erector Set of DNA.

But even Boyer, adventurous as he was, blanched at lunging straight for insulin. He wanted an easier test case, a more pliant peak to scale before attempting the Everest of molecules. He focused on another protein—

---

* Minkowski does not recollect this, but others present in the lab have written about the urine-as-treacle experiment.

somatostatin—also a hormone, but with little commercial potential. Its main advantage was size. Insulin was a daunting fifty-one amino acids in length—twenty-one in one chain and thirty in the other. Somatostatin was its duller, shorter cousin, with just fourteen.

To synthesize the somatostatin gene from scratch, Boyer recruited two chemists from the City of Hope hospital in Los Angeles—Keiichi Itakura and Art Riggs—both veterans of DNA synthesis.* Swanson was bitterly opposed to the whole plan. Somatostatin, he feared, would turn into a distraction; he wanted Boyer to move to insulin directly. Genentech was living in borrowed space on borrowed money. Scratched even a millimeter below its surface, the "pharma company" was, in truth, a rented cubicle in an office space in San Francisco with an offshoot in a microbiology lab at UCSF, which, in turn, was about to subcontract two chemists at yet another lab to make genes—a pharmaceutical Ponzi scheme. Still, Boyer convinced Swanson to give somatostatin a chance. They hired an attorney, Tom Kiley, to negotiate the agreements among UCSF, Genentech, and the City of Hope. Kiley had never heard the term *molecular biology*, but felt confident because of his track record of representing unusual cases; before Genentech, his most famous former client had been Miss Nude America.

Time too felt borrowed at Genentech. Boyer and Swanson knew that two reigning wizards of genetics had also entered the race to make insulin. At Harvard, Walter Gilbert, the DNA chemist who would share the Nobel Prize with Berg and Sanger, was leading a formidable team of scientists to synthesize insulin using gene cloning. And at UCSF, in Boyer's own backyard, another team was racing toward the gene cloning. "I think it was on our minds most of the time . . . most days," one of Boyer's collaborators recalled. "I thought about it all the time: Are we going to hear an announcement that Gilbert has been successful?"

By the summer of 1977, working frantically under Boyer's anxious eye, Riggs and Itakura had assembled all the reagents for the synthesis of somatostatin. The gene fragments had been created and inserted into a bacterial plasmid. The bacteria had been transformed, grown, and prepped for the production of the protein. In June, Boyer and Swanson flew to Los Angeles to witness the final act. The team gathered in Riggs's lab in the

---

* They later added other collaborators, including Richard Scheller, from Caltech. Boyer put two researchers, Herbert Heyneker and Francisco Bolivar, on the project. The City of Hope added another DNA chemist, Roberto Crea.

morning. They leaned over to watch the molecular detectors check for the appearance of somatostatin in the bacteria. The counters blinked on, then off. Silence. Not even the faintest blip of a functional protein.

Swanson was devastated. The next morning, he developed acute indigestion and was sent to the emergency room. The scientists, meanwhile, recovered over coffee and doughnuts, poring through the experimental plan, troubleshooting. Boyer, who had worked with bacteria for decades, knew that microbes often digest their own proteins. Perhaps somatostatin had been destroyed by the bacteria—a microbe's last stand against being co-opted by human geneticists. The solution, he surmised, would be to add another trick to the bag of tricks: they would hook the somatostatin gene to another bacterial gene to make a conjoined protein, then cleave off the somatostatin after. It was a genetic bait and switch: the bacteria would think they were making a bacterial protein, but would end up (secretly) secreting a human one.

It took another three months to assemble the decoy gene, with somatostatin now Trojan-horsed within another bacterial gene. In August 1977, the team reassembled at Riggs's lab for the second time. Swanson nervously watched the monitors flicker on, and momentarily turned his face away. The detectors for the protein crackled again in the background. As Itakura recalled, "We have about ten, maybe fifteen samples. Then we look at the printout of the radioimmunoassay, and the printout show[s] clearly that the gene is expressed." He turned to Swanson. "Somatostatin is there."

℗

Genentech's scientists could barely stop to celebrate the success of the somatostatin experiment. One evening, one new human protein; by the next morning, the scientists had regrouped and made plans to attack insulin. The competition was fierce, and rumors abounded: Gilbert's team had apparently cloned the native human gene out of human cells and were readying to make the protein in buckets. Or the UCSF competitors had synthesized a few micrograms of protein and were planning to inject the human hormone into patients. Perhaps somatostatin *had* been a distraction. Swanson and Boyer suspected ruefully that they had taken a wrong turn and been left behind in the insulin race. Dyspeptic even during the best of times, Swanson edged toward another bout of anxiety and indigestion.

Ironically, it was Asilomar—the very meeting that Boyer had so vociferously disparaged—that came to their rescue. Like most university laboratories with federal funding, Gilbert's lab at Harvard was bound by the Asilomar restrictions on recombinant DNA. The restrictions were especially severe because Gilbert was trying to isolate the "natural" human gene and clone it into bacterial cells. In contrast, Riggs and Itakura, following the lead with somatostatin, had decided to use a chemically synthesized version of the insulin gene, building it up nucleotide by nucleotide from scratch. A synthetic gene—DNA created as a naked chemical—fell into the gray zone of Asilomar's language and was relatively exempt. Genentech, as a privately funded company, was also relatively exempt from the federal guidelines.* The combination of factors proved to be a crucial advantage for the company. As one worker recalled, "Gilbert was, as he had for many days past, trudging through an airlock, dipping his shoes in formaldehyde on his way into the chamber in which he was obliged to conduct his experiments. Out at Genentech, we were simply synthesizing DNA and throwing it into bacteria, none of which even required compliance with the NIH guidelines." In the world of post-Asilomar genetics, "being natural" had turned out to be a liability.

℗

Genentech's "office"—the glorified booth in San Francisco—was no longer adequate. Swanson began scouring the city for lab space for his nascent company. In the spring of 1978, having searched up and down the Bay Area, he found a suitable site. Stretched across a tawny, sun-scorched flank of hillside a few miles south of San Francisco, the place was called Industrial City, although it was hardly industrial and barely a city. Genentech's lab was ten thousand square feet of a raw warehouse on 460 Point San Bruno Boulevard, set amid storage silos, dump sites, and airport-freight hangars. The back half of the warehouse housed a storage facility

---

* Genentech's strategy for the synthesis of insulin was also critical to its relative exemption from Asilomar's protocols. In the human pancreas, insulin is normally synthesized as a single contiguous protein and then cut into two pieces, leaving just a narrow cross-linkage. Genentech, in contrast, had chosen to synthesize the two chains of insulin, A and B, as separate, individual proteins and link them together afterward. Since the two separate chains used by Genentech were not "natural" genes, the synthesis did not fall under the federal moratorium that restricted the creation of recombinant DNA with "natural" genes.

for a distributor of porn videos. "You'd go through the back of Genentech's door and there would be all these movies on shelves," one early recruit wrote. Boyer hired a few additional scientists—some barely out of graduate school—and began to install equipment. Walls were constructed to divide the vast space. A makeshift lab was created by slinging black tarp across part of the roof. The first "fermenter" to grow gallons of microbial sludge—an upscale beer vat—arrived that year. David Goeddel, the company's third employee, walked around the warehouse in sneakers and a black T-shirt that read CLONE OR DIE.

Yet no human insulin was in sight. In Boston, Swanson knew, Gilbert had upped his war effort—literally. Fed up with the constraints on recombinant DNA at Harvard (on the streets of Cambridge, young protesters were carrying placards against gene cloning), Gilbert had gained access to a high-security biological-warfare facility in England and dispatched a team of his best scientists there. The conditions in the military facility were absurdly stringent. "You totally change your clothes, shower in, shower out, have gas masks available so that if the alarm goes off you can sterilize the entire laboratory," Gilbert recalled. The UCSF team, in turn, sent a student to a pharmaceutical lab in Strasbourg, France, hoping to create insulin at the well-secured French facility.

Gilbert's group oscillated at the brink of success. In the summer of 1978, Boyer learned that Gilbert's team was about to announce the successful isolation of the human insulin gene. Swanson braced himself for another breakdown—his third. To his immense relief, the gene that Gilbert had cloned was not human but *rat* insulin—a contaminant that had somehow tainted the carefully sterilized cloning equipment. Cloning had made it easy to cross the barriers between species—but that same breach meant that a gene from one species could contaminate another in a biochemical reaction.

In the narrow cleft of time between Gilbert's move to England and the mistaken cloning of rat insulin, Genentech forged ahead. It was an inverted fable: an academic Goliath versus a pharmaceutical David, one lumbering, powerful, handicapped by size, the other nimble, quick, adept at dancing around rules. By May 1978, the Genentech team had synthesized the two chains of insulin in bacteria. By July, the scientists had purified the proteins out of the bacteria debris. In early August, they snipped off the attached bacterial proteins and isolated the two individual chains. Late at night on August 21, 1978, Goeddel joined the protein chains together in a test tube to create the first molecules of recombinant insulin.

Ⓘ

In September 1978, two weeks after Goeddel had created insulin in a test tube, Genentech applied for a patent for insulin. Right at the onset, the company faced a series of unprecedented legal challenges. Since 1952, the United States Patent Act had specified that patents could be issued on four distinct categories of inventions: methods, machines, manufactured materials, and compositions of matter—the "four M's," as lawyers liked to call the categories. But how could insulin be pigeonholed into that list? It was a "manufactured material," but virtually every human body could evidently manufacture it without Genentech's ministrations. It was a "composition of matter," but also, indisputably, a natural product. Why was patenting insulin, the protein or its gene, different from patenting any other part of the human body—say, the nose or cholesterol?

Genentech's approach to this problem was both ingenious and counterintuitive. Rather than patenting insulin as "matter" or "manufacture," it concentrated its efforts, boldly, on a variation of "method." Its application claimed a patent for a "DNA vehicle" to carry a gene into a bacterial cell, and thereby produce a recombinant protein in a microorganism. The claim was so novel—no one had ever produced a recombinant human protein in a cell for medicinal use—that the audacity paid off. On October 26, 1982, the US Patent and Trademark Office (USPTO) issued a patent to Genentech to use recombinant DNA to produce a protein such as insulin or somatostatin in a microbial organism. As one observer wrote: "effectively, the patent claimed, as an invention, [all] genetically modified microorganisms." The Genentech patent would soon become one of the most lucrative, and most hotly disputed, patents in the history of technology.

Ⓘ

Insulin was a major milestone for the biotechnology industry, and a blockbuster drug for Genentech. But it was not, notably, the medicine that would catapult gene-cloning technology to the forefront of public imagination.

In April 1982, a ballet dancer in San Francisco, Ken Horne, visited a dermatologist, complaining of an inexplicable cluster of symptoms. Horne had felt weak for months and developed a cough. He had bouts of intractable diarrhea, and weight loss had hollowed his cheeks and made his neck muscles stand out like leather straps. His lymph nodes had swollen. And now—he pulled his shirt up to demonstrate—reticulated bumps were emerging on his skin, purple-blue of all colors, like hives in a macabre cartoon film.

Horne's case was not isolated. Between May and August 1982, as the coasts sweltered in a heat wave, similarly bizarre medical cases were reported in San Francisco, New York, and Los Angeles. At the CDC in Atlanta, a technician was asked to fill nine requests for pentamidine, an unusual antibiotic reserved to treat *Pneumocystis* pneumonia. These requisitions made no sense: PCP was a rare infection that typically afflicted cancer patients with severely depleted immune systems. But these applications were for young men, previously in excellent health, whose immune systems had suddenly been pitched into inexplicable, catastrophic collapse.

Horne, meanwhile, was diagnosed with Kaposi's sarcoma—an indolent skin tumor found among old men in the Mediterranean. But Horne's case, and the other nine cases reported in the next four months, bore little resemblance to the slow-growing tumors previously described as Kaposi's in the scientific literature. These were fulminant, aggressive cancers that spread rapidly through the skin and into the lungs, and they seemed to have a predilection for gay men living in New York and San Francisco. Horne's case mystified medical specialists, for now, as if to intersect puzzle upon puzzle, he developed *Pneumocystis* pneumonia and meningitis as well. By late August, an epidemiological disaster was clearly appearing out of thin air. Noting the preponderance of gay men afflicted, doctors began to call it GRID—gay-related immune deficiency. Many newspapers accusingly termed it the "gay plague."

By September, the fallacy of that name had become evident: symptoms of immunological collapse, including *Pneumocystis* pneumonia and strange variants of meningitis, had now begun to sprout up among three patients with hemophilia A. Hemophilia, recall, was the bleeding illness of the English royals—caused by a single mutation in the gene for a crucial clotting factor in blood, called factor VIII. For centuries, patients with hemophilia had lived in constant fear of a bleeding crisis: a nick in the skin could snowball into disaster. By the mid-1970s, though, hemophiliacs were being treated with injections of concentrated factor VIII. Distilled out of thousands of liters of human blood, a single dose of the clotting factor was equivalent to a hundred blood transfusions. A typical patient with hemophilia was thus exposed to the condensed essence of blood from thousands of donors. The emergence of the mysterious immunological collapse among patients with multiple blood transfusions pinpointed the cause of the illness to a blood-borne factor that had contaminated the supply of factor VIII—possibly a novel virus. The syndrome was renamed acquired immunodeficiency syndrome—AIDS.

Ⅰ

In the spring of 1983, against the backdrop of the early AIDS cases, Dave Goeddel at Genentech began to focus on cloning the factor VIII gene. As with insulin, the logic behind the cloning effort was immediately evident: rather than purifying the missing clotting factor out of liters of human blood, why not create the protein artificially, using gene cloning? If factor VIII could be produced through gene-cloning methods, it would be virtually free of any human contaminants, thereby rendering it inherently safer than any blood-derived protein. Waves of infections and deaths might be prevented among hemophiliacs. It was Goeddel's old T-shirt slogan brought to life—"clone or die."

Goeddel and Boyer were not the only geneticists musing about cloning factor VIII. As with the cloning of insulin, the effort had evolved into a race, although with different competitors. In Cambridge, Massachusetts, a team of researchers from Harvard, led by Tom Maniatis and Mark Ptashne, were also racing toward the factor VIII gene, having formed their own company, named the Genetics Institute—colloquially called GI. The factor VIII project, both teams knew, would challenge the outer limits of gene-cloning technology. Somatostatin had 14 amino acids; insulin had 51. Factor VIII had 2,350. The leap in size between somatostatin and factor VIII was 160-fold—almost equivalent to the jump in distance between Wilbur Wright's first airborne circle at Kitty Hawk and Lindbergh's journey across the Atlantic.

The leap in size was not just a quantitative barrier; to succeed, the gene cloners would need to use new cloning technologies. Both the somatostatin and insulin genes had been created from scratch by stitching together bases of DNA—A added chemically to the G and the C and so forth. But the factor VIII gene was far too large to be created using DNA chemistry. To isolate the factor VIII gene, both Genentech and GI would need to pull the native gene out of human cells, spooling it out as if extracting a worm from the soil.

①

But the "worm" would not come out easily, or intact, from the genome. Most genes in the human genome are, recall, interrupted by stretches of DNA called introns, which are like garbled stuffers placed in between parts of a message. Rather than the word *genome*, the actual gene reads *gen . . . . om . . . . . . e.* The introns in human genes are often enormous, stretching across vast lengths of DNA, making it virtually impossible to clone a gene directly (the intron-containing gene is too long to fit into a bacterial plasmid).

Maniatis found an ingenious solution: he had pioneered the technology to build genes out of RNA templates using reverse transcriptase, the enzyme that could build DNA from RNA. The use of reverse transcriptase made gene cloning vastly more efficient. Reverse transcriptase made it possible to clone a gene *after* the intervening stuffer sequences had been snipped off by the cell's splicing apparatus. The cell would do all the work; even long, unwieldy, intron-interrupted genes such as factor VIII would be processed by the cell's gene-splicing apparatus and could thus be cloned from cells.

By the late summer of 1983, using all the available technologies, both teams had managed to clone the factor VIII gene. It was now a furious race to the finish. In December 1983, still running shoulder to shoulder, both groups announced that they had assembled the entire sequence and inserted the gene into a plasmid. The plasmid was then introduced into hamster-derived ovary cells known for their ability to synthesize vast quantities of proteins. In January 1984, the first cargoes of factor VIII began to appear in the tissue-culture fluid. In April, exactly two years after the first AIDS clusters had been reported in America, both Genentech and GI announced that they had purified recombinant factor VIII in test tubes—a blood-clotting factor untainted by human blood.

In March 1987, Gilbert White, a hematologist, conducted the first clinical trial of the hamster-cell-derived recombinant factor VIII at the Center for Thrombosis in North Carolina. The first patient to be treated was

G.M., a forty-three-year-old man with hemophilia. As the initial drops of intravenous liquid dripped into his veins, White hovered anxiously around G.M.'s bed, trying to anticipate reactions to the drug. A few minutes into the transfusion, G.M. stopped speaking. His eyes were closed; his chin rested on his chest. "Talk to me," White urged. There was no response. White was about to issue a medical alert when G.M. turned around, made the sound of a hamster, and burst into laughter.

℗

News of G.M.'s successful treatment spread through a desperate community of hemophiliacs. AIDS among hemophiliacs had been a cataclysm within a cataclysm. Unlike gay men, who had quickly organized a concerted, defiant response to the epidemic—boycotting bathhouses and clubs, advocating safe sex, and campaigning for condoms—hemophiliacs had watched the shadow of the illness advance with numb horror: they could hardly boycott blood. Between April 1984 and March 1985, until the first test for virally contaminated blood was released by the FDA, every hemophiliac patient admitted to a hospital faced the terrifying choice of bleeding to death or becoming infected with a fatal virus. The infection rate among hemophiliacs during this period was staggering: among those with the severe variant of the disease, 90 percent would acquire HIV through contaminated blood.

Recombinant factor VIII arrived too late to save the lives of most of these men and women. Nearly all the HIV-infected hemophiliacs from the initial cohort would die of the complications of AIDS. Even so, the production of factor VIII from its gene broke important conceptual ground—although it was tinged with peculiar irony. The fears of Asilomar had been perfectly inverted. In the end, a "natural" pathogen had unleashed havoc on human populations. And the strange artifice of gene cloning—inserting human genes into bacteria and then manufacturing proteins in hamster cells—had emerged as potentially the safest way to produce a medical product for human use.

℗

It is tempting to write the history of technology through products: the wheel; the microscope; the airplane; the Internet. But it is more illuminating to write the history of technology through transitions: linear motion to circular motion; visual space to subvisual space; motion on land to motion in air; physical connectivity to virtual connectivity.

The production of proteins from recombinant DNA represented one such crucial transition in the history of medical technology. To understand the impact of this transition—from gene to medicine—we need to understand the history of medicinal chemicals. Stripped to its bare essence, a medicinal chemical—a drug—is nothing more than a molecule that enables a therapeutic change in human physiology. Medicines can be simple chemicals—water, in the right context and at the right dose, is a potent drug—or they can be complex, multidimensional, many-faced molecules. They are also astoundingly rare. Although there are seemingly thousands of drugs in human usage—aspirin alone comes in dozens of variants—the number of molecular *reactions* targeted by these drugs is a minuscule fraction of the total number of reactions. Of the several million variants of biological molecules in the human body (enzymes, receptors, hormones—and so forth), only about 250—0.025 percent—are therapeutically modulated by our current pharmacopeia. If human physiology is visualized as a vast global telephone network with interacting nodes and networks, then our current medicinal chemistry touches only a fraction of a fraction of its complexity; medicinal chemistry is a pole operator in Wichita tinkering with a few lines in the network's corner.

The paucity of medicines has one principal reason: specificity. Nearly every drug works by binding to its target and enabling or disabling it—turning molecular switches on or off. To be useful, a drug must bind to its switches—but to only a selected set of switches; an indiscriminate drug is no different from a poison. Most molecules can barely achieve this level of discrimination—but proteins have been designed explicitly for this purpose. Proteins, recall, are the hubs of the biological world. They are the enablers and the disablers, the machinators, the regulators, the gatekeepers, the operators, of cellular reactions. They *are* the switches that most drugs seek to turn on and off.

Proteins are thus poised to be some of the most potent and most discriminating medicines in the pharmacological world. But to make a protein, one needs its gene—and here recombinant DNA technology provided the crucial missing stepping-stone. The cloning of human genes allowed scientists to manufacture proteins—and the synthesis of proteins opened the possibility of targeting the millions of biochemical reactions in the human body. Proteins made it possible for chemists to intervene on previously impenetrable aspects of our physiology. The use of recombinant DNA to produce proteins thus marked a transition not just between one gene and one medicine, but between genes and a novel universe of drugs.

℗

On October 14, 1980, Genentech sold 1 million of its shares to the public, provocatively listing itself at the stock exchange under the trading symbol GENE. This initial sale would rank among the most dazzling debuts of any technology company in Wall Street history: within a few hours, the company had generated $35 million in capital. By then, the pharmaceutical giant Eli Lilly had acquired the license to produce and sell recombinant insulin—called *Humulin*, to distinguish it from cow and pig insulin—and was rapidly expanding its market. Sales rose from $8 million in 1983 to $90 million in 1996 to $700 million in 1998. Swanson—"a short, chunky chipmunk-cheeked thirty-six-year-old," as *Esquire* magazine described him—was now a millionaire several times over, as was Boyer. A graduate student who had held on to a few throwaway shares for helping to clone the somatostatin gene over the summer of 1977 woke up one morning and found himself a newly minted multimillionaire.

In 1982, Genentech began to produce human growth hormone—HGH—used to treat certain variants of dwarfism. In 1986, biologists at the company cloned alpha interferon, a potent immunological protein used to treat blood cancers. In 1987, Genentech made recombinant TPA, a blood thinner to dissolve the clots that occur during a stroke or a heart attack. In 1990, it launched efforts to create vaccines out of recombinant genes, beginning with a vaccine against hepatitis B. In December 1990, Roche Pharmaceuticals acquired a majority stake in Genentech for $2.1 billion. Swanson stepped down as the chief executive; Boyer left his position as vice president in 1991.

In the summer of 2001, Genentech launched its physical expansion into the largest biotech research complex in the world—a multiacre stretch of glass-wrapped buildings, rolling greens, and Frisbee-playing research students that is virtually indistinguishable from any university campus. At the center of the vast complex sits a modest bronze statue of a man in a suit gesticulating over a table to a scientist in flared jeans and a leather vest. The man is leaning forward. The geneticist looks puzzled and is gazing distantly over the man's shoulder.

Swanson, unfortunately, was not present for the formal unveiling of the statue commemorating his first meeting with Boyer. In 1999, at age fifty-two, he was diagnosed with glioblastoma multiforme, a brain tumor. He died on December 6, 1999, at home in Hillsborough, a few miles from Genentech's campus.

PART FOUR

"THE PROPER STUDY OF MANKIND IS MAN"

Human Genetics

(1970–2005)

①

①

Know then thyself, presume not God to scan;
The proper study of mankind is man.

—Alexander Pope, *Essay on Man*

How beauteous mankind is! O brave new world,
That has such people in't!

—William Shakespeare,
The Tempest, act 5, scene 1

①

The Miseries of My Father

ALBANY: *How have you known the miseries of your father?*
EDGAR: *By nursing them, my lord.*
—William Shakespeare, *King Lear*, act 5, scene 3

In the spring of 2014, my father had a fall. He was sitting on his favorite rocking chair—a hideous, off-kilter contraption that he had commissioned from a local carpenter—when he tipped over the back and fell off (the carpenter had devised a mechanism to make the chair rock, but had forgotten to add a mechanism to stop the chair from rocking over). My mother found him facedown on the veranda, his hand tucked under his body unnaturally, like a snapped wing. His right shoulder was bathed in blood. She could not pull his shirt over his head, so she took a pair of scissors to it, while he screamed in pain from his wound, and in deeper agony at having a perfectly intact piece of clothing ripped to shreds before his eyes. "You could have tried to save it," he later groused as they drove to the emergency room. It was an ancient quarrel: *his* mother, who had never had five shirts for all five boys at a time, would have found a way to rescue it. You could take a man out of Partition, but you could not take Partition out of the man.

He had gashed the skin on his forehead and broken his right shoulder. He was—like me—a terrible patient: impulsive, suspicious, reckless, anxious about confinement, and deluded about his recovery. I flew to India to see him. By the time I arrived home from the airport, it was late at night. He was lying in bed, looking vacantly at the ceiling. He seemed to have aged suddenly. I asked him if he knew what day it was.

"April twenty-fourth," he said correctly.

"And the year?"

"Nineteen forty-six," he said, then corrected himself, groping for the memory: "Two thousand six?"

It was a fugitive memory. I told him it was 2014. Nineteen forty-six, I noted privately, had been another season of catastrophe—the year that Rajesh had died.

Over the next days, my mother nursed him back to health. His lucidity ebbed back and some of his long-term memory returned, although his short-term memory was still significantly impaired. We determined that the rocking-chair accident was not as simple as it had sounded. He had not tipped backward but had attempted to get up from the chair, then lost his balance and shot forward, unable to catch himself. I asked him to walk across the room and noticed that his gait had an ever-so-slight shuffle. There was something robotic and constrained in his movements, as if his feet were made of iron, and the floor had turned magnetic. "Turn around quickly," I said, and he almost fell forward again.

Late that night, another indignity occurred: he wet his bed. I found him in the bathroom, bewildered and ashamed, clutching his underwear. In the Bible, Ham's descendants are cursed because he stumbles on his father, Noah, drunken and naked, his genitals exposed, lying in a field in the half-light of dawn. In the modern version of that story, you encounter your father, demented and naked, in the half-light of the guest bathroom—and see the curse of your own future, illuminated.

The urinary incontinence, I learned, had been occurring for a while. It had begun with the feeling of urgency—the inability to hold back once the bladder was half-full—and progressed to bed-wetting. He had told his doctors about it, and they had waved it off, vaguely attributing it to a swollen prostate. It's all old age, they had told him. He was eighty-two. Old men fall. They lose their memory. They wet their beds.

The unifying diagnosis came to us in a flash of shame the next week when he had an MRI of his brain. The ventricles of the brain, which bathe the brain in fluid, were swollen and dilated, and the tissue of the brain had been pushed out to the edges. The condition is called normal pressure hydrocephalus (NPH). It is thought to result from the abnormal flow of fluid around the brain, causing a buildup in the ventricles—somewhat akin to the "hypertension of the brain," the neurologist explained. NPH is characterized by an inexplicable classic triad of symptoms—gait instability, urinary incontinence, dementia. My father had not fallen by accident. He had fallen ill.

Over the next few months, I learned everything I could about the condition. The illness has no known cause. It runs in families. One variant of

the illness is genetically linked to the X chromosome, with a disproportionate predominance for men. In some families, it occurs in men as young
as twenty or thirty. In other families, only the elderly are affected. In some,
the pattern of inheritance is strong. In others, only occasional members
have the illness. The youngest documented familial cases are in children
four or five years old. The oldest patients are in their seventies and eighties.

It is, in short, quite likely to be a genetic disease—although not "genetic"
in the same sense as sickle-cell anemia or hemophilia. No single gene governs the susceptibility to this bizarre illness. Multiple genes, spread across
multiple chromosomes, specify the formation of the aqueducts of the brain
during development—just as multiple genes, spread across multiple chromosomes, specify the formation of the wing in a fruit fly. Some of these
genes, I learned, govern the anatomical configurations of the ducts and
vessels of the ventricles (as an analogue, consider how "pattern-formation"
genes can specify organs and structures in flies). Others encode the molecular channels that transmit fluids between the compartments. Yet other
genes encode proteins that regulate the absorption of fluids from the brain
into the blood, or vice versa. And since the brain and its ducts grow in the
fixed cavity of the skull, genes that determine the size and shape of the
skull also indirectly affect the proportions of the channels and the ducts.

Variations in any of these genes may alter the physiology of the aqueducts and ventricles, changing the manner in which fluid moves through the
channels. Environmental influences, such as aging or cerebral trauma, interpose further layers of complexity. There is no one-to-one mapping of one
gene and one illness. Even if you inherit the entire set of genes that causes
NPH in one person, you may still need an accident or an environmental
trigger to "release" it (in my father's case, the trigger was most likely his age).
If you inherit a particular combination of genes—say, those that specify a
particular rate of fluid absorption with those that specify a particular size of
the aqueducts—you might have an increased risk of succumbing to the illness. It is a Delphic boat of a disease—determined not by one gene, but by
the relationship between genes, and between genes and the environment.

"How does an organism transmit the information needed to create
form and function to its embryo?" Aristotle had asked. The answer to
that question, viewed through model organisms such as peas, fruit flies,
and bread molds, had launched the discipline of modern genetics. It had
resulted, ultimately, in that monumentally influential diagram that forms
the basis of our understanding of information flow in living systems:

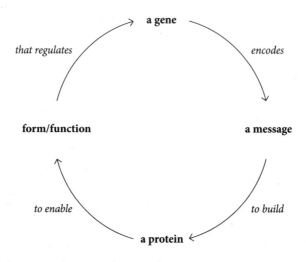

But my father's illness offers yet another lens by which we might view how hereditary information influences the form, function, and fate of an organism. Was my father's fall the consequence of his genes? Yes and no. His genes created a propensity for an outcome, rather than the outcome itself. Was it a product of his environment? Yes and no. It was the chair, after all, that had done it—but he had sat on that same chair, without event, for the good part of a decade before an illness had tipped him (literally) over an edge. Was it chance? Yes: Who knew that certain pieces of furniture, moved at certain angles, are designed to jettison you forward? Was it an accident? Yes, but his physical instability virtually guaranteed a fall.

The challenge of genetics, as it moved from simple organisms to the *human* organism, was to confront new ways to think about the nature of heredity, information flow, function, and fate. How do genes intersect with environments to cause normalcy versus disease? For that matter, what *is* normalcy versus disease? How do the variations in genes cause variations in human form and function? How do multiple genes influence a single outcome? How can there be so much uniformity among humans, yet such diversity? How can the variants in genes sustain a common physiology, yet also produce unique pathologies?

The Birth of a Clinic

I start with the premise that all human disease is genetic.

—Paul Berg

In 1962, a few months after the DNA "triplet code" had been deciphered by Nirenberg and his colleagues in Bethesda, the *New York Times* published an article about the explosive future of human genetics. Now that the code had been "cracked," the *Times* projected, human genes would become malleable to intervention. "It is safe to say that some of the biological 'bombs' that are likely to explode before long as a result of [breaking the genetic code] will rival even the atomic variety in their meaning for man. Some of them might be: determining the basis of thought . . . developing remedies for afflictions that today are incurable, such as cancer and many of the tragic inherited disorders."

Skeptics might, however, have been forgiven for their lack of enthusiasm; the biological "bomb" of human genetics had, thus far, burst forth with a rather underwhelming whimper. The astonishing growth spurt of molecular genetics between 1943 and 1962—from Avery's experiment to the solution of the structure of DNA, and the mechanisms of gene regulation and repair—had produced a progressively detailed mechanistic vision of the gene. Yet the gene had barely touched the human world. On one hand, the Nazi eugenicists had so definitively scorched the earth of human genetics that the discipline had been leached of scientific legitimacy and rigor. On the other, simpler model systems—bacteria, flies, worms—had proved to be vastly more tractable to experimental studies than humans. When Thomas Morgan traveled to Stockholm to collect the Nobel Prize for his contributions to genetics in 1934, he was pointedly dismissive about the medical relevance of his work. "The most important contribution to medicine that genetics has made is, in my opinion, intellectual," Morgan wrote. The word *intellectual* was not meant as a compliment, but as an insult. Genetics, Morgan noted, was unlikely to have even a marginal impact on human health in the near future. The notion that a doctor "may then want to call in his genetic friends for consultation," as Morgan put it, seemed like a silly, far-fetched fantasy.

Yet the entry, or rather reentry, of genetics into the human world *was* the product of medical necessity. In 1947, Victor McKusick, a young internist at Johns Hopkins University in Baltimore, saw a teenage patient with spots on his lips and tongue and multiple internal polyps. McKusick was intrigued by the symptoms. Other members of the family were also affected by similar symptoms, and familial cases with similar features had been published in the literature. McKusick described the case in the *New England Journal of Medicine*, arguing that the cluster of seemingly diffuse symptoms—tongue spots, polyps, bowel obstruction, and cancer—were all the product of a mutation in a single gene.

McKusick's case—later classified as Peutz-Jeghers syndrome, after the first clinicians who described it—launched his lifelong interest in the study of the links between genetics and human diseases. He began by studying human diseases in which the influence of genes was the simplest and strongest—where one gene was known to cause one disease. The best-established examples of such illnesses in humans, though few, were unforgettable: hemophilia among the English royals and sickle-cell anemia in African and Caribbean families. By digging through old papers from the medical libraries at Hopkins, McKusick discovered that a London doctor working in the early 1900s had reported the first example of a human disease apparently caused by a single genetic mutation.

In 1899, Archibald Garrod, an English pathologist, had described a bizarre illness that ran in families and became manifest within days of childbirth. Garrod had first observed it in a child at the Sick Hospital in London. Several hours after the boy had been born, his diaper linens had turned black with a peculiar stain of urine. By meticulously tracking down all such afflicted patients and their relatives, Garrod discovered that the disease ran in families and persisted through adulthood. In adults, sweat darkened spontaneously, sending rivulets of deep brown stains under the arms of shirts. Earwax, even, turned red as it touched the air, as if it had rusted on contact.

Garrod guessed that some inherited factor had been altered in these patients. The boy with the dark urine, Garrod reasoned, must have been born with an alteration in a unit of inheritance that had changed some metabolic function in cells, resulting in a difference in the composition of urine. "The phenomena of obesity and the various tints of hair, skin and eyes" can all be explained by variations in units of inheritance causing "chemical diversities" in human bodies, Garrod wrote. The prescience was remarkable. Even as the concept of the "gene" was being rediscovered by Bateson in England (and nearly a decade before the word *gene* had been coined), Garrod had conceptually visualized a human gene and explained human variation as "chemical diversities" encoded by units of inheritance. Genes make us human, Garrod had reasoned. And mutations make us different.

Inspired by Garrod's work, McKusick launched a systematic effort to create a catalog of genetic diseases in humans—an "encyclopedia of phenotypes, genetic traits and disorders." An exotic cosmos opened up before him; the range of human diseases governed by individual genes was vaster and stranger than he had expected. In Marfan syndrome, originally described by a French pediatrician in the 1890s, a gene controlling the structural integrity of the skeleton and blood vessels was mutated. Patients grew unusually tall, with elongated arms and fingers, and had a propensity to die of sudden ruptures of the aorta or the heart's valves (for decades, some medical historians have asserted that Abraham Lincoln had an undiagnosed variant of the syndrome). Other families were afflicted by osteogenesis imperfecta, a disease caused by a mutation in a gene for collagen, a protein that forms and strengthens bone. Children with this disease were born with brittle bones that, like dry plaster, could crumble at the slightest provocation; they might fracture their legs spontaneously or awake one morning with dozens of broken ribs (often mistaken for child abuse, the cases were brought to medical attention after police investigations). In 1957, McKusick founded the Moore Clinic at Johns Hopkins. Named after Joseph Earle Moore, the Baltimore physician who had spent his life working on chronic illnesses, the clinic would focus on hereditary disorders.

McKusick turned into a walking repository of knowledge of genetic syndromes. There were patients, unable to process chloride, who were afflicted by intractable diarrhea and malnourishment. There were men prone to heart attacks at twenty; families with schizophrenia, depression, or aggression; children born with webbed necks or extra fingers or the permanent odor of fish. By the mid-1980s, McKusick and his students had cataloged 2,239 genes linked with diseases in humans, and

3,700 diseases linked to single genetic mutations. By the twelfth edition of his book, published in 1998, McKusick had discovered an astounding 12,000 gene variants linked to traits and disorders, some mild and some life threatening.

Emboldened by their taxonomy of single-gene—"monogenic"—diseases, McKusick and his students ventured into diseases caused by the convergent influence of multiple genes—"polygenic" syndromes. Polygenic diseases, they found, came in two forms. Some were caused by the presence of whole extra chromosomes. In Down syndrome, first described in the 1860s, children are born with an extra copy of chromosome twenty-one, which has three-hundred-odd genes strung on it.* Multiple organs are affected by the extra copy of the chromosome. Men and women with the syndrome are born with flattened nasal bridges, wide faces, small chins, and altered folds in the eyes. They have cognitive deficits, accelerated heart disease, hearing loss, infertility, and an increased risk for blood cancers; many children die in infancy or childhood, and only a few survive to late adulthood. Most notably, perhaps, children with Down syndrome have an extraordinary sweetness of temperament, as if in inheriting an extra chromosome they had acquired a concomitant loss of cruelty and malice (if there is any doubt that genotypes can influence temperament or personality, then a single encounter with a Down child can lay that idea to rest).

The final category of genetic diseases that McKusick characterized was the most complex—polygenic illnesses caused by multiple genes scattered diffusely throughout the genome. Unlike the first two categories, populated by rare and strange syndromes, these were familiar, pervasive, highly prevalent chronic illnesses—diabetes, coronary artery disease, hypertension, schizophrenia, depression, infertility, obesity.

These illnesses lay on the opposite end of the One Gene–One Disease paradigm; they were Many Genes–Many Diseases. Hypertension, for instance, came in thousands of varieties and was under the influence of hundreds of genes, each exerting a minor additive effect on blood pressure and vascular integrity. Unlike Marfan or Down syndrome, where a single potent mutation or a chromosomal aberration was necessary and sufficient to cause the disease, the effect of any individual gene in poly-

---

* The abnormal chromosomal number in Down syndrome was discovered by Jérôme Lejeune in 1958.

genic syndromes was dulled. The dependence on environmental variables—diet, age, smoking, nutrition, prenatal exposures—was stronger. The phenotypes were variable and continuous, and the patterns of inheritance complex. The genetic component of the disease only acted as one trigger in a many-triggered gun—necessary, but not sufficient to cause the illness.

⚭

Four important ideas emerged from McKusick's taxonomy of genetic diseases. First, McKusick realized that mutations in a single gene can cause diverse manifestations of disease in diverse organs. In Marfan syndrome, for instance, a mutation in a fiberlike structural protein affects all connective tissues—tendons, cartilage, bones, and ligaments. Marfan patients have recognizably abnormal joints and spines. Less recognizable, perhaps, are the cardiovascular manifestations of Marfan disease: the same structural protein that supports tendons and cartilage also supports the large arteries and valves of the heart. Mutations in that gene thus lead to the catastrophic heart failures and aortic ruptures. Patients with Marfan syndrome often die in their youth because their blood vessels have been ruptured by the flow of blood.

Second, the precise converse, surprisingly, was also true: multiple genes could influence a single aspect of physiology. Blood pressure, for instance, is regulated through a variety of genetic circuits, and abnormalities in one or many of these circuits all result in the same disease—hypertension. It is perfectly accurate to say "hypertension is a genetic disease," but to also add, "There is no gene for hypertension." Many genes tug and push the pressure of blood in the body, like a tangle of strings controlling a puppet's arms. If you change the length of any of these individual strings, you change the configuration of the puppet.

McKusick's third insight concerned the "penetrance" and "expressivity" of genes in human diseases. Fruit fly geneticists and worm biologists had discovered that certain genes only become actualized into phenotypes depending upon environmental triggers or random chance. A gene that causes facets to appear in the fruit fly eye, for instance, is temperature dependent. Another gene variant changes the morphology of a worm's intestine—but only does so in about 20 percent of worms. "Incomplete penetrance" meant that even if a mutation was present in the genome, its

263

capacity to *penetrate* into a physical or morphological feature was not always complete.

McKusick found several examples of incomplete penetrance in human diseases. For some disorders, such as Tay-Sachs disease, penetrance was largely complete: the inheritance of the gene mutation virtually guaranteed the development of the disease. But for other human diseases, the actual effect of a gene on the disorder was more complex. In breast cancer, as we shall later learn, inheritance of the mutant *BRCA1* gene increases the risk of breast cancer dramatically—but not all women with the mutation will develop breast cancer, and different mutations in that gene have different levels of penetrance. Hemophilia, the bleeding disorder, is clearly the result of a genetic abnormality, but the extent to which a patient with hemophilia experiences bleeding episodes varies widely. Some have monthly life-threatening bleeds, while others rarely bleed at all.

⟟

The fourth insight is so pivotal to this story that I have separated it from the others. Like the fly geneticist Theodosius Dobzhansky, McKusick understood that mutations are just variations. The statement sounds like a bland truism, but it conveys an essential and profound truth. A mutation, McKusick realized, is a statistical entity, not a pathological or moral one. A mutation doesn't imply disease, nor does it specify a gain or loss of function. In a formal sense, a mutation is defined only by its deviation from the norm (the opposite of "mutant" is not "normal" but "wild type"—i.e., the type or variant found more commonly in the wild). A mutation is thus a statistical, rather than normative, concept. A tall man parachuted into a nation of dwarfs is a mutant, as is a blond child born in a country of brunettes—and both are "mutants" in *precisely* the same sense that a boy with Marfan syndrome is a mutant among non-Marfan, i.e., "normal," children.

By itself, then, a mutant, or a mutation, can provide no real information about a disease or disorder. The definition of disease rests, rather, on the specific disabilities caused by an *incongruity* between an individual's genetic endowment and his or her current environment—between a mutation, the circumstances of a person's existence, and his or her goals for survival or success. It is not mutation that ultimately causes disease, but mismatch.

The mismatch can be severe and debilitating—and in such cases, the disease becomes identical to the disability. A child with the fiercest variant of autism who spends his days rocking monotonously in a corner, or scratching his skin into ulcers, possesses an unfortunate genetic endowment that is mismatched to nearly any environment or any goals. But another child with a different—and rarer—variant of autism may be functional in most situations, and possibly *hyper*functional in some (a chess game, say, or a memory contest). His illness is situational; it lies more evidently in the incongruity of his specific genotype and his specific circumstances. Even the nature of the "mismatch" is mutable: since the environment is constantly subject to change, the definition of disease has to change with it. In the land of the blind, the sighted man is king. But flood that land with a toxic, blinding light—and the kingdom reverts to the blind.

McKusick's belief in this paradigm—the focus on disability rather than abnormalcy—was actualized in the treatment of patients in his clinic. Patients with dwarfism, for instance, were treated by an interdisciplinary team of genetic counselors, neurologists, orthopedic surgeons, nurses, and psychiatrists trained to focus on specific disabilities of persons with short stature. Surgical interventions were reserved to correct specific deformities as they arose. The goal was not to restore "normalcy"—but vitality, joy, and function.

McKusick had rediscovered the founding principles of modern genetics in the realm of human pathology. In humans, as in wild flies, genetic variations abounded. Here too genetic variants, environments, and gene-environment interactions ultimately collaborated to cause phenotypes—except in this case, the "phenotype" in question was disease. Here too some genes had partial penetrance and widely variable expressivity. One gene could cause many diseases, and one disease could be caused by many genes. And here too "fitness" could not be judged in absolutes. Rather, the lack of fitness—*illness*, in colloquial terms—was defined by the relative mismatch between an organism and environment.

⊕

"The imperfect is our paradise," Wallace Stevens wrote. If the entry of genetics into the human world carried one immediate lesson, it was this: the imperfect was not just our paradise; it was also, inextricably, our mortal world. The degree of human genetic variation—and the depth of its influence on human pathology—was unexpected and surprising. The world was vast and various. Genetic diversity was our natural state—not just in

265

isolated pockets in faraway places, but everywhere around us. Seemingly homogeneous populations were, in fact, strikingly heterogeneous. We had seen the mutants—and they were us.

Nowhere, perhaps, was the increased visibility of "mutants" more evident than in that reliable barometer of American anxieties and fantasies—comic strips. In the early 1960s, human mutants burst ferociously into the world of comic characters. In November 1961, Marvel Comics introduced the *Fantastic Four*, a series about four astronauts who, trapped inside a rocket ship—like Hermann Muller's fruit flies in bottles—are exposed to a shower of radiation and acquire mutations that bestow supernatural powers on them. The success of the *Fantastic Four* prompted the even more successful *Spider-Man*, the saga of young science whiz Peter Parker, who is bitten by a spider that has swallowed "a fantastic amount of radioactivity." The spider's mutant genes are transmitted to Parker's body presumably by horizontal transfer—a human version of Avery's transformation experiment—thus endowing Parker with the "agility and proportionate strength of an arachnid."

While *Spider-Man* and *Fantastic Four* introduced the mutant superhero to the American public, the *X-Men*, launched in September 1963, brought the mutant story to its psychological crescendo. Unlike its predecessors, *X-Men*'s central plot concerned a conflict between mutants and normal humans. The "normals" had grown suspicious of the mutants, and the mutants, under fear of surveillance and the threat of mob violence, had retreated to a cloistered School for Gifted Youngsters designed to protect and rehabilitate them—a Moore Clinic for comic-book mutants. The most remarkable feature of *X-Men* was not its growing, multifarious menagerie of mutant characters—a wolf man with steel claws or a woman able to summon English weather on command—but the reversed roles of the victim and the victimizer. In the typical comic book of the fifties, humans ran and hid from the terrifying tyranny of monsters. In *X-Men*, the *mutants* were forced to run and hide from the terrifying tyranny of normalcy.

℗

These concerns—about imperfection, mutation, and normalcy—leaped out of the pages of comic books into a two-foot-by-two-foot incubator in the spring of 1966. In Connecticut, two scientists working on the genetics of mental retardation, Mark Steele and Roy Breg, aspirated a few milliliters of fluid containing fetal cells from the amniotic sac of a pregnant woman. They grew the fetal cells in a petri dish, stained the chromosomes, and then analyzed them under a microscope.

None of these individual techniques was novel. Fetal cells from the amnion had first been examined to predict gender (XX versus XY chromosomes) in 1956. Amniotic fluid had safely been aspirated in the early 1890s, and the staining of chromosomes dated from Boveri's original work on sea urchins. But the advancing front of human genetics changed the stakes of these procedures. Breg and Steele realized that well-established genetic syndromes with evident chromosomal abnormalities—Down, Klinefelter, Turner—could be diagnosed *in utero*, and that pregnancy could voluntarily be terminated if fetal chromosomal abnormalities were detected. Two rather trivial and relatively safe medical procedures—amniocentesis and abortion—could thus be combined into a technology that would vastly exceed the sum of the individual parts.

We know little of the first women to pass through the crucible of this procedure. What remains—in the barest sketches of case reports—are stories of young mothers faced with terrifying choices, and their grief, bewilderment, and its reprieve. In April 1968, a twenty-nine-year-old woman, J.G., was seen at the New York Downstate Medical Center in Brooklyn. Her family had been crisscrossed with a hereditary variant of Down syndrome. Her grandfather and her mother were both carriers. Six years earlier, late in pregnancy, she had miscarried one child—a girl—with Down syndrome. In the summer of 1963, a second girl was born, a healthy child. Two years later, in the spring of 1965, she gave birth to another child—a boy. He was diagnosed with Down syndrome, mental retardation, and severe congenital abnormalities, including two open holes in his heart. The boy had lived for five and a half months. Much of that brief life had been miserable. After a series of heroic surgical attempts to correct his congenital defects, he had died of heart failure in the intensive care unit.

Five months into her fourth pregnancy, with this haunted history in the backdrop, J.G. came to her obstetrician and requested prenatal testing. An unsuccessful amniocentesis was performed in early April. On April 29, with the third trimester rapidly approaching, a second amniocentesis was attempted. This time, sheets of fetal cells grew out in the incubator. Chromosomal analysis revealed a male fetus with Down syndrome.

On May 31, 1968, on the very last week that abortion was still medically permissible, J.G. decided to terminate the pregnancy. The remains of the fetus were delivered on June 2. It bore the cardinal characteristics of Down syndrome. The mother "withstood the procedure without complications," the case report states, and she was discharged home two days after. Nothing more is known about the mother or her family. The first "therapeutic abortion," performed entirely on the basis of a genetic test, entered human history shrouded in secrecy, anguish, and grief.

The floodgates of prenatal testing and abortion were thrown open in the summer of 1973 by an unexpected maelstrom of forces. In September 1969, Norma McCorvey, a twenty-one-year-old carnival barker living in Texas, became pregnant with her third child. Penniless, often homeless, and out of work, she sought an abortion to terminate the unwanted pregnancy, but was unable to find a clinic to perform the procedure legally or, for that matter, sanitarily. The one place she found, she later revealed, was a closed clinic in an abandoned building, "with dirty instruments scattered around the room, and . . . dried blood on the floor."

In 1970, two attorneys brought her case against the state to a Texas court, arguing that McCorvey had a legal right to her abortion. The nominal defendant was Henry Wade, the Dallas district attorney. McCorvey had switched her name for the legal proceedings to a bland pseudonym—Jane Roe. The case—*Roe v. Wade*—moved through the Texas courts and climbed to the US Supreme Court in 1970.

The Supreme Court heard oral arguments for *Roe v. Wade* between 1971 and 1972. In January 1973, in a historic decision, the court ruled for McCorvey. Writing the majority opinion, Henry Blackmun, associate justice of the Supreme Court, decreed that states could no longer outlaw abortions. A woman's right to privacy, Blackmun wrote, was "broad enough to encompass [her] decision whether or not to terminate her pregnancy."

Yet a "woman's right to privacy" was not absolute. In an acrobatic attempt to counterbalance a pregnant woman's rights against the growing "personhood" of the fetus, the Court found that the state could not limit abortions during the first trimester of pregnancy but that as the fetus matured, its personhood became progressively protected by the state, and abortions could be restricted. The division of pregnancy into trimesters was a biologically arbitrary, but legally necessary, invention. As the legal scholar Alexander Bickel described it, "The individual's [i.e., mother's] interest, here, overrides society's interest in the first three months and, subject only to health regulations, also in the second; in the third trimester, society is preeminent."

The power unleashed by *Roe* reverberated swiftly through medicine. *Roe* may have handed reproductive control to women, but it had largely handed the control of the fetal genome to medicine. Before *Roe*, prenatal genetic testing had inhabited an uncertain limbo: amniocentesis was permitted, but the precise legal stature of abortion was unknown. But with first- and second-trimester abortion legalized, and the primacy of medical judgment acknowledged, genetic testing was poised to diffuse widely through clinics and hospitals around the nation. Human genes had become "actionable."

The effects of widespread testing and abortion were soon evident. In some states, the incidence of Down syndrome fell between 20 and 40 percent between 1971 and 1977. Among high-risk women in New York City, more pregnancies were terminated than carried to full term in 1978.* By the mid-1970s, nearly a hundred chromosomal disorders, and twenty-three metabolic diseases, were detectable by genetic testing *in utero*, including Turner and Klinefelter syndromes and Tay-Sachs and Gaucher's disease. "Tiny fault after tiny fault," medicine was sifting its way "through the risk of several hundred known genetic diseases," one geneticist wrote. "Genetic diagnosis," as one historian described it, "became a medical industry." "The selective abortion of affected fetuses" had transformed into "the primary intervention of genomic medicine."

Invigorated by its capacity to intervene on human genes, genetic medicine entered a period of such headiness that it could even begin to rewrite its own past. In 1973, a few months after *Roe v. Wade*, McKusick published a new edition of his textbook on medical genetics. In a chapter on the "prenatal detection of hereditary diseases," Joseph Dancis, the pediatrician, wrote:

> In recent years the feeling has grown among both physicians and the general public that we must be concerned not simply with ensuring the birth of a baby, but one who will not be a liability to society, to its parents, or to itself. The "right to be born" is being qualified by another right: to

* Across the world too, the legalization of abortion opened the floodgates for prenatal testing. In 1967, a parliamentary act legalized abortion in Britain, and prenatal testing rates and pregnancy termination rates increased dramatically in the 1970s.

have a reasonable chance of a happy and useful life. This shift in attitude is shown by, among other things, the widespread movement for the reform or even the abolition of abortion law.

Dancis had gently, but deftly, inverted history. The abortion movement, in Dancis's formulation, had not pushed the frontiers of human genetics forward by enabling doctors to terminate fetuses with genetic disorders. Rather, *human genetics* had pulled the reluctant cart of the abortion movement behind it—by shifting the "attitude" toward the treatment of devastating congenital diseases and thus softening the stance against abortion. In principle, Dancis continued, any illness with a sufficiently powerful genetic link could be intervened upon by prenatal testing and selective abortion. The "right to be born" could be rephrased as a right to be born with the right kind of genes.

⊕

In June 1969, a woman named Hetty Park gave birth to a daughter with infantile polycystic kidney disease. Born with malformed kidneys, the child died five hours after birth. Devastated, Park and her husband sought the counsel of a Long Island obstetrician, Herbert Chessin. Assuming, incorrectly, that the child's disease was not genetic (in fact, infantile PKD, like cystic fibrosis, results from two copies of mutated genes inherited from the child's parents), Chessin reassured the parents and sent them home. In Chessin's opinion, the chance that Park and her husband would have another child born with the same illness was negligible—possibly nil. In 1970, following Chessin's counsel, the Parks conceived again and gave birth to another daughter. Unfortunately, Laura Park was also born with polycystic kidney disease. She suffered multiple hospitalizations and then died of complications of kidney failure at age two and a half.

In 1979, as opinions such as Joseph Dancis's began to appear regularly in the medical and popular literature, the Parks sued Herbert Chessin, arguing that he had given them incorrect medical advice. Had the Parks known the true genetic susceptibilities of their child, they argued, they would have chosen not to conceive Laura. Their daughter was the victim of a flawed estimation of normalcy. Perhaps the most extraordinary feature of the case was the description of the harm. In traditional legal battles concerning medical error, the defendant (usually the physician) stood accused of the wrongful causation of death. The Parks argued that Chessin, their obstetrician, was guilty of the equal and opposite sin: "the wrongful causation of life." In a landmark judgment, the court agreed with the Parks. "Potential parents have a right to choose not to have a child when it can be reasonably established that the child would be deformed," the judge opined. One commentator noted, "The court asserted that the right of a child to be born free of [genetic] anomalies is a fundamental right."

"Interfere, Interfere, Interfere"

After millennia in which most people have produced babies in happy ignorance of the risks they run, we may all have to start acting with the severe responsibility of genetic foresight. . . . We never had to think about medicine like this before.

—Gerald Leach,
"Breeding Better People," 1970

No newborn should be declared human until it has passed certain tests regarding its genetic endowment.
—Francis Crick

Joseph Dancis was not just rewriting the past; he was also announcing the future. Even a casual reader of the extraordinary claim—that every parent had to shoulder the duty to create babies "who will not be a liability to society," or that the right to be born without "genetic anomalies" was a fundamental right—might have detected the cry of a rebirth within it. This was eugenics being reincarnated, if more politely, in the late half of the twentieth century. "Interfere, interfere, interfere," Sidney Webb, the British eugenicist, had urged in 1910. A little more than six decades later, the legalization of abortion and the growing science of genetic analysis had provided the first formal framework for a novel kind of genetic "interference" on humans—a new form of eugenics.

This—as its proponents were quick to point out—was not your Nazi grandfather's eugenics. Unlike American eugenics of the 1920s, or the more virulent European strain of the 1930s, there were no enforced sterilizations, no compulsory confinements, and no exterminations in gas chambers. Women were not sent away to isolation camps in Virginia. Ad hoc judges were not called in to classify men and women as "imbeciles," "morons," and "idiots," nor was chromosome number decided as a matter of personal taste. The genetic tests that formed the basis of fetal selection were, its proponents insisted, objective, standardized, and scientifically rigorous. The correlation between the test and the development of the subsequent medical syndrome was nearly absolute: *all* children born with an extra copy of chromosome twenty-one or a missing copy of the X chromosome, say, manifested at least some of the cardinal features of Down or Turner syndrome respectively. Most important, prenatal testing and selective abortion were performed with no state mandate, no centralized directive, and with full freedom of choice. A woman could choose to be tested or not, choose to know the results or not, and choose to terminate or continue her pregnancy even after testing positive for a fetal abnormality. This was eugenics in its benevolent avatar. Its champions called it neo-eugenics or newgenics.

A crucial distinction between newgenics and old eugenics was the use of *genes* as units of selection. For Galton, for American eugenicists such as Priddy, and for Nazi eugenicists, the only mechanism to ensure genetic selection was through the selection of physical or mental attributes—i.e., through phenotypes. But these attributes are complex, and their link to genes cannot be simply captured. "Intelligence," for instance, may have a genetic component, but it is much more evidently a consequence of genes, environments, gene-environment interactions, triggers, chance, and opportunities. Selecting "intelligence," therefore, cannot guarantee that genes for intelligence will be selected any more than selecting "richness" guarantees that a propensity for accumulating wealth will be selected.

In contrast to Galton's and Priddy's method, the major advance of newgenics, its proponents insisted, was that scientists were no longer selecting phenotypes as surrogates for the underlying genetic determinants. Now, geneticists had the opportunity to select *genes* directly—by examining the genetic composition of a fetus.

℗

To its many enthusiasts, neo-eugenics had shed the menacing guise of its past and emerged anew out of a scientific chrysalis. Its scope broadened even further in the mid-1970s. Prenatal testing and selective abortion had enabled a privatized form of "negative eugenics"—a means to select *against* certain genetic disorders. But coupled to this was the desire to instigate an equally expansive, laissez-faire form of "positive eugenics"—a means to select *for* favorable genetic attributes. As Robert Sinsheimer, a geneticist, described it, "The old eugenics was limited to the numerical enhancement of the best of our existing gene pool. The new eugenics would in principle allow the *conversion* of all the unfit to the highest genetic level."

In 1980, Robert Graham, a millionaire entrepreneur who had developed shatterproof sunglasses, endowed a sperm bank in California that would preserve sperm from men of the "highest intellectual caliber," to be accessed only to inseminate healthy, intelligent women. Called the Repository for Germinal Choice, the bank sought sperm from Nobel laureates across the world. The physicist William Shockley, the inventor of the silicon transistor, was among the few scientists who agreed to donate. Perhaps predictably, Graham ensured that his own sperm was added to the bank on the pretext that—although the committee in Stockholm was yet to recognize it—he was a "future Nobel laureate," a genius-in-waiting. However ardent its fantasies, Graham's cryogenic utopia was not embraced by the public. Over the next decade, only fifteen children would be born from sperm banked at the Repository. The long-term achievements of most of these children remain unknown, although none, thus far, appears to have won another Nobel Prize.

Although Graham's "genius bank" was ridiculed and eventually disbanded, its early advocacy of "germinal choice"—that individuals should be free to pick and choose the genetic determinants of their offspring—was hailed by several scientists. A sperm bank of selected genetic geniuses was obviously a crude idea—but selecting "genius genes" in sperm, on the other hand, was considered a perfectly tenable prospect for the future.

But how might sperm (or eggs, for that matter) be selected to carry specific enhanced genotypes? Could new genetic material be introduced into the human genome? Although the precise contours of the technology that would enable positive eugenics were yet unknown, several scientists considered this a mere technological hurdle that would be solved in the near future. The geneticist Hermann Muller, evolutionary biologists Ernst Mayr and Julian Huxley, and the population biologist James Crow were among the vociferous proponents of positive eugenics. Until the birth of eugenics, the only mechanism to select for beneficial human genotypes had been natural selection—governed by the brutal logic of Malthus and

Darwin: the struggle for survival and the slow, tedious emergence of survivors. Natural selection, Crow wrote, was "cruel, blundering and inefficient." In contrast, artificial genetic selection and manipulation could be based on "health, intelligence or happiness." Support from scientists, intellectuals, writers, and philosophers poured into the movement. Francis Crick staunchly backed neo-eugenics, as did James Watson. James Shannon, director of the National Institutes of Health, told Congress that genetic screening was not just a "moral obligation of the medical profession, but a serious social responsibility as well."

As newgenics rose in prominence nationally and internationally, its founders tried valiantly to dissociate the new movement from its ugly past—in particular, from the Hitlerian shadows of Nazi eugenics. German eugenics had fallen into the abyss of Nazi horrors, neo-eugenicists argued, because of two cardinal errors—its scientific illiteracy and its political illegitimacy. Junk science had been used to prop up a junk state, and the junk state had fostered junk science. Neo-eugenics would sidestep these pitfalls by sticking to two lodestone values—scientific rigor and choice.

Scientific rigor would ensure that the perversities of Nazi eugenics would not contaminate neo-eugenics. Genotypes would be evaluated objectively, without interference or mandates from the state, using strict scientific criteria. And choice would be conserved at every step, guaranteeing that eugenic selections—such as prenatal testing and abortion—would occur only with full freedom.

Yet to its critics, newgenics was riddled with some of the same fundamental flaws that had cursed eugenics. The most resonant criticism of neo-eugenics emerged, unsurprisingly, from the very discipline that had breathed life into it—human genetics. As McKusick and his colleagues were discovering with increasing lucidity, the interactions between human genes and illnesses were vastly more complicated than newgenics might have anticipated. Down syndrome and dwarfism offered instructive case studies. For Down syndrome, where the chromosomal abnormality was distinct and easily identifiable, and where the link between the genetic lesion and the medical symptoms was highly predictable, prenatal testing and abortion might seem justifiable. But even with Down syndrome, as with dwarfism, the variation between individual patients carrying the same mutation was striking. Most men and women with Down syndrome experienced deep physical, developmental, and cognitive disabilities. But some, undeniably, were highly functional—leading near-independent lives requiring minimal interventions. Even a whole extra chromosome—as significant a genetic lesion as was conceivable in human cells—could not be a singular determinant of disability; it lived in the context of other genes and was modified by environmental inputs and by the genome at large. Genetic illness and genetic wellness were not discrete neighboring countries; rather, wellness and illness were continuous kingdoms, bounded by thin, often transparent, borders.

The situation became even more complex with polygenic illnesses—schizophrenia or autism, say. Although schizophrenia was well-known to have a strong genetic component, early studies suggested that multiple genes on multiple chromosomes were intimately involved. How might negative selection exterminate all these independent determinants? And what if some of the gene variants that caused mental disorders in some genetic or environmental contexts were the very variants that produced enhanced ability in other contexts? Ironically, William Shockley—the most prominent donor to Graham's genius bank—was himself afflicted by a syndrome of paranoia, aggression, and social withdrawal that several biographers have suggested was a form of high-functioning autism. What if—poring through Graham's bank in some future era—the selected "genius specimens" were found to possess the very genes that, in alternative situations, might be identified as disease enabling (or vice versa: What if "disease-causing" gene variants were also *genius* enabling?)?

McKusick, for one, was convinced that "overdeterminism" in genetics, and its indiscriminate application to human selection, would result in the creation of what he called the "genetic-commercial" complex. "Near the end of his terms of office, President Eisenhower warned against the dangers of the military-industrial complex," McKusick said. "It is appropriate to warn of a potential hazard of the genetic-commercial complex. The increasing availability of tests for presumed genetic quality or poor quality could lead the commercial sector and the Madison Avenue publicist to bring subtle or not so subtle pressure on couples to make value judgments in the choice of their gametes for reproduction."

In 1976, McKusick's concerns still seemed largely theoretical. Although the list of human diseases influenced by genes had grown exponentially, most of the actual genes were yet to be identified. Gene-cloning and gene-sequencing technologies, both invented in the late 1970s, made it conceivable that such genes could be successfully identified in humans, leading to predictive diagnostic tests. But the human genome has 3 billion base pairs—while a typical disease-linked gene mutation might result in the alteration of just *one* base pair in the genome. Cloning and sequencing all genes in the genome to find that mutation was inconceivable. To find a disease-linked gene, the gene would need to be somehow mapped, or localized, to a smaller part of the genome. But that, precisely, was the missing piece of technology: although genes that caused diseases seemed abundant, there was no easy way to find them in the vast expanse of the human genome. As one geneticist described it, human genetics was stuck in the ultimate "needle in a haystack problem."

A chance meeting in 1978 would offer a solution to the "needle in a haystack" problem of human genetics, enabling geneticists to map and clone human disease-linked genes. The meeting, and the discovery that followed it, would mark one of the turning points in the study of the human genome.

A Village of Dancers, an Atlas of Moles

Glory be to God for dappled things.
—Gerard Manley Hopkins, "Pied Beauty"

We suddenly came upon two women, mother and daughter, both tall, thin, almost cadaverous, both bowing, twisting, grimacing.

—George Huntington

In 1978, two geneticists, David Botstein, from MIT, and Ron Davis, from Stanford, traveled to Salt Lake City to serve on a graduate review committee for the University of Utah. The meeting was held at Alta, high on the Wasatch Mountains, a few miles from the city. Botstein and Davis sat through the presentations taking notes—but one talk struck a particular chord with both of them. A graduate student, Kerry Kravitz, and his adviser, Mark Skolnick, were painstakingly mapping the inheritance of a gene that causes hemochromatosis, a hereditary illness. Known by physicians since antiquity, hemochromatosis is caused by a mutation in a gene that regulates iron absorption from the intestines. Patients with hemochromatosis absorb enormous amounts of iron, resulting in a body that is slowly choked by iron deposits. The liver asphyxiates on iron; the pancreas stops working. The skin turns bronze and then ashen gray. Organ by organ, the body transforms into mineral, like the Tin Man in *The Wizard of Oz*, ultimately leading to tissue degeneration, organ failure, and death.

The problem that Kravitz and Skolnick had decided to solve concerned a fundamental conceptual gap in genetics. By the mid-1970s, thousands of genetic diseases had been identified—hemochromatosis, hemophilia, and sickle-cell anemia among them. Yet, discovering the genetic nature of an illness is not the same as identifying the actual gene that causes that illness. The pattern of inheritance of hemochromatosis, for instance, clearly suggests that a single gene governs the disease, and that the mutation is recessive—i.e., two defective copies of the gene (one from each parent) are necessary to cause the illness. But the pattern of inheritance tells us nothing about what the hemochromatosis gene is or what it does.

Kravitz and Skolnick proposed an ingenious solution to identify the hemochromatosis gene. The first step to finding a gene is to "map" it to a particular chromosomal location: once a gene has been physically located on a particular stretch of a chromosome, standard cloning techniques can be used to isolate the gene, sequence it, and test its function. To map the hemochromatosis gene, Kravitz and Skolnick reasoned, they would use the one property that all genes possess: they are linked to each other on chromosomes.

Consider the following thought experiment. Say the hemochromatosis gene sits on chromosome seven, and the gene that governs hair texture—straight versus kinked or curly or wavy—is its immediate neighbor on the same chromosome. Now assume that somewhere in distant evolutionary history, the defective hemochromatosis gene arose in a man with curly hair. Every time this ancestral gene is passed from parent to child, the curly-haired gene travels with it: both are bound on the same chromosome, and since chromosomes rarely splinter, the two gene variants inevitably associate with each other. The association may not be obvious in a single generation, but over multiple generations, a statistical pattern begins to emerge: curly-haired children in this family tend to have hemochromatosis.

Kravitz and Skolnick had used this logic to their advantage. By studying Mormons in Utah with cascading, many-branched family trees, they had discovered that the hemochromatosis gene was genetically linked to an immune-response gene that exists in hundreds of variants. Prior work had mapped the immune-response gene to chromosome six—and so the hemochromatosis gene had to be located on that chromosome.

Careful readers might object that the example above was loaded: the gene for hemochromatosis happened to be conveniently linked to an easily identifiable, highly variant trait on the same chromosome. But surely such traits were fleetingly rare. That Skolnick's gene of interest happened to be sitting, cheek by jowl, with a gene that encoded an immune-response protein that existed in many easily detectable variants was surely a lucky aberration. To achieve this kind of mapping for any other gene, wouldn't the human genome have to be littered with strings of variable, easily identifiable markers—lamplit signposts planted conveniently along every mile of chromosome?

But Botstein knew that such signposts might exist. Over centuries of evolution, the human genome has diverged enough to create thousands of minute variations in DNA sequence. These variants are called *polymorphisms*—"many forms"—and they are exactly like alleles or variants, except they need not be in genes themselves; they might exist in the long stretches of DNA between genes, or in introns.

These variants can be imagined as molecular versions of eye or skin color, existing in thousands of varied forms in the human population. One family might carry an A<u>C</u>AAGTCC at a particular location on a chromosome, while another might have A<u>G</u>AAGTCC at that same location—a one-base-pair difference.* Unlike hair color or the immune response, these variants are invisible to the human eye. The variations need not enable a change in phenotype, or even alter a function of a gene. They cannot be distinguished using standard biological or physical traits—but they can be discerned using subtle molecular techniques. A DNA-cutting enzyme that recognizes A<u>C</u>AAG, but not A<u>G</u>AAG, for instance, might discriminate one sequence variant and not the other.

Ⓘ

When Botstein and Davis had first discovered DNA polymorphisms in yeast and bacterial genomes in the 1970s, they had not known what to make of them. At the same time, they had also identified a few such polymorphisms scattered across human genomes—but the extent and location of such variations in humans was still unknown. The poet Louis MacNeice once wrote about feeling "the drunkenness of things being various." The thought of tiny molecular variations peppered randomly through the genome—like freckles across a body—might have provoked a certain pleasure in a drunken human geneticist, but it was hard to imagine how this information might be useful. Perhaps the phenomenon was perfectly beautiful and perfectly useless—a map of freckles.

But as Botstein listened to Kravitz that morning in Utah, he was struck by a compelling idea: if such variant genetic signposts existed in the human genome, then by linking a genetic trait to one such variant, *any gene could*

---

* In 1978, two other researchers, Y. Wai Kan and Andree Dozy, had found a polymorphism of DNA near the sickle-cell gene—and used it to follow the inheritance of the sickle-cell gene in patients. Maynard Olson and colleagues also described gene mapping methods using polymorphisms in the late 1970s.

be mapped to an approximate chromosomal location. A map of genetic freckles was not useless at all; it could be deployed to chart the basic anatomy of genes. The polymorphisms would act like an internal GPS system for the genome; a gene's location could be pinpointed by its association, or linkage, to one such variant. By lunchtime, Botstein was nearly frantic with excitement. Skolnick had spent more than a decade hunting down the immune-response marker to map the hemochromatosis gene. "We can give you markers . . . markers spread all over the genome," he told Skolnick.

The real key to human gene mapping, Botstein had realized, was not finding the gene, but finding the humans. If a large-enough family bearing a genetic trait—any trait—could be found, and if that trait could be correlated with any of the variant markers spread across the genome, then gene mapping would become a trivial task. If all the members of a family affected by cystic fibrosis inevitably "co-inherited" some variant DNA marker, call it Variant-X, located on the tip of chromosome seven, then the cystic fibrosis gene had to sit in proximity to this location.

Botstein, Davis, Skolnick, and Ray White, a human geneticist, published their idea about gene mapping in the *American Journal of Human Genetics* in 1980. "We describe a new basis for the construction of a genetic . . . map of the human genome," Botstein wrote. It was an odd study, tucked into the middle pages of a relatively obscure journal and festooned with statistical data and mathematical equations, reminiscent of Mendel's classic paper.

It would take some time for the full implication of the idea to sink in. The crucial insights of genetics, I said before, are always transitions—from statistical traits to inheritable units, from genes to DNA. Botstein had also made a crucial conceptual transition—between human genes as inherited biological characteristics, and their physical maps on chromosomes.

①

Nancy Wexler, a psychologist, heard about Botstein's gene-mapping proposal in 1978 while corresponding with Ray White and David Housman, the MIT geneticist. She had a poignant reason to pay attention. In the summer of 1968, when Wexler was twenty-two, her mother, Leonore Wexler, was chastised by a policeman for walking erratically while crossing a street in Los Angeles. Leonore had suffered inexplicable bouts of depression but had never been considered physically ill. Two of Leonore's brothers, Paul and Seymour, once members of a swing band in New York, had been diagnosed with a rare genetic syndrome called Huntington's disease, in the 1950s. Another brother, Jessie, a salesman who liked to perform magic tricks, had found his fingers dancing uncontrollably during his performances. He too was diagnosed with the same illness. Their father, Abraham Sabin, had died of Huntington's disease in 1929. Leonore saw a neurologist and was diagnosed with Huntington's disease in May 1968.

Named after the Long Island doctor who first described the condition in the 1870s, Huntington's disease was once called Huntington's chorea—*chorea* from the Greek word for "dance." The "dance," of course, is the opposite of dance, a joyless and pathological caricature, the ominous manifestation of dysregulated brain function. Typically, patients who inherit the dominant Huntington's gene—only one copy is sufficient to precipitate the disease—are neurologically intact for the first three or four decades of their life. They might experience occasional mood swings or subtle signs of social withdrawal. Then minor, barely discernible twitches appear. Objects become hard to grasp. Wineglasses and watches slip between fingers, and movements dissolve into jerks and spasms. Finally, the involuntary "dance" begins, as if set to devil's music. The hands and legs move of their own accord, tracing writhing, arclike gestures separated by staccato, rhythmic jolts—"like watching a giant puppet show . . . jerked by an unseen puppeteer." The late stage of the disease is marked by deep cognitive decline and near-complete loss of motor function. Patients die of malnourishment, dementia, and infections—yet "dancing" to the last.

Part of the macabre denouement of Huntington's is the late onset of the illness. Those carrying the gene only discover their fate in their thirties or forties—i.e., after they have had their own children. The disease thus persists in human populations. Since every patient with Huntington's disease has one normal copy and one mutant copy of the gene, every child born to him or her has a fifty-fifty chance of being affected. For these children, life devolves into "a grim roulette—a waiting game for the onset of symptoms," as Nancy Wexler described it. One patient wrote about the strange terror of this limbo: "I don't know the point where the grey zone ends and a much darker fate awaits. . . . So I play the terrible waiting game, wondering about the onset and the impact."

_①

Milton Wexler, Nancy's father, a clinical psychologist in Los Angeles, broke the news of their mother's diagnosis to his two daughters in 1968. Nancy and Alice were still asymptomatic, but they each carried a 50 percent chance of being affected, with no genetic test for the disease. "Each one of you has a one-in-two chance of getting the disease," Milton Wexler told his daughters. "And if you get it, your kids have a one-in-two chance of getting it."

"We were all hanging on to each other and sobbing," Nancy Wexler recalled. "The passivity of just waiting for this to come and kill me was unbearable."

That year, Milton Wexler launched a nonprofit foundation, called the Hereditary Disease Foundation, dedicated to funding research on Huntington's chorea and other rare inherited diseases. Finding the Huntington's gene, Wexler reasoned, would be the first step toward diagnosis, future treatments, and cures. It would give his daughters a chance to predict and plan for their future illness.

Leonore Wexler, meanwhile, gradually descended into the chasm of her disease. Her speech began to slur uncontrollably. "New shoes would wear out the moment you put them on her feet," her daughter recalled. "In one nursing home, she sat in a chair in the narrow space between her bed and the wall. No matter where the chair was put, the force of her continual movements edged it against the wall, until her head began bashing into the plaster. . . . We tried to keep her weight up; for some unknown reason, people with Huntington's disease do better when they are heavy, although their constant motion makes them thin. . . . Once she polished off a pound of Turkish delight in half an hour with a grin of mischievous delight. But she never gained weight. I gained weight. I ate to keep her company; I ate to keep from crying."

Leonore died on May 14, 1978, on Mother's Day. In October 1979, Nancy Wexler, of the Hereditary Disease Foundation, David Housman, Ray White, and David Botstein organized a workshop at the NIH to focus on the best strategy to pursue mapping the gene. Botstein's gene mapping method was still largely theoretical—thus far, no human gene had been successfully mapped with it—and the likelihood of using the method to map the Huntington's gene was remote. Botstein's technique was, after all, crucially dependent on the association between a disease and markers: the more patients, the stronger the association, the more refined the genetic map. Huntington's chorea—with only a few thousand patients scattered across the United States—seemed perfectly mismatched to this gene-mapping technique.

Yet Nancy Wexler could not shake the image of gene maps from her mind. A few years earlier, Milton Wexler had heard from a Venezuelan neurologist of two neighboring villages, Barranquitas and Lagunetas, on the shores of Lake Maracaibo in Venezuela with a striking prevalence of Huntington's disease. In a fuzzy, black-and-white home movie filmed by the neurologist, Milton Wexler had seen more than a dozen villagers wandering on the streets, their limbs shaking uncontrollably. There were scores of Huntington's patients in the village. If Botstein's technique had any chance of working, Nancy Wexler reasoned, she would need to access the genomes of the Venezuelan cohort. It was in Barranquitas, several thousand miles from Los Angeles, that the gene for her family's illness would most likely be found.

In July 1979, Wexler set off to Venezuela to hunt the Huntington's gene. "There have been a few times in my life when I felt certain that something was really right, times when I couldn't sit still," Wexler wrote.

①

At first glance, a visitor to Barranquitas might notice nothing unusual about its inhabitants. A man walks by on a dusty road, followed by a band of shirtless children. A thin, dark-haired woman in a floral dress appears from a tin-roof shed and makes her way to the market. Two men sit across from each other, conversing and playing cards.

The initial impression of normalcy changes quickly. There is something in the man's walk that seems profoundly unnatural. A few steps, and his body begins to move with jerking, staccato gestures, while his hand traces sinuous arcs in midair. He twitches and lunges sideways, then corrects himself. Occasionally, his facial muscles contort into a frown. The woman's hands also twist and writhe, tracing airy half circles around her body. She looks emaciated and drools. She has progressive dementia. One of the two men in conversation flings out his arm violently, then the talk resumes, as if nothing has happened.

When the Venezuelan neurologist Américo Negrette first arrived in Barranquitas in the 1950s, he thought that he had stumbled upon a village of alcoholics. He soon realized that he was wrong: all the men and women with dementia, facial twitches, muscle wasting, and uncontrolled movement had a heritable neurological syndrome, Huntington's disease.

In the United States, the syndrome is fleetingly rare—only one in ten thousand have the disease. In some parts of Barranquitas and nearby Lagunetas, in contrast, more than one in twenty men and women were afflicted with the disease.

ⅅ

Wexler landed in Maracaibo in July 1979. She hired a team of eight local workers, ventured into the barrios along the lake, and began to document the pedigrees of affected and unaffected men and women (although trained as a clinical psychologist, Wexler had, by then, become one of the world's leading experts on choreas and neurodegenerative illnesses). "It was an impossible place to conduct research," her assistant recalled. A makeshift ambulatory clinic was set up so that neurologists could identify the patients, characterize the disease, and provide information and supportive care. Wexler was particulary interested in finding men and women with two copies of the mutated Huntington's disease gene—i.e., "homozygotes". To find such individuals, she needed a family where both parents were affected. One morning, a local fisherman brought a crucial clue: he knew of a boating shanty, about two hours along the lake, where many families were afflicted by *el mal*. Would Wexler like to venture through the swamps to the village?

She would. The next day, Wexler and two assistants set off on a boat toward the *pueblo de agua*, the village on stilts. The heat was sweltering. They paddled for hours through the backwaters—and then, as they rounded the bend of an inlet, they saw a woman with a brown-print dress sitting cross-legged on a porch. The arrival of the boat startled the woman. She rose to go inside the house and was suddenly struck, midway, by the jerking choretic movements characteristic of Huntington's disease. A continent away from home, Wexler had come face-to-face with that achingly recognizable dance. "It was a clash of total bizarreness and total familiarity," she recalled. "I felt connected and alienated. I was overcome."

Moments later, as Wexler paddled into the heart of the village, she found another couple lying on two hammocks. Both were shaking and dancing. They had fourteen children. As Wexler collected information about the children and *their* children, the documented lineage grew rapidly. In a few months, she had established a list containing hundreds of

285

men, women, and children with Huntington's disease. Over the next months, Wexler returned to the sprawl of villages with a team of trained nurses and physicians to collect vial upon vial of blood. They assiduously collected and assembled a family tree of the Venezuelan kindred. The blood was then shipped to the laboratory of James Gusella, at the Massachusetts General Hospital in Boston, and to Michael Conneally, a population geneticist at Indiana University.

In Boston, Gusella purified DNA from blood cells and cut it with a barrage of enzymes, looking for a variant that might be genetically linked to Huntington's disease. Conneally's group analyzed the data to quantify the statistical link between the DNA variants and the disease. The three-part team expected to plod along slowly—they had to sift through thousands of polymorphic variants—but they were immediately surprised. In 1983, barely three years after the blood had arrived, Gusella's team stumbled on a single piece of variant DNA, located on a stretch of chromosome four, that was strikingly associated with the disease. Notably, Gusella's group had also collected blood from a much smaller American cohort with Huntington's disease. Here too the illness seemed to be weakly associated with a DNA signpost located on chromosome four. With two independent families demonstrating such a powerful association, there could be little doubt about a genetic link.

In August 1983, Wexler, Gusella, and Conneally published a paper in *Nature* definitively mapping the Huntington's disease gene to a distant outpost of chromosome four—4p16.3. It was a strange region of the genome, largely barren, with a few unknown genes within it. For the team of geneticists, it was like the sudden landing of a boat on a derelict beachhead, with no known landmarks in sight.

Ⓒ

To map a gene to its chromosomal location using linkage analysis is to zoom in from outer space into the genetic equivalent of a large metropolitan city: it produces a vastly refined understanding of the location of the gene, but it is still a long way from identifying the gene itself. Next, the gene map is refined by identifying more linkage markers, progressively narrowing the location of a gene to smaller and yet smaller chunks of the chromosome. Districts and subdistricts whiz by; neighborhoods and blocks appear.

The final steps are improbably laborious. The piece of the chromosome carrying the suspected culprit gene is divided into parts and subparts. Each of these parts is isolated from human cells, inserted into yeast or bacterial chromosomes to make millions of copies, and thereby cloned. These cloned pieces are sequenced and analyzed, and the sequenced fragments scanned to determine if they contain a potential gene. The process is repeated and refined, every fragment sequenced and rechecked, until a piece of the candidate gene has been identified in a single DNA fragment. The ultimate test is to sequence the gene in normal and affected patients to confirm that the fragment is altered in patients with the hereditary illness. It is like moving door-to-door to identify a culprit.

⟐

On a bleak February morning in 1993, James Gusella received an e-mail from his senior postdoc with a single word in it: "Bingo." It signaled an arrival—a landing. Since 1983, when the Huntington's gene had been mapped to chromosome four, an international team of six lead investigators and fifty-eight scientists (organized, nurtured and funded through the Hereditary Disease Foundation) had spent the bleakest of decades hunting for the gene on that chromosome. They had tried all sorts of shortcuts to isolate the gene. Nothing had worked. Their initial burst of luck had run out. Frustrated, they had resorted to gene-by-gene plodding. In 1992, they had gradually zeroed in on one gene, initially named *IT15*—"interesting transcript 15." It was later renamed *Huntingtin*.

IT15 was found to encode an enormous protein—a biochemical behemoth containing 3,144 amino acids, larger than nearly any other protein in the human body (insulin has a mere 51 amino acids). That morning in February, Gusella's postdoc had sequenced the *IT15* gene in a cohort of normal controls and patients with Huntington's disease. As she counted the bands in the sequencing gel, she found an obvious difference between patients and their unaffected relatives. The candidate gene had been found.

Wexler was about to leave for yet another trip to Venezuela to collect samples when Gusella called her. She was overwhelmed. She could not stop weeping. "We've got it, we've got it," she told an interviewer. "It's been a long day's journey into the night."

①

The *Huntingtin* protein is found in neurons and in testicular tissue. In mice, it is required for the development of the brain. The mutation that causes the disease is even more mysterious. The normal gene sequence contains a highly repetitive sequence, CAGCAGCAGCAG . . . a molecular singsong that stretches for seventeen such repeats on average (some people have ten, while others may have up to thirty-five). The mutation found in Huntington's patients is peculiar. Sickle-cell anemia is caused by the alteration of a single amino acid in the protein. In Huntington's disease, the mutation is not an alteration of one amino acid or two, but an increase in the number of repeats, from less than thirty-five in the normal gene to more than forty in the mutant. The increased number of repeats lengthens the size of the *Huntingtin* protein. The longer protein is thought to be aggregated into pieces in neurons, and these pieces accumulate in tangled spools inside cells, possibly leading to their death and dysfunction.

The origin of this strange molecular "stutter"—the alteration of a repeat sequence—still remains a mystery. It might be an error made in gene copying. Perhaps the DNA replication enzyme adds extra CAGs to the repetitive stretches, like a child who writes an additional *s* while spelling *Mississippi*. A remarkable feature of the inheritance of Huntington's disease is a phenomenon called "anticipation": in families with Huntington's disease, the number of repeats gets amplified over generations, resulting in fifty or sixty repeats in the gene (the child, having misspelled *Mississippi* once, keeps adding more *s*'s). As the repeats increase, the disease accelerates in severity and onset, affecting younger and younger members. In Venezuela, even boys and girls as young as twelve years old are now afflicted, some of them carrying strings of seventy or eighty repeats.

①

Davis and Botstein's technique of mapping genes based on their physical positions on chromosomes—later called positional cloning—marked a transformative moment in human genetics. In 1989, the technique was used to identify a gene that causes cystic fibrosis, a devastating illness that affects the lungs, pancreas, bile ducts, and intestines. Unlike the mutation that causes Huntington's disease, which is fleetingly rare in most populations (except the unusual cluster of patients in Venezuela), the mutated variant of the cystic fibrosis is common: one in twenty-five men and women of European descent carries the mutation. Humans with a single copy of the mutant gene are largely asymptomatic. If two such asymptomatic carriers conceive a child, chances are one in four that the child will be born with both mutant genes. The consequence of inheriting two mutant copies of the CF gene can be fatal. Some of the mutations have a nearly 100 percent penetrance. Until the 1980s, the average life span of a child carrying two such mutant alleles was twenty years.

That cystic fibrosis had something to do with salt and secretions had been suspected for centuries. In 1857, a Swiss almanac for children's songs and games warned about the health of a child whose "brow tastes salty when kissed." Children with the disease were known to secrete such enormous quantities of salt through their sweat glands that their sweat-drenched clothes, hung on wires to dry, would corrode the metal, like seawater. The secretions of the lung were so viscous that they blocked the airways with gobs of mucus. The phlegm-clogged airways became breeding grounds for bacteria, causing frequent, lethal pneumonias, among the most common causes of death. It was a horrific life—a body drowning in its own secretions—that often culminated in a horrific death. In 1595, a professor of anatomy at Leiden wrote of a child's death: "Inside the pericardium, the heart was floating in a poisonous liquid, sea green in colour. Death had been caused by the pancreas which was oddly swollen. . . . The little girl was very thin, worn out by hectic fever—a fluctuating but persistent fever." It is virtually certain that he was describing a case of cystic fibrosis.

In 1985, Lap-Chee Tsui, a human geneticist working in Toronto, found an "anonymous marker," one of Botstein's DNA variants along the genome, that was linked to the mutant CF gene. The marker was quickly pinpointed on chromosome seven, but the CF gene was still lost somewhere in the genetic wilderness of that chromosome. Tsui began to hunt for the CF gene by progressively narrowing the region that might contain it. The hunt was joined by Francis Collins, a human geneticist at the University of Michigan, and by Jack Riordan, also in Toronto. Collins had made an ingenious modification to the standard gene-hunting technique. In gene mapping, one usually "walked" along a chromosome—cloning one bit, then the next, one contiguous, overlapping stretch after another.

It was painstakingly laborious, like climbing a rope by placing one fist directly upon the other. Collins's method allowed him to move up and down the chromosome with a greatly outstretched reach. He called it chromosome "jumping."

By the spring of 1989, Collins, Tsui, and Riordan had used chromosome jumping to narrow the gene hunt to a few candidates on chromosome seven. The task was now to sequence the genes, confirm their identity, and define the mutation that affected the function of the CF gene. On a rain-drenched evening late that summer, while both Tsui and Collins were attending a gene-mapping workshop in Bethesda, they stood penitently by a fax machine, waiting for news of the gene sequence from a postdoc researcher in Collins's lab. As the machine spit out sheaves of paper with garbles of sequence, ATGCCGGTC . . . Collins watched the revelation materialize out of thin air: only one gene was persistently mutated in both copies in affected children, while their unaffected parents carried a single copy of the mutation.

The CF gene codes a molecule that channels salt across cellular membranes. The most common mutation is a deletion of three bases of DNA that results in the removal, or deletion, of just one amino acid from the protein (in the language of genes, three bases of DNA encode a single amino acid). This deletion creates a dysfunctional protein that is unable to move chloride—one component of sodium chloride, i.e., common salt—across membranes. The salt in sweat cannot be absorbed back into the body, resulting in the characteristically salty sweat. Nor can the body secrete salt and water into the intestines, resulting in the abdominal symptoms.*

---

* The high prevalence of the mutant cystic fibrosis gene in European populations has puzzled human geneticists for decades. If CF is such a lethal disease, then why was the gene not culled out by evolutionary selection? Recent studies posit a provocative theory: the mutant cystic fibrosis gene may provide a selective *advantage* during cholera infection. Cholera in humans causes a severe, intractable diarrhea that is accompanied by the acute loss of salt and water; this loss can lead to dehydration, metabolic disarray, and death. Humans with *one* copy of the mutant CF gene have a slightly diminished capacity to lose salt and water through their membranes and are thus relatively protected from the most devastating complications of cholera (this can be demonstrated using genetically engineered mice). Here too a mutation in a gene can have a dual and circumstantial effect—potentially beneficial in one copy, and lethal in two copies. Humans with one copy of the mutant CF gene may thus have survived cholera epidemics in Europe. When two such people reproduced, they had a one-in-four chance of creating a child with two mutant genes—i.e., a child with CF—but the selective advantage was strong enough to maintain the mutant CF gene in the population.

The cloning of the CF gene was a landmark achievement for human geneticists. Within a few months, a diagnostic test for the mutant allele became available. By the early 1990s, carriers could be screened for the mutation, and the disease could routinely be diagnosed *in utero*, allowing parents to consider aborting affected fetuses, or to monitor children for early manifestations of the disease. "Carrier couples"—in which both parents happen to possess at least one copy of the mutant gene—could choose not to conceive a child, or to adopt children. Over the last decade, the combination of targeted parental screening and fetal diagnosis has reduced the prevalence of children born with cystic fibrosis by about 30 to 40 percent in populations where the frequency of the mutant allele is the highest. In 1993, a New York hospital launched an aggressive program to screen Ashkenazi Jews for three genetic diseases, including cystic fibrosis, Gaucher's disease, and Tay-Sachs disease (mutations in these genes are more prevalent in the Ashkenazi population). Parents could freely choose to be screened, to undergo amniocentesis for prenatal diagnosis, and to terminate a pregnancy if the fetus was found to be affected. Since the launch of the program, not a single baby with any of these genetic diseases has been born at that hospital.

Ⓓ

It is important to conceptualize the transformation in genetics that occurred between 1971—the year that Berg and Jackson created the first molecule of recombinant DNA—and 1993, the year that the Huntington's disease gene was definitively isolated. Even though DNA had been identified as the "master molecule" of genetics by the late 1950s, no means then existed to sequence, synthesize, alter, or manipulate it. Aside from a few notable exceptions, the genetic basis of human disease was largely unknown. Only a few human diseases—sickle-cell anemia, thalassemia, and hemophilia B—had been definitively mapped to their causal genes. The only human genetic interventions available clinically were amniocentesis and abortion. Insulin and clotting factors were being isolated from pig organs and human blood; no medicine had been created by genetic engineering. A human gene had never intentionally been expressed outside a human cell. The prospect of changing an organism's genome by introducing foreign genes, or by deliberately mutating its native genes, was far outside the reach of any technology. The word *biotechnology* did not exist in the Oxford dictionary.

Two decades later, the transformation of the landscape of genetics was remarkable: human genes had been mapped, isolated, sequenced, synthesized, cloned, recombined, introduced into bacterial cells, shuttled into viral genomes, and used to create medicines. As the physicist and historian Evelyn Fox Keller described it: once "molecular biologists [had discovered] techniques by which they themselves could manipulate [DNA]," there "emerged a technological know-how that decisively altered our historical sense of the immutability of 'nature.'"

"Where the traditional view had been that 'nature' spelt destiny, and 'nurture' freedom, now the roles appeared to be reversed. . . . We could more readily control the former [i.e., genes], than the latter [i.e., the environment]—not simply as a long-term goal but as an immediate prospect."

In 1969, on the eve of the revelatory decade, Robert Sinsheimer, the geneticist, wrote an essay about the future. The capacity to synthesize, sequence, and manipulate genes would unveil "a new horizon in the history of man."

"Some may smile and may feel that this is but a new version of the old dream of the perfection of man. It is that, but it is something more. The old dreams of the cultural perfections of man were always sharply constrained by his inherent, inherited imperfections and limitations. . . . We now glimpse another route—the chance to ease and consciously perfect far beyond our present vision this remarkable product of two billion years of evolution."

Other scientists, anticipating this biological revolution, had been less sanguine about it. As the geneticist J. B. S. Haldane had described it in 1923, once the power to control genes had been harnessed, "no beliefs, no values, no institutions are safe."

"To Get the Genome"

A-hunting we will go, a-hunting we will go!
We'll catch a fox and put him in a box,
And then we'll let him go.
> —Children's rhyme from the
> eighteenth century

Our ability to read out this sequence of our own genome has
the makings of a philosophical paradox. Can an intelligent
being comprehend the instructions to make itself?
> —John Sulston

Scholars of Renaissance shipbuilding have often debated the nature of the technology that spurred the explosive growth of transoceanic navigation in the late 1400s and 1500s, ultimately leading to the discovery of the New World. Was it the capacity to build larger ships—galleons, carracks, and fluyts—as one camp insists? Or was it the invention of new navigation technologies—a superior astrolabe, the navigator's compass, and the early sextant?

In the history of science and technology too, breakthroughs seem to come in two fundamental forms. There are scale shifts—where the crucial advance emerges as a result of an alteration of size or scale alone (the moon rocket, as one engineer famously pointed out, was just a massive jet plane pointed vertically at the moon). And there are conceptual shifts—in which the advance arises because of the emergence of a radical new concept or idea. In truth, the two modes are not mutually exclusive, but reinforcing. Scale shifts enable conceptual shifts, and new concepts, in turn, demand new scales. The microscope opened a door to a subvisual world. Cells and intracellular organelles were revealed, raising questions about the inner anatomy and physiology of a cell, and demanding yet more powerful microscopes to understand the structures and functions of these subcellular compartments.

Between the mid-1970s and the mid-1980s, genetics had witnessed many conceptual shifts—gene cloning, gene mapping, split genes, genetic engineering, and new modes of gene regulation—but no radical shifts in scale. Over the decade, hundreds of individual genes had been isolated, sequenced, and cloned by virtue of functional characteristics—but no comprehensive catalog of all genes of a cellular organism existed. In principle, the technology to sequence an entire organismal genome had been invented, but the sheer size of the effort had made scientists balk. In 1977, when Fred Sanger had sequenced the genome of the phiX virus, with 5,386 bases of DNA, that number represented the outer limit of gene-sequencing capability. The human genome contains 3,095,677,412 base pairs—representing a scale shift of 574,000-fold.

①

The potential benefit of a comprehensive sequencing effort was particularly highlighted by the isolation of disease-linked genes in humans. Even as the mapping and identification of crucial human genes was being celebrated in the popular press in the early 1990s, geneticists—and patients—were privately voicing concerns about the inefficiency and laboriousness of the process. For Huntington's disease, it had taken no less than twenty-five years to move from one patient (Nancy Wexler's mother) to the gene (one hundred and twenty-one years, if you count Huntington's original case history of the disease). Hereditary forms of breast cancer had been known since antiquity, yet the most common breast-cancer-associated gene, *BRCA1*, was only identified in 1994. Even with new technologies, such as chromosome jumping, that had been used to isolate the cystic fibrosis gene, finding and mapping genes was frustratingly slow. "There was no shortage of exceptionally clever people trying to find genes in the human," John Sulston, the worm biologist, noted, "but they were wasting their time theorizing about the bits of the sequence that might be necessary." The gene-by-gene approach, Sulston feared, would eventually come to a standstill.

James Watson echoed the frustration with the pace of "single-gene" genetics. "But even with the immense power of recombinant DNA methodologies," he argued, "the eventual isolation of most disease genes still seemed in the mid 1980s beyond human capability." What Watson sought was the sequence of the entire human genome—all 3 billion base pairs of it, starting with the first nucleotide and ending with the last. Every known human gene, including all of its genetic code, all the regulatory sequences, every intron and exon, and all the long stretches of DNA between genes and all protein-coding segments, would be found in that sequence. The sequence would act as a template for the annotation of genes discovered in the future: if a geneticist found a novel gene that increases the risk for breast cancer, for instance, she should be able to decipher its precise location and sequence by mapping it to the master sequence of the human genome. And the sequence would also be the "normal" template against which abnormal genes—i.e., mutations—could be annotated: by comparing that breast-cancer-associated gene between affected and unaffected women, the geneticist would be able to map the mutation responsible for causing the disease.

①

Impetus for sequencing the entire human genome came from two other sources. The one-gene-at-a-time approach worked perfectly for "monogenetic" diseases, such as cystic fibrosis and Huntington's disease. But most common human diseases do not arise from single-gene mutations. These are not genetic illnesses as much as *genomic* illnesses: multiple genes, spread diffusely throughout the human genome, determine the risk for the illness. These diseases cannot be understood through the action of a single gene. They can only be understood, diagnosed, or predicted by understanding the interrelationships between several independent genes.

The archetypal genomic disease is cancer. That cancer is a disease of genes had been known for more than a century: in 1872, Hilário de Gouvêa, a Brazilian ophthalmologist, had described a family in which a rare form of eye cancer, called retinoblastoma, coursed tragically through multiple generations. Families certainly share much more than genes: bad habits, bad recipes, neuroses, obsessions, environments, and behaviors—but the familial pattern of the illness suggested a genetic cause. De Gouvêa proposed an "inherited factor" as the cause of these rare eye tumors. Halfway across the globe and seven years prior, an unknown botanist-monk named Mendel had published a paper on inherited factors in peas—but de Gouvêa had never encountered Mendel's paper or the word *gene*.

By the late 1970s, a full century after de Gouvêa, scientists began to converge on the uncomfortable realization that cancers arose from normal cells that had acquired mutations in growth-controlling genes.* In normal cells, these genes act as powerful regulators of growth: hence a wound in the skin, having healed itself, typically stops healing and does not morph into a tumor (or in the language of genetics: genes tell the cells in a wound when to start growing, and when to stop). In cancer cells, geneticists realized, these pathways were somehow disrupted. Start genes were jammed *on*, and stop genes were flicked *off*; genes that altered metabolism and identity of a cell were corrupted, resulting in a cell that did not know how to stop growing.

That cancer was the result of alterations of such *endogenous* genetic pathways—a "distorted version of our normal selves," as Harold Varmus, the cancer biologist, put it—was ferociously disquieting: for decades, scientists had hoped that some pathogen, such as a virus or bacterium, would be implicated as the universal cause of cancer, and might potentially be eliminated via a vaccine or antimicrobial therapy. The intimacy of the relationship between cancer genes and normal genes threw open a central challenge of cancer biology: How might the mutant genes be restored to their *off* or *on* states, while allowing normal growth to proceed unperturbed? This was—and still remains—the defining goal, the perennial fantasy, and the deepest conundrum, of cancer therapy.

Normal cells could acquire these cancer-causing mutations through four mechanisms. The mutations could be caused by environmental insults, such as tobacco smoke, ultraviolet light, or X-rays—agents that attack DNA and change its chemical structure. Mutations could arise from spontaneous errors during cell division (every time DNA is replicated in a cell, there's a minor chance that the copying process generates an error—an A switched to a T, G, or C, say). Mutant cancer genes could be inherited from parents, thereby causing hereditary cancer syndromes such as retinoblastoma and breast cancer that coursed through families.

---

* The twisted intellectual journey, with its false leads, exhausting trudges, and inspired shortcuts, that ultimately revealed that cancer was caused by the corruption of *endogenous* human genes deserves a book in its own right.

In the 1970s, the reigning theory of carcinogenesis was that all, or most, cancers were caused by viruses. Pathbreaking experiments performed by several scientists, including Harold Varmus and J. Michael Bishop at UCSF, revealed, surprisingly, that these viruses typically caused cancer by tampering with *cellular genes*—called *proto-oncogenes*. The vulnerabilities, in short, were already present within the human genome. Cancer occurs when these genes are mutated, thereby unleashing dysregulated growth.

Or the genes could be carried into the cells via viruses, the professional gene carriers and gene swappers of the microbial world. In all four cases, the result converged on the same pathological process: the inappropriate activation or inactivation of genetic pathways that controlled growth, causing the malignant, dysregulated cellular division that was characteristic of cancer.

That one of the most elemental diseases in human history happens to arise from the corruption of the two most elemental processes in biology is not a co-incidence: cancer co-opts the logic of both evolution and heredity; it is a pathological convergence of Mendel and Darwin. Cancer cells arise via mutation, survival, natural selection, and growth. And they transmit the instructions for malignant growth to their daughter cells via their genes. As biologists realized in the early 1980s, cancer, then, was a "new" kind of genetic disease—the result of heredity, evolution, environment, and chance all mixed together.

Ⓞ

But how many such genes were involved in causing a typical human cancer? One gene per cancer? A dozen? A hundred? In the late 1990s, at Johns Hopkins University, a cancer geneticist named Bert Vogelstein decided to create a comprehensive catalog of nearly all the genes implicated in human cancers. Vogelstein had already discovered that cancers arise from a step-by-step process involving the accumulation of dozens of mutations in a cell. Gene by gene, a cell slouches toward cancer—acquiring one, two, four, and then dozens of mutations that tip its physiology from controlled growth to dysregulated growth.

To cancer geneticists, these data clearly suggested that the one-gene-at-a-time approach would be insufficient to understand, diagnose, or treat cancer. A fundamental feature of cancer was its enormous genetic diversity: two specimens of breast cancer, removed from two breasts of the same woman at the same time, might have vastly different spectra of mutations—and thereby behave differently, progress at different rates, and respond to different chemotherapies. To understand cancer, biologists would need to assess the entire genome of a cancer cell.

If the sequencing of cancer genomes—not just individual cancer genes—was necessary to understand the physiology and diversity of cancers, then it was all the more evident that the sequence of the normal genome had to be completed first. The human genome forms the normal counterpart to the cancer genome. A genetic mutation can be described only in the context of a normal or "wild-type" counterpart. Without that template of normalcy, one had little hope that the fundamental biology of cancer could be solved.

①

Like cancer, heritable mental illnesses were also turning out to involve dozens of genes. Schizophrenia, in particular, sparked a furor of national attention in 1984, when James Huberty, a man known to have paranoid hallucinations, strolled casually into a McDonald's in San Diego on a July afternoon and shot and killed twenty-one people. The day before the massacre, Huberty had left a desperate message with a receptionist at a mental health clinic, pleading for help, then waited for hours by his phone. The return phone call never came; the receptionist had mistakenly spelled his name *Shouberty* and neglected to copy his number. The next morning, still afloat in a paranoid fugue, he had left home with a loaded semiautomatic wrapped in a checkered blanket, having told his daughter that he was "going hunting humans."

The Huberty catastrophe occurred seven months after an enormous National Academy of Sciences (NAS) study published data definitively linking schizophrenia to genetic causes. Using the twin method pioneered by Galton in the 1890s, and by Nazi geneticists in the 1940s, the NAS study found that identical twins possessed a striking 30 to 40 percent concordance rate for schizophrenia. An earlier study, published by the geneticist Irving Gottesman in 1982, had found an even more provocative correlation of between 40 and 60 percent in identical twins. If one twin was diagnosed with schizophrenia, then the chance of the other twin developing the illness was fifty times higher than the risk of schizophrenia in the general population. For identical twins with the severest form of schizophrenia, Gottesman had found the concordance rate was 75 to 90 percent: nearly *every* identical twin with one of the severest variants of schizophrenia had been found to have a twin with the same illness. This high degree of concordance between identical twins suggested a powerful genetic influence on schizophrenia. But notably, both the NAS and the Gottesman study found that the concordance rate fell sharply between nonidentical twins (to about 10 percent).

To a geneticist, such a pattern of inheritance offers important clues about the underlying genetic influences on an illness. Suppose schizophrenia is caused by a single, dominant, highly penetrant mutation in one gene. If one identical twin inherits that mutant gene, then the other will invariably inherit that gene. Both will manifest the disease, and the concordance between the twins should approach 100 percent. Fraternal twins and siblings should, on average, inherit that gene about half the time, and the concordance between them should fall to 50 percent.

In contrast, suppose schizophrenia is not one disease but a family of diseases. Imagine that the cognitive apparatus of the brain is a complex mechanical engine, composed of a central axle, a main gearbox, and dozens of smaller pistons and gaskets to regulate and fine-tune its activity. If the main axle breaks, and the gearbox snaps, then the entire "cognition engine" collapses. This is analogous to the severe variant of schizophrenia: a combination of a few highly penetrant mutations in genes that control neural communication and development might cause the axle and the gears to collapse, resulting in severe deficits of cognition. Since identical twins inherit identical genomes, both will invariably inherit mutations in the axle and the gearbox genes. And since the mutations are highly penetrant, the concordance between identical twins will still approach 100 percent.

But now imagine that the cognition engine can also malfunction if several of the smaller gaskets, spark plugs, and pistons do not work. In this case, the engine does not fully collapse; it sputters and gasps, and its dysfunction is more situational: it worsens in the winter. This, by analogy, is the milder variant of schizophrenia. The malfunction is caused by a *combination* of mutations, each with low penetrance: these are gasket-and-piston and spark-plug genes, exerting more subtle control on the overall mechanism of cognition.

Here too identical twins, possessing identical genomes, will inherit, say, all five variants of the genes together—but since the penetrance is incomplete, and the triggers more situational, the concordance between identical twins might fall to only 30 or 50 percent. Fraternal twins and siblings, in contrast, will share only a few of these gene variants. Mendel's laws guarantee that all five variants will rarely be inherited *in toto* by two siblings. The concordance between fraternal twins and siblings will fall even more sharply—to 5 or 10 percent.

This pattern of inheritance is more commonly observed in schizophrenia. That identical twins share only a 50 percent concordance—i.e., if one twin is affected, then the other twin is affected only 50 percent of the time—clearly demonstrates that some other triggers (environmental factors or chance events) are required to tip the predisposition over an edge. But when a child of a schizophrenic parent is adopted at birth by a nonschizophrenic family, the child still has a 15 to 20 percent risk of developing the illness—about twentyfold higher than the general population—demonstrating that the genetic influences can be powerful and autonomous despite enormous variations in environments. These patterns strongly suggest that schizophrenia is a complex, polygenic illness, involving multiple variants, multiple genes, and potential environmental or chance triggers. As with cancer and other polygenic diseases, then, a gene-by-gene approach is unlikely to unlock the physiology of schizophrenia.

Ⓓ

Populist anxieties about genes, mental illness, and crime were fanned further with the publication in the summer of 1985 of *Crime and Human Nature: The Definitive Study of the Causes of Crime*, an incendiary book written by James Q. Wilson, a political scientist, and Richard Herrnstein, a behavioral biologist. Wilson and Herrnstein argued that particular forms of mental illness—most notably schizophrenia, especially in its violent, disruptive form—were highly prevalent among criminals, likely to be genetically ingrained, and likely to be the cause of criminal behavior. Addiction and violence also had strong genetic components. The hypothesis seized popular imagination. Postwar academic criminology had been dominated by "environmental" theories of crime—i.e., criminals were the products of bad influences: "bad friends, bad neighborhoods, bad labels." Wilson and Herrnstein acknowledged these factors, but added the most controversial fourth—"bad genes." The soil was not contaminated, they suggested; the seed was. *Crime and Human Nature* steamrolled into a major media phenomenon: twenty major news outlets—the *New York Times*, *Newsweek*, and *Science* among them—reviewed or featured it. *Time* reinforced its essential message in its headline: "Are Criminals Born, Not Made?" *Newsweek*'s byline was more blunt: "Criminals Born and Bred."

Wilson and Herrnstein's book was met with a barrage of criticism. Even die-hard believers of the genetic theory of schizophrenia had to admit that the etiology of the illness was largely unknown, that acquired influences had to play a major triggering role (hence the 50—not 100—percent concordance among identical twins), and that the vast majority of schizophrenics lived in the terrifying shadow of their illness but had no history of criminality whatsoever.

But to a public frothing with concern about violence and crime in the eighties, the idea that the human genome might contain the answers not just to medical illnesses, but to social maladies such as deviance, alcoholism, violence, moral corruption, perversion, or addiction, was potently seductive. In an interview in the *Baltimore Sun*, a neurosurgeon wondered if the "crime-prone" (such as Huberty) could be identified, quarantined, and treated *before* they had committed crimes—i.e., via genetic profiling of precriminals. A psychiatric geneticist commented on the impact that identifying such genes might have on the public discourse around crime, responsibility, and punishment. "The link [to genetics] is quite clear. . . . We would be naïve not to think that one aspect of [curing crime] will be biological."

①

Set against this monumental backdrop of hype and expectation, the first conversations to approach human genome sequencing were remarkably deflating. In the summer of 1984, Charles DeLisi, a science administrator from the Department of Energy (DOE) convened a meeting of experts to evaluate the technical feasibility of human genome sequencing. Since the early 1980s, DOE researchers had been investigating the effects of radiation on human genes. The Hiroshima/Nagasaki bombings of 1945 had sprayed hundreds of thousands of Japanese citizens with varying doses of radiation, including twelve thousand surviving children, now in their forties and fifties. How many mutations had occurred in these children, in what genes, and over what time? Since radiation-induced mutations would likely be randomly scattered through the genome, a gene-by-gene search would be futile. In December 1984, another meeting of scientists was called to evaluate whether whole-genome sequencing might be used to detect genetic alterations in radiation-exposed children. The conference was held at Alta, in Utah—the same mountain town where Botstein and Davis had originally conceived the idea of mapping human genes using linkage and polymorphisms.

On the surface, the Alta meeting was a spectacular failure. Scientists realized that the sequencing technology available in the mid-1980s was nowhere close to being able to map mutations across a human genome. But the meeting was a crucial platform to jump-start a conversation about comprehensive gene sequencing. A fleet of meetings on genome sequencing followed—in Santa Cruz in May 1985 and in Santa Fe in March 1986. In the late summer of 1986, James Watson convened perhaps the most decisive of these meetings at Cold Spring Harbor, provocatively titling it "The Molecular Biology of *Homo sapiens*." As with Asilomar, the serenity of the campus—on a placid, crystalline bay, with rolling hills tipping into the water—contrasted with the fervent energy of the discussions.

A host of new studies was presented at the meeting that suddenly made genome sequencing seem within technological reach. The most important technical breakthrough, perhaps, came from Kary Mullis, a biochemist studying gene replication. To sequence genes, it is crucial to have enough starting material of DNA. A single bacterial cell can be grown into hundreds of millions of cells, thereby supplying abundant amounts of bacterial DNA for sequencing. But it is difficult to grow hundreds of millions of human cells. Mullis had discovered an ingenious shortcut. He made a copy of a human gene in a test tube using DNA polymerase, then used that copy to make copies of the copy, then copied the multiple copies for dozens of cycles. Each cycle of copying amplified the DNA, resulting in an exponential increase in the yield of a gene. The technique was eventually called the polymerase chain reaction, or PCR, and would become crucial for the Human Genome Project.

Eric Lander, a mathematician turned biologist, told the audience about new mathematical methods to find genes related to complex, multigenic diseases. Leroy Hood, from Caltech, described a semiautomated machine that could speed up Sanger's sequencing method by ten- or twentyfold.

Earlier, Walter Gilbert, the DNA-sequencing pioneer, had prepared an edge-of-napkin calculation of the costs and personnel involved. To sequence all 3 billion base pairs of human DNA, Gilbert estimated, would take about fifty thousand person years and cost around $3 billion—one dollar per base. As Gilbert, with characteristic panache, strode across the floor to inscribe the number on a chalkboard, an intense debate broke out in the audience. "Gilbert's number"—which would turn out to be startlingly accurate—had reduced the genome project to tangible realities. Indeed, put in perspective, the cost was not even particularly large: at its peak, the Apollo program had employed nearly four hundred thousand people, with a total cumulative cost of about $100 billion. If Gilbert was right, the human genome could be had for less than one-thirtieth of the moon landing. Sydney Brenner later joked that the sequencing of the human genome would perhaps ultimately be limited not by cost or technology, but only by the severe monotony of its labor. Perhaps, he speculated, genome sequencing should be doled out as a punishment to criminals and convicts—1 million bases for robbery, 2 million for homicide, 10 million for murder.

As dusk fell on the bay that evening, Watson spoke to several scientists about an unfolding personal crisis of his. On May 27, the night before the conference, his fifteen-year-old-son, Rufus Watson, had escaped from a psychiatric facility in White Plains. He was later found wandering in the woods, near the train tracks, captured, and brought back to the facility. A few months earlier, Rufus had tried to break a window at the World Trade Center to jump off the building. He had been diagnosed with schizophrenia. To Watson, a firm believer in the genetic basis for the disease, the Human Genome Project had come home—literally. There were no animal models for schizophrenia, nor any obviously linked polymorphisms that would allow geneticists to find the relevant genes. "The only way to give Rufus a life was to understand why he was sick. And the only way we could do that was to get the genome."

①

But which genome "to get"? Some scientists, including Sulston, advocated a graded approach—starting with simple organisms, such as baker's yeast, the worm, or the fly, and then scaling the ladder of complexity and size to the human genome. Others, such as Watson, wanted to leap directly into the human genome. After a prolonged internal debate, the scientists reached a compromise. The sequencing of the genomes of simple organisms, such as worms and flies, would begin at first. These projects would carry the names of their respective organisms: the Worm Genome Project, or the Fruit Fly Genome Project—and they would fine-tune the technology of gene sequencing. The sequencing of human genes would continue in parallel. The lessons learned from simple genomes would be applied to the much larger and more complex human genome. This larger endeavor—the comprehensive sequencing of the entire human genome—was termed the Human Genome Project.

The NIH and DOE, meanwhile, jostled to control the Human Genome Project. By 1989, after several congressional hearings, a second compromise was reached: the National Institutes of Health would act as the official "lead agency" of the project, with the DOE contributing resources and strategic management. Watson was chosen as its head. International collaborators were swiftly added: the Medical Research Council of the United Kingdom and the Wellcome Trust joined the effort. In time, French, Japanese, Chinese, and German scientists would join the Genome Project.

In January 1989, a twelve-member council of advisers met in a conference room in Building 31 on the far corner of the NIH campus in Bethesda. The council was chaired by Norton Zinder, the geneticist who had helped draft the Asilomar moratorium. "Today we begin," Zinder announced. "We are initiating an unending study of human biology. Whatever it's going to be, it will be an adventure, a priceless endeavor. And when it's done, someone else will sit down and say, 'It's time to begin.'"

①

On January 28, 1983, on the eve of the launch of the Human Genome Project, Carrie Buck died in a nursing home in Waynesboro, Pennsylvania. She was seventy-six years old. Her birth and death had bookended the near century of the gene. Her generation had borne witness to the scientific resurrection of genetics, its forceful entry into public discourse, its perversion into social engineering and eugenics, its postwar emergence as the central theme of the "new" biology, its impact on human physiology and pathology, its powerful explanatory power in our understanding of illness, and its inevitable intersection with questions of fate, identity, and choice. She had been one of the earliest victims of the misunderstandings of a powerful new science. And she had watched that science transform our understanding of medicine, culture, and society.

What of her "genetic imbecility"? In 1930, three years after her Supreme Court–mandated sterilization, Carrie Buck was released from the Virginia State Colony and sent to work with a family in Bland County, Virginia. Carrie Buck's only daughter, Vivian Dobbs—the child who had been examined by a court and declared "imbecile"—died of enterocolitis in 1932. During the eight-odd years of her life, Vivian had performed reasonably well in school. In Grade 1B, for instance, she received A's and B's in deportment and spelling, and a C in mathematics, a subject that she had always struggled with. In April 1931, she was placed on the honor roll. What remains of the school report cards suggests a cheery, pleasant, happy-go-lucky child whose performance was no better, and no worse, than that of any other schoolchild. Nothing in Vivian's story bears an even remote suggestion of an inherited propensity for mental illness or imbecility—the diagnosis that had sealed Carrie Buck's fate in court.

The Geographers

So Geographers in Afric-maps,
With Savage-Pictures fill their Gaps;
And o'er uninhabitable Downs
Place Elephants for want of Towns.
— Jonathan Swift, "On Poetry"

More and more, the Human Genome Project, supposedly
one of mankind's noblest undertakings, is resembling a
mud-wrestling match.

— Justin Gillis, 2000

It is fair to say that the first surprise for the Human Genome Project had nothing to do with genes. In 1989, as Watson, Zinder, and their colleagues were gearing up to launch the Genome Project, a little-known neurobiologist at NIH, Craig Venter, proposed a shortcut to genome sequencing.

Pugnacious, single-minded, and belligerent, a reluctant student with middling grades, a surfing and sailing addict, and a former serviceman in the Vietnam War, Venter had an ability to lunge headlong into unknown projects. He had trained in neurobiology and had spent much of his scientific life studying adrenaline. In the mid-eighties, working at the NIH, Venter had become interested in sequencing genes expressed in the human brain. In 1986, he had heard of Leroy Hood's rapid-sequencing machine and rushed to buy an early version for his laboratory. When it arrived, he called it "my future in a crate." He had an engineer's tinkering hands, and a biochemist's love of mixing solutions. Within months, Venter had become an expert in rapid genome sequencing using the semiautomated sequencer.

Venter's strategy for genome sequencing relied on a radical simplification. While the human genome contains genes, of course, the vast part of the genome is devoid of genes. The enormous stretches of DNA between genes, called intergenic DNA, are somewhat akin to the long highways between Canadian towns. And as Phil Sharp and Richard Roberts had demonstrated, a gene is itself broken up into segments, with long spacers, called introns, interposed between the protein-coding segments.

Intergenic DNA and introns—spacers between genes and stuffers within genes—do not encode any protein information.* Some of these stretches contain information to regulate and coordinate the expression of genes in time and space; they encode on and off switches appended to genes. Other stretches encode no known function. The structure of the human genome can thus be likened to a sentence that reads—

This . . . . . . is the . . . . . . str . . . uc . . . . . . ture . . . , , , . . . of . . . your . . .
(. . . gen . . . ome . . .) . . .

—where the words correspond to the genes, the ellipses correspond to the spacers and stuffers, and the occasional punctuation marks demarcate the regulatory sequences of genes.

Ⓓ

Venter's first shortcut was to ignore the spacers and stuffers of the human genome. Introns and intergenic DNA did not carry protein information, he reasoned, so why not focus on the "active," protein-encoding parts? And—piling shortcut on shortcut—he proposed that perhaps even these active parts could be assessed even faster if only fragments of genes were sequenced. Convinced that this fragmented-gene approach would work, Venter had begun to sequence hundreds of such gene fragments from brain tissue.

To continue our analogy between genomes and sentences in English, it was as if Venter had decided to find shards of words in a sentence—*struc,*

* Stretches of DNA associated with a gene called promoters can be likened to "on" switches for that gene. These sequences encode information about when and where to activate a gene (thus hemoglobin is only turned on in red blood cells). In contrast, other stretches of DNA encode information about when and where to turn a gene "off" (thus lactose-digesting genes are turned off in a bacterial cell unless lactose becomes the dominant nutrient). It is remarkable that the system of "on" and "off" gene switches, first discovered in bacteria, is conserved throughout biology.

your, and *geno*—in the human genome. He might not learn the content of the entire sentence with this method, he knew, but perhaps he could deduce enough from the shards to understand the crucial elements of human genes.

Watson was appalled. Venter's "gene-fragment" strategy was indubitably faster and cheaper, but to many geneticists, it was also sloppy and incomplete, since it produced only fragmentary information about the genome.* The conflict was deepened by an unusual development. In the summer of 1991, as Venter's group began to drudge up sequences of human gene fragments from the brain, the NIH technology transfer office contacted Venter about patenting the novel gene fragments. For Watson, the dissonance was embarrassing: now, it seemed, one arm of the NIH was filing for exclusive rights to the same information that another arm was trying to discover and make freely available.

But by what logic could genes—or, in Venter's case, "active" fragments of genes—be patented? At Stanford, Boyer and Cohen, recall, had patented a *method* to "recombine" pieces of DNA to create genetic chimeras. Genentech had patented a *process* to express proteins such as insulin in bacteria. In 1984, Amgen had filed a patent for the isolation of the blood-production hormone erythropoietin using recombinant DNA—but even that patent, carefully read, involved a scheme for the production and isolation of a distinct protein with a distinct function. No one had ever patented a gene, or a piece of genetic information, for its own sake. Was a human gene not like any other body part—a nose or the left arm—and therefore fundamentally unpatentable? Or was the discovery of new genetic information so novel that it would merit ownership and patentability? Sulston, for one, was firmly opposed to the idea of gene patents. "Patents (or so I had believed) are designed to protect inventions," he wrote. "There was no 'invention' involved in finding [gene fragments] so how could they be patentable?" "It's a quick and dirty land grab," one researcher wrote dismissively.

The controversy around Venter's gene patents reached an even more fervent pitch because the gene fragments were being sequenced at random, without the ascription of any function to most of the genes. Since

---

* Venter's strategy of sequencing protein-encoding and RNA-encoding portions of the genome would, in the end, prove to be an invaluable resource for geneticists. Venter's method revealed parts of the genome that were "active," thereby allowing geneticists to annotate these active parts against the whole genome.

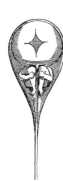

This homunculus, wrapped inside human sperm, was drawn by Nicolaas Hartsoeker in 1694. Like many other biologists in his time, Hartsoeker believed in "spermism," the theory that the information to create a fetus was transmitted by the miniature human form lodged inside sperm.

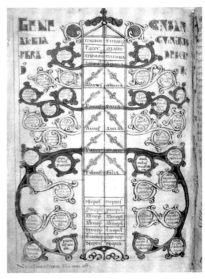

In medieval Europe, "trees of lineage" were often created to mark the ancestors and descendants of noble families. These trees were used to stake claims on peerage and property, or to seek marital arrangements between families (in part, to decrease the chances of consanguineous marriages between cousins). The word *gene*—at the top left corner—was used in the sense of genealogy or descent. The modern connotation of *gene*, as a unit of hereditary information, appeared centuries later in 1909.

Charles Darwin (here in his seventies) and his "tree of life" sketch, showing organisms radiating out from a common ancestral organism (note the doubt-ridden phrase "I think," scribbled above the diagram). Darwin's theory of evolution by variation and natural selection demanded a theory of heredity via genes. Close readers of Darwin's theory realized that evolution could work only if there were indivisible, but mutable, particles of heredity that transmit information between parents and offspring. Yet Darwin, having never read Gregor Mendel's paper, never found an adequate formulation of such a theory during his lifetime.

Gregor Mendel holds a flower, possibly from a pea plant, in his monastery garden in Brno (now in the Czech Republic). Mendel's seminal experiments in the 1850s and '60s identified indivisible particles of information as carriers of hereditary information. Mendel's paper (1865) was largely ignored for four decades, and then transformed the science of biology.

William Bateson's "rediscovery" of Mendel's work in 1900 converted him into a believer in genes. Bateson coined the term *genetics* in 1905 to describe the study of heredity. Wilhelm Johannsen (*left*) coined the term *gene* to describe a unit of heredity. Johannsen visited Bateson at his house in Cambridge, England; the two became close collaborators and vigorous defenders of the gene theory.

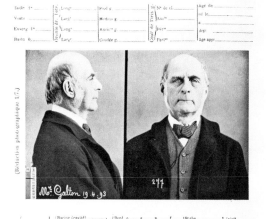

Francis Galton, aged 71, photographed as a criminal on his visit to Bertillon's Criminal Identification Laboratory in Paris, 1893.

Francis Galton—mathematician, biologist, and statistician—put himself on one of his own "anthropometry cards," in which he tabulated a person's height, weight, facial features, and other characteristics. Galton resisted Mendel's theory of genes. He also believed that the selective breeding of humans with the "best" features would lead to the creation of an improved human race. *Eugenics*, a term coined by Galton for the science of human emancipation through the manipulation of heredity, would soon morph into a macabre form of social and political control.

The Nazi doctrine of "racial hygiene" prompted a vast state-sponsored effort to cleanse the human race through sterilization, confinement, and murder. Twin studies were used to prove the power of hereditary influences, and men, women, and children were exterminated based on an assumption that they carried defective genes. The Nazis extended their eugenic efforts to exterminate Jews, Gypsies, dissidents, and homosexuals. Here, Nazi scientists measure the height of twins, and demonstrate family history charts to Nazi recruits.

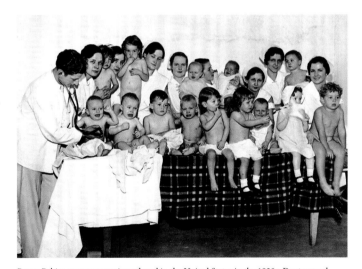

Better Babies contests were introduced in the United States in the 1920s. Doctors and nurses examined children (all white) for the best genetic features. Better Babies contests generated passive support for eugenics in America by showcasing the healthiest babies as products of genetic selection.

A "eugenics tree" cartoon from the United States argues for the "self-direction of human evolution." Medicine, surgery, anthropology, and genealogy are the "roots" of the tree. Eugenic science hoped to use these foundational principles to select fitter, healthier, and more accomplished humans.

In the 1920s, Carrie Buck and her mother, Emma Buck, were sent to the Virginia State Colony for Epileptics and Feebleminded, where women classified as "imbeciles" were routinely sterilized. The photograph, obtained on the pretext of capturing a casual moment between mother and daughter, was staged to provide evidence of the resemblance between Carrie and Emma, and thus proof of their "hereditary imbecility."

At Columbia University, and subsequently at Caltech University in the 1920s and '30s, Thomas Morgan used fruit flies to demonstrate that genes were physically linked to each other, presciently predicting that a single, chainlike molecule carried genetic information. Linkage between genes would eventually be used to generate genetic maps in humans and lay the foundation for the Human Genome Project. This is Morgan in his Caltech Fly Room, surrounded by the milk bottles in which he bred his maggots and flies.

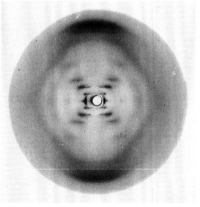

Rosalind Franklin looks down a microscope at King's College in London in the 1950s. Franklin used X-ray crystallography to photograph and study the structure of DNA. Photograph 51 is the clearest of Franklin's photographs of a DNA crystal. The photo suggested a double-helix structure, although the precise orientations of the bases A, C, T, and G were not clear from it.

James Watson and Francis Crick demonstrate their model of DNA as a double helix in Cambridge in 1953. Watson and Crick solved the structure of DNA by realizing that the A in one strand was paired against the T in the other, and the G against the C.

At the Moore Clinic in Baltimore in the 1950s, Victor McKusick created a vast catalog of human mutations. He found that one phenotype—short stature, or "dwarfism"—could be caused by mutations in several disparate genes. Conversely, diverse phenotypes could be caused by mutations in a single gene.

Nancy Wexler's mother and uncles were diagnosed with Huntington's disease, a lethal neurodegenerative disease that spurs involuntary sinuous or jerking movements. The diagnosis launched her personal hunt for the gene that causes the illness. Wexler found a cluster of several patients with Huntington's disease in Venezuela, all likely descended from a founder with the disease. Huntington's disease was one of the first human diseases to be definitively linked to a single gene using modern gene-mapping methods.

Students protest a genetics meeting in the 1970s. The novel technologies of gene sequencing, gene cloning, and recombinant DNA raised anxieties that new forms of eugenics would be used to create a "perfect race." The link to Nazi eugenics was not forgotten.

Herb Boyer (*left*) and Robert Swanson founded Genentech in 1976 to produce medicines out of genes. The drawing on the blackboard shows the scheme to produce insulin using recombinant DNA technology. The first such proteins were produced in enormous bacterial incubators under Swanson's watchful eye.

Paul Berg speaks to Maxine Singer and Norton Zinder at the Asilomar meeting in 1975, while Sydney Brenner takes notes in the background. Following the discovery of technologies to create genetic hybrids between genes (recombinant DNA) and produce millions of copies of these hybrids in bacterial cells (gene cloning), Berg and others proposed a "moratorium" on certain recombinant DNA work until the risks had been adequately assessed.

Frederick Sanger examines a DNA sequencing gel. Sanger's invention of a technique to sequence DNA (i.e., read the precise stretch of letters—A, C, T, and G—in a gene's sequence) revolutionized our understanding of genes, and set the stage for the Human Genome Project.

Jesse Gelsinger poses in Philadelphia a few months before his death in 1999. Gelsinger was one of the first patients to be treated with gene therapy. A virus was designed to deliver the correct form of a mutated gene into his liver, but Gelsinger had a brisk immunological response to the virus, resulting in organ failure and death. Gelsinger's "biotech death" would spur nationwide responses to ensure the safety of gene-therapy trials.

The February 2001 cover of *Science* magazine announced the draft sequence of the human genome.

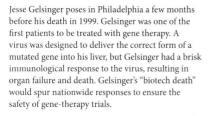

Craig Venter (*left*), President Bill Clinton, and Francis Collins announce the draft sequence of the human genome on June 26, 2000, at the White House.

Even without subtle techniques to alter human genomes, the capacity to assess a child's genome *in utero* has led to vast dysgenic efforts around the world. In parts of China and India, the assessment of male versus female gender by amniocentesis, and the selective abortion of female fetuses, has skewed the sex ratio to 0.8 females to 1 male, and caused unprecedented alterations of population and family structures.

Faster and more accurate gene-sequencing machines (housed inside gray boxlike containers) linked to supercomputers that analyze and annotate genetic information can now sequence individual human genomes in months. Variations of this technique can be used to sequence the genome of a multicelled embryo or a fetus, enabling preimplantation genetic diagnosis and *in utero* diagnosis of future illness.

Jennifer Doudna (*right*), a biologist and RNA researcher at Berkeley, is among those working on a system to deliver targeted, intentional mutations in genes. In principle, the system can be used to "edit" the human genome, although the technology still remains to be perfected and assessed for safety and fidelity. If intentional genetic changes were introduced into sperm, egg, or human embryonic stem cells, the technology would portend the genesis of humans with altered genes.

Venter's approach often resulted in incomplete shards of genes being sequenced, the nature of the information was necessarily garbled. At times, the shards were long enough to deduce the function of a gene—but more commonly, no real understanding could be ascribed to these fragments. "Could you patent an elephant by describing its tail? What about patenting an elephant by describing three discontinuous parts of its tail?" Eric Lander argued. At a congressional hearing on the genome project, Watson erupted with vehemence: "virtually any monkey" could generate such fragments, he argued. Walter Bodmer, the English geneticist, warned that if the Americans granted gene-fragment patents to Venter, the British would start their own rival patenting effort. Within weeks, the genome would be balkanized—carved up into a thousand territorial colonies carrying American, British, and German flags.

On June 10, 1992, fed up with the interminable squabbles, Venter left the NIH to launch his own private gene-sequencing institute. At first the organization was called Institute for Genome Research, but Venter cannily spotted the flaw in the name: its acronym, IGOR, carried the unfortunate association with a cross-eyed Gothic butler apprenticed to Frankenstein. Venter changed it to *The* Institute for Genomic Research—or TIGR, for short.

①

On paper—or, at least, on scientific papers—TIGR was a phenomenal success. Venter collaborated with scientific luminaries, such as Bert Vogelstein and Ken Kinzler, to discover new genes associated with cancer. More important, Venter kept battering at the technological frontiers of genome sequencing. Uniquely sensitive to his critics, he was also uniquely responsive to them: in 1993, he expanded his sequencing efforts beyond gene fragments to full genes, and to genomes. Working with a new ally, Hamilton Smith, the Nobel Prize–winning bacteriologist, Venter decided that he would sequence the entire genome of a bacterium that causes lethal human pneumonias—*Haemophilus influenzae*.

Venter's strategy was an expansion of the gene-fragment approach he had used with the brain—except with an important twist. This time he would shatter the bacterial genome into a million pieces using a shotgun-like device. He would then sequence hundreds of thousands of fragments at random, then use their overlapping segments to assemble them to solve the entire genome. To return to our sentence analogy, imagine trying to assemble a word using the following word fragments: *stru, uctu, ucture, structu,* and *ucture.* A computer can use the overlapping segments to assemble the full word: *structure.*

The solution depends on the presence of the overlapping sequences: if an overlap does not exist, or some part of the word gets omitted, it becomes impossible to assemble the correct word. But Venter was confident that he could use this approach to shatter and reassemble most of the genome. It was a Humpty Dumpty strategy: all the king's men would solve the jigsaw puzzle by fitting the parts together again. The technique, called "shotgun" sequencing, had been used by Fred Sanger, the inventor of gene sequencing, in the 1980s—but Venter's attack on the *Haemophilus* genome was the most ambitious application of this method in its history.

Venter and Smith launched the *Haemophilus* project in the winter of 1993. By July 1995, it was complete. "The final [paper] took forty drafts," Venter later wrote. "We knew this paper was going to be historic, and I was insistent that it be as near perfect as possible."

It was a marvel: the Stanford geneticist Lucy Shapiro wrote about how members of her lab had stayed up all night reading the *H. flu* genome, "thrilled by the first glimpse at the complete gene content of a living species." There were genes to generate energy, genes to make coat proteins, genes to manufacture proteins, to regulate food, to evade the immune system. Sanger himself wrote to Venter to describe the work as "magnificent."

①

While Venter was sequencing bacterial genomes at TIGR, the Human Genome Project was undergoing drastic internal changes. In 1993, after a series of quarrels with the head of the NIH, Watson stepped down as the project's director. He was swiftly replaced by Francis Collins, the Michigan geneticist known for cloning the cystic fibrosis gene in 1989.

If the Genome Project had not found Collins in 1993, it might have found it necessary to invent him: he was almost preternaturally matched to its peculiar challenges. A devout Christian from Virginia, an able communicator and administrator, a first-rate scientist, Collins was measured, cautious, and diplomatic; to Venter's furious little yacht constantly tilting against the winds, Collins was a transoceanic liner, barely registering the tumult around him. By 1995, as TIGR had roared forward with the

Haemophilus genome, the Genome Project had concentrated its efforts on refining the basic technologies for gene sequencing. In contrast to TIGR's strategy, which was to shred the genome to pieces, sequence at random, and reassemble the data post hoc, the Genome Project had chosen a more orderly approach—assembling and organizing the genomic fragments into a physical map ("Who is next to whom?"), confirming the identity and the overlaps of the clones, and then sequencing the clones in order.

To the early leaders of the Human Genome Project, this clone-by-clone assembly was the only strategy that made any sense. A mathematician-turned-biologist-turned-gene-sequencer, Eric Lander, whose opposition to shotgun sequencing could almost be described as an aesthetic revulsion, liked the idea of sequencing the complete genome piece by piece, as if solving an algebra problem. Venter's strategy, he worried, would inevitably leave potholes in the genome. "What if you took a word, broke it apart, and tried to reconstruct it from the parts?" Lander asked. "That might work if you can find every piece of the word, or if every fragment overlaps. But what if some letters of the word are missing?" The word you might construct out of the available alphabets might convey precisely the *opposite* meaning to the real word; what if you found just the letters "p . . . u . . . n . . . y" in "profundity"?

Proponents of the public Genome Project also feared the false intoxication of a half-finished genome: if gene sequencers left 10 percent of the genome incomplete, the full sequence would never be completed. "The real challenge of the Human Genome Project wasn't starting the sequence. It was *finishing* the sequence of the genome. . . . If you left holes in the genome, but gave yourself the impression of completion, then no one would have the patience to finish the full sequence. Scientists would clap, dust their hands, pat their backs and move on. The draft would always remain a draft," Lander later said.

The clone-by-clone approach required more money, deeper investments in infrastructure, and the one factor that seemed to have gone missing among genome researchers: patience. At MIT, Lander had assembled a formidable team of young scientists—mathematicians, chemists, engineers, and a group of caffeinated twentysomething computer hackers. Phil Green, a mathematician from the University of Washington, was developing algorithms to plod their way methodically through the genome. The British team, funded by the Wellcome Trust, was developing its own platforms for analysis and assembly. More than a dozen groups around the world worked to collect and assemble the data.

①

In May 1998, the ever-moving Venter tacked sharply windward yet again. Although TIGR's shotgun-sequencing efforts had been undeniably successful, Venter still chafed under the organizational structure of the institute. TIGR had been set up as a strange hybrid—a nonprofit institute nestled inside a for-profit company, called Human Genome Sciences (HGS). Venter found this Russian-doll organizational system ridiculous. He argued relentlessly with his bosses. Venter decided to sever his ties with TIGR. He formed yet another new company, which would focus entirely on human genome sequencing. Venter called the new company Celera, a contraction of "accelerate."

A week before a pivotal Human Genome Project meeting at Cold Spring Harbor, Venter met Collins, between flights, at the Red Carpet Lounge at Dulles Airport. Celera was about to launch an unprecedented push to sequence the human genome using shotgun sequencing, Venter announced matter-of-factly. It had bought two hundred of the most sophisticated sequencing machines, and was prepared to run them to the ground to finish the sequence in record time. Venter agreed to make much of the information available as a public resource—but with a menacing clause: Celera would seek to patent the three hundred most important genes that might act as targets for drugs for diseases such as breast cancer, schizophrenia, and diabetes. He laid out an ambitious timeline. Celera hoped to have the whole human genome assembled by 2001, beating the projected deadline for the publicly funded Human Genome Project by four years. He got up abruptly and caught the next flight to California.

Stung into action, the Wellcome Trust doubled its funding for the public effort. Congress threw open the sluices of federal funding, sending $60 million in sequencing grants to seven American centers. Maynard Olson and Robert Waterston acted as strategic leaders and coordinators of the public project and provided key advice to continue the systematic assembly of the genome.

①

In December 1998, the Worm Genome Project scored a decisive victory. John Sulston, Robert Waterson and other researchers working on the genome brought news that the worm (*C. elegans*) genome had been completely sequenced using the clone-by-clone approach favored by proponents of the Human Genome Project.

If the *Haemophilus* genome had nearly brought geneticists to their knees with amazement and wonder in 1995, then the worm genome—the first complete sequence of a multicellular organism—demanded a full-fledged genuflection. Worms are vastly more complex than *Haemophilus*—and vastly more similar to humans. They have mouths, guts, muscles, a nervous system—and even a rudimentary brain. They touch; they feel; they move. They turn their heads away from noxious stimuli. They socialize. Perhaps they register something akin to worm anxiety when their food runs out. Perhaps they feel a fleeting pulse of joy when they mate.

C. elegans was found to have 18,891 genes.* Thirty-six percent of the encoded proteins were similar to proteins found in humans. The rest—about 10,000—had no known similarities to known human genes; these 10,000 genes were either unique to worms, or, much more likely, a potent reminder of how little humans knew of human genes (many of these genes would, indeed, later be found to have human counterparts). Notably, only 10 percent of the encoded genes were similar to genes found in bacteria. Ninety percent of the nematode genome was dedicated to the unique complexities of organism building—demonstrating, yet again, the fierce starburst of evolutionary innovation that had forged multicellular creatures out of single-celled ancestors several million years ago.

As was the case with human genes, a single worm gene could have multiple functions. A gene called *ceh-13*, for instance, organizes the location of cells in the developing nervous system, allows the cells to migrate to the anterior parts of the worm's anatomy, and ensures that the vulva of the worm is appropriately created. And conversely, a single "function" might

* Estimating the number of genes in any organism is complicated and requires some fundamental assumptions about the nature and structure of a gene. Before the advent of whole-genome sequencing, genes were identified by their function. However, whole-genome sequencing does not consider the function of a gene; it is like identifying all the words and letters in an encyclopedia without reference to what any of these words or letters *mean*. The number of genes is estimated by examining the genome sequence and identifying stretches of DNA sequence that *look like* genes—i.e., they contain some regulatory sequences and encode an RNA sequence or resemble other genes found in other organisms. However, as we learn more about the structures and functions of genes, this number is bound to change. Currently, worms are believed to have about 19,500 genes, but that number will continue to evolve as we understand more about genes.

be specified by multiple genes: the creation of a mouth in worms requires the coordinated function of multiple genes.

The discovery of ten thousand new proteins, with more than ten thousand new functions, would have amply justified the novelty of the project—yet the most surprising feature of the worm genome was not protein-encoding genes, but the number of genes that made RNA messages, but no protein. These genes—called "noncoding" (because they do not encode proteins)—were scattered through the genome, but they clustered on certain chromosomes. There were hundreds of them, perhaps thousands. Some noncoding genes were of known function: the ribosome, the giant intracellular machine that makes proteins, contains specialized RNA molecules that assist in the manufacture of proteins. Other noncoding genes were eventually found to encode small RNAs—called micro-RNAs—which regulate genes with incredible specificity. But many of these genes were mysterious and ill defined. They were not dark matter, but shadow matter, of the genome—visible to geneticists, yet unknown in function or significance.

℗

What is a gene, then? When Mendel discovered the "gene" in 1865, he knew it only as an abstract phenomenon: a discrete determinant, transmitted intact across generations, that specified a single visible property or phenotype, such as flower color or seed texture in peas. Morgan and Muller deepened this understanding by demonstrating that genes were physical—*material*—structures carried on chromosomes. Avery advanced this understanding of genes by identifying the chemical form of that material: genetic information was carried in DNA. Watson, Crick, Wilkins, and Franklin solved its molecular structure as a double helix, with two paired, complementary strands.

In the 1930s, Beadle and Tatum solved the mechanism of gene action by discovering that a gene "worked" by specifying the structure of a protein. Brenner and Jacob identified a messenger intermediate—an RNA copy—that is required for the translation of genetic information into a protein. Monod and Jacob added to the dynamic conception of genes by demonstrating that genes can be turned on and off by increasing or decreasing this RNA message, using regulatory switches appended to each gene.

The comprehensive sequencing of the worm genome extended and modified these insights on the concept of a gene. A gene specifies a function in an organism, yes—but a single gene can specify more than a single function. A gene need not provide instructions to build a protein: it can be used to encode RNA alone, and no proteins. It need not be a contiguous piece of DNA: it can be split into parts. It has regulatory sequences appended to it, but these sequences need not be immediately adjacent to a gene.

Already, comprehensive genome sequencing had opened the door to an unexplored universe in organismal biology. Like an infinitely recursive encyclopedia—whose entry under *encyclopedia* has to be updated constantly—the sequencing of a genome had shifted our conception of genes, and therefore, of the genome itself.

☉

The *C. elegans* genome—published to universal scientific acclaim in a special issue of *Science* magazine, with a picture of the subcentimeter nematode emblazoned on its cover in December 1998—was a powerful vindication for the Human Genome Project. A few months after the worm genome announcement, Lander had exciting news of his own: the Human Genome Project had completed one-quarter of the sequence of the human genome. In a dark, dry, vaultlike warehouse on an industrial lot near Kendall Square in Cambridge, Massachusetts, 125 semiautomated sequencing machines, shaped like enormous gray boxes, were reading about two hundred DNA letters every second (Sanger's virus, which had taken him three years to sequence, would have been completed in twenty-five seconds). The sequence of an entire human chromosome—chromosome twenty-two—had been fully assembled and was awaiting final confirmation. In October 1999, the project would cross a memorable sequencing landmark: its one-billionth human base pair (a G-C, as it turned out), of the total 3 billion.

Celera, meanwhile, had no intention of lagging behind in this arms race. Flush with funds from private investors, it had doubled its output of gene sequences. On September 17, 1999, barely nine months after the publication of the worm genome, Celera opened a vast genome conference at the Fontainebleau Hotel in Miami with its own strategic counterpunch: it had sequenced the genome of the fruit fly, *Drosophila melanogaster*. Working with the fruit fly geneticist Gerry Rubin and a team of geneticists from Berkeley and Europe, Venter's team had assembled the fly genome in a record spurt of eleven months—faster than any prior gene-sequencing project. As Venter, Rubin, and Mark Adams rose to the podium to give talks, the leap of the advance became clear: in the nine decades since Thomas Morgan had begun his work on fruit flies, geneticists had identified about 2,500 genes. Celera's draft sequence contained all 2,500 known genes—and, in a single swoop, had added 10,500 new ones. In the hushed, reverential minute that followed the end of the presentations, Venter did not hesitate to drive a switchblade through his competitors' spines: "Oh, and by the way, we [have] just started sequencing human DNA, and it looks as if [the technical hurdles are] going to be less of a problem than they had been with the fly."

In March 2000, *Science* published the sequence of the fruit fly genome in yet another special issue of the journal, this time with a 1934 engraving of a male and a female fruit fly on its cover. Even the most strident critics of shotgun sequencing were sobered by the quality and depth of the data. Celera's shotgun strategy had left some important gaps in the sequence—but significant sections of the fly genome were complete. Comparisons between human, worm, and fly genes revealed several provocative patterns. Of the 289 human genes known to be involved with a disease, 177 genes—more than 60 percent—had a related counterpart in the fly. There were no genes for sickle-cell anemia or hemophilia—flies do not have red blood cells or form clots—but genes involved in colon cancer, breast cancer, Tay-Sachs disease, muscular dystrophy, cystic fibrosis, Alzheimer's disease, Parkinson's disease, and diabetes, or close counterparts of those genes, were present. Although separated by four legs, two wings, and several million years of evolution, flies and humans shared core pathways and genetic networks. As William Blake had suggested in 1794, the diminutive fly had turned out to be "a man like me."

The most bewildering feature of the fly genome was also a matter of size. Or more accurately, it was the proverbial revelation that size does not matter. Contrary to the expectations of even the most seasoned fly biologists, the fly was found to have just 13,601 genes—5,000 *fewer* genes than a worm. Less had been used to build more: out of just 13,000 genes was created an organism that mates, grows old, gets drunk, gives birth, experiences pain, has smell, sight, taste, and touch, and shares our insatiable desire for ripe summer fruit. "The lesson is that the complexity apparent [in flies] is not achieved by the sheer number of genes," Rubin said. "The human genome . . . is likely to be an amplified version of a fly genome. . . . The evolution of additional complex attributes is essentially an *organizational* one: a matter of novel interactions that derive from the temporal and spatial segregation of fairly similar components."

As Richard Dawkins puts it, "All animals probably have a relatively similar repertoire of proteins that need to be 'called forth' at any particular time. . . ." The difference between a more complex organism and a simpler one, "between a human and a nematode worm, is not that humans have more of those fundamental pieces of apparatus, but that they can call them into action in more complicated sequences and in a more complicated range of spaces." It was not the size of the ship, yet again, but the way the planks were configured. The fly genome was its own Delphic boat.

⊕

In May 2000, with Celera and the Human Genome Project sprinting neck and neck toward a draft sequence of the human genome, Venter received a phone call from his friend Ari Patrinos, from the Department of Energy. Patrinos had contacted Francis Collins and asked him to stop by Patrinos's town house for an evening drink. Would Venter consider joining? There would be no aides, advisers, or journalists, no entourage of investors or funders. The conversation would be entirely private, and the conclusions would remain strictly confidential.

Patrinos's phone call to Venter had been orchestrated for several weeks. News of the arms race between Celera and the Human Genome Project had filtered through political channels and reached the White House. President Clinton, with his unfailing nose for public relations, realized that news of the contest could escalate into an embarrassment for the government, especially if Celera was the first to announce victory. Clinton had sent his aides a note with a terse, two-word dictum—"Fix this!"—appended to the margin. Patrinos was the appointed "fixer."

A week later, Venter and Collins met in the rec room in the basement of Patrinos's town house in Georgetown. The atmosphere was understandably chilly. Patrinos waited for the mood to thaw, then delicately broached the subject of the meeting: Would Collins and Venter consider a joint announcement of the sequencing of the human genome?

Both Venter and Collins had come mentally prepared for such an offer. Venter mulled over the possibility and acquiesced—but with several caveats. He agreed to a joint ceremony at the White House to celebrate the draft sequence, and back to back publications in *Science*. He made no commitments about timelines. This, as one journalist would later describe it, was the most "carefully scripted draw."

That initial meeting in Ari Patrinos's basement room became the first of several private meetings between Venter, Collins, and Patrinos. Over the next three weeks, Collins and Venter warily choreographed the general outline of the announcement: President Clinton would open the event, followed by Tony Blair, and by talks from Collins and Venter. In effect, Celera and the Human Genome Project would be declared joint victors in the race to sequence the human genome. The White House was swiftly informed of the possibility of the announcement and acted quickly to secure a date. Venter and Collins returned to their respective groups and agreed on June 26, 2000.

①

At 10:19 a.m. on the morning of June 26, Venter, Collins, and the president gathered at the White House to reveal the "first survey" of the human genome to a large group of scientists, journalists, and foreign dignitaries (in truth, neither Celera nor the Genome Project had completed their sequences—but both groups had decided to continue with the announcement as a symbolic gesture; even as the White House was unveiling the supposed "first survey" of the genome, scientists at Celera and the Genome Project were keying frantically at their terminals, trying to string the sequence together into a meaningful whole). Tony Blair joined the meeting from London by satellite. Norton Zinder, Richard Roberts, Eric Lander, and Ham Smith sat in the audience, joined by James Watson in a crisp white suit.

Clinton spoke first, comparing the map of the human genome to the Lewis and Clark map of the continent:

"Nearly two centuries ago, in this room, on this floor, Thomas Jefferson and a trusted aide spread out a magnificent map, a map Jefferson had long prayed he would get to see in his lifetime. . . . It was a map that defined the contours and forever expanded the frontiers of our continent and our imagination. Today the world is joining us here in the East Room to behold the map of even greater significance. We are here to celebrate the completion of the first survey of the entire human genome. Without a doubt, this is the most important, most wondrous map ever produced by humankind."

Venter, the last to speak, could not resist reminding his audience that this "map" had also been achieved, in parallel, by a private expedition led by a private explorer: "At twelve thirty today, at a joint press conference with the public genome effort, Celera Genomics will describe the first assembly of the human genetic code from whole genome shotgun method. . . . The method used by Celera has determined the genetic code of five individuals. We have sequenced the genome of three females and two males who have identified themselves as Hispanic, Asian, Caucasian, or African-American."

☉

Like so many truces, the fragile armistice between Venter and Collins barely outlived its tortuous birth. In part, the conflict centered on old quarrels. Although the status of its gene patents was still uncertain, Celera had decided to monetize its sequencing project by selling subscriptions to its database to academic researchers and pharmaceutical companies (big pharma companies, Venter had astutely reasoned, might want to know gene sequences to discover new drugs—especially ones that target particular proteins). But Venter also wanted to publish Celera's human genome sequence in a major scientific journal—*Science*, for example—which required the company to deposit its gene sequences in a public repository (a scientist cannot publish a scientific paper for the general public while insisting that its essential data is secret). Justifiably, Watson, Lander, and Collins were bitingly critical of Celera's attempt to straddle the commercial and academic worlds. "My greatest success," Venter told an interviewer, "was I managed to get hated by both worlds."

The Genome Project, meanwhile, was struggling with technical hurdles. Having sequenced vast parts of the human genome using the clone-by-clone approach, the project was now poised at a critical juncture: it had to assemble the pieces to complete the puzzle. But that task—seemingly modest on a theoretical level—represented a daunting computational problem. Substantial portions of the sequence were still missing. Not every part of the genome was amenable to cloning and sequencing, and assembling nonoverlapping segments was vastly more complicated than had been anticipated, like solving a puzzle containing pieces that had fallen into the cracks of furniture. Lander recruited yet another team of scien-

319

tists to help him—David Haussler, a computer scientist at the University of California, Santa Cruz, and his forty-year-old protégé, James Kent, a former programmer–turned-molecular biologist. In a fit of inspired frenzy, Haussler convinced the university to buy a hundred desktop PCs so that Kent could write and run tens of thousands of lines of code in parallel, icing his wrists at night so that he could start coding every morning.

At Celera too the genome assembly problem was proving to be frustrating. Parts of the human genome are full of strange repetitive sequences—"equivalent to a big stretch of blue sky in a jigsaw puzzle," as Venter described it. Computational scientists in charge of assembling the genome worked week upon week to put the gene fragments in order, but the complete sequence was still missing.

By the winter of 2000, both projects neared completion—but the communications between the groups, strained at its best moments, had fallen apart. Venter accused the Genome Project of "a vendetta against Celera." Lander wrote to the editors of *Science* protesting Celera's strategy of selling the sequence database to subscribers and restricting parts of it to the public, while trying to publish yet other selected parts of the data in a journal; Celera was trying to "have its genome and sell it too." "In the history of scientific writing since the 1600s," Lander complained, "the disclosure of data has been linked to the announcement of a discovery. That's the *basis* of modern science. In pre-modern times, you could say: 'I've found an answer, or I've made lead into gold, proclaim the discovery, and then refuse to show the results.' But the whole point of professional scientific journals is disclosure and credit." Worse, Collins and Lander accused Celera of using the Human Genome Project's published sequence as a "scaffold" to assemble its own genome—molecular plagiarism (Venter retorted that the idea was ludicrous; Celera had deciphered all the other genomes with no help from such "scaffolds"). Left to its own devices, Lander announced, Celera's data was nothing more than a "genome tossed salad."

As Celera edged toward the final draft of its paper, scientists made frantic appeals for the company to deposit its results in the publicly available repository of sequences, named GenBank. Ultimately, Venter agreed to provide free access to academic researchers—but with several important constraints. Dissatisfied with the compromise, Sulston, Lander, and Collins chose to send their paper to a rival journal, *Nature*.

On February 15 and 16, 2001, the Human Genome Project consortium and Celera published their papers in *Nature* and *Science*, respectively. Both were enormous studies, nearly spanning the lengths of the two journals (at sixty-six thousand words, the Human Genome Project paper was the largest study published in *Nature*'s history). Every great scientific paper is a conversation with its own history—and the opening paragraphs of the *Nature* paper were written with full cognizance of its moment of reckoning:

"The rediscovery of Mendel's laws of heredity in the opening weeks of the 20th century sparked a scientific quest to understand the nature and content of genetic information that has propelled biology for the last hundred years. The scientific progress made [since that time] falls naturally into four main phases, corresponding roughly to the four quarters of the century."

"The first established the cellular basis of heredity: the chromosomes. The second defined the molecular basis of heredity: the DNA double helix. The third unlocked the informational basis of heredity [i.e., the genetic code], with the discovery of the biological mechanism by which cells read the information contained in genes, and with the invention of the recombinant DNA technologies of cloning and sequencing by which scientists can do the same."

The sequence of the human genome, the project asserted, marked the starting point of the "fourth phase" of genetics. This was the era of "genomics"—the assessment of the entire genomes of organisms, including humans. There is an old conundrum in philosophy that asks if an intelligent machine can ever decipher its own instruction manual. For humans, the manual was now complete. Deciphering it, reading it, and understanding it would be quite another matter.

The Book of Man
(in Twenty-Three Volumes)

Is man no more than this? Consider him well.
—William Shakespeare, *King Lear*, act 3, scene 4

There are mountains beyond mountains.
—Haitian proverb

- It has 3,088,286,401 letters of DNA (give or take a few; a more recent estimate is about 3.2 billion letters).

- Published as a book with a standard-size font, it would contain just four letters . . . AGCTTGCAGGGG . . . and so on, stretching, inscrutably, page upon page, for over 1.5 million pages—sixty-six times the size of the *Encyclopaedia Britannica*.

- It is divided into twenty-three pairs of chromosomes—forty-six in all—in most cells in the body. All other apes, including gorillas, chimpanzees, and orangutans, have twenty-four pairs. At some point in hominid evolution, two medium-size chromosomes in some ancestral ape fused to form one. The human genome departed cordially from the ape genome several million years ago, acquiring new mutations and variations over time. We lost a chromosome, but gained a thumb.

- It encodes about 20,687 genes in total—only 1,796 more than worms, 12,000 fewer than corn, and 25,000 fewer genes than rice or wheat. The difference between "human" and "breakfast cereal" is not a matter of gene numbers, but of the sophistication of gene networks. It is not what we have; it is how we use it.

- It is fiercely inventive. It squeezes complexity out of simplicity. It orchestrates the activation or repression of certain genes in only certain cells and at certain times, creating unique contexts and partners for each gene in time and space, and thus produces near-infinite functional variation out of its limited repertoire. And it mixes and matches gene modules—called exons—within single genes to extract even further combinatorial diversity out of its gene repertoire. These two strategies—gene regulation and gene splicing—appear to be used more extensively in the human genome than in the genomes of most organisms. More than the enormity of gene numbers, the diversity of gene types, or the originality of gene function, it is the *ingenuity* of our genome that is the secret to our complexity.

- It is dynamic. In some cells, it reshuffles its own sequence to make novel variants of itself. Cells of the immune system secrete "antibodies"—missilelike proteins designed to attach themselves to invading pathogens. But since pathogens are constantly evolving, antibodies must also be capable of changing; an evolving pathogen demands an evolving host. The genome accomplishes this counterevolution by reshuffling its genetic elements—thereby achieving astounding diversity (*s . . . tru . . . c . . . t . . . ure* and *g . . . en . . . ome* can be reshuffled to form an entirely new word *c . . . ome . . . t*). The reshuffled genes generate the diversity of antibodies. In these cells, every genome is capable of giving rise to an entirely different genome.

- Parts of it are surprisingly beautiful. On a vast stretch on chromosome eleven, for instance, there is a causeway dedicated entirely to the sensation of smell. Here, a cluster of 155 closely related genes encodes a series of protein receptors that are professional smell sensors. Each receptor binds to a unique chemical structure, like a key to a lock, and generates a distinctive sensation of smell in the brain—spearmint, lemon, caraway, jasmine, vanilla, ginger, pepper. An elaborate form of gene regulation ensures that only one odor-receptor gene is chosen from this cluster and expressed in a single smell-sensing neuron in the nose, thereby enabling us to discriminate thousands of smells.

- Genes, oddly, comprise only a minuscule fraction of it. An enormous proportion—a bewildering 98 percent—is not dedicated to genes per se, but to enormous stretches of DNA that are interspersed between genes (intergenic DNA) or within genes (introns). These long stretches encode no RNA, and no protein: they exist in the genome either because they regulate gene expression, or for reasons that we do not yet understand, or because of no reason whatsoever (i.e., they are "junk" DNA). If the genome were a line stretching across the Atlantic Ocean between North America and Europe, genes would be occasional specks of land strewn across long, dark tracts of water. Laid end to end, these specks would be no longer than the largest Galápagos island or a train line across the city of Tokyo.

- It is encrusted with history. Embedded within it are peculiar fragments of DNA—some derived from ancient viruses—that were inserted into the genome in the distant past and have been carried passively for millennia since then. Some of these fragments were once capable of actively "jumping" between genes and organisms, but they have now been largely inactivated and silenced. Like decommissioned traveling salesmen, these pieces are permanently tethered to our genome, unable to move or get out. These fragments are vastly more common than genes, resulting in yet another major idiosyncrasy of our genome: much of the human genome is not particularly human.

- It has repeated elements that appear frequently. A pesky, mysterious three-hundred-base-pair sequence called Alu appears and reappears millions of times, although its origin, function, or significance is unknown.

- It has enormous "gene families"—genes that resemble each other and perform similar functions—which often cluster together. Two hundred closely related genes, clustered in archipelagoes on certain chromosomes, encode members of the "Hox" family, many of which play crucial roles in determining the fate, identity, and structure of the embryo, its segments, and its organs.

- It contains thousands of "pseudogenes"—genes that were once functional but have become nonfunctional, i.e., they give rise to no protein or RNA. The carcasses of these inactivated genes are littered throughout its length, like fossils decaying on a beach.

- It accommodates enough variation to make each one of us distinct, yet enough consistency to make each member of our species profoundly different from chimpanzees and bonobos, whose genomes are 96 percent identical to ours.

- Its first gene, on chromosome one, encodes a protein that senses smell in the nose (again: those ubiquitous olfactory genes!). Its last gene, on chromosome X, encodes a protein that modulates the interaction between cells of the immune system. (The "first" and "last" chromosomes are arbitrarily assigned. The first chromosome is labeled first because it is the longest.)

- The ends of its chromosomes are marked with "telomeres." Like the little bits of plastic at the ends of shoelaces, these sequences of DNA are designed to protect the chromosomes from fraying and degenerating.

- Although we fully understand the genetic code—i.e., how the information in a single gene is used to build a protein—we comprehend virtually nothing of the *genomic* code—i.e., how multiple genes spread across the human genome coordinate gene expression in space and time to build, maintain, and repair a human organism. The genetic code is simple: DNA is used to build RNA, and RNA is used to be build a protein. A triplet of bases in DNA specifies one amino acid in the protein. The genomic code is complex: appended to a gene are sequences of DNA that carry information on when and where to express the gene. We do not know why certain genes are located in particular geographic locations in the genome, and how the tracts of DNA that lie between genes regulate and coordinate gene physiology. There are codes beyond codes, like mountains beyond mountains.

- It imprints and erases chemical marks on itself in response to alterations in its environment—thereby encoding a form of cellular "memory" (there is more to come on this topic).

- It is inscrutable, vulnerable, resilient, adaptable, repetitive, and unique.

- It is poised to evolve. It is littered with the debris of its past.

- It is designed to survive.

- It resembles us.

PART FIVE

THROUGH THE
LOOKING GLASS

The Genetics of Identity and "Normalcy"

(2001–2015)

℗

①

How nice it would be if we could only get through into Looking-glass House! I'm sure it's got, oh! such beautiful things in it!

—Lewis Carroll, *Alice in Wonderland*

①

"So, We's the Same"

Medicine, the sociologist Everett Hughes once observed wryly, perceives the world through "mirror writing." Illness is used to define wellness. Abnormalcy marks the boundaries of normalcy. Deviance demarcates the limits of conformity. This mirror writing can result in an epically perverse vision of the human body. An orthopedist thus begins to think of bones as sites of fractures; a brain, in a neurologist's imagination, is a place where memories are lost. There is an old story, probably apocryphal, of a Boston surgeon who did lose his memory and could only recall his friends by the names of the various operations he had performed on them.

Throughout much of the history of human biology, as the science writer Matt Ridley noted, genes too were largely perceived in mirror writing— identified by the abnormality or disease caused when they mutated. Hence the *cystic fibrosis* gene, the *Huntington's* gene, the *breast-cancer-causing BRCA1* gene, and so forth. To a biologist the nomenclature is absurd: the function of the *BRCA1* gene is not to cause breast cancer when mutated, but to repair DNA when normal. The sole function of the "benign" breast cancer gene *BRCA1* is to make sure that DNA is repaired when it is damaged. The hundreds of millions of women without a family history of breast cancer inherit this benign variant of the *BRCA1* gene. The mutant variant or allele—call it *m-BRCA1*—causes a change in the structure of the *BRCA1* protein such that it fails to repair damaged DNA. Hence cancer-causing mutations arise in the genome when *BRCA1* malfunctions.

The gene called *wingless* in fruit flies encodes a protein whose real function is not to make wingless insects, but to encode instructions to build wings. Naming a gene *cystic fibrosis* (or *CF*), as Ridley observed, is "as absurd as defining the organs of the body by the diseases they get: livers are there to cause cirrhosis, hearts to cause heart attacks, and brains to cause strokes."

The Human Genome Project allowed geneticists to invert this mirror writing on itself. The comprehensive catalog of every normal gene in the human genome—and the tools generated to produce such a catalog—made it possible, in principle, to approach genetics from the front side of its mirror: it was no longer necessary to use pathology to define the borders of normal physiology. In 1988, a National Research Council document on the Genome Project made a crucial projection about the future of genomic research: "Encoded in the DNA sequence are fundamental determinants of those mental capacities—learning, language, memory—essential to human culture. Encoded there as well are the mutations and variations that cause or increase the susceptibility to many diseases responsible for much human suffering."

Vigilant readers may have noted that the two sentences signaled the twin ambitions of a new science. Traditionally, human genetics had concerned itself largely with pathology—with "diseases responsible for much human suffering." But armed with new tools and methods, genetics could also roam freely to explore aspects of human biology that had hitherto seemed impenetrable to it. Genetics had crossed over from the strand of pathology to the strand of normalcy. The new science would be used to understand history, language, memory, culture, sexuality, identity, and race. It would, in its most ambitious fantasies, try to become the science of normalcy: of health, of identity, of destiny.

The shift in the trajectory of genetics also signals a shift in the story of the gene. Until this point, the organizing principle of our story has been historical: the journey from the gene to the Genome Project has moved through a relatively linear chronology of conceptual leaps and discoveries. But as human genetics shifted its glance from pathology to normalcy, a strictly chronological approach could no longer capture the diverse dimensions of its inquiry. The discipline shifted to a more *thematic* focus, organizing itself around distinct, albeit overlapping, arenas of inquiry concerning human biology: the genetics of race, gender, sexuality, intelligence, temperament, and personality.

The expanded dominion of the gene would vastly deepen our understanding of the influence of genes on our lives. But the attempt to confront human normalcy through genes would also force the science of genetics to confront some of the most complex scientific and moral conundrums in its history.

Ⅸ

To understand what genes tell us about human beings, we might begin by trying to decipher what genes tell us about the origins of human beings. In the mid-nineteenth century, before the advent of human genetics, anthropologists, biologists, and linguists fought furiously over the question of human origin. In 1854, a Swiss-born natural historian named Louis Agassiz became the most ardent proponent of a theory called *polygenism*, which suggested that the three major human races—Whites, Asians, and Negroes, as he liked to categorize them—had arisen independently, from separate ancestral lineages, several million years ago.

Agassiz was arguably the most distinguished racist in the history of science—"racist" both in the original sense of the word, a believer in the inherent differences between human races, and in an operational sense, a believer that some races were fundamentally superior to others. Repulsed by the horror that he might share a common ancestor with Africans, Agassiz maintained that each race had its unique forefather and foremother and had arisen independently and forked out independently over space and time. (The name *Adam*, he suggested, arose from the Hebrew word for "one who blushes," and only a white man could detectably blush. There had to be multiple Adams, Agassiz concluded—blushers and nonblushers—one for each race.)

In 1859, Agassiz's theory of multiple origins was challenged by the publication of Darwin's *Origin of Species*. Although *Origin* pointedly dodged the question of human origin, Darwin's notion of evolution by natural selection was obviously incompatible with Agassiz's separate ancestry of all human races: if finches and tortoises had cascaded out from a common ancestor, why would humans be different?

As academic duels go, this one was almost comically one-sided. Agassiz, a grandly bewhiskered Harvard professor, was among the most prominent natural historians of the world, while Darwin, a doubt-ridden, self-taught parson-turned-naturalist from the "other" Cambridge, was still virtually unknown outside England. Still, recognizing a potentially fatal confrontation when he saw one, Agassiz issued a scalding rebuttal to Darwin's book. "Had Mr. Darwin or his followers furnished a single fact to show that individuals change, in the course of time, in such a manner as to produce, at last, species . . . the state of the case might be different," he thundered.

But even Agassiz had to concede his theory of separate ancestors for separate races ran the risk of being challenged not by a "single fact" but a multitude of facts. In 1848, stone diggers in a limestone quarry in the Neander Valley in Germany had accidentally unearthed a peculiar skull that resembled a human skull but also possessed substantial differences, including a larger cranium, a recessed chin, powerfully articulated jawbones, and a pronounced outward jut of the brow. At first, the skull was dismissed as the remnant of a freak that had suffered an accident—a madman stuck in a cave—but over the next decades, a host of similar skulls and bones were disgorged from gorges and caves scattered across Europe and Asia. The bone-by-bone reconstruction of these specimens pointed to a strongly built, prominently browed species that walked upright on somewhat bowed legs—an ornery wrestler with a permanent frown. The hominid was called a Neanderthal, after the site of its original location.

Initially, many scientists believed that the Neanderthals represented an ancestral form of modern humans, one link in the chain of missing links between humans and apes. In 1922, for instance, an article in *Popular Science Monthly* called the Neanderthal "an early time in the evolution of man." Accompanying the text was a variant of a now-familiar image of human evolution, with gibbonlike monkeys transmuting into gorillas, gorillas into upright-walking Neanderthals, and so forth, until humans were formed. But by the 1970s and 1980s, the Neanderthal-as-human-ancestor hypothesis had been debunked and replaced by a much stranger idea—that early modern humans *coexisted* with Neanderthals. The "chain of evolution" drawings were revised to reflect that gibbons, gorillas, Neanderthals, and modern humans were not progressive stages of human evolution, but had all emerged from a common ancestor. Further anthropological evidence suggested that modern humans—then called Cro-Magnons—had arrived on the Neanderthal's scene around forty-five thousand years ago, most likely by migrating into parts of Europe where Neanderthals lived. We now know that Neanderthals had become extinct forty thousand years ago, having overlapped with early modern humans for about five thousand years.

Cro-Magnons are, indeed, our closer, truer ancestors, possessing the smaller skull, flattened face, receded brow, and thinner jaw of contemporary humans (the politically correct phrase for the anatomically correct Cro-Magnons is European Early Modern Human or EEMH). These early modern humans intersected with Neanderthals, at least in parts of Europe, and likely competed with them for resources, food, and space. Neanderthals were our neighbors and rivals. Some evidence suggests that we interbred with them and that in competing with them for food and resources we may have contributed to their extinction. We loved them—and, yes, we killed them.

①

But the distinction between Neanderthals and modern humans returns us, full circle, to our original questions: How old are humans, and where did we come from? In the 1980s, a biochemist at the University of California, Berkeley, named Allan Wilson began to use genetic tools to answer these questions.* Wilson's experiment began with a rather simple idea. Imagine being thrown into a Christmas party. You don't know the host or the guests. A hundred men, women, and children are milling around, drinking punch, and suddenly a game begins: You are asked to arrange the crowd by family, relatedness, and descent. You cannot ask for names or ages. You are blindfolded; you are not allowed to construct family trees by looking at facial resemblances or studying mannerisms.

To a geneticist, this is a tractable problem. First, he recognizes the existence of hundreds of natural variations—mutations—scattered through each individual genome. The closer the relatedness of individuals, the closer the spectrum of their variants or mutations shared by them (identical twins share the entire genome; fathers and mothers contribute, on average, half to their children, and so forth). If these variants can be sequenced and identified in each individual, lineage can be solved imme-

---

* Wilson drew his crucial insight from two giants of biochemistry, Linus Pauling and Émile Zuckerkandl, who had proposed an entirely novel way to think of the genome—not just as a compendium of information to build an individual organism, but as a compendium of information for an organism's evolutionary history: a "molecular clock." The Japanese evolutionary biologist Motoo Kimura also developed this theory.

diately: relatedness is a function of mutatedness. Just as facial features or skin color or height are shared among related individuals, variations are more commonly shared within families than across families (indeed, facial features and heights are shared *because* genetic variations are shared among individuals).

And what if the geneticist is also asked to find the family with the most generations present, without knowing the ages of any of the individuals at the party? Suppose one family is represented by a great-grandfather, grandfather, father, and son at the celebration; four generations are present. Another family also has four attendees—a father and his identical triplets, representing just two generations. Can we identify the family with the most generations in the crowd with no prior knowledge about faces or names? Merely counting the number of members in a family will not do: the father-and-triplet family, and the great-grandfather and his multigenerational descendants, each have the same number of family members: four.

Genes and mutations provide a clever solution. Since mutations accumulate over generations—i.e., over intergenerational time—the family with the greatest *diversity* in gene variations is the one with the most generations. The triplets have exactly the same genome; their genetic diversity is minimal. The great-grandfather and great-grandson pair, in contrast, have related genomes—but their genomes have the most differences. Evolution is a metronome, ticktocking time through mutations. Genetic diversity thus acts as a "molecular clock," and variations can organize lineage relationships. The intergenerational time between any two family members is proportional to the extent of genetic diversity between them.

Wilson realized that this technique could be applied not just across a family, but across an entire population of organisms. Variations in genes could be used to create a map of relatedness. And genetic diversity could be used to measure the oldest populations within a species: a tribe that has the most genetic diversity within it is older than a tribe with little or no diversity.

Wilson had almost solved the problem of estimating the age of any species using genomic information—except for a glitch. If genetic variation was produced only by mutation, Wilson's method would be absolutely fail-safe. But genes, Wilson knew, are present in two copies in most human cells, and they can "cross over" between paired chromosomes, generating variation and diversity by an alternative method. This method of generating variation would inevitably confound Wilson's study. To construct an ideal genetic lineage, Wilson realized, he needed a stretch of the human genes that was intrinsically resistant to reassortment and crossing over—a lonely, vulnerable corner of the genome where change can occur only through the accumulation of mutations, thereby allowing that genomic segment to act as the perfect molecular clock.

But where might he find such a vulnerable stretch? Wilson's solution was ingenious. Human genes are stored in chromosomes in the nucleus of the cell, but with one exception. Every cell possesses a subcellular structure called a mitochondrion that is used to generate energy. Mitochondria have their own mini-genome, with only thirty-seven genes, about one six-thousandth the number of genes on human chromosomes. (Some scientists propose that mitochondria originated from some ancient bacteria that invaded single-celled organisms. These bacteria formed a symbiotic alliance with the organism; they provided energy, but used the organism's cellular environment for nutrition, metabolism, and self-defense. The genes lodged within mitochondria are left over from this ancient symbiotic relationship; indeed, human mitochondrial genes resemble bacterial genes more than human ones.)

The mitochondrial genome rarely recombines and is only present in a single copy. Mutations in mitochondrial genes are passed intact across generations, and they accumulate over time without crossing over, making the mitochondrial genome an ideal genetic timekeeper. Crucially, Wilson realized, this method of age reconstruction was entirely self-contained and independent of bias: it made no reference to the fossil record, to linguistic lineages, geologic strata, geographical maps, or anthropological surveys. Living humans are endowed with the evolutionary history of our species in our genomes. It is as if we permanently carry a photograph of each of our ancestors in our wallets.

Between 1985 and 1995, Wilson and his students learned to apply these techniques to human specimens (Wilson died of leukemia in 1991, but his students continued his work). The results of these studies were startling for three reasons. First, when Wilson measured the overall diversity of the human mitochondrial genome, he found it to be surprisingly small—less diverse than the corresponding genomes of chimpanzees. Modern humans, in other words, are substantially younger and substantially more homogenous than chimpanzees (every chimp might look like every other chimp to human eyes, but to a discerning chimpanzee, it is humans that are vastly more alike). Calculating backward, the age of humans was estimated to be about two hundred thousand years—a minor blip, a ticktock, in the scale of evolution.

Where did the first modern humans come from? By 1991, Wilson could use his method to reconstruct the lineage relationship between various populations across the globe and calculate the relative age of any population using genetic diversity as his molecular clock. As gene-sequencing and annotation technologies evolved, geneticists refined this analysis—broadening its scope beyond mitochondrial variations and studying thousands of individuals across hundreds of populations around the world.

In November 2008, a seminal study led by Luigi Cavalli-Sforza, Marcus Feldman, and Richard Myers from Stanford University characterized 642,690 genetic variants in 938 individuals drawn from 51 subpopulations across the world. The second startling result about human origins emerges from this study: modern humans appear to have emerged exclusively from a rather narrow slice of earth, somewhere in sub-Saharan Africa, about one hundred to two hundred thousand years ago, and then migrated northward and eastward to populate the Middle East, Europe, Asia, and the Americas. "You get less and less variation the further you go from Africa," Feldman wrote. "Such a pattern fits the theory that the first modern humans settled the world in stepping-stone fashion after leaving Africa less than 100,000 years ago. As each small group of people broke away to found a new region, it took only a sample of the parent population's genetic diversity."

The oldest human populations—their genomes peppered with diverse and ancient variations—are the San tribes of South Africa, Namibia, and Botswana, and the Mbuti Pygmies, who live deep in the Ituri forest in the Congo. The "youngest" humans, in contrast, are the indigenous North Americans who left Europe, and crossed into the Seward peninsula in Alaska through the icy cleft of the Bering Strait, some fifteen to thirty thousand years ago. This theory of human origin and migration, corroborated by fossil specimens, geological data, tools from archaeological digs, and linguistic patterns, has overwhelmingly been accepted by most human geneticists. It is called the Out of Africa theory, or the Recent Out of Africa model (the *recent* reflecting the surprisingly modern evolution of modern humans, and its acronym, ROAM, a loving memento to an ancient peripatetic urge that seems to rise directly out of our genomes).

①

The third important conclusion of these studies requires some conceptual background. Consider the genesis of a single-celled embryo produced by the fertilization of an egg by a sperm. The *genetic* material of this embryo comes from two sources: paternal genes (from sperm) and maternal genes (from eggs). But the *cellular* material of the embryo comes exclusively from the egg; the sperm is no more than a glorified delivery vehicle for male DNA—a genome equipped with a hyperactive tail.

Aside from proteins, ribosomes, nutrients, and membranes, the egg also supplies the embryo with specialized structures called *mitochondria*. These mitochondria are the energy-producing factories of the cell; they are so anatomically discrete and so specialized in their function that cell biologists call them "organelles"—i.e., mini-organs resident within cells. Mitochondria, recall, carry a small, independent genome that resides within the mitochondrion itself—not in the cell's nucleus, where the twenty-three pairs of chromosomes (and the 21,000-odd human genes) can be found.

The exclusively female origin of all the mitochondria in an embryo has an important consequence. All humans—male or female—must have inherited their mitochondria from their mothers, who inherited *their* mitochondria from their mothers, and so forth, in an unbroken line of female ancestry stretching indefinitely into the past. (A woman also carries the mitochondrial genomes of all her future descendants in her cells; ironically, if there is such a thing as a "homunculus," then it is exclusively female in origin—technically, a "femunculus"?)

Now imagine an ancient tribe of two hundred women, each of whom bears one child. If the child happens to be a daughter, the woman dutifully passes her mitochondria to the next generation, and, through her daughter's daughter, to a third generation. But if she has only a son and no daughter, the woman's mitochondrial lineage wanders into a genetic blind alley and becomes extinct (since sperm do not pass their mitochondria to the embryo, sons cannot pass their mitochondrial genomes to their children). Over the course of the tribe's evolution, tens of thousands of such mitochondrial lineages will land on lineal dead ends by chance, and be snuffed out. And here is the crux: if the founding population of a species is small enough, and if enough time has passed, the number of surviving maternal lineages will keep shrinking, and shrinking further, until only a few are left. If half of the two hundred women in our tribe have sons, and only sons, then one hundred mitochondrial lineages will dash against the glass pane of male-only heredity and vanish in the next generation. Another half will dead-end into male children in the second generation, and so forth. By the end of several generations, all the descendants of the tribe, male or female, might track their mitochondrial ancestry to just a few women.

For modern humans, that number has reached *one*: each of us can trace our mitochondrial lineage to a single human female who existed in Africa about two hundred thousand years ago. She is the common mother of our species. We do not know what she looked like, although her closest modern-day relatives are women of the San tribe from Botswana or Namibia.

I find the idea of such a founding mother endlessly mesmerizing. In human genetics, she is known by a beautiful name—Mitochondrial Eve.

①

In the summer of 1994, as a graduate student interested in the genetic origin of the immune system, I traveled along the Rift Valley, from Kenya to Zimbabwe, past the basin of the Zambezi River to the flat plains of South Africa. It was the evolutionary journey of humans in reverse. The final station of the journey was an arid mesa in South Africa, roughly equidistant from Namibia and Botswana, where some of the San tribes once lived. It was a place of lunar desolation—a flat, dry tabletop of land decapitated by some geophysically vengeful force and perched above the plains below. By then, a series of thefts and losses had whittled my possessions down to virtually nothing: four pairs of boxers, which I often doubled up and wore as shorts, a box of protein bars, and bottled water. Naked we come, the Bible suggests; I was almost there.

With a little imagination, we can reconstruct the history of humans using that windblown mesa as a starting point. The clock begins about two hundred thousand years ago, when a population of early modern humans begins to inhabit this site, or some such site in its vicinity (the evolutionary geneticists Brenna Henn, Marcus Feldman, and Sarah Tishkoff have pinpointed the origin of human migration farther west, near the coast of Namibia). We know virtually nothing about the culture and habits of this ancient tribe. They left no artifacts—no tools, no drawings, no cave dwellings—except the most profound of all remnants: their genes, stitched indelibly into our own.

The population was likely quite small, even minuscule by contemporary standards—no more than about six thousand or ten thousand individuals. The most provocative estimate is a bare seven hundred—about the number of humans that might inhabit a single city block or a village. Mitochondrial Eve may have lived among them, bearing at least one daughter, and at least one granddaughter. We do not know when, or why, these individuals stopped interbreeding with other hominids—but we do know that they began breeding with each other with relative exclusivity about two hundred millennia ago. ("Sexual intercourse began," the poet Philip Larkin once wrote, "in nineteen sixty-three." He was off by about two hundred thousand years.) Perhaps they were isolated here by climate changes, or stranded by geographic barriers. Perhaps they fell in love.

①

From here, they went west, as young men often do, and then traveled north.* They clambered through the gash of the Rift Valley or ducked into the canopies of the humid rain forests around the Congo basin, where the Mbuti and Bantu now live.

The story is not as geographically contained or as neat as it sounds. Some populations of early modern humans are known to have wandered back into the Sahara—a lush landscape then, crisscrossed with finger lakes and rivers—and eddied backward into local pools of humanoids, coexisted and even interbred with them, perhaps generating evolutionary backcrosses. As Christopher Stringer, the paleoanthropologist, described it, "In terms of modern humans, this means that . . . some modern humans have got more archaic genes than others. That does seem to be so. So it leads us on to ask again: what is a modern human? Some of the most fascinating ongoing research topics in the next year or two will be homing in on the DNA that some of us have acquired from Neanderthals. . . . Scientists will look at that DNA and ask, is it functional? Is it actually doing something in the bodies of those people? Is it affecting brains, anatomy, physiology, and so on?"

But the long march went on. Some seventy-five thousand years ago, a group of humans arrived at the northeastern edge of Ethiopia or Egypt, where the Red Sea narrows to a slitlike strait between the shrugged shoulder of Africa and the downward elbow of the Yemeni peninsula. There was no one there to part the ocean. We do not know what drove these men

---

* If the origin of this group was in south*western* Africa, as some recent studies suggest, then these humans traveled largely *east* and *north*.

and women to fling themselves across the water, or how they managed to cross it (the sea was shallower then, and some geologists have wondered whether chains of sandbar islands spanned the strait along which our ancestors hopscotched their way to Asia and Europe). A volcano had erupted in Toba, Indonesia, about seventy thousand years ago, spewing enough dark ash into the skies to launch a decades-long winter that might have precipitated a desperate search for new food and land.

Others have proposed that multiple dispersals, prompted by smaller catastrophes, may have taken place at various times in human history. One dominant theory suggests that at least two independent crossings occurred. The earliest crossing occurred 130,000 years ago. The migrants landed in the Middle East and took a "beachcomber" route through Asia, hugging the coast toward India and then fanning out southward toward Burma, Malaysia, and Indonesia. A later crossing happened more recently, about sixty thousand years ago. These migrants moved north into Europe, where they encountered Neanderthals. Either route used the Yemeni peninsula as its hub. This is the true "melting pot" of the human genome.

What is certain is that every perilous ocean-crossing left hardly any survivors—perhaps as few as six hundred men and women. Europeans, Asians, Australians, and Americans are the descendants of these drastic bottlenecks, and this corkscrew of history too has left its signature in our genomes. In a genetic sense, nearly all of us who emerged out of Africa, gasping for land and air, are even more closely yoked than previously imagined. We were on the same boat, brother.

①

What does this tell us about race and genes? A great deal. First, it reminds us that the racial categorization of humans is an inherently limited proposition. Wallace Sayre, the political scientist, liked to quip that academic disputes are often the most vicious because the stakes are so overwhelmingly low. By similar logic, perhaps our increasingly shrill debates on race should begin with the recognition that the actual range of human genomic variation is strikingly low—lower than in many other species (lower, remember, than in chimpanzees). Given our rather brief tenure on earth as a species, we are much more alike than unlike each other. It is an inevitable consequence of the bloom of our youth that we haven't even had time to taste the poisoned apple.

Yet, even a young species possesses history. One of the most penetrating powers of genomics is its ability to organize even closely related genomes into classes and subclasses. If we go hunting for discriminatory features and clusters, then we will, indeed, find features and clusters to discriminate. Examined carefully, the variations in the human genome *will* cluster in geographic regions and continents, and along traditional boundaries of race. Every genome bears the mark of its ancestry. By studying the genetic characteristics of an individual you can pinpoint his or her origin to a certain continent, nationality, state, or even a tribe with remarkable accuracy. It is, to be sure, an apotheosis of small differences—but if this is what we mean by "race," then the concept has not just survived the genomic era, it has been amplified by it.

The problem with racial discrimination, though, is not the inference of a person's race from their genetic characteristics. It is quite the opposite: it is the inference of a person's characteristics from their race. The question is not, can you, given an individual's skin color, hair texture, or language, infer something about their ancestry or origin. That is a question of biological systematics—of lineage, of taxonomy, of racial geography, of biological discrimination. Of course you can—and genomics has vastly refined that inference. You can scan any individual genome and infer rather deep insights about a person's ancestry, or place of origin. But the vastly more controversial question is the converse: Given a racial identity—African or Asian, say—can you infer anything about an individual's characteristics: not just skin or hair color, but more complex features, such as intelligence, habits, personality, and aptitude? *Genes can certainly tell us about race, but can race tell us anything about genes?*

To answer this question, we need to measure how genetic variation is distributed across various racial categories. Is there more diversity *within* races or *between* races? Does knowing that someone is of African versus European descent, say, allow us to refine our understanding of their genetic traits, or their personal, physical, or intellectual attributes in a meaningful manner? Or is there so much variation within Africans and Europeans that *intraracial* diversity dominates the comparison, thereby making the category "African" or "European" moot?

We now know precise and quantitative answers to these questions. A number of studies have tried to quantify the level of genetic diversity of the human genome. The most recent estimates suggest that the vast proportion of genetic diversity (85 to 90 percent) occurs *within* so-called races (i.e., within Asians or Africans) and only a minor proportion (7 percent) between racial groups (the geneticist Richard Lewontin had estimated a similar distribution as early as 1972). Some genes certainly vary sharply between racial or ethnic groups—sickle-cell anemia is an Afro-Caribbean and Indian disease, and Tay-Sachs disease has a much higher frequency in Ashkenazi Jews—but for the most part, the genetic diversity within any racial group dominates the diversity between racial groups—not marginally, but by an enormous amount. This degree of intraracial variability makes "race" a poor surrogate for nearly any feature: in a genetic sense, an African man from Nigeria is so "different" from another man from Namibia that it makes little sense to lump them into the same category.

For race and genetics, then, the genome is a strictly one-way street. You can use genome to predict where X or Y came from. But, knowing where A or B came from, you can predict little about the person's genome. Or: *every genome carries a signature of an individual's ancestry—but an individual's racial ancestry predicts little about the person's genome.* You can sequence DNA from an African-American man and conclude that his ancestors came from Sierra Leone or Nigeria. But if you encounter a man whose great-grandparents came from Nigeria or Sierra Leone, you can say little about the features of this particular man. The geneticist goes home happy; the racist returns empty-handed.

As Marcus Feldman and Richard Lewontin put it, "Racial assignment loses any general biological interest. For the human species, racial assignment of individuals does not carry any general implication about genetic differentiation." In his monumental study on human genetics, migration, and race published in 1994, Luigi Cavalli-Sforza, the Stanford geneticist, described the problem of racial classification as a "futile exercise" driven by cultural arbitration rather than genetic differentiation. "The level at which we stop our classification is completely arbitrary. . . . We can identify 'clusters' of populations . . . [but] since every level of clustering would determine a different partition . . . there is no biological reason to prefer a particular one." Cavalli-Sforza continued, "The evolutionary explanation is simple. There is great genetic variation in populations, even in small ones. This individual variation has accumulated over long periods, because most [genetic variations] antedate the separation into continents, and perhaps even the origin of the species, less than half a million years ago. . . . There has therefore been too little time for the accumulation of substantial divergence."

That extraordinary last statement was written to address the past: it is a measured scientific retort to Agassiz and Galton, to the American eugenicists of the nineteenth century, and to the Nazi geneticists of the twentieth. Genetics unleashed the specter of scientific racism in the nineteenth century. Genomics, thankfully, has stuffed it back into its bottle. Or, as Aibee, the African-American maid, tells Mae Mobley plainly in *The Help*, "So, we's the same. Just a different color."

⊕

In 1994, the very year that Luigi Cavalli-Sforza published his comprehensive review of race and genetics, Americans were convulsing with anxiety around a very different kind of book about race and genes. Written by Richard Herrnstein, the behavioral psychologist, and Charles Murray, a political scientist, *The Bell Curve* was, as the *Times* described it, "a flamethrowing treatise on class, race and intelligence." *The Bell Curve* offered a glimpse of how easily the language of genes and race can be distorted, and how potently those distortions can reverberate through a culture that is obsessed with heredity and race.

As public flame-throwers go, Herrnstein was an old hand: his 1985 book, *Crime and Human Nature*, had ignited its own firestorm of controversy by claiming that ingrained characteristics, such as personality and temperament, were linked to criminal behavior. A decade later, *The Bell Curve* made an even more incendiary set of claims. Murray and Herrnstein proposed that intelligence was also largely ingrained—i.e., genetic—and that it was unequally segregated among races. Whites and Asians possessed higher IQs on average, and Africans and African-Americans possessed lower IQs. This difference in "intellectual capacity," Murray and Herrnstein claimed, was largely responsible for the chronic underperformance of African-Americans in social and economic spheres. African-Americans were lagging behind in the United States not because of systematic flaws in our social contracts, but because of systematic flaws in their mental constructs.

To understand *The Bell Curve*, we need to begin with a definition of "intelligence." Predictably, Murray and Herrnstein chose a narrow definition of intelligence—one that brings us back to nineteenth-century biometrics and eugenics. Galton and his disciples, we might recall, were obsessed with the measurement of intelligence. Between 1890 and 1910, dozens of tests were devised in Europe and America that purported to measure intelligence in some unbiased and quantitative manner. In 1904, Charles Spearman, a British statistician, noted an important feature of these tests: people who did well in one test generally tended to do well in another test. Spearman hypothesized that this positive correlation existed because all the tests were obliquely measuring some mysterious common factor. This factor, Spearman proposed, was not knowledge itself, but the capacity to *acquire and manipulate* abstract knowledge. Spearman called it "general intelligence." He labeled it *g*.

By the early twentieth century, *g* had caught the imagination of the public. First, it captivated early eugenicists. In 1916, the Stanford psychologist Lewis Terman, an avid supporter of the American eugenics movement, created a standardized test to rapidly and quantitatively assess general intelligence, hoping to use the test to select more intelligent humans for eugenic breeding. Recognizing that this measurement varied with age during childhood development, Terman advocated a new metric to quantify age-specific intelligence. If a subject's "mental age" was the same as his or her physical age, their "intelligence quotient," or IQ, was defined as exactly 100. If a subject lagged in mental age compared to physical age, the IQ was less than a hundred; if she was more mentally advanced, she was assigned an IQ above 100.

A numerical measure of intelligence was also particularly suited to the demands of the First and Second World Wars, during which recruits had to be assigned to wartime tasks requiring diverse skills based on rapid, quantitative assessments. When veterans returned to civilian life after the wars, they found their lives dominated by intelligence testing. By the early 1940s, such tests had become accepted as an inherent part of American culture. IQ tests were used to rank job applicants, place children in school, and recruit agents for the Secret Service. In the 1950s, Americans commonly listed their IQs on their résumés, submitted the results of a test for a job application, or even chose their spouses based on the test. IQ scores were pinned on the babies who were on display in Better Babies contests (although how IQ was measured in a two-year-old remained mysterious).

These rhetorical and historical shifts in the concept of intelligence are worth noting, for we will return to them in a few paragraphs. General intelligence (*g*) originated as a *statistical* correlation between tests given under particular circumstances to particular individuals. It morphed into the notion of "general intelligence" because of a *hypothesis* concerning the nature of human knowledge acquisition. And it was codified into "IQ" to serve the particular exigencies of war. In a cultural sense, the definition of g was an exquisitely self-reinforcing phenomenon: those who possessed it, anointed as "intelligent" and given the arbitration of the quality, had every incentive in the world to propagate its definition. Richard Dawkins, the evolutionary biologist, once defined a meme as a cultural unit that spreads virally through societies by mutating, replicating, and being selected. We might imagine g as such a self-propagating unit. We might even call it a "selfish g."

It takes counterculture to counter culture—and it was only inevitable, perhaps, that the sweeping political movements that gripped America in the 1960s and 1970s would shake the notions of general intelligence and IQ by their very roots. As the civil rights movement and feminism highlighted chronic political and social inequalities in America, it became evident that biological and psychological features were not just inborn but likely to be deeply influenced by context and environment. The dogma of a single form of intelligence was also challenged by scientific evidence. Developmental psychologists such as Louis Thurstone (in the fifties) and Howard Gardner (in the late seventies) argued that "general intelligence" was a rather clumsy way to lump together many vastly more context-specific and subtle forms of intelligence, such as visuospatial, mathematical, or verbal intelligence. A geneticist, revisiting this data, might have concluded that g—the measurement of a hypothetical quality invented to serve a particular context—might be a trait barely worth linking to genes, but this hardly dissuaded Murray and Herrnstein. Drawing heavily from an earlier article by the psychologist Arthur Jensen, Murray and Herrnstein set out to prove that g was heritable, that it varied between ethnic groups, and—most crucially—that the racial disparity was due to inborn genetic differences between whites and African-Americans.

①

Is g heritable? In a certain sense, yes. In the 1950s, a series of reports suggested a strong genetic component. Of these, twin studies were the most definitive. When identical twins who had been reared together—i.e., with shared genes and shared environments—were tested in the early fifties, psychologists had found a striking degree of concordance in their IQs, with a correlation value of 0.86.* In the late eighties, when identical twins who were separated at birth and reared separately were tested, the correlation fell to 0.74—still a striking number.

But the heritability of a trait, no matter how strong, may be the result of multiple genes, each exerting a relatively minor effect. If that was the case, identical twins would show strong correlations in g, but parents and children would be far less concordant. IQ followed this pattern. The correlation between parents and children living together, for instance, fell to 0.42. With parents and children living apart, the correlation collapsed to 0.22. Whatever the IQ test was measuring, it was a heritable factor, but one also influenced by many genes and possibly strongly modified by environment—part nature and part nurture.

The most logical conclusion from these facts is that while some combination of genes and environments can strongly influence g, this combination will rarely be passed, intact, from parents to their children. Mendel's laws virtually guarantee that the particular permutation of genes will scatter apart in every generation. And environmental interactions are so difficult to capture and predict that they cannot be reproduced over time. Intelligence, in short, is *heritable* (i.e., influenced by genes), but not easily *inheritable* (i.e., moved down intact from one generation to the next).

Had Murray and Herrnstein reached these conclusions, they would have published an accurate, if rather uncontroversial, book on the inheritance of intelligence. But the molten centerpiece of *The Bell Curve* is not the heritability of IQ—but its racial distribution. Murray and Herrnstein began by reviewing 156 independent studies that had compared IQs between races. Taken together, these studies had found an average IQ of 100 for whites (by definition, the average IQ of the index population has to be 100) and 85 for African-Americans—a 15-point difference. Murray and Herrnstein tried, somewhat valiantly, to ferret out the possibility that the tests were biased against African-Americans. They limited the tests to only those administered after 1960, and only given outside the South, hoping to curtail endemic biases—but the 15-point difference persisted.

Could the difference in black-white IQ scores be a result of socioeconomic status? That impoverished children, regardless of race, perform

---

* More recent estimates have pinned the correlation between identical twins to 0.6–0.7. When the 1950s data was reexamined in subsequent decades by several psychologists, including Leon Kamin, the methodologies used were found to be suspect, and the initial estimates called into question. .

worse in IQ tests had been known for decades. Indeed, of all the hypotheses concerning the difference in racial IQs, this was, by far, the most plausible: that the vast part of the black-white difference may be the consequence of the overrepresentation of poor African-American children. In the 1990s, the psychologist Eric Turkheimer strongly validated this theory by demonstrating that genes play a rather minor role in determining IQ in severely impoverished circumstances. If you superpose poverty, hunger, and illness on a child, then these variables dominate the influence on IQ. Genes that control IQ only become significant if you remove these limitations.

It is easy to demonstrate an analogous effect in a lab: If you raise two plant strains—one tall and one short—in undernourished circumstances, then both plants grow short regardless of intrinsic genetic drive. In contrast, when nutrients are no longer limiting, the tall plant grows to its full height. Whether genes or environment—nature or nurture—dominates in influence depends on context. When environments are constraining, they exert a disproportionate influence. When the constraints are removed, genes become ascendant.*

The effects of poverty and deprivation offered a perfectly reasonable cause for the *overall* black-white difference in IQ, but Murray and Herrnstein dug deeper. Even correcting for socioeconomic status, they found, the black-white score difference could not be fully eliminated. If you plot a curve of IQ of whites and African-Americans across increasing socioeconomic status, the IQ increases in both cases, as expected. Wealthier children certainly score better than their poorer counterparts—both in white and in African-American populations. Yet, the difference between the IQ scores across races persists. Indeed, paradoxically, the difference *increases* as you increase the socioeconomic status of whites and African-Americans. The difference in IQ scores between wealthy whites and wealthy African-Americans is even more pronounced: far from narrowing, the gap *widens* at the top brackets of income.

①

Quarts and quarts of ink have been spilled in books, magazines, scientific journals, and newspapers analyzing, cross-examining, and debunk-

---

* There can hardly be a more cogent genetic argument for equality. It is impossible to ascertain any human's genetic potential without first equalizing environments.

ing these results. In a blistering article written for the *New Yorker*, for instance, the evolutionary biologist Stephen Jay Gould argued that the effect was far too mild, and the variation within tests was far too great, to make any statistical conclusions about the difference. The Harvard historian Orlando Patterson, in the slyly titled "For Whom the Bell Curves," reminded readers that the frayed legacies of slavery, racism, and bigotry had deepened the cultural rifts between whites and African-Americans so dramatically that biological attributes across races could not be compared in a meaningful way. Indeed, the social psychologist Claude Steele demonstrated that when black students are asked to take an IQ test under the pretext that they are being tested to try out a new electronic pen, or a new way of scoring, they perform well. Told that they are being tested for "intelligence," however, their scores collapse. The real variable being measured, then, is not intelligence but an aptitude for test taking, or self-esteem, or simply ego or anxiety. In a society where black men and women experience routine, pervasive, and insidious discrimination, such a propensity could become fully self-reinforcing: black children do worse at tests because they've been told that they are worse at tests, which makes them perform badly in tests and furthers the idea that they are less intelligent—*ad infinitum*.

But the final fatal flaw in *The Bell Curve* is something far simpler, a fact buried so inconspicuously in a single throwaway paragraph in an eight-hundred-page book that it virtually disappears. If you take African-Americans and whites with identical IQ scores, say 105, and measure their performance in various *subtests* for intelligence, black children often score better in certain sets (tests of short-term memory and recall, for instance), while whites often score better in others (tests of visuospatial and perceptual changes). In other words, the way an IQ test is configured profoundly affects the way different racial groups, and their gene variants, perform on it: alter the weights and balances within the same test, and you alter the measure of intelligence.

The strongest evidence for such a bias comes from a largely forgotten study performed by Sandra Scarr and Richard Weinberg in 1976. Scarr studied transracial adoptees—black children adopted by white parents—and found that these children had an average IQ of 106, at least as high as white children. By analyzing carefully performed controls, Scarr concluded that "intelligence" was not being enhanced, but performance on particular subtests of intelligence.

We cannot shrug this proposition away by suggesting that the current construction of the IQ test must be correct since it predicts performance in the real world. Of course it does—because the concept of IQ is powerfully self-reinforcing: it measures a quality imbued with enormous meaning and value whose job it is to propagate itself. The circle of its logic is perfectly closed and impenetrable. Yet the actual configuration of the test is relatively arbitrary. You do not render the word *intelligence* meaningless by shifting the balance in a test—from visuospatial perception to short-term recall, say—but you *do* shift the black-white IQ score discrepancy. And therein lies the rub. The tricky thing about the notion of *g* is that it pretends to be a biological quality that is measurable and heritable, while it is actually strongly determined by cultural priorities. It is—to simplify it somewhat—the most dangerous of all things: a meme masquerading as a gene.

If the history of medical genetics teaches us one lesson, it is to be wary of precisely such slips between biology and culture. Humans, we now know, are largely similar in genetic terms—but with enough variation within us to represent true diversity. Or, perhaps more accurately, we are culturally or biologically inclined to magnify variations, even if they are minor in the larger scheme of the genome. Tests that are explicitly designed to capture variances in abilities will likely capture variances in abilities—and these variations may well track along racial lines. But to call the score in such a test "intelligence," especially when the score is uniquely sensitive to the configuration of the test, is to insult the very quality it sets out to measure.

Genes cannot tell us how to categorize or comprehend human diversity; environments can, cultures can, geographies can, histories can. Our language sputters in its attempt to capture this slip. When a genetic variation is statistically the most common, we call it *normal*—a word that implies not just superior statistical representation but qualitative or even moral superiority (Merriam-Webster's dictionary has no less than eight definitions of the word, including "occurring naturally" and "mentally and physically healthy"). When the variation is rare, it is termed a *mutant*—a word that implies not just statistical uncommonness, but qualitative inferiority, or even moral repugnance.

And so it goes, interposing linguistic discrimination on genetic variation, mixing biology and desire. When a gene variant reduces an organism's fitness in a particular environment—a hairless man in Antarctica—we call the phenomenon *genetic illness*. When the same variant increases fitness in a different environment, we call the organism *genetically enhanced*. The synthesis of evolutionary biology and genetics reminds us that these judgments are meaningless: *enhancement* or *illness* are words that measure the fitness of a particular genotype to a particular environment; if you alter the environment, the words can even reverse their meanings. "When nobody read," the psychologist Alison Gopnik writes, "dyslexia wasn't a problem. When most people had to hunt, a minor genetic variation in your ability to focus attention was hardly a problem, and may even have been an advantage [enabling a hunter to maintain his focus on multiple and simultaneous targets, for instance]. When most people have to make it through high school, the same variation can become a life-altering disease."

℗

The desire to categorize humans along racial lines, and the impulse to superpose attributes such as intelligence (or criminality, creativity, or violence) on those lines, illustrates a general theme concerning genetics and categorization. Like the English novel, or the face, say, the human genome can be lumped or split in a million different ways. But whether *to* split or lump, *to* categorize or synthesize, is a choice. When a distinct, heritable biological feature, such as a genetic illness (e.g., sickle-cell anemia), is the ascendant concern, then examining the genome to identify the locus of that feature makes absolute sense. The narrower the definition of the heritable feature or the trait, the more likely we will find a genetic locus for that trait, and the more likely that the trait will segregate within some human subpopulation (Ashkenazi Jews in the case of Tay-Sachs disease, or Afro-Caribbeans for sickle-cell anemia). There's a reason that marathon running, for instance, is becoming a genetic sport: runners from Kenya and Ethiopia, a narrow eastern wedge of one continent, dominate the race not just because of talent and training, but also because the marathon is a narrowly defined test for a certain form of extreme fortitude. Genes that enable this fortitude (e.g., particular combinations of gene variants that produce distinct forms of anatomy, physiology, and metabolism) will be naturally selected.

Conversely, the more we widen the definition of a feature or trait (say, intelligence, or temperament), the less likely that the trait will correlate with single genes—and, by extension, with races, tribes, or subpopulations. Intelligence and temperament are not marathon races: there are no fixed criteria for success, no start or finish lines—and running sideways or backward, might secure victory.

The narrowness, or breadth, of the definition of a feature is, in fact, a question of identity—i.e., how we define, categorize, and understand humans (ourselves) in a cultural, social, and political sense. The crucial missing element in our blurred conversation on the definition of race, then, is a conversation on the definition of identity.

The First Derivative of Identity

For several decades, anthropology has participated in the general deconstruction of "identity" as a stable object of scholarly inquiry. The notion that individuals craft their identity through social performances, and hence that their identity is not a fixed essence, fundamentally drives current research into gender and sexuality. The notion that collective identity emerges out of political struggle and compromise underlies contemporary studies of race, ethnicity and nationalism.

—Paul Brodwin, "Genetics, Identity, and the
Anthropology of Essentialism"

Methinks you are my glass, and not my brother.
—William Shakespeare,
The Comedy of Errors, act 5, scene 1

On October 6, 1942, five years before my father's family left Barisal, my mother was born twice in Delhi. Bulu, her identical twin, came before her, placid and beautiful. My mother, Tulu, emerged several minutes later, squirming and crying murderously. The midwife must, fortunately, have known enough about infants to recognize that the most beautiful are often the most damned: the quiet twin, on the edge of listlessness, was quite severely undernourished and had to be swaddled in blankets and revived. The first few days of my aunt's life were the most tenuous. She could not suckle at the breast, the story runs (perhaps apocryphally), and there were no infant bottles to be found in Delhi in the forties, so she was fed through a cotton wick dipped in milk, and then

352

from the caul of a cowrie shell shaped like a spoon. A nurse was hired to tend to her. When the breast milk began to run dry at seven months, my mother was quickly weaned to let her sister have its last remnants. Right from the onset, then, my mother and her twin were living experiments in genetics—emphatically identical in nature and emphatically divergent in nurture.

My mother—the "younger" of the two by minutes—was boisterous. She had a slippery, mercurial temper. She was carefree and fearless, fast to learn and willing to make mistakes. Bulu was physically timid. Her mind was more agile, her tongue sharper, her wit more lancing. Tulu was gregarious. She made friends easily. She was impervious to insults. Bulu was reserved and restrained, quieter and more brittle. Tulu liked theater and dancing. Bulu was a poet, a writer, a dreamer.

Yet the contrasts only highlighted the similarities between the twins. Tulu and Bulu looked strikingly similar: they had the same freckled skin, almond-shaped face, and high cheekbones, unusual among Bengalis, and the slight downward tilt of the outer edge of the eye, the trick that Italian painters used to make Madonnas seem to exude a mysterious empathy. They shared the inner language that twins often share. They had jokes that only the other twin understood.

Over the years, their lives drifted apart. Tulu married my father in 1965 (he had moved to Delhi three years earlier). It was an arranged marriage, but also a risky one. My father was a penniless immigrant in a new city, saddled with a domineering mother and a half-mad brother who lived at home. To my mother's overly genteel West Bengali relatives, my father's family was the very embodiment of East Bengali hickness: when his brothers sat down to lunch, they would pile their rice in mounds and punch volcanic holes in it for gravy, as if marking the insatiable, perpetual hunger of their village days in the form of craters on their plates. Bulu's marriage seemed a vastly safer prospect by comparison. In 1966, she was engaged to a young lawyer, the eldest son of a well-established clan in Calcutta. In 1967, Bulu married him and moved to his family's sprawling, decrepit mansion in South Calcutta, with a garden already choked by weeds.

By the time I was born, in 1970, the sisters' fortunes had started to move in unexpected directions. In the late 1960s, Calcutta began its steady descent into hell. Its economy was fraying, its tenuous infrastructure heaving under the weight of waves of immigration. Internecine political movements broke out frequently, shuttering the streets and businesses for weeks. As the city convulsed between cycles of violence and apathy, Bulu's new family hemorrhaged its savings to keep itself afloat. Her husband kept up the pretense of a job, leaving home every morning with the requisite briefcase and tiffin box—but who needed a lawyer in a city without laws? Eventually, the family sold the mildewing house, with its grand veranda and inner courtyard, and moved into a modest two-room flat—just a few miles from the house that had sheltered my grandmother on her first night in Calcutta.

My father's fate, in contrast, mirrored that of his adoptive city. Delhi, the capital, was India's overnourished child. Bolstered by the nation's aspirations to build a mega-metropolis, fattened by subsidies and grants, its roads widened and its economy expanded. My father rose through the ranks of a Japanese multinational firm, clambering swiftly from lower- to upper-middle class. Our neighborhood, once girded by forests of thornbushes overrun with wild dogs and goats, was soon transformed into one of the most affluent pockets of real estate in the city. We vacationed in Europe. We learned to eat with chopsticks and swam in hotel pools in the summer. When the monsoons hit Calcutta, the mounds of garbage on the streets clogged the drains and turned the city into a vast infested swamp. One such stagnant pond, festering with mosquitoes, was deposited yearly outside Bulu's house. She called it her own "swimming pool."

There is something in that comment—a lightness—that was symptomatic. You might imagine that the sharp vicissitudes of fortune had reshaped Tulu and Bulu in drastically different ways. On the contrary: over the years, their physical resemblance had dwindled to the point of vanishing, but something ineffable about them—an approach, a temperament—remained remarkably similar and even amplified in its convergence. Despite the growing economic rift between the two sisters, they shared an optimism about the world, a curiosity, a sense of humor, an equanimity that borders on nobility but comes with no pride. When we traveled abroad, my mother would bring home a collection of souvenirs for Bulu—a wooden toy from Belgium, fruit-flavored chewing gum from America that smelled of no earthly fruit, or a glass trinket from Switzerland. My aunt would read travel guides of the countries that we had visited. "I've been there too," she would say, arranging the souvenirs in a glass case, with no trace of bitterness in her voice.

There is no word, or phrase, in the English language for that moment in a son's consciousness when he begins to understand his mother—not just superficially, but with the immersive clarity with which he understands himself. My experience of this moment, somewhere in the depths of my childhood, was perfectly dual: as I understood my mother, I also learned to understand her twin. I knew, with luminous certainty, when she would laugh, what made her feel slighted, what would animate her, or where her sympathies or affinities might lie. To see the world through the eyes of my mother was to also see it through the eyes of her twin, except, perhaps, with lenses tinted in slightly different colors.

What had converged between my mother and her sister, I began to realize, was not personality but its tendency—its first derivative, to borrow a mathematical term. In calculus, the first derivative of a point is not its position in space, but its propensity to change its position; not where an object *is*, but how it *moves* in space and time. This shared quality, unfathomable to some, and yet self-evident to a four-year-old, was the lasting link between my mother and her twin. Tulu and Bulu were no longer recognizably identical—but they shared the first derivative of identity.

℗

Anyone who doubts that genes can specify identity might well have arrived from another planet and failed to notice that the humans come in two fundamental variants: male and female. Cultural critics, queer theorists, fashion photographers, and Lady Gaga have reminded us—accurately—that these categories are not as fundamental as they might seem, and that unsettling ambiguities frequently lurk in their borderlands. But it is hard to dispute three essential facts: that males and females are anatomically and physiologically different; that these anatomical and physiological differences are specified by genes; and that these differences, interposed against cultural and social constructions of the self, have a potent influence on specifying our identities as individuals.

That genes have anything to do with the determination of sex, gender, and gender identity is a relatively new idea in our history. The distinction between the three words is relevant to this discussion. By *sex*, I mean the anatomic and physiological aspects of male versus female bodies. By *gender*, I am referring to a more complex idea: the psychic, social, and cultural roles that an individual assumes. By *gender identity*, I mean an individual's sense of self (as female versus male, as neither, or as something in between).

For millennia, the basis of the anatomical dissimilarities between men and women—the "anatomical dimorphism" of sex—was poorly understood. In AD 200, Galen, the most influential anatomist in the ancient world, performed elaborate dissections to try to prove that male and female reproductive organs were analogs of each other, with the male organs turned inside out and the female's turned outside in. The ovaries, Galen argued, were just internalized testicles retained inside the female body because females lacked some "vital heat" that could extrude the organs. "Turn outward the woman's [organs] and double the man's, and you will find the same," he wrote. Galen's students and followers stretched this analogy, quite literally, to its absurd point, reasoning that the uterus was the scrotum ballooning inward, and that the fallopian tubes were the seminal vesicles blown up and expanded. The theory was memorialized in a medieval verse, an anatomical mnemonic for medical students:

> *Though they of different sexes be*
> *Yet on the whole, they're the same as we*
> *For those that have the strictest searchers been*
> *Find women are just men turned outside in.*

But what force was responsible for turning men "inside out," or women "outside in," like socks? Centuries before Galen, the Greek philosopher Anaxagoras, writing around 400 BC, claimed that gender, like New York real estate, was determined entirely by location. Like Pythagoras, Anaxagoras believed that the essence of heredity was carried by male sperm, while the female only "shaped" male semen in the womb to produce the fetus. The inheritance of gender also followed this pattern. Semen produced in the left testicle gave rise to male children, while semen produced in the right testicle gave rise to females. The specification of gender continued in the womb, extending the left-right spatial code sparked off during ejaculation. A male fetus was deposited, with exquisite specificity, in the right horn of the uterus. A female, conversely, was nurtured in the left horn.

It is easy to laugh Anaxagoras's theory off as anachronistic and bizarre. Its peculiar insistence on left and right placement—as if gender were determined by some sort of cutlery arrangement—clearly belongs to another era. But the theory was revolutionary for its time, for it made two crucial advances. First, it recognized that the determination of gender was essentially random—and so a random cause (the left or right origin of sperm) would need to be invoked to explain it. And second, it reasoned that once established, the original random act had to be amplified and consolidated to fully engender gender. The developmental plan of the fetus was crucial. Right-sided sperm found its way to the right side of the uterus, where it was further specified into a male fetus. Left-sided sperm was segregated to the left side to make a female child. Gender determination was a chain reaction, set off by a single step but then amplified by the location of the fetus into the full-fledged dimorphism between men and women.

And there, for the most part, sex determination sat, for centuries. Theories abounded, but conceptually they were variants of Anaxagoras's idea—that sex was determined by an essentially random act, consolidated and amplified by the environment of the egg or fetus. "Sex is not inherited," one geneticist wrote in 1900. Even Thomas Morgan, perhaps the most prominent proponent of the role of genes in development, proposed that sex could not be determined through genes. In 1903, Morgan wrote that sex was likely determined by multiple environmental inputs rather than a single genetic one: "The egg, as far as sex is concerned, appears to be in a sort of balanced state, and the conditions to which it is exposed . . . may determine which sex it will produce. It may be a futile attempt to try to discover any one influence that has a deciding influence for all kinds of eggs."

☉

In the winter of 1903, the very year that Morgan had published his casual dismissal of a genetic theory of sex determination, Nettie Stevens, a graduate student, performed a study that would transform the field. Stevens was born to a carpenter in Vermont in 1861. She took courses to become a schoolteacher, but by the early 1890s, had saved enough money from her tutoring jobs to attend Stanford University in California. She chose to attend graduate school in biology in 1900—an unusual choice for a woman in her time—and, even more unusually, chose to perform fieldwork at the zoological station in faraway Naples, where Theodor Boveri had collected his urchin eggs. She learned Italian so that she could speak the lingo of the local fishermen who brought her eggs from the shores. From Boveri, she learned to stain eggs to identify chromosomes—the strange blue-stained filaments that resided in cells.

Boveri had demonstrated that cells with altered chromosomes could not develop normally—and so hereditary instructions for development had to be carried within chromosomes. But could the genetic determinant for sex also be carried by chromosomes? In 1903, Stevens chose a simple organism—the common mealworm—to investigate the correlation between an individual worm's chromosomal makeup and its sex. When Stevens used Boveri's chromosome-staining method on male and female worms, the answer leaped out of the microscope: a variation in just one chromosome correlated perfectly with the worm's sex. Mealworms have twenty chromosomes in all—ten pairs (most animals have paired chromosomes; humans have twenty-three pairs). Cells from female worms inevitably possessed ten matched pairs. Cells from male worms, in contrast, had two unpaired chromosomes—a small, nublike band and a larger chromosome. Stevens suggested that the presence of the small chromosome was sufficient to determine sex. She termed it the *sex chromosome*.

To Stevens, this suggested a simple theory of sex determination. When sperm was created in the male gonad, it was made in two forms—one bearing the nublike male chromosome, and another bearing the normal-size female chromosome—in roughly equal ratios. When sperm bearing the male chromosome—i.e., "male sperm"—fertilized the egg, the embryo was born male. When "female sperm" fertilized an egg, the result was a female embryo.

Stevens's work was corroborated by that of her close collaborator, the cell biologist Edmund Wilson, who simplified Stevens's terminology, calling the male chromosome Y, and the female X. In chromosomal terms, male cells were XY, and females were XX. The egg contains a single X chromosome, Wilson reasoned. When a sperm carrying a Y chromosome fertilizes an egg, it results in an XY combination, and *maleness* is determined. When a sperm carrying an X chromosome meets a female egg, the result is XX, which determines *femaleness*. Sex was not determined by right or left testicles, but by a similarly random process—by the nature of the genetic payload of the first sperm to reach and fertilize an egg.

℗

The XY system discovered by Stevens and Wilson had an important corollary: if the Y chromosome carried all the information to determine maleness, then that chromosome had to carry genes to make an embryo male. At first, geneticists expected to find dozens of male-determining genes on the Y chromosome: sex, after all, involves the exacting coordination of multiple anatomical, physiological, and psychological features, and it was hard to imagine that a single gene could be capable of performing such diverse functions all by itself. Yet, careful students of genetics knew that the Y chromosome was an inhospitable place for genes. Unlike any other chromosome, the Y is "unpaired"—i.e., it has no sister chromosome and no duplicate copy, leaving every gene on the chromosome to fend for itself. A mutation in any other chromosome can be repaired by copying the intact gene from the other chromosome. But a Y chromosome gene cannot be fixed, repaired, or recopied from another chromsome; it has no backup or guide (there is, however, a unique internal system to repair genes in the Y chromosome). When the Y chromosome is assailed by mutations, it lacks a mechanism to recover information. The Y is thus pockmarked with the potshots and scars of history. It is the most vulnerable spot in the human genome.

As a consequence of this constant genetic bombardment, the human Y chromosome began to jettison information millions of years ago. Genes that were truly valuable for survival were likely shuffled to other parts of the genome where they could be stored securely; genes with limited value were made obsolete, retired, or replaced; only the most essential genes were retained (some of these genes were duplicated in the Y chromosome itself—but even this strategy does not solve the problem completely). As information was lost, the Y chromosome itself shrank—whittled down piece by piece by the mirthless cycle of mutation and gene loss. That the Y chromosome is the smallest of all chromosomes is not a coincidence: it is largely a victim of planned obsolescence (in 2014, scientists discovered that a few extremely important genes may be permanently lodged in the Y).

In genetic terms, this suggests a peculiar paradox. Sex, one of the most complex of human traits, is unlikely to be encoded by multiple genes. Rather, a single gene, buried rather precariously in the Y chromosome, must be the master regulator of maleness.* Male readers of that last paragraph should take notice: we barely made it.

⊙

In the early 1980s, a young geneticist in London named Peter Goodfellow began to hunt for the sex-determining gene on the Y chromosome. A die-hard soccer enthusiast—scruffy, bone-thin, taut, with an unmistakable East Anglian drawl and a "punk meets new romantic" dress sense—Goodfellow intended to use the gene-mapping methods pioneered by Botstein and Davis to narrow the search to a small region of the Y chro-

---

* With such steep liabilities, it is a genuine wonder that the XY system of gender determination exists in the first place. Why did mammals evolve a mechanism of sex determination burdened with such obvious pitfalls? Why carry the sex-determination gene in, of all places, an unpaired, hostile chromosome, where it's most likely to be assailed by mutations?

To answer the question, we need to step back and ask a more fundamental question: Why was sexual reproduction invented? Why, as Darwin wondered, should new beings "be produced by the union of two sexual elements, instead of by a process of parthenogenesis"?

Most evolutionary biologists agree that sex was created to enable rapid genetic reassortment. No quicker way exists, perhaps, to mix genes from two organisms than by mixing their eggs and sperm. And even the genesis of spermatozoa and egg cells causes genes to be shuffled through the gene recombination. The powerful reassortment of genes during sexual reproduction increases variation. Variation, in turn, increases an organism's fitness and survival in the face of a constantly changing environment. The phrase *sexual reproduction*, then, is a perfect misnomer. The evolutionary purpose of sex is not "reproduction": organisms can make superior facsimiles—re-productions—of themselves in the absence of sex. Sex was invented for quite the opposite reason: to enable *recombination*.

But "sexual reproduction" and "sex determination" are not the same. Even if we recognize the many advantages of sexual reproduction, we might still ask why most mammals use the XY system for gender determination. Why, in short, the Y? We do not know. The XY system for gender determination was clearly invented in evolution several million years ago. In birds, reptiles, and some insects, the system is reversed: the female carries two different chromosomes, while the male carries two identical chromosomes. And in yet other animals, such as some reptiles and fish, gender is determined by the temperature of the egg, or the size of an organism relative to its competitors. These systems of gender determination are thought to predate the XY system of mammals. But why the XY system was fixed in mammals—and why it is still in use—remains a mystery. Having two sexes has some evident advantages: males and females can carry out specialized functions and occupy different roles in breeding. But having two sexes does not require a Y chromosome *per se*. Perhaps evolution stumbled on the Y chromosome as a quick and dirty solution for sex determination—confining the male-determinant gene in a separate chromosome and putting a powerful gene in it to control maleness is certainly a workable solution. Some geneticists believe that the Y may continue shrinking, while others argue that it will only shrink to a point, retaining SRY and other essential genes.

mosome. But how could a "normal" gene be mapped without the existence of a variant phenotype, or an associated disease? Cystic fibrosis and Huntington's disease genes had been mapped to their chromosomal locations by tracking the link between the disease-causing gene and signposts along the genome. In both cases, affected siblings carrying the gene also carried the signpost, while unaffected siblings did not. But where might Goodfellow find a human family with a variant gender—a third sex—that was genetically transmitted, and carried by some siblings but not others?

℗

In fact, such humans existed—although identifying them was a much more complicated task than anticipated. In 1955, Gerald Swyer, an English endocrinologist investigating female infertility, had discovered a rare syndrome that made humans biologically female but chromosomally male. "Women" born with "Swyer syndrome" were anatomically and physiologically female throughout childhood, but did not achieve female sexual maturity in early adulthood. When their cells were examined, geneticists discovered that these "women" had XY chromosomes in all their cells. Every cell was chromosomally male—yet the person built from these cells was anatomically, physiologically, and psychologically female. A "woman" with Swyer syndrome had been born with the male chromosomal pattern (i.e., XY chromosomes) in all of her cells, but had somehow failed to signal "maleness" to her body.

The most likely scenario behind Swyer syndrome was that the master-regulatory gene that specifies maleness had been inactivated by a mutation, leading to femaleness. At MIT, a team led by the geneticist David Page had used such sex-reversed women to map the male-determinant gene to a relatively narrow region of the Y chromosome. The next step was the most laborious—the gene-by-gene sifting to find the correct candidate among the dozens of genes in that general location. Goodfellow was making slow, steady progress when he received devastating news. In the summer of 1989, he learned that Page had landed on the male-determinant gene. Page called the gene ZFY, for its presence in the Y chromosome.

Initially, ZFY seemed like the perfect candidate: it was located in the right region of the Y chromosome, and its DNA sequence suggested that it could act as a master switch for dozens of other genes. But when Goodfellow looked carefully, the shoe wouldn't fit: when ZFY was sequenced in women with Swyer syndrome, it was completely normal. There was no mutation that would explain the disruption of the male signal in these women.

With *ZFY* disqualified, Goodfellow returned to his search. The gene for maleness had to be in the region identified by Page's team: they must have come close, but just missed it. In 1989, rooting about close to the *ZFY* gene, Goodfellow found another promising candidate—a small, nondescript, tightly packed, intronless gene called *SRY*. Right at the onset, it seemed like the perfect candidate. The normal *SRY* protein was abundantly expressed in the testes, as one might expect for a sex-determination gene. Other animals, including marsupials, also carried variants of the gene on their Y chromosomes—and thus only males inherited the gene. The most striking proof that *SRY* was the correct gene came from the analysis of human cohorts: the gene was indisputably mutated in females with Swyer syndrome, and nonmutated in their unaffected siblings.

But Goodfellow had one last experiment to clinch the case—the most dramatic of his proofs. If the *SRY* gene was the singular determinant of "maleness," what if he forcibly *activated* the gene in female animals? Would females be forced to turn into males? When Goodfellow and Robin Lovell-Badge inserted an extra copy of the *SRY* gene into female mice, their offspring were born with XX chromosomes in every cell (i.e., genetically female), as expected. Yet the mice developed as anatomically male—including growing a penis and testicles, mounting females, and performing every behavior characteristic of male mice. By flicking a single genetic switch, Goodfellow had switched an organism's sex—creating Swyer syndrome in reverse.

①

Is all of sex just one gene, then? Almost. Women with Swyer syndrome have male chromosomes in every cell in the body—but with the maleness-determining gene inactivated by a mutation, the Y chromosome is literally emasculated (not in a pejorative but in a purely biological sense).* The

---

* The opposite scenario is also notable. In rare circumstances, the SRY gene is translocated *to the X chromosome*, resulting in 46 XX humans (i.e. chromosomal females) who carry the male-determination gene – the obverse of Swyer syndrome. Humans with this genetic arrangement can have normal male anatomy; some have smaller or undescended testes. Notably, children typically identify as male. Here, again, the SRY gene dominates anatomy, physiology and gender identity, although it clearly requires the correct context of other genes to fully execute its function.

presence of the Y chromosome in the cells of women with Swyer syndrome does disrupt some aspects of the anatomical development of females. In particular, breasts do not form properly, and ovarian function is abnormal, resulting in low levels of estrogen. But these women feel absolutely no disjunction in their physiology. Most aspects of female anatomy are formed perfectly normally: the vulva and vagina are intact, and a urinary outlet is attached to them with textbook fidelity. Astonishingly, even the *gender identity* of women with Swyer syndrome is unambiguous: just one gene flicked off and they "become" women. Although estrogen is undoubtedly required to enable the development of secondary sexual characteristics and reinforce some anatomical aspects of femininity in adults, women with Swyer syndrome are typically never confused about gender or gender identity. As one woman wrote, "I definitely identify with female gender roles. I've always considered myself one hundred percent female. . . . I played on a boy's soccer team for a while—I have a twin brother; we look nothing alike—but I was definitely a girl on a boy's team. I didn't fit in well: I suggested that we name our team 'the butterflies.'"

Women with Swyer syndrome are not "women trapped in men's bodies." They are women trapped in *women's* bodies that are chromosomally male (except for just one gene). A mutation in that single gene, *SRY*, creates a (largely) female body—and, more crucially, a wholly female self. It is as artless, as plain, as binary, as leaning over the nightstand and turning a switch on or off.*

Ⅸ

If genes determine sexual anatomy so unilaterally, then how do genes affect gender identity? On the morning of May 5, 2004, David Reimer, a

---

* What about "intersexuality" – i.e., the fact that some humans are born with reproductive anatomy or physiology that does not fit the typical definitions of male and female bodies? Does intersexuality contradict the idea of a strong binary genetic switch that controls sexual anatomy and physiology? No. SRY, note, lies atop a cascade of events that produces males versus females: it turns genes *on* and *off*, and these genes, in turn, activate and repress other networks of genes to produce diffuse aspects of reproductive and sexual anatomy and physiology. Variations in these downstream networks, intersecting with variations in exposures and environments (hormones, say), might produce variations in reproductive anatomy – *even though a strong binary switch sits atop the cascade.* We will revisit this theme – of hierarchies in genetic networks, with strong, autonomous drivers on top, and more subtle integrators and effectors organized below – several times in subsequent pages.

thirty-eight-year-old man in Winnipeg, walked into the parking lot of a grocery store and killed himself with a sawed-off shotgun. Born in 1965 as Bruce Reimer—chromosomally, and genetically, male—David had been the victim of a ghoulish attempt at circumcision by an inept surgeon, resulting in a severely damaged penis in early infancy. Reconstructive surgery was impossible, and so Bruce's parents had rushed him to see John Money, a psychiatrist at Johns Hopkins University, known internationally for his interest in gender and sexual behavior. Money evaluated the child and, as part of an experiment, asked Bruce's parents to have their son castrated and raise him as a girl. Desperate to give their son a "normal" life, his parents capitulated. They changed his name to Brenda.

Money's experiment on David Reimer—for which he never asked or received permission from the university or hospital—was an attempt to test a theory widely fashionable in academic circles in the sixties. The notion that gender identity was not innate and was crafted through social performance and cultural mimicry ("you are who you act; nurture can overcome nature") was in its full prime in that era—and Money was among its most ardent and most vocal proponents. Casting himself as the Henry Higgins of sexual transformation, Money advocated "sexual reassignment," the reorientation of sexual identity through behavioral and hormonal therapy—a decades-long process invented by him that allowed his experimental subjects to emerge with their identities sanguinely switched. Based on Money's advice, "Brenda" was dressed and treated as a girl. Her hair was grown long. She was given female dolls and a sewing machine. Her teachers and friends were never informed about the switch.

Brenda had an identical twin—a boy named Brian—who was brought up as a male child. As part of the study, Brenda and Brian visited Money's clinic in Baltimore at frequent intervals throughout their childhood. As preadolescence approached, Money prescribed estrogen supplements to feminize Brenda. The surgical construction of an artificial vagina was scheduled to complete her anatomical transformation into a woman. Money published a steady stream of highly cited papers touting the extraordinary success of the sexual reassignment. Brenda was adjusting to her new identity with perfect equanimity, he proposed. Her twin, Brian, was a "rough and tumble" boy, while Brenda was an "active little girl." Brenda would ease into womanhood with scarcely any hurdles, Money declared. "Gender identity is sufficiently incompletely differentiated at birth to permit successful assignment of a genetic male as a girl."

In reality, nothing could have been further from the truth. At age four, Brenda took scissors and shredded the pink and white dresses she had been forced to wear. She lapsed into fits of fury when told to walk or talk like a girl. Pinioned to an identity that she found evidently false and discordant, she was anxious, depressed, confused, anguished, and often frankly enraged. In her school reports, Brenda was described as "tomboyish" and "dominant," with "abundant physical energy." She refused to play with dolls or other girls, preferring her brother's toys (the only time she played with her sewing machine was when she sneaked a screwdriver out of her father's toolbox and took the machine meticulously apart, screw by screw). Perhaps most confoundingly to her young classmates, Brenda went to the girl's bathroom dutifully—but then preferred to urinate with her legs spread wide, standing up.

After fourteen years, Brenda brought this grotesque charade to an end. She refused the vaginal operation. She stopped the estrogen pills, underwent a bilateral mastectomy to excise her breast tissue, and began injecting testosterone to revert back to male. She—*he*—changed her name to David. He married a woman in 1990, but the relationship was tormented from the start. Bruce/Brenda/David—the boy who became a girl who became a man—continued to ricochet between devastating bouts of anxiety, anger, denial, and depression. He lost his job. The marriage failed. In 2004, shortly after a bitter altercation with his wife, David killed himself.

David Reimer's case was not unique. In the 1970s and 1980s, several other cases of sexual reassignment—the attempted conversion of chromosomally male children into females through psychological and social conditioning—were described, each troubled and troubling in its own right. In some cases, the gender dysphoria was not as acute as David's—but the wo/men often suffered haunting pangs of anxiety, anger, dysphoria, and disorientation well into adulthood. In one particularly revealing case, a woman—called C—came to see a psychiatrist in Rochester, Minnesota. Dressed in a frilly, floral blouse and a rough cowhide jacket—"my leather-and-lace look," as she described it—C had no problems with some aspects of her duality, yet had trouble reconciling her "sense of herself as fundamentally female." Born and raised as a girl in the 1940s, C recalled being a tomboy in school. She had never thought of herself as physically male, but had always felt a kinship with men ("I feel like I have the brain of a man"). She married a man in her twenties and lived with him—until a chance ménage à trois involving a woman kindled her fantasies about women. Her husband married the other woman, and C left him and entered a series of lesbian relationships. She oscillated between periods of equanimity and depression. She joined a church and discovered a nurturing spiritual community—except for a pastor who railed against her homosexuality and recommended therapy to "convert" her.

At forty-eight, goaded by guilt and fear, she finally sought psychiatric assistance. During the medical examination, her cells were sent for chromosomal analysis, and she was found to have XY chromosomes in her cells. Genetically speaking, C was male. She later discovered that s/he had been born with ambiguous, underdeveloped genitals, although chromosomally male. Her mother had consented to reconstructive surgery to transform her into a female. Sexual reassignment had begun when she was six months old, and she had been given hormones at puberty on the pretext of curing a "hormonal imbalance." Throughout her childhood and adolescence, C did not have the faintest spasm of doubt about her gender.

C's case illustrates the importance of thinking carefully about the link between gender and genetics. Unlike David Reimer, C was not confused about the performance of gender *roles*: she wore female clothes in public, maintained a heterosexual marriage (for a while, at least), and acted within the range of cultural and social norms to pass as female for forty-eight years. Yet despite her guilt about her sexuality, crucial aspects of her identity—kinship, fantasy, desire, and erotic drive—remained fastened to maleness. C had been able to learn many of the essential features of her acquired gender through social performance and mimesis, but she couldn't unlearn the psychosexual drives of her genetic self.

In 2005, a team of researchers at Columbia University validated these case reports in a longitudinal study of "genetic males"—i.e., children born with XY chromosomes—who had been assigned to female gender at birth, typically because of the inadequate anatomical development of their genitals. Some of the cases were not as anguished as David Reimer's or C's—but an overwhelming number of males assigned to female gender roles reported experiencing moderate to severe gender dysphoria during childhood. Many had suffered anxiety, depression, and confusion. Many had voluntarily changed genders back to male upon adolescence and adulthood. Most notably, when "genetic males" born with ambiguous genitals were brought up as *boys*, not girls, not a single case of gender dysphoria or gender change in adulthood was reported.

These case reports finally put to rest the assumption, still unshakably prevalent in some circles, that gender identity can be created or programmed entirely, or even substantially, by training, suggestion, behavioral enforcement, social performance, or cultural interventions. It is now clear that genes are vastly more influential than virtually any other force in shaping sex identity and gender identity—although in limited circumstances a few attributes of gender can be learned through cultural, social, and hormonal reprogramming. Since even hormones are ultimately "genetic"—i.e., the direct or indirect products of genes—then the capacity to reprogram gender using purely behavioral therapy and cultural reinforcement begins to tip into the realm of impossibility. Indeed, the growing consensus in medicine is that, aside from exceedingly rare exceptions, children should be assigned to their *chromosomal* (i.e., genetic) sex regardless of anatomical variations and differences—with the option of switching, if desired, later in life. As of this writing, none of these children have opted to switch from their gene-assigned sexes.

①

How can we reconcile this idea—of a single genetic switch that dominates one of the most profound dichotomies in human identity—with the fact that human gender identity in the real world appears in a continuous spectrum? Virtually every culture has recognized that gender does not exist in discrete half-moons of black and white, but in a thousand shades of gray. Even Otto Weininger, the Austrian philosopher famous for his misogyny, conceded, "Is it really the case that all women and men are marked off sharply from each other . . . ? There are transitional forms between the metals and nonmetals; between chemical combinations and simple mixtures, between animals and plants, between phanerogams and cryptogams, and between mammals and birds. . . . The improbability may henceforth be taken for granted of finding in Nature a sharp cleavage between all that is masculine on the one side and all that is feminine on the other."

In genetic terms, though, there is no contradiction: master switches and hierarchical organizations of genes are perfectly compatible with continuous curves of behavior, identity, and physiology. The *SRY* gene indubitably controls sex determination in an on/off manner. Turn *SRY* on, and an animal becomes anatomically and physiologically male. Turn it off, and the animal becomes anatomically and physiologically female.

But to enable more profound aspects of gender determination and gender identity, *SRY* must act on dozens of targets—turning *them* on and off, activating some genes and repressing others, like a relay race that moves a baton from hand to hand. These genes, in turn, integrate inputs from the self and the environment—from hormones, behaviors, exposures, social performance, cultural role-playing, and memory—to engender gender. What we call gender, then, is an elaborate genetic and developmental cascade, with *SRY* at the tip of the hierarchy, and modifiers, integrators, instigators, and interpreters below. This geno-developmental cascade specifies gender identity. To return to an earlier analogy, genes are single lines in a recipe that specifies gender. The *SRY* gene is the first line in the recipe: "Start with four cups of flour." If you fail to start with the flour, you will certainly not bake anything close to a cake. But infinite variations fan out of that first line—from the crusty baguette of a French bakery to the eggy mooncakes of Chinatown.

☉

The existence of a transgender identity provides powerful evidence for this geno-developmental cascade. In an anatomical and physiological sense, sex identity is quite binary: just one gene governs sex identity, resulting in the striking anatomical and physiological dimorphism that we observe between males and females. But gender and gender identity are far from binary. Imagine a gene—call it *TGY*—that determines how the brain responds to *SRY* (or some other male hormone or signal). One child might inherit a *TGY* gene variant that is highly resistant to the action of *SRY* on the brain, resulting in a body that is anatomically male, but a brain that does not read or interpret that male signal. Such a brain might recognize itself as psychologically female; it might consider itself neither male or female, or imagine itself belonging to a third gender altogether.

These men (or women) have something akin to a Swyer syndrome of *identity*: their chromosomal and anatomical gender is male (or female), but their chromosomal/anatomical state does not generate a synonymous signal in their brains. In rats, notably, such a syndrome can be caused by changing a single gene in the brains of female embryos or exposing embryos to a drug that blocks the signaling of "femaleness" to the brain. Female mice engineered with this altered gene or treated with this drug have all the anatomical and physiological features of femaleness, but perform the activities associated with male mice, including mounting females: these animals might be anatomically female, but they are behaviorally male.

⌖

The hierarchical organization of this genetic cascade illustrates a crucial principle about the link between genes and environments in general. The perennial debate rages on: nature or nurture, genes or environment? The battle has gone on for so long, and with such animosity, that both sides have capitulated. Identity, we are now told, is determined by nature *and* nurture, genes *and* environment, intrinsic *and* extrinsic inputs. But this too is nonsense—an armistice between fools. If genes that govern gender identity are hierarchically organized—starting with *SRY* on top and then fanning out into thousands of rivulets of information below—then whether nature predominates or nurture is not absolute, but depends quite acutely on the level of organization one chooses to examine.

At the top of the cascade, nature works forcefully and unilaterally. Up top, gender is quite simple—just one master gene flicking on and off. If we learned to toggle that switch—by genetic means or with a drug—we could control the production of men or women, and they would emerge with male versus female identity (and even large parts of anatomy) quite intact. At the bottom of the network, in contrast, a purely genetic view fails to perform; it does not provide a particularly sophisticated understanding of gender or its identity. Here, in the estuarine plains of crisscrossing information, history, society, and culture collide and intersect with genetics, like tides. Some waves cancel each other, while others reinforce each other. No force is particularly strong—but their combined effect produces the unique and rippled landscape that we call an individual's identity.

The Last Mile

*Like sleeping dogs, unknown twins might be better left
alone.*

—William Wright, *Born That Way*

Whether *sex* identity is innate or acquired in the one-in-two-thousand
babies born with ambiguous genitals does not typically incite national
debates about inheritance, preference, perversity, and choice. Whether
sex*ual* identity—the choice and preference of sexual partners—is innate or acquired does, absolutely. For a while in the 1950s and 1960s, it
seemed that that discussion had been settled for good. The dominant theory among psychiatrists was that sexual preference—i.e., "straightness"
versus "gayness"—was acquired, not inborn. Homosexuality was characterized as a frustrated form of neurotic anxiety. "It is the consensus of
many contemporary psychoanalytic workers that permanent homosexuals, like all perverts, are neurotics," the psychiatrist Sándor Lorand wrote
in 1956. "The homosexual's real enemy," wrote another psychiatrist in the
late sixties, "is not so much his perversion but [his] ignorance of the possibility that he can be helped, plus his psychic masochism which leads
him to shun treatment."

In 1962, Irving Bieber, a prominent New York psychiatrist known for
his attempts to convert gay men to straightness, wrote the enormously
influential *Homosexuality: A Psychoanalytic Study of Male Homosexuals*.
Bieber proposed that male homosexuality was caused by the distorted dynamics of a family—by the fatal combination of a smothering mother who
was often "close-binding and [sexually] intimate," if not overtly seductive,
to her son, and by a detached, distant, or "emotionally hostile" father.
Boys responded to these forces by exhibiting neurotic, self-destructive,
and crippling behaviors ("a homosexual is a person whose heterosexual
function is crippled, like the legs of a polio victim," Bieber famously said

in 1973). Ultimately, in some such boys, a subconscious desire to identify with the mother and to emasculate the father became manifest as a choice to embrace a lifestyle that fell outside the norm. The sexual "polio victim" adopts a pathological style of being, Bieber argued, just as victims of polio might grasp a pathological style of walking. By the late 1980s, the notion that homosexuality represented the choice of a deviant lifestyle had sclerosed into dogma, leading Dan Quayle, then vice president, to sanguinely proclaim in 1992 that "homosexuality is more of a choice than a biological situation. . . . It is a wrong choice."

In July 1993, the discovery of the so-called gay gene would incite one of the most vigorous public discussions about genes, identity, and choice in the history of genetics. The discovery would illustrate the power of the gene to sway public opinion and almost fully invert the terms of the discussion. In *People* magazine (not, we might note, a particularly strident voice for radical social change), the columnist Carol Sarler wrote that October, "What do we say of the woman who will opt for an abortion rather than raising a gentle, caring boy who might—only *might*, mind you—grow up to love another gentle caring boy? We say that she is a warped, dysfunctional monster who—if forced to have the child—will make the child's life hell. We say that no child should be forced to have her as a parent."

The phrase "gentle, caring boy"—chosen to illustrate a child's inborn propensity rather than an adult's perverted preferences—exemplified the inversion of the debate. Once *genes* had been implicated in the development of sexual preference, the gay child was instantly transformed to normal. His hateful enemies were the abnormal monsters.

①

It was boredom, more than activism, that prompted the search for the gay gene. Dean Hamer, a researcher at the National Cancer Institute, was not looking for controversy. He was not even looking for himself. Although openly gay, Hamer had never been particularly intrigued by the genetics of any form of identity, sexual or otherwise. He had spent much of his life comfortably ensconced in a "normally quiet US government laboratory . . . jumbled floor to ceiling with beakers and vials," studying the regulation of a gene called metallothionine—or MT—that is used by cells to respond to poisonous heavy metals, such as copper and zinc.

In the summer of 1991, Hamer flew to Oxford to present a scientific seminar on gene regulation. It was his standard research talk—well received, as usual—but when he opened the floor to discussion, he experienced the most desolate form of déjà vu: the questions seemed exactly the same as the queries raised by his talk a decade ago. When the next speaker, a competitor from another lab, presented data that affirmed and extended Hamer's work, Hamer found himself becoming even more bored and depressed. "I realized that even if I stuck with this research for another ten years, the best thing I could hope for was to build a three-dimensional replica of our little [genetic] model. It didn't seem like much of a lifetime goal."

In the lull between sessions, Hamer walked out in a daze, his mind churning. He stopped at Blackwell's, the cavernous bookstore on High Street, and descended into its concentric rooms, browsing through books on biology. He bought two books. The first was Darwin's *Descent of Man, and Selection in Relation to Sex*. Published in 1871, Darwin's book had set off a storm of controversy by claiming human descent from an apelike ancestor (in *Origin of Species*, Darwin had coyly skirted the question of human descent, but in *Descent of Man*, he had taken the question head-on).

Descent of Man is to biologists what *War and Peace* is to graduate students of literature: nearly every biologist claims to have read the book, or appears to know its essential thesis, but few have actually even opened its pages. Hamer had never read it either. To his surprise, Hamer found that Darwin had spent a substantial portion of the book discussing sex, the choice of sexual partners, and its influence on dominance behaviors and social organization. Darwin had clearly felt that heredity had a powerful effect on sexual behavior. Yet the genetic determinants of sexual behavior and preference—"the final cause of sexuality," as Darwin described it—had remained mysterious to him.

But the idea that sexual behavior, or any behavior, was linked to genes had fallen out of fashion. The second book, Richard Lewontin's *Not in Our Genes: Biology, Ideology, and Human Nature*, proposed a different view. Published in 1984, Lewontin had launched an attack on the idea that much of human nature was biologically determined. Elements of human behavior that are considered genetically determined, Lewontin argued, are often nothing more than arbitrary, and often manipulative, constructions of culture and society to reinforce power structures. "There is no acceptable evidence that homosexuality has any genetic basis. . . . The story has been manufactured out of whole cloth," Lewontin wrote. Darwin was broadly right about organismal evolution, he argued—but not about the evolution of human identity.

Which of these two theories was correct? To Hamer, at least, sexual orientation seemed far too fundamental to be entirely constructed by cultural forces. "Why was Lewontin, a formidable geneticist, so determined not to believe that behavior could be inherited?" Hamer wondered. "He could not disprove the genetics of behavior in the lab and so he wrote a political polemic against it? Maybe there was room for real science here." Hamer intended to give himself a crash course on the genetics of sexual behavior. He returned to his lab to start exploring—but there was little to be learned from the past. When Hamer searched a database of all scientific journals published since 1966 for articles concerning "homosexuality" and "genes," he found 14. Searching for the metallothionine gene, in contrast, brought up 654.

But Hamer did find a few tantalizing clues, even if they were half buried in the scientific literature. In the 1980s, a professor of psychology named J. Michael Bailey had tried to study the genetics of sexual orientation using a twin-study experiment. Bailey's methodology was classical: if sexual orientation was partly inherited, then a higher proportion of identical twins should both be gay compared to fraternal twins. By placing strategic advertisements in gay magazines and newspapers, Bailey had recruited 110 male twin pairs in which at least one twin was gay. (If this seems difficult today, imagine running this experiment in 1978, when few men were publicly out of the closet, and gay sex in certain states was punishable as a crime.)

When Bailey looked for concordance of gayness among twins, the results were striking. Among the fifty-six pairs of identical twins, both twins were gay in 52 percent.* Of the fifty-four pairs of nonidentical twins, 22 percent were both gay—lower than the fraction for identical twins, but

---

* A shared intrauterine environment, or exposures during gestation, might explain some of this concordance, but the fact that *non*identical twins share these environments, yet have a lower concordance compared to identical twins, argues against such theories. The genetic argument is also strengthened by the fact that gay siblings also have a higher rate of concordance than the general population (although lower than identical twins). Future studies might reveal a combination of environmental and genetic factors in the determination of sexual preference, but genes will likely remain important factors.

still significantly higher than the estimate of 10 percent gay in the overall population. (Years later, Bailey would hear of striking cases such as this: In 1971, two Canadian twin brothers were separated within weeks of birth. One was adopted by a prosperous American family. The other was raised in Canada by his natural mother under vastly different circumstances. The brothers, who looked virtually identical, knew nothing of each other's existence until they ran into each other, by accident, in a gay bar in Canada.)

Male homosexuality was not just genes, Bailey found. Influences such as families, friends, schools, religious beliefs, and social structure clearly modified sexual behavior—so much so that one identical twin identified as gay and the other as straight as much as 48 percent of the time. Perhaps external or internal triggers were required to release distinct patterns of sexual behavior. Undoubtedly, the pervasive and repressive cultural beliefs that surrounded homosexuality were potent enough to sway the choice of a "straight" identity in one twin but not the other. But the twin studies provided incontrovertible evidence that genes influenced homosexuality more strongly than, say, genes influenced the propensity for type 1 diabetes (the concordance rate among twins is only 30 percent), and almost as strongly as genes influence height (a concordance of about 55 percent).

Bailey had profoundly changed the conversation around sexual identity away from the 1960s rhetoric of "choice" and "personal preference" toward biology, genetics, and inheritance. If we did not think of variations in height or the development of dyslexia or type 1 diabetes as choices, then we could not think of sexual identity as a choice.

But was it one gene, or many genes? And what was the gene? Where was it located? To identify the "gay gene," Hamer needed a much larger study—preferably a study involving families in which sexual orientation could be tracked over multiple generations. To fund such a study, Hamer would need a new grant—but where on earth might a federal researcher studying metallothionine regulation find money to hunt for a gene that influences human sexuality?

℗

In early 1991, two developments enabled Hamer's hunt. The first was the announcement of the Human Genome Project. Even though the precise sequence of the human genome would not be known for another decade, the mapping of pivotal genetic signposts along the human genome made it vastly easier to hunt for any gene. Hamer's idea—of mapping genes related to homosexuality—would have been methodologically intractable in the 1980s. A decade later, with genetic markers strung like marquee lights along most chromosomes, it was at least conceptually within reach.

The second was AIDS. The illness had decimated the gay community in the late 1980s—and goaded by activists and patients, often through civil disobedience and militant protests, the NIH had eventually committed hundreds of millions of dollars to AIDS-focused research. Hamer's tactical genius was to piggyback the gay gene hunt on an AIDS-related study. He knew that Kaposi's sarcoma, a previously rare, indolent tumor, had been found at a strikingly high frequency among gay men with AIDS. Perhaps, Hamer reasoned, the risk factors for the progression of Kaposi's sarcoma were related to homosexuality—and if so, finding genes for one might lead to identifying genes for the other. The theory was spectacularly wrong: Kaposi's sarcoma would later be found to be caused by a virus, transmitted sexually and occur mainly in immunocompromised people, thus explaining its co-occurrence with AIDS. But it was tactically brilliant: in 1991, the NIH granted Hamer $75,000 for his new protocol, a study to find homosexuality-related genes.

Protocol #92-C-0078 was launched in the fall of 1991. By 1992, Hamer had attracted 114 gay men to his study. Hamer planned to use the cohort to create elaborate family trees to determine if sexual orientation ran in families, to describe the pattern of its inheritance, and to map the gene. But mapping the gay gene, Hamer knew, would become vastly easier if he could find *brother* pairs where both were known to be gay. Twins share the same genes, but brothers share only some sections of their genomes. If Hamer could find brothers who were gay, he would find the subsections of the genome shared by them, and thereby isolate the gay gene. Beyond family trees, then, Hamer needed samples of genes from such brothers. His budget allowed him to fly such siblings to Washington and provide a $45 stipend for a weekend. The brothers, often estranged, got a reunion. Hamer got a tube of blood.

By the late summer of 1992, Hamer had collected information about nearly one thousand family members and built family trees for each of the 114 gay men. In June, he sat down for the first glimpse of the data on his computer. Almost instantly, he felt the gratifying heave of validation: as with the Bailey study, the siblings in Hamer's study had a higher concordance in sexual orientation—about 20 percent, nearly twice the population rate of about 10 percent. The study had produced real data—but the gratification soon turned cold. As Hamer pored through the numbers, he could find no other insight. Beyond the concordance between gay siblings, he found no obvious pattern or trend.

Hamer was devastated. He tried organizing the numbers into groups and subgroups, but to no avail. He was about to throw the family trees, sketched on pieces of paper, back into their piles, when he stumbled on a pattern—an observation so subtle that only the human eye could have discerned it. By chance, while drawing the trees, he had placed the paternal relatives on the left, and maternal relatives on the right, for each family. Gay men were marked with red. And as he shuffled the papers, he instinctively discerned a trend: the red marks tended to cluster toward the right, while the unmarked men tended to cluster to the left. Gay men tended to have gay uncles—*but only on the maternal side.* The more Hamer hunted up and down the family trees for gay relatives—a "gay *Roots* project," as he called it—the more the trend intensified. Maternal cousins had higher rates of concordance—but not paternal cousins. Maternal cousins through *aunts* tended to have higher concordance than any other cousins.

The pattern ran generation on generation. To a seasoned geneticist, this trend meant the gay gene had to be carried on the X chromosome. Hamer could almost see it now in his mind's eye—an inherited element passing between generations like a shadowy presence, nowhere as penetrant as the typical cystic fibrosis or Huntington's gene mutations, but inevitably tracking the trail of the X chromosome. In a typical family tree, a great-uncle might be identified as potentially gay. (Family histories were often vague. The historical closet was substantially darker than the current sexual closet—but Hamer had collected data from occasional families where sexual identity was known for up to two or even three generations.) All the sons born from that uncle's brothers were straight—men do not pass on the X chromosome to their male children (in all human males, the X chromosome must come from the mother). But one of his *sister's* sons might be gay, and that son's sister's son might also be gay: a man shares parts of his X chromosome with his sister and with his sister's sons. And so forth: great-uncle, uncle, eldest nephew, nephew's sibling, sidestepping through generations, forward and across, like a knight's move in chess. Hamer had suddenly moved from a phenotype (sexual preference) to a potential location on a chromosome—a genotype. He had not identified the gay gene—but he had proved that a piece of DNA associated with sexual orientation could be physically mapped to the human genome.

But where on the X chromosome? Hamer now turned to forty gay sibling pairs from whom he had collected blood. Assume, for a moment, that the gay gene is indeed located on some small stretch of the X chromosome. Wherever the stretch is, the forty siblings would tend to share that particular chunk of DNA at a significantly higher frequency than siblings where one is gay and the other straight. Using signposts along the genome defined by the Human Genome Project, and careful mathematical analysis, Hamer began to sequentially narrow the stretch to shorter and shorter regions of the X chromosome. He ran through a series of twenty-two markers along the entire length of the chromosome. Notably, of the forty gay siblings, Hamer found that thirty-three brothers shared a small stretch of the X chromosome called Xq28. Random chance predicted that only half the brothers—i.e., twenty—would share that marker. The chance that thirteen extra brothers would carry the same marker was vanishingly small—less than one in ten thousand. Somewhere near Xq28, then, was a gene that determined male sexual identity.

①

Xq28 was an instant sensation. "The phone rang off the hook," Hamer recalled. "There were TV cameramen lined up outside the lab; the mailbox and e-mail overflowed." The *Daily Telegraph*, the conservative London newspaper, wrote that if science had isolated the gay gene, then "science could be used to eradicate it." "A lot of mothers are going to feel guilty," wrote another newspaper. "Genetic tyranny!" claimed yet another headline. Ethicists wondered whether parents would "genotype" their way out of having homosexual children by testing fetuses. Hamer's research "does identify a chromosomal region that could be analyzed for individual males," one writer wrote, "but the results of any test based on this research would, again, offer only probabilistic tools by which to estimate the sexual orientation of some men." Hamer was attacked left and right—literally. Antigay conservatives argued that by reducing homosexuality to genetics, Hamer had justified it biologically. Advocates of gay rights accused Hamer of furthering the fantasy of a "gay test" and thereby propelling new mechanisms of detection and discrimination.

Hamer's own approach was neutral, rigorous, and scientific—often corrosively so. He continued to refine his analysis, running the Xq28 association through a variety of tests. He asked if Xq28 might encode not a gene for homosexuality, but a "gene for sissyness" (only a gay man would dare to use that phrase in a scientific paper). It did not: men who shared Xq28 did not have any significant alterations in gender-specific behaviors, or in conventional aspects of masculinity. Could it be a gene for receptive anal intercourse ("Is it the bottoms up gene?" he asked)? Again, there was no correlation. Could it be related to rebelliousness? Or a gene for bucking repressive social customs? A gene for contrary behavior? Hypothesis after hypothesis was turned over, but there was no link. The exhaustive elimination of all possibilities left only one conclusion: male sexual identity was partially determined by a gene near Xq28.

①

Since Hamer's 1993 paper in *Science*, several groups have tried to validate Hamer's data. In 1995, Hamer's own team published a larger analysis that confirmed the original study. In 1999, a Canadian group tried to replicate Hamer's study on a small group of gay siblings but failed to find the link to Xq28. In 2005, in perhaps the largest study to date, 456 sibling pairs were studied. The link to Xq28 was not discovered, but links to chromosomes seven, eight, and ten were found. In 2015, in yet another detailed analysis of 409 additional sibling pairs, the link to Xq28 was validated again—albeit weakly—and the previously identified link to chromosome eight was reiterated.

Perhaps the most intriguing feature of all these studies is that, thus far, no one has isolated an actual gene that influences sexual identity. Linkage analysis does not identify a gene itself; it only identifies a chromosomal region where a gene might be found. After nearly a decade of intensive hunting, what geneticists have found is not a "gay gene" but a few "gay locations." Some genes that reside in these locations are indeed tantalizing candidates as regulators of sexual behavior—but none of these candidates has been experimentally linked to homo- or heterosexuality. One gene that sits in the Xq28 region, for instance, encodes a protein that is known to regulate the testosterone receptor, a well-known mediator of sexual behavior. But whether this gene is the long-sought gay gene on Xq28 remains unknown.

The "gay gene" might not even be a gene, at least not in the traditional sense. It might be a stretch of DNA that regulates a gene that sits near it or influences a gene quite far from it. Perhaps it is located in an intron— the sequences of DNA that interrupt genes and break them up into modules. Whatever the molecular identity of the determinant, this much is certain: sooner or later, we will discover the precise nature of the heritable elements that influence human sexual identity. Whether Hamer is right or wrong about Xq28 is immaterial. The twin studies clearly suggest that several determinants that influence sexual identity are part of the human genome, and as geneticists discover more powerful methods to map, identify, and categorize genes, we will inevitably find some of these determinants. Like gender, these elements will likely be hierarchically organized—with master regulators on top, and complex integrators and modifiers on the bottom. Unlike gender, though, sexual identity is unlikely to be governed by a single master regulator. Multiple genes with small effects—in particular, genes that modulate and integrate inputs from the environment—are much more likely to be involved in the determination of sexual identity. There will be no *SRY* gene for straightness.

℗

The publication of Hamer's article on the gay gene coincided with the forceful reemergence of the notion that genes could influence diverse behaviors, impulses, personalities, desires, and temperaments—an idea that had been out of intellectual fashion for nearly two decades. In 1971, in a book titled *Genes, Dreams and Realities*, Macfarlane Burnet, the renowned Anglo-Australian biologist, wrote, "It is self-evident that the genes that we are born with provide, along with the rest of our functional selves, the basis of our intelligence, temperament and personality." But by the midseventies, Burnet's conception had become far from "self-evident." The notion that genes, of all things, could predispose humans to acquiring particular "functional selves"—possessing particular variants of temperament, personality, and identity—had unceremoniously been drummed out of universities. "An environmentalist view . . . dominated psychological theory and research from the 1930s through the 1970s," Nancy Segal, the psychologist, wrote. "Other than being born with a general capacity to learn, human behavior was explained almost exclusively by forces outside the individual." A "toddler," as one biologist recalled, was perceived as a "random access memory onto which any number of operating systems could be downloaded by culture." The Silly Putty of a child's psyche was infinitely malleable; you could mold it into any shape and force it into any dress by changing the environment or reprogramming behavior (hence the stupefying credulity that enabled experiments, such as John Money's, to attempt definitive changes in gender using behavioral and cultural therapy). Another psychologist, entering a research program at Yale University in the 1970s to study human behaviors, was bewildered by the dogmatic stance against genetics in his new department: "Whatever back-porch wisdom we might have brought to New Haven about inherited traits [driving and influencing human behaviors] was the kind of bunk that we were paying Yale to purge." The environment was all about environments.

The return of the native—the emergence of the *gene* as a major driver for psychological impulses—was not as easy to orchestrate. In part, it required a fundamental reinvention of that classic workhorse of human genetics: the much maligned, much misunderstood twin study. Twin studies had been around since the Nazis—recall Mengele's macabre preoccupation with *Zwillinge*—but they had reached a conceptual gridlock. The problem with studying identical twins from the same family, geneticists knew, was the impossibility of unbraiding the twisted strands of nature and nurture. Reared in the same home, by the same parents, often schooled in the same classrooms by the same teachers, dressed, fed, and nurtured identically, these twins offered no self-evident way to separate the effects of genes versus the environment.

Comparing identical twins to fraternal twins partially solved the problem, since fraternal twins share the same environment, but share only half the genes, on average. But critics argued that such identical/fraternal comparisons were also intrinsically flawed. Perhaps identical twins are treated more similarly than fraternal twins by their parents. Identical twins, for instance, were known to have more similar patterns of nutrition and growth compared to fraternal twins—but was this nature or nurture? Or identical twins might react *against* each other to distinguish themselves from one another—my mother and her twin often self-consciously chose opposite shades of lipstick—but was that dissimilarity encoded by genes, or a reaction to genes?

①

In 1979, a scientist in Minnesota found a way out of the impasse. One evening in February, Thomas Bouchard, a behavioral psychologist, found a news article left by a student in Bouchard's mailbox. It was an unusual story: a pair of identical twins from Ohio had been separated at birth, adopted by different families, and experienced a remarkable reunion at age thirty. These brothers were obviously part of a fleetingly rare group—identical twins given up for adoption and reared apart—but they represented a powerful way to interrogate the effects of human genes. Genes had to be identical in these twins, but environments were often radically different. By comparing separated-at-birth twins against twins brought up in the same family, Bouchard could untwist the effects of genes and environments. The similarities between such twins could have nothing to do with nurture; they could only reflect hereditary influences—nature.

Bouchard began recruiting such twins for his study in 1979. By the late eighties, he had assembled the world's largest cohort of reared-apart and reared-together twins. Bouchard called it the Minnesota Study of Twins Reared Apart (or "MISTRA"). In the summer of 1990, his team presented a comprehensive analysis as a lead article in *Science* magazine.* The team had collected data from fifty-six reared-apart identical twins and thirty reared-apart fraternal twins. In addition, data from an earlier study, containing 331 reared-together twins (both identical and fraternal), was included. The twin sets came from a broad range of socioeconomic classes, with frequent discordances between two individual twins (one reared in a poor family, another adopted by a wealthy family). Physical and racial environments were also broadly different. To assess environments, Bouchard made the twins compile meticulous records of their homes, schools, offices, behaviors, choices, diets, exposures, and lifestyles. To determine indicators of "cultural class," Bouchard's team ingeniously recorded whether the family possessed a "telescope, an unabridged dictionary, or original artwork."

The punch line of the paper was presented in a single table—unusual for *Science*, where papers typically contain dozens of figures. Over nearly eleven years, the Minnesota group had subjected the twins to battery upon battery of detailed physiological and psychological tests. In test upon test, the similarities between twins remained striking and consistent. The correlations between physical features had been expected: the number of fingerprint ridges on the thumb, for instance, was virtually

---

* Earlier versions of the paper appeared in 1984 and 1987.

identical, with a correlational value of 0.96 (a value of 1 reflects complete concordance or absolute identity). IQ testing also revealed a strong correlation of about 0.70, corroborating previous studies. But even the most mysterious and profound aspects of personality, preferences, behaviors, attitudes, and temperament, tested broadly using multiple independent tests, showed strong correlations between 0.50 and 0.60—virtually identical to the correlation between identical twins that had been reared together. (As a sense of the strength of this association, consider that the correlation between height and weight in human populations lies between 0.60 and 0.70, and between educational status and income is about 0.50. The concordance among twins for type 1 diabetes, an illness considered unequivocally genetic, is only 0.35.)

The most intriguing correlations obtained by the Minnesota study were also among the most unexpected. Social and political attitudes between twins reared apart were just as concordant as those between twins reared together: liberals clustered with liberals, and orthodoxy was twinned with orthodoxy. Religiosity and faith were also strikingly concordant: twins were either both faithful or both nonreligious. Traditionalism, or "willingness to yield to authority," was significantly correlated. So were characteristics such as "assertiveness, drive for leadership, and a taste for attention."

Other studies on identical twins continued to deepen the effect of genes on human personality and behavior. Novelty seeking and impulsiveness were found to have striking degrees of correlation. Experiences that one might have imagined as intensely personal were, in fact, shared between twins. "Empathy, altruism, sense of equity, love, trust, music, economic behavior, and even politics are partially hardwired." As one startled observer wrote, "A surprisingly high genetic component was found in the ability to be enthralled by an esthetic experience such as listening to a symphonic concert." Separated by geographic and economic continents, when two brothers, estranged at birth, were brought to tears by the same Chopin nocturne at night, they seemed to be responding to some subtle, common chord struck by their genomes.

℗

Bouchard had measured characteristics where measurable—but it is impossible to convey the strange feel of this similarity without citing actual examples. Daphne Goodship and Barbara Herbert were twins from

England. They had been born to an unmarried Finnish exchange student in 1939, and their mother had given them up for adoption before returning to Finland. The twins were raised separately—Barbara, the daughter of a lower-middle-class municipal gardener, and Daphne, the daughter of a prominent upper-class metallurgist. Both lived near London—although, given the rigidity of class structure in 1950s England, they might as well have been brought up on different planets.

Yet in Minnesota, Bouchard's staff was repeatedly struck by the similarities between the twins. Both laughed uncontrollably, erupting into peals of giggles with minimal provocation (the staff called them the "giggle twins"). They played pranks on the staff, and on each other. Both were five feet three inches tall, and both had crooked fingers. Both had gray-brown hair; both had dyed it an unusual shade of auburn. They tested identically on IQ tests. Both had fallen down the stairs as children and broken their ankles; both had a consequent fear of heights, and yet both, despite some clumsiness, had taken lessons in ballroom dancing. Both had met their future husbands through dancing lessons.

Two men—both renamed Jim after adoption—had been separated from each other thirty-seven days after birth and had grown up eighty miles apart in an industrial belt in northern Ohio. Both had struggled through school. "Both drove Chevrolets, both chain-smoked Salems, and both loved sports, especially stock-car racing, but both *disliked* baseball. . . . Both Jims had married women named Linda Both had owned dogs that they had named Toy. . . . One had a son named James Allan; the other's son was named James Alan. Both Jims had undergone vasectomies, and both had slightly high blood pressure. Each had become overweight at roughly the same time and had leveled off at about the same age. Both suffered migraine headaches that lasted approximately half a day and that did not respond to any medication."

Two other women, also separated at birth, emerged from separate airplanes wearing seven rings each. A pair of male twins—one brought up Jewish, in Trinidad, and the other Catholic, in Germany—wore similar clothes, including blue oxford shirts with epaulets and four pockets, and shared peculiar obsessive behaviors, such as carrying wads of Kleenex in their pockets and flushing toilets twice, once before and once after using them. Both had invented fake sneezes, which they deployed strategically—as "jokes"—to diffuse tense moments of conversation. Both had violent, explosive tempers and unexpected spasms of anxiety.

A pair of twins had an identical manner of rubbing their noses, and—even though they had never met—had each invented a new word to describe the odd habit: *squidging*. Two sisters in Bouchard's study shared the same pattern of anxiety and despair. As teenagers, they confessed, they had been haunted by the same nightmare: of feeling suffocated in the middle of the night because their throats were being stuffed with various—but typically metallic—things: "door-knobs, needles and fishhooks."

Several features *were* quite different in the reared-apart twins. Daphne and Barbara looked similar, but Barbara was twenty pounds heavier (although, notably, despite the twenty pounds, their heart rates and blood pressures were the same). The German twin, of the Catholic/Jewish set, had been a staunch German nationalist as a young man, while his twin brother had spent his summers at a kibbutz. Yet, both shared a fervor, a rigidity of belief, even if the beliefs themselves were almost diametrically opposed. The picture that emerged from the Minnesota study was not that reared-apart twins were identical, but that they shared a powerful tendency toward similar or convergent behaviors. What was common to them was not identity, but its first derivative.

①

In the early 1990s, Richard Ebstein, a geneticist in Israel, read studies about the subtypes of human temperaments. Ebstein was intrigued: some of these studies had shifted our understanding of personality and temperament—away from culture and environment and toward genes. But like Hamer, Ebstein wanted to identify the actual genes that determined variant forms of behavior. Genes had been linked to temperaments before: the extraordinary, otherworldly sweetness of children with Down syndrome had long been noted by psychologists, and other genetic syndromes had been linked with outbursts of violence and aggression. But Ebstein was not interested in the outer bounds of pathology; he was interested in normal variants of temperament. Extreme genetic changes could evidently cause extreme variants of temperament. But were there "normal" gene variants that influenced normal subtypes of personality?

To find such genes, Ebstein knew, he would have to begin with rigorous definitions of the subtypes of personality that he wished to link to genes. In the late 1980s, psychologists studying variations in human temperament had proposed that a questionnaire, containing just one hundred true/false questions, could effectively split personalities along four archetypal dimensions: *novelty seeking* (impulsive versus cautious), *reward dependent* (warm versus detached), *risk avoidant* (anxious versus calm), and *persistent* (loyal versus fickle). Twin studies suggested that each of these personality types had a strong genetic component: identical twins had reported more than 50 percent concordance in their scores on these questionnaires.

Ebstein was particularly intrigued by one of the subtypes. Novelty seekers—or "neophiles"—were characterized as "impulsive, exploratory, fickle, excitable and extravagant" (think Jay Gatsby, Emma Bovary, Sherlock Holmes). In contrast, "neophobes" were "reflective, rigid, loyal, stoic, slow-tempered and frugal" (think Nick Carraway, the always-suffering Charles Bovary, the always-bested Dr. Watson). The most extreme novelty seekers—the greatest among the Gatsbys—seemed virtually addicted to stimulation and excitement. Scores aside, even their test-*taking* behavior was temperamental. They might leave questions unanswered. They might pace the room, trying to look for ways to get out. They might be frequently, hopelessly, maddeningly bored.

Ebstein collected a cohort of 124 volunteers and asked them to fill out standard questionnaires to measure novelty-seeking behavior (do you "often try things just for fun and thrills, even if most people think it's a waste of time"? Or "how often do things based on how [you] feel at the moment, without thinking about how they have been done in the past"?). He then used molecular and genetic techniques to determine the genotypes within his cohort with a limited panel of genes. The most extreme novelty seekers, he discovered, had a disproportionate representation of one genetic determinant: a variant of a dopamine-receptor gene called *D4DR*. (This kind of analysis is broadly termed an *association study*, since it identifies genes via their association with a particular phenotype—extreme impulsivity in this case.)

Dopamine, a neurotransmitter—a molecule that transmits chemical signals between neurons in the brain—is especially involved in the recognition of "reward" to the brain. It is one of the most potent neurochemical signals that we know: a rat, given a lever to electrically stimulate the dopamine-responsive reward center in the brain, will stimulate itself to death because it neglects to eat and drink.

D4DR acts as the "docking station" for dopamine, from which the signal is relayed to a dopamine-responsive neuron. Biochemically, the variant associated with novelty seeking, "*D4DR-7* repeat," dulls the response to dopamine, perhaps thereby heightening the requirement for external stimulation to reach the same level of reward. It is like a half-stuck switch, or a velvet-stuffed receiver: it needs a stronger push, or a louder voice, to be turned on. Novelty seekers try to amplify the signal by stimulating their brains with higher and higher forms of risk. They are like habituated drug users, or like the rats in the dopamine-reward experiment—except the "drug" is a brain chemical that signals excitement itself.

Ebstein's original study has been corroborated by several other groups. Interestingly, as one might suspect from the Minnesota twin studies, *D4DR* does not "cause" a personality or temperament. Instead, it causes a *propensity* toward a temperament that seeks stimulation or excitement— the first derivative of impulsivity. The precise nature of stimulation varies from one context to the next. It can produce the most sublime qualities in humans—exploratory drive, passion, and creative urgency—but it can also spiral toward impulsivity, addiction, violence, and depression. The *D4DR-7* repeat variant has been associated with bursts of focused creativity, and also with attention deficit disorder—a seeming paradox until you understand that both can be driven by the same impulse. The most provocative human studies have cataloged the geographic distribution of the *D4DR* variant. Nomadic and migratory populations have higher frequencies of the variant gene. And the farther one moves from the original site of human dispersal from Africa, the more frequently the variant seems to appear as well. Perhaps the subtle drive caused by the *D4DR* variant drove the "out-of-Africa" migration, by throwing our ancestors out to sea. Many attributes of our restless, anxious modernity, perhaps, are products of a restless, anxious gene.

Yet studies on the *D4DR* variant have been difficult to replicate across different populations and in differing contexts. Some of this, undoubtedly, is because novelty-seeking behaviors depend on age. Perhaps predictably, by age fifty or so, much of the exploratory impulse and its variance has been extinguished. Geographic and racial variations also affect the influence of *D4DR* on temperament. But the most likely reason for the lack of reproducibility is that the effect of the *D4DR* variant is relatively weak. One researcher estimates that the effect of *D4DR* explains only about 5 percent of the variance in novelty-seeking behavior among individuals. *D4DR* is likely only one of many genes—as many as ten—that determine this particular aspect of personality.

①

Gender. Sexual preference. Temperament. Personality. Impulsivity. Anxiety. Choice. One by one, the most mystical realms of human experience have become progressively encircled by genes. Aspects of behavior relegated largely or even exclusively to cultures, choices, and environments, or to the unique constructions of self and identity, have turned out to be surprisingly influenced by genes.

But the real surprise, perhaps, is that we should be surprised at all. If we accept that variations in genes can influence diffuse aspects of human pathology, then we can hardly be astonished that variations in genes can also influence equally diffuse aspects of *normalcy*. There is a fundamental symmetry to the idea that the mechanism by which genes cause disease is precisely analogous to the mechanism by which genes cause normal behavior and development. "How nice it would be if we could only get through into Looking-glass House!" says Alice. Human genetics has traveled through its looking-glass house—and the rules on one side have turned out to be exactly the same as the rules on the other.

How can we describe the influence of genes on normal human form and function? The language should have a familiar ring to it; it is the very language that was once used to describe the link between genes and illness. The variations that you inherit from your parents, mixed and matched, specify variations in cellular and developmental processes that ultimately result in variations in physiological states. If these variations affect master-regulatory genes at the tip of a hierarchy, the effect can be binary and strong (male versus female; short statured versus normal). More commonly, the variant/mutant genes lie in lower rungs of cascades of information and can only cause alterations in propensities. Often, dozens of genes are required to create these propensities or predispositions.

These propensities intersect with diverse environmental cues and chance to effect diverse outcomes—including variations in form, function, behavior, personality, temperament, identity, and fate. Mostly, they do so only in a probabilistic sense—i.e., only by altering weights and balances, by shifting likelihoods, by making certain outcomes more or less probable.

Yet these shifts in likelihoods are sufficient to make us observably different. A change in the molecular structure of a receptor that signals "reward" to neurons in the brain might result in nothing more than a change in the length of time that one molecule engages with its receptor. The signal that emanates from that variant receptor might persist in a neuron for just one-half of a second longer. Yet that change is enough to tip one human being toward impulsivity, and his counterpart toward caution, or one man toward mania and another toward depression. Complex perceptions, choices, and feelings might result from such changes in physical and mental states. The length of a chemical interaction is thus transformed into, say, the longing for an emotional interaction. A man with a propensity toward schizophrenia interprets a fruit vendor's conversation as a plot to kill him. His brother, with a genetic propensity for bipolar disorder, perceives that same conversation as a grandiose fable about his future: even the fruit seller recognizes his incipient fame. One man's misery becomes another man's magic.

①

This much is easy. But how can we explain an *individual* organism's form, temperament, and choices? How do we go, say, from genetic propensities in the abstract to a concrete and particular personhood? We might describe this as the "last mile" problem of genetics. Genes can describe the form or fate of a complex organism in likelihoods and probabilities—but they cannot accurately describe the form or fate itself. A particular combination of genes (a genotype) might predispose you to a particular configuration of a nose or personality—but the precise shape or length of the nose that you acquire remains unknowable. A *pre*disposition cannot be confused with the disposition itself: one is a statistical probability; the other, a concrete reality. It is as if genetics can nearly beat its way to the door of human form, identity, or behavior—but it cannot traverse the final mile.

Perhaps we can reframe the last-mile problem of genes by contrasting two very different lines of investigation. Since the 1980s, human genetics has spent much of its time concerned with how identical twins separated at birth demonstrate all sorts of similarities. If separated-at-birth twins share a tendency toward impulsivity, depression, cancer, or schizophrenia, then we know that the genome must contain information that encodes predispositions for these characteristics.

But it requires quite the opposite line of thought to understand how a predisposition is transformed into a disposition. To answer that, we need to ask the converse question: Why do identical twins raised in identical homes and families end up with *different* lives and become such different beings? Why do identical genomes become manifest in such dissimilar personhoods, with nonidentical temperaments, personalities, fates, and choices?

For nearly three decades since the eighties, psychologists and geneticists have tried to catalog and measure subtle differences that might explain the divergent developmental fates of identical twins brought up in the same circumstances. But all attempts at finding concrete, measurable, and systematic differences have invariably fallen short: twins share families, live in the same homes, typically attend the same school, have virtually identical nutrition, often read the same books, are immersed in the same culture, and share similar circles of friends—and yet are unmistakably different.

What causes the difference? Forty-three studies, performed over two decades, have revealed a powerful and consistent answer: "unsystematic, idiosyncratic, serendipitous events." Illnesses. Accidents. Traumas. Triggers. A missed train; a lost key; a suspended thought. Fluctuations in molecules that cause fluctuations in genes, resulting in slight alterations in forms.* Rounding a bend in Venice and falling into a canal. Falling in love. Randomness. Chance.

Is that an infuriating answer? After decades of musing, have we reached the conclusion that fate is, well . . . fate? That being happens through . . . be-ing? I find that formulation illuminatingly beautiful. Prospero, rag-

---

* Perhaps the most provocative recent study on chance, identity, and genetics comes from the laboratory of Alexander van Oudenaarden, a worm biologist at MIT. Van Oudenaarden used the worm as a model to ask one of the most difficult questions about chance and genes: Why do two animals that have the same genome and inhabit the same environment—perfect twins—have different fates? Van Oudenaarden examined a mutation of a gene, *skn-1*, that is "incompletely penetrant"—i.e., one worm with the mutation manifests a phenotype (cells are formed in the gut), while its twin worm, with the same mutation, does not manifest the phenotype (the cells are not formed). What determines the difference between the two twin worms? Not genes, since both worms share the same *skn-1* gene mutation, and not environments, since both are reared and housed in exactly the same conditions. How, then, might the same genotype cause an incompletely penetrant phenotype? Van Oudenaarden found that the expression level of a single regulatory gene, called *end-1*, is the crucial determinant. The expression of *end-1*—i.e., the number of molecules of RNA made during a particular phase of worm development—varies between worms, most likely due to random or stochastic effects—i.e., chance. If the expression exceeds a threshold, the worm manifests the phenotype; if it is below the level, the worm manifests a different phenotype. Fate reflects random *fluctuations in a single molecule in a worm's body*. For further details, see Arjun Raj et al., "Variability in gene expression underlies incomplete penetrance," *Nature* 463, no. 7283 (2010): 913–18.

ing against the deformed monster Caliban in *The Tempest*, describes him as "a devil, a born devil, on whose nature, nurture can never stick." The most monstrous of Caliban's flaws is that his intrinsic nature cannot be rewritten by any external information: his nature will not allow nurture to stick. Caliban is a genetic automaton, a windup ghoul—and this makes him vastly more tragic and more pathetic than anything human.

It is a testament to the unsettling beauty of the genome that it can make the real world "stick." Our genes do not keep spitting out stereotypical responses to idiosyncratic environments: if they did, we too would devolve into windup automatons. Hindu philosophers have long described the experience of "being" as a web—*jaal*. Genes form the threads of the web; the detritus that sticks is what transforms every individual web into a being. There is an exquisite precision in that mad scheme. Genes must carry out programmed responses to environments—otherwise, there would be no conserved form. But they must also leave exactly enough room for the vagaries of chance to stick. We call this intersection "fate." We call our responses to it "choice." An upright organism with opposable thumbs is thus built from a script, but built to go off script. We call one such unique variant of one such organism a "self."

The Hunger Winter

Identical twins have exactly the same genetic code as each other. They share the same womb, and usually they are brought up in very similar environments. When we consider this, it doesn't seem surprising that if one of the twins develops schizophrenia, the chance that his or her twin will also develop the illness is very high. In fact, we have to start wondering why it isn't higher. Why isn't the figure 100 percent?

—Nessa Carey, *The Epigenetics Revolution*

Genes have had a glorious run in the 20th century. . . . They have carried us to the edge of a new era in biology, one that holds out the promise of even more astonishing advances. But these very advances will necessitate the introduction of other concepts, other terms and other ways of thinking about biological organization, thereby loosening the grip that genes have had on the imagination of the life sciences.

—Evelyn Fox Keller, *An Anthropology of Biomedicine*

A question implicit in the last chapter must be answered: If the "self" is created through the chance interactions among events and genes, then how are these interactions actually recorded? One twin falls on ice, fractures a knee, and develops a callus, while the other does not. One sister marries a rising executive in Delhi, while the other moves to a crumbling household in Calcutta. Through what mechanism are these "acts of fate" registered within a cell or a body?

The answer had been standard for decades: through genes. Or to be more precise, by turning genes on and off. In Paris, in the 1950s, Monod and Jacob had demonstrated that when bacteria switch their diet from glucose to lactose, they turn glucose-metabolizing genes off and turn lactose-metabolizing genes on (these genes are turned "on" and "off" by master-regulatory factors—activators and repressors—also called transcription factors.) Nearly thirty years later, biologists working on the worm had found that signals from neighboring cells—events of fate, as far as an individual cell is concerned—are also registered by the turning on and off of master-regulatory genes, leading to alterations in cell lineages. When one twin falls on ice, wound-healing genes are turned on. These genes enable the wound to harden into the callus that marks the site of the fracture. Even when a complex memory is recorded in the brain, genes must be turned on and off. When a songbird encounters a new song from another bird, a gene called *ZENK* is turned up in the brain. If the song isn't right—if it's a song from a different species, or a flat note—then *ZENK* is not turned on at the same level, and the song is not released.

But does the activation or repression of genes in cells and bodies (in response to environmental inputs: a fall, an accident, a scar) leave some sort of permanent mark or stamp on the genome? What happens when an organism reproduces: do the marks or stamps on the genome get transmitted to another organism? Can the information from the environment be transmitted across generations?

①

We are now about to enter one of the most controversial arenas in the history of the gene, and some historical context is essential. In the 1950s, Conrad Waddington, an English embryologist, tried to understand the mechanisms by which environmental signals might affect a cell's genome. In embryonic development, Waddington saw the genesis of thousands of diverse cell types—neurons, muscle cells, blood, sperm—out of a single fertilized cell. In an inspired analogy, Waddington likened embryonic differentiation to a thousand marbles sent tumbling down a sloping landscape full of crags, nooks, and crevices. As every cell charted its unique path down the "Waddington landscape," he proposed, it got trapped in some particular channel or crag in the landscape, thereby limiting the type of cell that it could become.

Waddington was particularly intrigued by the manner in which a cell's environment might affect the use of its genes. He termed the phenomenon "*epi*-genetics"—literally "above genes".* Epigenetics, Waddington wrote, concerns "the interaction of genes with their environment [. . .] that brings their phenotype into being."

. ⬭

A macabre human experiment provided evidence for Waddington's theory, although its denouement would not be obvious for generations. In September 1944, amid the most vengeful phase of World War II, German troops occupying the Netherlands banned the export of food and coal to its northern parts. Trains were stopped and roads blockaded. Travel on the waterways was frozen. The cranes, ships, and quays of the port of Rotterdam were blown up with explosives, leaving a "tortured and bleeding Holland," as one radio broadcaster described it.

Heavily crisscrossed by waterways and barge traffic, Holland was not just tortured and bleeding. She was also hungry. Amsterdam, Rotterdam, Utrecht, and Leiden depended on regular transportation of supplies for food and fuel. By the early winter of 1944, wartime rations reaching the provinces north of the Waal and Rhine Rivers dwindled to a bare trickle, and the population edged toward famine. By December, the waterways were reopened, but now the water was frozen. Butter disappeared first, then cheese, meat, bread, and vegetables. Desperate, cold, and famished, people dug up tulip bulbs from their yards, ate vegetable skins, and then graduated to birch bark, leaves, and grass. Eventually, food intake fell to about four hundred calories a day—the equivalent of three potatoes. A human being "only [consists of] a stomach and certain instincts," one man wrote. The period, still etched in the national memory of the Dutch, would be called the Hunger Winter, or Hongerwinter.

---

* Waddington initially used the term "epigenesis" as a verb, not a noun, to describe the process by which the embryo develops out of a single cell ("epigenesis" referred to the genesis of an embryo through different kinds of cells—neurons, skin cells— arising sequentially out of the original fertilized cell). In time, however, "epigenetics" would be used to refer to the ways in which cells or organisms might acquire features without changing the gene sequence—i.e., through gene regulation. The more contemporary use refers to chemical or physical changes in DNA that affect gene regulation without changing the DNA sequence. Some scientists believe that "epigenetics" should only be reserved to refer to changes that are *heritable*—i.e., passed from cell to cell, or organism to organism. The shifting meaning of the word "epigenetics" has created enormous confusion within the field.

The famine raged on until 1945. Tens of thousands of men, women, and children died of malnourishment; millions survived. The change in nutrition was so acute and abrupt that it created a horrific natural experiment: as the citizens emerged from the winter, researchers could study the effect of a sudden famine on a defined cohort of people. Some features, such as malnourishment and growth retardation, were expected. Children who survived the Hongerwinter also potentially suffered chronic health issues associated with malnourishment: depression, anxiety, heart disease, gum disease, osteoporosis, and diabetes. (Audrey Hepburn, the wafer-thin actress, was one such survivor, and she would be afflicted by a multitude of chronic illnesses throughout her life.)

In the 1980s, however, a more intriguing pattern emerged: when the children born to women who were pregnant during the famine grew up, they too had higher rates of obesity and heart disease. This finding too might have been anticipated. Exposure to malnourishment *in utero* is known to cause changes in fetal physiology. Nutrient-starved, a fetus alters its metabolism to sequester higher amounts of fat to defend itself against caloric loss, resulting, paradoxically, in late-onset obesity and metabolic disarray. But the oddest result of the Hongerwinter study would take yet another generation to emerge. In the 1990s, when the *grandchildren* of men and women exposed to the famine were studied, they too had higher rates of obesity and heart disease (some of these health issues are still being evaluated). The acute period of starvation had somehow altered genes not just in those directly exposed to the event; the message had been transmitted to their grandchildren. Some heritable factor, or factors, must have been imprinted into the genomes of the starving men and women and crossed at least two generations. The Hongerwinter had etched itself into national memory, but it had penetrated genetic memory as well.[*]

①

But what was "genetic memory"? How—beyond genes themselves—was gene memory encoded? Waddington did not know about the Hongerwinter study—he had died, largely unrecognized, in 1975—but geneticists

---

[*] Some scientists have argued that the Dutch famine study is inherently biased: parents with metabolic disorders (obesity, say) might alter the dietary choices of their children, or change their habits in some non-genetic manner. The factor being "transmitted" across generations, critics claim, is not a genetic signal but a cultural or dietary choice.

cannily saw the connection between Waddington's hypothesis and multi-generational illnesses of the Dutch cohort. Here too, a "genetic memory" was evident: the children and grandchildren of famine-starved individuals tended to develop metabolic illnesses, as if their genomes carried some recollection of their grandparents' metabolic travails. Here too, the factor responsible for the "memory" could not be an alteration of the gene sequence: the hundreds of thousands of men and women in the Dutch cohort could not have mutated their genes over the span of three generations. And here too, an interaction between "the genes and the environment" had changed a phenotype (i.e., the propensity for developing an illness). Something must have been stamped onto the genome by virtue of its exposure to the famine—some permanent, heritable mark—that was now being transmitted across generations.

If such a layer of information could be interposed on a genome, it would have unprecedented consequences. First, it would challenge an essential feature of classical Darwinian evolution. Conceptually, a key element of Darwinian theory is that genes do not—*cannot*—remember an organism's experiences in a permanently heritable manner. When an antelope strains its neck to reach a tall tree, its genes do not record that effort, and its children are not born as giraffes (the direct transmission of an adaptation into a heritable feature, remember, was the basis for Lamarck's flawed theory of evolution by adaptation). Rather, giraffes arise via spontaneous variation and natural selection: a tall-necked mutant appears in an ancestral tree-grazing animal, and during a period of famine, this mutant survives and is naturally selected. August Weismann had formally tested the idea that an environmental influence could permanently alter genes by chopping off the tails of five generations of mice—and yet, mice in the sixth generation had been born with perfectly intact tails. Evolution can craft perfectly adapted organisms, but not in an intentional manner: it is not just a "blind watchmaker," as Richard Dawkins once famously described it, but also a forgetful one. Its sole driver is survival and selection; its only memory is mutation.

Yet the grandchildren of the Hongerwinter had somehow acquired the memory of their grandparents' famine—not through mutations and selection, but via an environmental message that had somehow transformed into a heritable one. A genetic "memory" in this form could act as a wormhole for evolution. A giraffe's ancestor might be able to make a giraffe—not by trudging through the glum Malthusian logic of mutation, survival, and selection, but by simply straining its neck, and registering and imprinting a memory of that strain in its genome. A mouse with an excised tail would be able to bear mice with shortened tails by transmitting that information to its genes. Children raised in stimulating environments could produce more stimulated children. The idea was a restatement of Darwin's gemmule formulation: the particular experience, or history, of an organism would be signaled straight to its genome. Such a system would act as a rapid-transit system between an organism's adaptation and evolution. It would unblind the watchmaker.

Waddington, for one, had yet another stake in the answer—a personal one. An early, fervent convert to Marxism, he imagined that discovering such "memory-fixing" elements in the genome might be crucial not just to the understanding of human embryology, but also to his political project. If cells could be indoctrinated or de-indoctrinated by manipulating their gene memories, perhaps humans could be indoctrinated as well (recall Lysenko's attempt to achieve this with wheat strains, and Stalin's attempts to erase the ideologies of human dissidents). Such a process might undo cellular identity and allow cells to run *up* the Waddington landscape—turning back from an adult cell to an embryonic cell, thus reversing biological time. It might even undo the fixity of human memory, of identity—of choice.

℗

Until the late 1950s, epigenetics was more fantasy than reality: no one had witnessed a cell layering its history or identity above its genome. In 1961, two experiments performed less than six months, and less than twenty miles, from each other would transform the understanding of genes and lend credence to Waddington's theory.

In the summer of 1958, John Gurdon, a graduate student at Oxford University, began to study the development of frogs. Gurdon had never been a particularly promising student—he once scored 250th in a class of 250 in a science exam—but he had, as he once described it, an "aptitude for doing things on a small scale." His most important experiment involved the smallest of scales. In the early fifties, two scientists in Philadelphia had emptied an unfertilized frog egg of all its genes, sucking out the nucleus and leaving just the cellular husk behind, then injected the genome of another frog cell into the emptied egg. This was like evacuating a nest, slinking a false bird in, and asking if the bird developed normally. Did the "nest"—i.e., the egg cell, devoid of all its own genes—have all the factors in it to create an embryo out of an injected genome from another cell? It did. The Philadelphia researchers produced an occasional tadpole from an egg injected with the genome of a frog cell. It was an extreme form of parasitism: the egg cell became merely a host, or a vessel, for the genome of a normal cell and allowed that genome to develop into a perfectly normal adult animal. The researchers called their method nuclear transfer, but the process was extremely inefficient. In the end, they largely abandoned the approach.

Gurdon, fascinated by those rare successes, pushed the boundaries of that experiment. The Philadelphia researchers had injected nuclei from young embryos into the enucleated eggs. In 1961, Gurdon began to test whether injecting the genome from the cell of an *adult* frog intestine could also give rise to a tadpole. The technical challenges were immense. First, Gurdon learned to use a tiny beam of ultraviolet rays to lance the nucleus of an unfertilized frog egg, leaving the cytoplasm intact. Then, like a diver slicing into water, he punctured the egg membrane with a fire-sharpened needle, barely ruffling the surface, and blew in the nucleus from an adult frog cell in a tiny puff of liquid.

The transfer of an adult frog nucleus (i.e., all its genes) into an empty egg worked: perfectly functional tadpoles were born, and each of these tadpoles carried a perfect replica of the genome of the adult frog. If Gurdon transferred the nuclei from multiple adult cells drawn from the same frog into multiple evacuated eggs, he could produce tadpoles that were perfect clones of each other, and clones of the original donor frog. The process could be repeated *ad infinitum*: clones made from clones from clones, all carrying exactly the same genotype—reproductions without reproduction.

Gurdon's experiment incited the imagination of biologists—not the least because it seemed like a science-fiction fantasy brought to life. In one experiment, he produced eighteen clones from the intestinal cells of a single frog. Placed into eighteen identical chambers, they were like eighteen doppelgängers, inhabiting eighteen parallel universes. The scientific principle at stake was also provocative: the genome of an adult cell, having reached its full maturity, had been bathed briefly in the elixir of an egg cell and then emerged fully rejuvenated as an embryo. The egg cell, in short, had everything necessary—all the regulatory factors needed to drive a genome backward through developmental time into a functional embryo. In time, variations on Gurdon's method would be generalized to other animals. It would lead, famously, to the cloning of Dolly, the sheep, the only higher organism reproduced without reproduction (the biologist John Maynard Smith would later remark that the only other "observed case of a mammal produced without sex wasn't entirely convincing." He was referring to Jesus Christ). In 2012, Gurdon was awarded the Nobel Prize for his discovery of nuclear transfer.*

But for all the remarkable features of Gurdon's experiment, it was his *lack* of success that was just as revealing. Adult intestinal cells could certainly give rise to tadpoles, but despite Gurdon's laborious technical ministrations, they did so with great reluctance: his success rate at turning adult cells into tadpoles was abysmal. This demanded an explanation beyond classical genetics. The DNA sequence in the genome of an adult frog, after all, is identical to the DNA sequence of an embryo or a tadpole. Is it not the fundamental principle of genetics that all cells contain the same genome, and it is the manner in which these genes are *deployed* in different cells, turned on and off based on cues, that controls the development of an embryo into an adult?

But if genes are genes are genes, then why was the genome of an adult cell so inefficiently coaxed backward into an embryo? And why, as others discovered, were nuclei from younger animals more pliant to this age reversal than

* Gurdon's technique—of evacuating the egg and inserting a fully fertilized nucleus— has already found a novel clinical application. Some women carry mutations in mitochondrial genes—i.e., genes that are carried within mitochondria, the energy-producing organelles that live inside cells. All human embryos, recall, inherit their mitochondria exclusively from the female egg—i.e., from their mothers (the sperm does not contribute any mitochondria). If the mother carries a mutation in a mitochondrial gene, then all her children might be affected by that mutation; mutations in these genes, which often affect energy metabolism, can cause muscle wasting, heart abnormalities, and death. In a provocative series of experiments in 2009, geneticists and embryologists proposed a daring new method to tackle these maternal mitochondrial mutations. After the egg had been fertilized by the father's sperm, the nucleus was injected into an egg with intact ("normal") mitochondria from a normal donor. Since the mitochondria were derived from the *donor*, the maternal mitochondrial genes were intact, and the babies born would no longer carry the maternal mutations. Humans born from this procedure thus have *three* parents. The fertilized nucleus, formed by the union of the "mother" and "father" (parents 1 and 2), contributes virtually all the genetic material. The third parent—i.e., the egg donor—contributes only mitochondria, and the mitochondrial genes. In 2015, after a protracted national debate, Britain legalized the procedure, and the first cohorts of "three-parent children" are now being born. These children represent an unexplored frontier of human genetics (and of the future). Obviously, no comparable animals exist in the natural world.

those from older animals? Again, as with the Hongerwinter study, something must have been progressively imprinted on the adult cell's genome—some cumulative, indelible mark—that made it difficult for that genome to move back in developmental time. That mark could not live in the sequence of genes themselves, but had to be etched above them: it had to be *epi*genetic. Gurdon returned to Waddington's question: What if every cell carries an imprint of its history and its identity in its genome—a form of cellular memory?

⊕

Gurdon had visualized an epigenetic mark in an abstract sense, but he hadn't physically seen such an imprint on the frog genome. In 1961, Mary Lyon, a former student of Waddington's, found a visible example of an epigenetic change in an animal cell. The daughter of a civil servant and a schoolteacher, Lyon began her graduate work with the famously cantankerous Ron Fisher in Cambridge, but soon fled to Edinburgh to finish her degree, and then to a laboratory in the quiet English village of Harwell, twenty miles from Oxford, to launch her own research group.

At Harwell, Lyon studied the biology of chromosomes, using chemical dyes to visualize them. To her astonishment, she found that every paired chromosome stained with chromosomal dyes looked identical—except the two X chromosomes in females. One of two X chromosomes in every cell in female mice was inevitably more darkly stained. The *genes* in the darkly stained chromosome were unchanged: the actual sequence of DNA was identical between both chromosomes. What had changed, however, was their *activity*: the genes in that shrunken chromosome did not generate RNA, and therefore the entire chromosome was "silent." It was as if one chromosome had been purposely decommissioned—turned off. The inactivated X chromosome was chosen randomly, Lyon found: in one cell, it might be the paternal X, while its neighbor might inactivate the maternal X chromosome. This pattern was a universal feature of all cells that had two X chromosomes—i.e., every cell in the female body.

What purpose does X inactivation serve? Since females have two X chromosomes, while males have only one, female cells inactivate one X chromosome to equalize the "dose" of genes from the two X chromosomes. This random inactivation of the X has an important biological consequence: the female body is a mosaic of two types of cells. For the most part, this random silencing of one X chromosome is invisible—unless one of the X chromosomes (from the father, say) happens to carry a gene variant that produces a visible trait. In that case, one cell might express that variant, while its neighboring cell would lack that function—producing a mosaic-like effect. In cats, for instance, one gene for coat color lives on the X chromosome. The random inactivation of the X chromosome causes one cell to have a color pigment, while its neighbor has a different color. Epigenetics, not genetics, solves the conundrum of a female tortoiseshell cat. (If humans carried the skin color gene on their X chromosomes, then a female child of a dark-skinned and light-skinned couple would be born with patches of light and dark skin.)

How can a cell "silence" an entire chromosome? This process must involve not just the activation or inactivation of one or two genes based on an environmental cue; here an entire chromosome—including all its genes—was being shut off for the lifetime of a cell. The most logical guess, proposed in the 1970s, was that cells had somehow appended a permanent chemical stamp—a molecular "cancellation sign"—to the DNA in that chromosome. Since the genes themselves were intact, such a mark had to be above genes—i.e., *epi*genetic.

In the late 1970s, scientists working on gene-silencing discovered that the attachment of a small molecule—a methyl group—to some parts of DNA was correlated with a gene's turning off. One of the chief instigators of this process was later found to be an RNA molecule, called *XIST*. The RNA molecule "coats" parts of the chromosome and is thought to be crucial to the silencing of that chromosome. These methyl tags decorated the strands of DNA, like charms on a necklace, and they were recognized as shutdown signals for certain genes.

Φ

Methyl tags were not the only charms hanging off the DNA necklace. In 1996, working at Rockefeller University in New York, a biochemist named David Allis found yet another system to etch permanent marks on genes.* Rather than stamping the marks directly on genes, this second system placed its marks on proteins, called histones, that act as the packaging material for genes.

---

* The idea that histones might regulate genes had originally been proposed by Vincent Allfrey, a biochemist at Rockefeller University in the 1960s. Three decades later—and, as if to close a circle, at that very same institution—Allis's experiments would vindicate Allfrey's "histone hypothesis."

Histones hang tightly to DNA and wrap it into coils and loops, forming scaffolds for the chromosome. When scaffolding changes, the activity of a gene can change—akin to altering the properties of a material by changing the way that it is packaged (a skein of silk packed into a ball has very different properties from that same skein stretched into a rope). A "molecular memory" could potentially be stamped on a gene—this time, indirectly, by attaching the signal to proteins (there is enormous debate within the field of epigenetics whether some—or any—histone modifications carry consequential effects on the activity of a gene, or whether some of these histone changes are merely "bystanders" or side-effects of a gene's activity.) The heritability and stability of these histone marks, and the mechanism to ensure that the marks appear in the right genes at the right time, are still under investigation—but simple organisms, such as yeast and worms, can seemingly transmit these histone marks across several generations.

①

The silencing and activation of genes via protein regulators (called transcription factors)—the "master-conductors" of the symphony of genes in cells—has been established since the 1950s. But these conductors can potentially *recruit* other proteins—call them helpers—to place permanent chemical imprints on genes. They even ensure that the tags are maintained on the genome.* The tags can thus be added, erased, amplified, diminished, and toggled on and off in response to cues from a cell or from its environment.†

These marks function like notes written above a sentence, or like marginalia recorded in a book—pencil lines, underlined words, scratch marks, crossed-out letters, subscripts, and endnotes—that modify the context of

---

* A master-regulator gene can maintain its actions on its target genes, largely autonomously, through a mechanism called "positive feedback".

† Tim Bestor, the geneticist, and some of his colleagues have argued that DNA methylation marks are primarily used to inactivate ancient virus-like elements that are buried in the human genome, to inactivate of the X chromosomes (*à la* Lyon) and to differentially mark certain genes in sperm, but not eggs (or vice versa) so that an organism knows and "remembers" which genes had originated in the father versus the mother—a phenomenon called "imprinting". Notably, Bestor does not believe that environmental stimuli have a significant effect on the genome. Rather, epigenetic marks are used to regulate gene expression during development and imprinting.

the genome without changing the actual words. Every cell in an organism inherits the same book, but by scratching out particular sentences and appending others, by "silencing" and "activating" particular words, by emphasizing certain phrases, each cell could potentially write a unique novel from the same basic script. We might visualize genes in the human genome, with their appended chemical marks, thus:

> ... *This* .... *is* ... *the* ......,,,....... <u>struc</u> ... ture , ...
> *of* ... *Your* ...... *Gen* ... *ome* ...

As before, the words in the sentence correspond to the genes. The ellipses and punctuation marks denote the introns, the intergenic regions, and regulatory sequences. The boldface and capitalized letters and the underlined words are epigenetic marks appended to the genome to impose a final layer of meaning.

This was the reason that Gurdon, despite all his experimental ministrations, had rarely been able to coax an adult intestinal cell backward in developmental time to become an embryonic cell and then a full-fledged frog: the genome of the intestinal cell had been tagged with too many epigenetic "notes" for it to be easily erased and transformed into the genome of an embryo. Like human memories that persist despite attempts to alter them, the chemical scribbles overwritten on the genome can be changed—but not easily. These notes are designed to persist so that a cell can lock its identity into place. Only embryonic cells have genomes that are pliant enough to acquire many different kinds of identities—and can thus generate all the cell types in the body. Once the cells of the embryo have taken up fixed identities—turned into intestinal cells or blood cells or nerve cells, say—there is rarely any going back (hence Gurdon's difficulty in making a tadpole out of a frog's intestinal cell). An embryonic cell might be able to write a thousand novels from the same script. But Young Adult Fiction, once scripted, cannot easily be reformatted into Victorian Romance.

Ⓘ

The interplay between gene regulators and epigenetics partially solves the riddle of a cell's individuality—but perhaps it can also solve the more tenacious riddle of an *individual's* individuality. "Why are twins different?" we had asked earlier. Well, because idiosyncratic events are recorded through idiosyncratic marks in their bodies. But "recorded" in what manner? Not in the actual sequence of genes: if you sequence the genomes of a pair of identical twins every decade for fifty years, you get the same sequence over and over again. But if you sequence the *epigenomes* of a pair of twins over the course of several decades, you find substantial differences: the pattern of methyl groups attached to the genomes of blood cells or neurons is virtually identical between the twins at the start of the experiment, begins to diverge slowly over the first decade, and becomes substantially different over fifty years.*

Chance events—injuries, infections, infatuations; the haunting trill of that particular nocturne; the smell of that particular madeleine in Paris—impinge on one twin and not the other. Regulatory proteins turn genes "on" and "off" in response to these events, and epigenetic marks are gradually layered above genes.† How these epigenetic marks functionally impact the activity of genes remains to be determined—but some experiments suggest that these marks, in conjunction with transcription factors, can help orchestrate the activity of genes.

In his remarkable story "Funes the Memorious," the Argentine writer Jorge Luis Borges described a young man who awakes from an accident to discover that he has acquired "perfect" memory. Funes remembers every detail of every moment in his life, every object, every encounter—the "shape of every cloud . . . the marble grain of a leather-bound book." This extraordinary ability does not make Funes more powerful; it paralyzes him. He is inundated by memories that he cannot silence; the memories overwhelm him, like the constant noise from a crowd that he cannot silence. Borges finds Funes lying in a cot in the darkness, unable to contain the hideous influx of information and forced to shut the world out.

A cell without the capacity to selectively silence parts of its genome devolves into Funes the Memorious (or, as in the story, Funes the Incapaci-

---

* More recent studies, and more powerful methods of methylation analysis, have shown smaller differences between twins. The field remains contentious and is actively changing.

† The permanence of epigenetic marks, and the nature of memory recorded by these marks, has been questioned by the geneticist Mark Ptashne. In Ptashne's view, shared by several other geneticists, master-regulatory proteins—previously described as molecular "on" and "off" switches—orchestrate the activation or repression of genes. Epigenetic marks are laid down as a *consequence* of gene activation or repression, and may play an accompanying role in regulating gene activation and repression, but the main orchestration of gene expression occurs by virtue of these master-regulatory proteins.

tated). The genome contains the memory to build every cell in every tissue in every organism—memory so overwhelmingly profuse and diverse that a cell devoid of a system of selective repression and reactivation would become overwhelmed by it. As with Funes, the capacity to use any memory functionally depends, paradoxically, on the ability to silence memory. Perhaps an epigenetic system exists to allow the genome to function. Much about this system remains to be discovered. Different genomes, in different cells, seem to be modified by diverse chemical marks in response to various stimuli (including environments). But whether these marks contribute to the activity of genes, how they do so—and what their functions might be—is still hotly, often viciously debated among geneticists.

⏀

Perhaps the most startling demonstration of the power of master-regulator proteins interacting with epigenetic marks to reset cellular memory arises from an experiment performed by the Japanese stem-cell biologist Shinya Yamanaka in 2006. Like Gurdon, Yamanaka was intrigued by the idea that chemical marks attached to genes in a cell might function as a record of its cellular identity. What if he could erase these marks? Would the adult cell revert to an original state—and turn into the cell of an embryo, reversing time, annihilating history, furling back toward innocence?

Like Gurdon, again, Yamanaka began his attempt to reverse a cell's identity with a normal cell from an adult mouse—this one from a fully grown mouse's skin. Gurdon's experiment had proved that factors present in an egg—proteins and RNA—could erase the marks of an adult cell's genome and thereby reverse the fate of a cell and produce a tadpole from a frog cell. Yamanaka wondered whether he could identify and isolate these factors from an egg cell, then use them as molecular "erasers" of cellular fate. After a decades-long hunt, Yamanaka narrowed the mysterious factors down to proteins encoded by just four genes. He then introduced the four genes into an adult mouse's skin cell.

To Yamanaka's astonishment, and to the subsequent amazement of scientists around the world, the introduction of these four genes into a mature skin cell caused a small fraction of the cells to transform into something resembling an embryonic stem cell. This stem cell could give rise to skin, of course, but also to muscle, bones, blood, intestines, and nerve cells. In fact, it could give rise to all cell types found in an entire organism. When

Yamanaka and his colleagues analyzed the progression (or rather regression) of the skin cell to the embryo-like cell, they uncovered a cascade of events. Circuits of genes were activated or repressed. The metabolism of the cell was reset. Then, epigenetic marks were erased and rewritten. The cell changed shape and size. Its wrinkles unmarked, its stiffening joints made supple, its youth restored, the cell could now climb *up* Waddington's slope. Yamanaka had expunged a cell's memory, reversed biological time.

The story comes with a twist. One of the four genes used by Yamanaka to reverse cellular fate is called *c-myc*. *Myc*, the rejuvenation factor, is no ordinary gene: it is one of the most forceful regulators of cell growth and metabolism known in biology. Activated abnormally, it can certainly coax an adult cell back into an embryo-like state, thereby enabling Yamanaka's cell-fate reversal experiment (this function requires the collaboration of the three other genes found by Yamanaka). But *myc* is also one of the most potent cancer-causing genes known in biology; it is also activated in leukemias and lymphomas, and in pancreatic, gastric, and uterine cancer. As in some ancient moral fable, the quest for eternal youthfulness appears to come at a terrifying collateral cost. The very genes that enable a cell to peel away mortality and age can also tip its fate toward malignant immortality, perpetual growth, and agelessness—the hallmarks of cancer.

⊕

We can now try to understand the Dutch Hongerwinter, and its multigenerational effects, in mechanistic terms that involve both genes and regulatory proteins interacting with the genome. The acute starvation of men and women during those brutal months in 1945 indubitably altered the expression of genes involved in metabolism and storage. The first changes were transient—no more, perhaps, than the turning on and turning off of genes that respond to nutrients in the environment.

But as the landscape of metabolism was frozen and reset by prolonged starvation—as transience hardened into permanence—more durable changes were imprinted in the genome. Hormones fanned out between organs, signaling the potential long-term deprivation of food and auguring a broader reformatting of gene expression. Proteins intercepted these messages within cells. Genes were shut down, one by one, then imprints were stamped on DNA to close them down further. Like houses shuttering against a storm, entire gene programs were barricaded shut. Methylation marks were added to genes. Histones may have been chemically modified to record the memory of starvation.

Cell by cell, and organ by organ, the body was reprogrammed for survival. Ultimately, even the germ cells—sperm and egg—were marked (we do not know how, or why, sperm and egg cells carry the memory of a starvation response; perhaps ancient pathways in human DNA record starvation or deprivation in germ cells).* When children and grandchildren were born from these sperm and eggs, the embryos may have carried these marks, resulting in alterations in metabolism that remained etched in their genomes decades after the Hongerwinter. Historical memory was thus transformed into cellular memory.

⨀

A note of caution: epigenetics is also on the verge of transforming into a dangerous idea. Epigenetic modifications of genes can potentially superpose historical and environmental information on cells and genomes—but this capacity is speculative, limited, idiosyncratic, and unpredictable: a parent with an experience of starvation produces children with obesity and *overnourishment*, while a father with the experience of tuberculosis, say, does not produce a child with an altered response to tuberculosis. Most epigenetic "memories" are the consequence of ancient *evolutionary* pathways, and cannot be confused with our longing to affix desirable legacies on our children.

As with genetics in the early twentieth century, epigenetics is now being used to justify junk science and enforce stifling definitions of normalcy. Diets, exposures, memories, and therapies that purport to alter heredity are eerily reminiscent of Lysenko's attempt to "re-educate" wheat using shock therapy. Mothers are being asked to minimize anxiety during their pregnancy—lest they taint all their children, and *their* children, with traumatized mitochondria. Lamarck is being rehabilitated into the new Mendel.

These glib notions about epigenetics should invite skepticism. Environmental information can certainly be etched on the genome.

---

* Experiments performed on worms and mice have also demonstrated the trans-generational effects of starvation, although it's unclear whether these effects persist or are attenuated over generations. Some of these studies have implicated small RNAs in the transmission of epigenetic information across generations.

But most of these imprints are recorded as "genetic memories" in the cells and genomes of *individual organisms*—not carried forward across generations. A man who loses a leg in an accident bears the imprint of that accident in his cells, wounds, and scars—but does not bear children with shortened legs. Nor has the uprooted life of my family seem to have burdened me, or my children, with any wrenching sense of estrangement.

Despite Menelaus's admonitions, the blood of our fathers *is* lost in us—and so, fortunately, are their foibles and sins. It is an arrangement that we should celebrate more than rue. Genomes and epigenomes exist to record and transmit likeness, legacy, memory, and history across cells and generations. Mutations, the reassortment of genes, and the erasure of memories counterbalance these forces, enabling unlikeness, variation, monstrosity, genius, and reinvention—and the refulgent possibility of new beginnings, generation upon generation.

☉

It is conceivable that an interplay of genes and epigenes coordinates human embryogenesis. Let us return, yet again, to Morgan's problem: the creation of a multicellular organism from a one-celled embryo. Seconds after fertilization, a quickening begins in the embryo. Proteins reach into the nucleus of the cell and start flicking genetic switches on and off. A dormant spaceship comes to life. Genes are activated and repressed, and these genes, in turn, encode yet other proteins that activate and repress other genes. A single cell divides to form two, then four, and eight cells. An entire layer of cells forms, then hollows out into the outer skin of a ball. Genes that coordinate metabolism, motility, cell fate, and identity fire "on." The boiler room warms up. The lights flicker on in the corridors. The intercom crackles alive.

Now a second layer of information—instigated by master-regulator proteins—stirs to life to ensure that gene expression is locked into place in each cell, enabling each cell to acquire and fix an identity. Chemical marks are selectively added to certain genes and erased from others, modulating the expression of the genes in that cell alone. Methyl groups are inserted and erased, and histones are modified.

The embryo unfurls step by step. Primordial segments appear, and cells take their positions along various parts of the embryo. New genes are activated that command subroutines to grow limbs and organs, and more chemical marks are appended on the genomes of individual cells. Cells are added to create organs and structures—forelegs, hind legs, muscles, kidneys, bones, eyes. Some cells die a programmed death. Genes that maintain function, metabolism, and repair are turned on. An organism emerges from a cell.

ⅅ

Do not be lulled by that description. Do not, gentle reader, be tempted to think—"My goodness, what a complicated recipe!"—and then rest assured that someone will not learn to understand or hack or manipulate that recipe in some deliberate manner.

When scientists underestimate complexity, they fall prey to the perils of unintended consequences. The parables of such scientific overreach are well-known: foreign animals, introduced to control pests, become pests in their own right; the raising of smokestacks, meant to alleviate urban pollution, releases particulate effluents higher in the air and exacerbates pollution; stimulating blood formation, meant to prevent heart attacks, thickens the blood and results in an increased risk of blood clots to the heart.

But when nonscientists *overestimate* complexity—"No one can possibly crack *this* code"—they fall into the trap of unanticipated consequences. In the early 1950s, a common trope among some biologists was that the genetic code would be so context dependent—so utterly determined by a particular cell in a particular organism and so horribly convoluted—that deciphering it would prove impossible. The truth turned out to be quite the opposite: just one molecule carries the code, and just one code pervades the biological world. If we know the code, we can intentionally alter it in organisms, and ultimately in humans. Similarly, in the 1960s, many doubted that gene-cloning technologies could so easily shuttle genes between species. By 1980, making a mammalian protein in a bacterial cell, or a bacterial protein in a mammalian cell, was not just feasible; it was, in Berg's words, rather "ridiculously simple." Species were specious. "Being natural" was "often just a pose."

The genesis of a human from genetic instructions is indubitably complex, but nothing about it forbids or restricts manipulation or distortion. When a social scientist emphasizes that gene-environment interactions—not genes alone—determine form, function, and fate, he is underestimating the power of master-regulatory genes that act nonconditionally and autonomously to determine complex physiological and anatomical states. And when a human geneticist says, "Genetics cannot be used to manipulate complex states and behaviors because these are usually controlled by dozens of genes," that geneticist is underestimating the capacity of one gene, such as a master regulator of genes, to "reset" entire states of being. If the activation of four genes can turn a skin cell into a pluripotent stem cell, if one drug can reverse the identity of a brain, and if a mutation in a single gene can switch sex and gender identity, then our genomes, and our selves, are much more pliable than we had imagined.

①

Technology, I said before, is most powerful when it enables transitions—between linear and circular motion (the wheel), or between real and virtual space (the Internet). Science, in contrast, is most powerful when it elucidates rules of organization—laws—that act as lenses through which to view and organize the world. Technologists seek to liberate us from the constraints of our current realities through those transitions. Science defines those constraints, drawing the outer limits of the boundaries of possibility. Our greatest technological innovations thus carry names that claim our prowess over the world: the *engine* (from *ingenium*, or "ingenuity") or the *computer* (from *computare*, or "reckoning together"). Our deepest scientific laws, in contrast, are often named after the limits of human knowledge: *uncertainty, relativity, incompleteness, impossibility.*

Of all the sciences, biology is the most lawless; there are few rules to begin with, and even fewer rules that are universal. Living beings must, of course, obey the fundamental rules of physics and chemistry, but life often exists on the margins and interstices of these laws, bending them to their near-breaking limit. The universe seeks equilibriums; it prefers to disperse energy, disrupt organization, and maximize chaos. Life is designed to combat these forces. We slow down reactions, concentrate matter, and organize chemicals into compartments; we sort laundry on Wednesdays. "It sometimes seems as if curbing entropy is our quixotic purpose in the universe," James Gleick wrote. We live in the loopholes of natural laws, seeking extensions, exceptions, and excuses. The laws of nature still mark the outer boundaries of permissibility—but life, in all its idiosyncratic, mad weirdness, flourishes by reading between the lines.

Even the elephant cannot violate the law of thermodynamics—although its trunk, surely, must rank as one of the most peculiar means of moving matter using energy.

①

The circular flow of biological information—

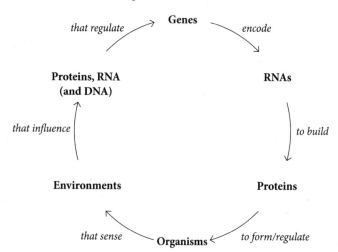

—is, perhaps, one of the few organizing rules in biology. Certainly the directionality of this flow of information has exceptions (retroviruses can pedal "backward" from RNA to DNA). And there are yet-undiscovered mechanisms in the biological world that might change the order or the components of information flow in living systems (RNA, for instance, is now known to be able to influence the regulation of genes). But the circular flow of biological information has been chalked out conceptually.

This flow of information is the closest thing that we might have to a biological law. When the technology to manipulate this law is mastered, we will move through one of the most profound transitions in our history. We will learn to read and write our selves, ourselves.

①

But before we leap into the genome's future, allow a quick diversion to its past. We do not know where genes come from, or how they arose. Nor can we know why *this* method of information transfer and data storage was chosen over all other possible methods in biology. But we can try to reconstruct the primordial origin of genes in a test tube. At Harvard, a soft-spoken biochemist named Jack Szostak has spent over two decades trying to create a self-replicating genetic system in a test tube—thereby reconstructing the origin of genes.

Szostak's experiment followed the work of Stanley Miller, the visionary chemist who had attempted to brew a "primordial soup" by mixing basic chemicals known to have existed in the ancient atmosphere. Working at the University of Chicago in the 1950s, Miller had sealed a glass flask and blown methane, carbon dioxide, ammonia, oxygen, and hydrogen into the flask through a series of vents. He had added hot steam and rigged an electrical spark to simulate bolts of lightning, then heated and cooled the flask cyclically to recapitulate the volatile conditions of the ancient world. Fire and brimstone, heaven and hell, air and water, were condensed in a beaker.

Three weeks later, no organism had crawled out of Miller's flask. But in the raw mixture of carbon dioxide, methane, water, ammonia, oxygen, hydrogen, heat, and electricity, Miller had found traces of amino acids— the building units of proteins—and trace amounts of the simplest sugars. Subsequent variations of the Miller experiment have added clay, basalt, and volcanic rock and produced the rudiments of lipids, fats, and even the chemical building blocks of RNA and DNA.

Szostak believes that genes emerged out of this soup through a fortuitous meeting between two unlikely partners. First, lipids created within the soup coalesced with each other to form *micelles*—hollow spherical membranes, somewhat akin to soap bubbles, that trap liquid inside and resemble the outer layers of cells (certain fats, mixed together in watery solutions, tend to naturally coalesce into such bubbles). In lab experiments, Szostak has demonstrated that such micelles can behave like protocells: if you add more lipids to them, these hollow "cells" begin to grow in size. They expand, move about, and extend thin extrusions that resemble the ruffling membranes of cells. Eventually they divide, forming two micelles out of one.

Second, while self-assembling micelles were being formed, chains of RNA arose from the joining together of nucleosides (A, C, G, U, or their chemical ancestors) to form strands. The vast bulk of these RNA chains had no reproductive capability: they had no ability to make copies of themselves. But among the billions of nonreplicating RNA molecules, one was accidentally created with the unique capacity to build an image of itself—or rather, generate a copy using its mirror image (RNA and DNA, recall, have inherent chemical designs that enable the generation of mirror-image molecules). This RNA molecule, incredibly, possessed the capacity to gather nucleosides from a chemical mix and string them together to form a new RNA copy. It was a self-replicating chemical.

The next step was a marriage of convenience. Somewhere on earth—Szostak thinks it might have been on the edge of a pond or a swamp—a self-copying RNA molecule collided with a self-replicating micelle. It was, conceptually speaking, an explosive affair: the two molecules met, fell in love, and launched a long conjugal entanglement. The self-replicating RNA began to inhabit the dividing micelle. The micelle isolated and protected the RNA, enabling special chemical reactions in its secure bubble. The RNA-molecule, in turn, began to encode information that was advantageous to the self-propagation not just of itself, but the entire RNA-micelle unit. Over time, the information encoded in the RNA-micelle complex allowed it to propagate more such RNA-micelle complexes.

"It is relatively easy to see how RNA-based protocells may have then evolved," Szostak wrote. "Metabolism could have arisen gradually, as . . . [the protocells learned to] synthesize nutrients internally from simpler and more abundant starting materials. Next, the organisms might have added protein synthesis to their bag of chemical tricks." RNA "proto-genes" may have learned to coax amino acids to form chains and thus build proteins—versatile, molecular machines that could make metabolism, self-propagation, and information transfer vastly more efficient.

①

When, and why, did discrete "genes"—modules of information—appear on a strand of RNA? Did genes exist in their modular form at the very beginning, or was there an intermediate or alternative form of information storage? Again, these questions are fundamentally unanswerable, but perhaps information theory can provide a crucial clue. The trouble with continuous, nonmodular information is that it is notoriously hard to manage. It tends to diffuse; it tends to become corrupted; it tends to tangle, dilute, and decay. Pulling one end causes another to unspool. If information bleeds into information, it runs a much greater risk of distortion: think of a vinyl record that acquires a single dent in the middle. Information that is "digitized," in contrast, is much easier to repair and recover. We can access and change one word in a book without having to reconfigure the entire library. Genes may have appeared for the same reason: discrete, information-bearing modules in one strand of RNA were used to encode instructions to fulfill discrete and individual functions.

The discontinuous nature of information would have carried an added benefit: a mutation could affect one gene, and only one gene, leaving the other genes unaffected. Mutations could now act on discrete modules of information rather than disrupting the function of the organism as a whole—thereby accelerating evolution. But that benefit came with a concomitant liability: too much mutation, and the information would be damaged or lost. What was needed, perhaps, was a backup copy—a mirror image to protect the original or to restore the prototype if damaged. Perhaps this was the ultimate impetus to create a *double-stranded* nucleic acid. The data in one strand would be perfectly reflected in the other and could be used to restore anything damaged; the yin would protect the yang. Life thus invented its own hard drive.

In time, this new copy—DNA—would become the master copy. DNA was an invention of the RNA world, but it soon overran RNA as a carrier of genes and became the dominant bearer of genetic information in living systems.* Yet another ancient myth—of the child consuming its father, of Cronus overthrown by Zeus—is etched into the history of our genomes.

---

* Some viruses still carry their genes in the form of RNA.

PART SIX

POST-GENOME

The Genetics of Fate and Future

$(2015- \ldots)$

☼

Those who promise us paradise on earth never produced anything but a hell.

—Karl Popper

It's only we humans who want to own the future, too.

—Tom Stoppard, *The Coast of Utopia*

The Future of the Future

Probably no DNA science is at once as hopeful, controversial, hyped, and even as potentially dangerous as the discipline known as gene therapy.

—Gina Smith, *The Genomics Age*

Clear the air! Clean the sky! Wash the wind! Take the stone from the stone, take the skin from the arm, take the muscle from the bone, and wash them. Wash the stone, wash the bone, wash the brain, wash the soul, wash them wash them!

—T. S. Eliot, *Murder in the Cathedral*

Let us return, for a moment, to a conversation on the ramparts of a fort. It is the late summer of 1972. We are in Sicily, at a scientific conference on genetics. It is late at night, and Paul Berg and a group of students have clambered up a hill overlooking the lights of a city. Berg's news—of the possibility of combining two pieces of DNA to create "recombinant DNA"—has sent tremors of wonder and anxiety through the meeting. At the conference, the students are concerned about the dangers of such novel DNA fragments: if the wrong gene is introduced into the wrong organism, the experiment might unleash a biological or ecological catastrophe. But Berg's interlocutors aren't only worried about pathogens. They have gone, as students often do, to the heart of the matter: they want to know about the prospects of human genetic engineering—of new genes being introduced permanently into the human genome. What about predicting the future from genes—and then altering that destiny through genetic manipulation? "They were already thinking several steps ahead," Berg later told me. "I was worried about the future, but they were worried about the future of the future."

For a while, the "future of the future" seemed biologically intractable. In 1974, barely three years after the invention of recombinant DNA technology, a gene-modified SV40 virus was used to infect early mouse embryonic cells. The plan was audacious. The virus-infected embryonic cells were mixed with the cells of a normal embryo to create a composite of cells, an embryological "chimera." These composite embryos were implanted into mice. All the organs and cells of the embryo emanated from that mix of cells—blood, brain, guts, heart, muscles, and, most crucially, the sperm and the eggs. If the virally infected embryonic cells formed some of the sperm and the egg cells of the newborn mice, then the viral genes would be transmitted from mouse to mouse vertically across generations, like any other gene. The virus, like a Trojan horse, might thus smuggle genes permanently into an animal's genome across multiple generations resulting in the first genetically modified higher organism.

The experiment worked at first—but it was stymied by two unexpected effects. First, although cells carrying viral genes clearly emerged in the blood, muscle, brain, and nerves of the mouse, the delivery of the viral genes into sperm and eggs was extremely inefficient. Try as they might, scientists could not achieve efficient "vertical" transmission of the genes across generations. And second, even though viral genes were present in the mouse cells, the *expression* of the genes was firmly shut down, resulting in an inert gene that did not make RNA or protein. Years later, scientists would discover that epigenetic marks had been placed on viral genes to silence them. We now know that cells have ancient detectors that recognize viral genes and stamp them with chemical marks, like cancellation signs, to prevent their activation.

The genome had, it seemed, already anticipated attempts to alter it. It was a perfect stalemate. There's an old proverb among magicians that it's essential to learn to make things reappear before one learns to make things disappear. Gene therapists were relearning that lesson. It was easy to slip a gene invisibly into a cell and into an embryo. The real challenge was to make it visible again.

①

Thwarted by these original studies, the field of gene therapy stagnated for another decade or so, until biologists stumbled on a critical discovery: embryonic stem cells, or ES cells. To understand the future of gene therapy in humans, we need to reckon with ES cells. Consider an organ such as the brain, or the skin. As an animal ages, cells on the surface of its skin grow, die, and slough off. This wave of cell death might even be catastrophic—after a burn, or a massive wound, for instance. To replace these dead cells, most organs must possess methods to regenerate their own cells.

Stem cells fulfill this function, especially after catastrophic cell loss. A stem cell is a unique type of cell that is defined by two properties. It can give rise to other functional cell types, such as nerve cells or skin cells, through differentiation. And it can renew *itself*—i.e., give rise to more stem cells, which can, in turn, differentiate to form the functional cells of an organ. A stem cell is somewhat akin to a grandfather that continues to produce children, grandchildren, and great-grandchildren, generation upon generation, without ever losing his own reproductive fecundity. It is the ultimate reservoir of regeneration for a tissue or an organ.

Most stem cells reside in particular organs and tissues and give rise to a limited repertoire of cells. Stem cells in the bone marrow, for instance, only produce blood cells. There are stem cells in the crypts of the intestine that are dedicated to the production of intestinal cells. But embryonic stem cells, or ES cells, which arise from the inner sheath of an animal's embryo, are vastly more potent; they can give rise to *every* cell type in the organism—blood, brains, intestines, muscles, bone, skin. Biologists use the word *pluripotent* to describe this property of ES cells.

ES cells also possess an unusual third characteristic—a quirk of nature. They can be isolated from the embryo of an organism and grown in petri dishes in the lab. The cells grow continuously in culture. Tiny, translucent spheres that can cluster into nestlike whirls under the microscope, they resemble a dissolving organ more than an organism in the making. Indeed, when the cells were first derived from mouse embryos in a laboratory in Cambridge, England, in the early eighties, they generated little interest among geneticists. "Nobody seems to be interested in my cells," the embryologist Martin Evans complained.

But the real power of an ES cell lies, yet again, in a transition: like DNA, like genes, and like viruses, it is the intrinsic duality of its existence that makes this cell such a potent biological tool. Embryonic stem cells behave like other experimentally amenable cells in tissue culture. They can be grown in petri dishes; they can be frozen in vials and thawed back to life. The cells can be propagated in liquid broth for generations, and genes can be inserted into their genomes or excised from their genomes with relative ease.

Yet, put the same cell into the right environment in the right context, and life literally leaps out of it. Mixed with cells from an early embryo and implanted into a mouse womb, the cells divide and form layers. They differentiate into all sorts of cells: blood, brain, muscle, liver—and even sperm and egg cells. These cells, in turn, organize themselves into organs and then become incorporated, miraculously, into a multilayered, multicellular organism—an actual mouse. Every experimental manipulation performed in the petri dish is thus carried forward into this mouse. The genetic modification of a cell in a dish "becomes" the genetic modification of an organism in a womb. It is a transition between lab and life.

The experimental ease permitted by embryonic stem cells also surmounted a second, and more intractable, problem. When viruses are used to deliver genes into cells, it is virtually impossible to control where the gene is inserted into the genome. At 3 billion base pairs of DNA, the human genome is about fifty thousand or a hundred thousand times the size of most viral genomes. A viral gene drops into the genome like a candy wrapper thrown from an airplane into the Atlantic: there is no way to predict where it might land. Virtually all viruses capable of gene integration, such as HIV or SV40, generally latch their genes randomly onto some spot in the human genome. For gene therapy, this random integration is an infernal nuisance. The viral genes might fall into a silent crevasse of the genome, never to be expressed. The genes might fall into an area of the chromosome that is actively silenced by the cell without much effort. Or worse, the integration might disrupt an essential gene or activate a cancer-causing gene, resulting in potential disasters.

With ES cells, however, scientists learned to make genetic changes not randomly, but in targeted positions in the genome, including within *the genes themselves*. You could choose to change the insulin gene and—through some rather basic but ingenious experimental manipulations—ensure that *only* the insulin gene was changed in the cells. And because the gene-modified ES cells could, in principle, generate all the cell types in a full mouse, you could be sure that a mouse with precisely that changed insulin gene would be born. Indeed, if the gene-modified ES cells eventually produced sperm and egg cells in the adult mice, then the gene would be transmitted from mouse to mouse across generations, thus achieving vertical hereditary transmission.

This technology had far-reaching implications. In the natural world, the only means to achieve a directional or intentional change in a gene is through random mutation and natural selection. If you expose an animal to X-rays, say, a genetic alteration might become permanently embedded in the genome—but there is no method to focus the attention of an X-ray to one particular gene. Natural selection must choose the mutation that confers the best fitness to the organism and thereby allows that mutation to become increasingly common in the gene pool. But in this scheme, neither mutation nor evolution has any intentionality or directionality. In nature, the engine that drives genetic alteration has no one in its driver's seat. The "watchmaker" of evolution, as Richard Dawkins reminds us, is inherently blind.

Using ES cells, however, a scientist could intentionally manipulate virtually any chosen gene and incorporate that genetic change permanently into the genome of an animal. It was mutation and selection in the same step—evolution fast-forwarded in a laboratory dish. The technology was so transformative that a new word had to be coined to describe these organisms: they were called *transgenic* animals—from "across genes." By the early 1990s, hundreds of strains of transgenic mice had been created in laboratories around the world to decipher the functions of genes. One mouse was made with a jellyfish gene inserted into its genome that allowed it to glow in the dark under blue lamps. Other mice, carrying variants of the growth hormone gene, grew twice the size of their normal counterparts. There were mice endowed with genetic alterations that forced them to develop Alzheimer's disease, epilepsy, or premature aging. Mice with activated cancer genes exploded with tumors, allowing biologists to use these mice as models for human malignancies. In 2014, researchers created a mouse carrying a mutation in a gene that controls the communication between neurons in the brain. These mice have substantially increased memory and superior cognitive function. They are the savants of the rodent world: they acquire memories faster, retain them longer, and learn new tasks nearly twice as fast as normal mice.

The experiments brimmed over with complex ethical implications. Could this technique be used in primates? In humans? Who would regulate the creation of animals with transgenes? What genes would be, or could be, introduced? What were the limits on transgenes?

Fortunately, technical barriers intervened before the ethical mayhem had a chance to become unmoored. Much of the original work on ES cells—including the production of transgenic organisms—had been per-formed using mouse cells. In the early 1990s, when several *human* em-bryonic stem cells were derived from early human embryos, scientists ran into an unexpected barrier. Unlike the mouse ES cells, which had proved so amenable to experimental manipulations, human ES cells did not be-have themselves in culture. "It may be the field's dirty little secret: human ES cells do not have the same capabilities as mouse ES cells," the biologist Rudolf Jaenisch said. "You can't clone them. You can't use them for gene targeting. . . . They are very different from mouse embryonic stem cells, which can do everything."

At least temporarily, the genie of transgenesis seemed contained.

①

Transgenic modification of human embryos was out of the question for a while—but what if gene therapists could settle for a less radical goal? Could viruses be used to deliver genes into human *nonreproductive* cells—i.e., to neurons, blood, or muscle cells? The problem of random integration into the genome would remain—and most crucially, no vertical transmis-sion of genes from one organism to the next would occur. But if the virally delivered genes could be put into the right kind of cells, they might still achieve their therapeutic purpose. Even that goal would represent a leap into the future of human medicine. It would be gene-therapy lite.

In 1988, a two-year-old girl named Ashanti DeSilva, or Ashi, for short, in North Olmsted, Ohio, began to develop peculiar symptoms. Children have dozens of transient ailments in infancy, as any parent knows, but Ashi's ill-nesses and symptoms were markedly abnormal: bizarre pneumonias and infections that seemed to persist, wounds that would not heal, and a white blood cell count that hovered consistently below normal. Much of Ashi's childhood was spent in and out of the hospital: at age two, a run-of-the-mill viral infection spun out of control, causing life-threatening internal bleeding and a prolonged hospitalization.

For a while, her doctors were mystified by her symptoms, attribut-ing her periodic illnesses loosely to an underdeveloped immune system that would eventually mature. But when the symptoms refused to abate as Ashi turned three, she underwent a barrage of tests. Her immunode-ficiency was attributed to her genes—to rare, spontaneous mutations in both copies of a gene called ADA on chromosome twenty. By then, Ashi had already suffered several near-death experiences. The physical toll on her body had been immense—but the emotional anguish that she had experienced was more pronounced: one morning, the four-year-old awoke and said, "Mommy, you shouldn't have had a child like me."

The ADA gene—short for "adenosine deaminase"—encodes an enzyme that converts adenosine, a natural chemical produced by the body, into a harmless product called inosine. In the absence of the ADA gene, the detoxification reaction fails to occur, and the body gets clogged with toxic by-products of adenosine metabolism. The cells that are most acutely poisoned are infection-fighting T cells—and in their absence, the immune system collapses rapidly. The illness is fleetingly rare—only one in 150,000 children is born with ADA deficiency—but it is even rarer still, because virtually all children die of the disease. ADA deficiency is part of a larger group of notorious illnesses called severe combined immunodeficiency, or SCID. The most famous SCID patient, a boy named David Vetter, had spent all twelve years of his life in a plastic chamber in a Texas hospital. The Bubble Boy, as David was called by the media, died in 1984, still imprisoned in his sterile plastic bubble, after a desperate attempt at a bone marrow transplant.

David Vetter's death gave pause to doctors who had hoped to use bone marrow transplants to treat ADA deficiency. The only other medicine, being tested in early clinical trials in the mideighties, was called PEG-ADA—the purified enzyme derived from cows and wrapped in an oily chemical sheath to make it long-lived in the blood (the normal ADA protein is too short-lived to be effective). But even PEG-ADA barely reversed the immunodeficiency. It had to be injected into the blood every month or so, to replace the enzyme degraded by the body. More ominously, PEG-ADA carried the risk of inducing antibodies against itself—depleting the levels of the enzyme even more acutely and precipitating a full catastrophe, making the solution infinitely worse than the original problem.

Could gene therapy correct ADA deficiency? After all, only one gene needed to be corrected, and the gene had already been identified and isolated. A vehicle, or vector, designed to deliver genes into human cells had also been identified. In Boston, Richard Mulligan, a virologist and geneticist, had designed a particular strain of retrovirus—a cousin of HIV's—that could potentially shuttle any gene into any human cell with relative safety. Retroviruses can be designed to infect many kinds of cells; their distinct capability is their ability to insert their own genome into the cell's genome, thereby permanently affixing their genetic material to that of a cell. By tweaking the technology, Mulligan had created partially crippled viruses that would infect cells and integrate into their genomes, but would not propagate the infection from cell to cell. Virus went in, but no virus came out. The gene fell into the genome but never popped out again.

①

In 1986, at the National Institutes of Health in Bethesda, a team of gene therapists, led by William French Anderson and Michael Blaese,* decided to use variants of Mulligan's vectors to deliver the ADA gene into children with ADA deficiency.† Anderson obtained the ADA gene from another lab and inserted it into the retroviral gene-delivery vector. In the early 1980s, Anderson and Blaese had run some preliminary trials hoping to use retroviral vectors to deliver the human ADA gene into blood-forming stem cells of mice and then monkeys. Once these stem cells had been infected by the virus carrying the ADA gene, Anderson hoped, they would form all the cellular elements of blood—including, crucially, T cells into which the now-functioning ADA gene had been delivered.

The results were far from promising: the level of gene delivery was abysmal. Of the five monkeys treated, only one—Monkey Roberts—had cells in the blood that showed long-term production of the human ADA protein from the virus-delivered gene. But Anderson was unfazed. "Nobody knows what may happen when new genes enter the body of a human being," he argued. "It's a total black box, despite what anyone tells you. . . . Test-tube and animal research can only tell us so much. Eventually you have to try it in a person."

On April 24, 1987, Anderson and Blaese applied to the NIH for per-

---

* Kenneth Culver was also a crucial member of this original team.

† In 1980, a UCLA scientist named Martin Cline attempted the first known gene therapy in humans. A hematologist by training, Cline chose to study beta-thalassemia, a genetic disease in which the mutation of a single gene, encoding a subunit of hemoglobin, causes severe anemia. Reasoning that he might be able to run his trials in foreign countries, where the use of recombinant DNA in humans was less constrained and regulated, Cline did not notify his hospital's review board, and ran his trials on two thalassemia patients in Israel and Italy. Cline's attempts were discovered by the NIH and UCLA. He was sanctioned by the NIH, found to be in breach of federal regulations, and ultimately resigned as the chair of his division. The complete data from his experiment were never formally published.

mission to launch their gene-therapy protocol. They proposed to extract bone marrow stem cells from the children with ADA deficiency, infect the cells with the virus in the lab, and transplant the modified cells back into the patients. Since the stem cells generate all the elements of blood—including B and T cells—the ADA gene would find its way into the T cells, where it was most required.

The proposal was sent to the Recombinant DNA Advisory Committee, or RAC, a consortium set up within the NIH in the wake of the Berg recommendations of the Asilomar meeting. Known for its tough oversight, the advisory committee was the gatekeeper for all experiments that involved recombinant DNA (the committee was so notoriously obstreperous that researchers called getting its approval being "taken through the Rack"). Perhaps predictably, the RAC rejected the protocol outright, citing poor animal data, the barely detectable level of gene delivery into stem cells, and the lack of a detailed experimental rationale, while noting that gene transfer into a human body had never been attempted before.

Anderson and Blaese returned to the lab to revamp their protocol. Begrudgingly, they admitted that the RAC's decision was correct. The barely detectable infection rate of bone marrow stem cells by the gene-carrying virus was clearly a problem, and the animal data was far from exhilarating. But if stem cells could not be used, how could gene therapy hope to succeed? Stem cells are the only cells in the body that can renew themselves and therefore provide a long-term solution to a gene deficiency. Without a source of self-renewing or long-lived cells, you might insert genes into the human body, but the cells carrying the genes would eventually die and vanish. There would be genes, but no therapy.

That winter, mulling over the problem, Blaese found a potential solution. What if, rather than delivering the genes into the blood-forming stem cells, they merely took *T cells* from the blood of ADA patients and put the virus into the cells? It would not be as radical or permanent an experiment as putting viruses into stem cells, but it would be far less toxic and much easier to achieve clinically. The T cells could be harvested from peripheral blood, not bone marrow, and the cells might live just long enough to make the ADA protein and correct the deficiency. Although the T cells would inevitably fade from the blood, the procedure could be repeated again and again. It would not qualify as definitive gene therapy, but it would still be a proof of principle—gene-therapy double-lite.

Anderson was reluctant; if he was to launch the first trial of human gene therapy, he wanted a definitive trial and a chance to stake a permanent claim on medical history. He resisted at first, but eventually, conceding Blaese's logic, relented. In 1990, Anderson and Blaese approached the committee again. Again, there was vicious dissent: the T cell protocol had even less supportive data than the original. Anderson and Blaese submitted modifications, and modifications to the modifications. Months passed without a decision. In the summer of 1990, after a prolonged series of debates, the committee agreed to let them proceed with the trial. "Doctors have been waiting for this day for a thousand years," the RAC's chairman, Gerard McGarrity, said. Most others in the committee were not as sanguine about the chances of success.

Anderson and Blaese searched hospitals around the country to find children with ADA deficiency for their trial. They came upon a small trove in Ohio: all of two patients with the genetic defect. One was a tall, dark-haired girl named Cynthia Cutshall. The second, Ashanti DeSilva, was a four-year-old daughter of a chemist and a nurse, both from Sri Lanka.

⊕

In September 1990, on an overcast morning in Bethesda, Van and Raja DeSilva, Ashi's parents, brought their daughter to the NIH. Ashi was now four years old—a shy, hesitant girl with a fringe of shiny hair, cut pageboy-style, and an apprehensive face that could suddenly be brightened by a smile. This was her first meeting with Anderson and Blaese. When they approached her, she looked away. Anderson brought her down to the hospital's gift shop and asked her to pick out a soft toy for herself. She chose a bunny.

Back at the Clinical Center, Anderson inserted a catheter into one of Ashi's veins, collected samples of her blood, and rushed it up to his lab. Over the next four days, 200 million retroviruses, in an enormous cloudy soup, were mixed with 200 million T cells drawn from Ashi's blood. Once infected, the cells grew in petri dishes, forming lush outcroppings of new cells, and even newer cells. They doubled by day and doubled by night in a silent, humid incubator in Building 10 of the Clinical Center—just a few hundred feet from the lab where Marshall Nirenberg, almost exactly twenty-five years before, had solved the genetic code.

Ashi DeSilva's gene-modified T cells were ready on September 14, 1990. That morning, Anderson ran out of his home at dawn, skipping breakfast, near nauseous with anticipation, and dashed up the steps to the lab on the third floor. The DeSilva family was already waiting for him; Ashi was standing by her mother, her elbows planted firmly on her seated mother's lap as if waiting for a dental exam. The morning was spent running more tests. The clinic was silent, except for the occasional footfall of research nurses running in and out. As Ashi sat on her bed in a loose yellow gown, a needle was put into one of her veins. She winced slightly but recovered: her veins had been cannulated dozens of times before.

At 12:52 p.m., a vinyl bag carrying the murky swirl of nearly 1 billion T cells infected by the ADA-gene-carrying retrovirus was brought up to the floor. Ashi looked apprehensively at the bag as the nurses hooked it to her vein. Twenty-eight minutes later, the bag had run dry, its last dregs emptied into Ashi. She played with a yellow sponge ball on her bed. Her vital signs were normal. Ashi's father was sent downstairs with a pile of quarters to buy candy from the vending machine on the ground floor. Anderson looked visibly relieved. "A cosmic moment has come and gone, with scarcely a sign of its magnitude," one observer wrote. It was celebrated, in style, with a bag of multicolored M&M's.

"Number one," Anderson said, pointing elatedly at Ashi as he wheeled her down the hallway after the transfusion had been completed. A few of his colleagues at the NIH were waiting outside the door to witness the coming of the first human to be transfused with gene-modified cells, but the crowd thinned quickly and the scientists vanished back to their labs. "It's like people say in downtown Manhattan," Anderson groused. "Jesus Christ himself could walk by and nobody would notice." The next day, Ashi's family returned home to Ohio.

Ⓘ

Did Anderson's gene-therapy experiment work? We do not know—and perhaps we will never know. Anderson's protocol was designed as a proof of principle for safety—i.e., could retrovirus-infected T cells be safely delivered into human bodies? It was not designed to test efficacy: Would this protocol cure ADA deficiency, even temporarily? Ashi DeSilva and Cynthia Cutshall, the first two patients on the study, received the gene-modified T cells, but they were allowed continued treatment with PEG-ADA, the artificial enzyme. Any effect of the gene therapy was thus confounded by that medicine.

Nonetheless, both DeSilva's and Cutshall's parents were convinced that the treatment worked. "It's not a big improvement," Cynthia Cutshall's mother admitted. "But to give you an example, she just got over one cold. Usually her colds end up in pneumonia. This one didn't. . . . That's a breakthrough for her." Ashi's father, Raja DeSilva, concurred: "With PEG, we'd seen a tremendous improvement. But even with [PEG-ADA], she had runny noses and a constant cold, was on antibiotics all the time. But by the second infusion of genes, in December, it began to change. We noticed because we were not using up so many boxes of tissues."

Despite Anderson's enthusiasm, and the anecdotal evidence from the families, many proponents of gene therapy, including Mulligan, were far from convinced that Anderson's trial had amounted to anything more than a publicity stunt. Mulligan, the most voluble critic of the trial from the very first, was particularly incensed by the claims of success when the data was insufficient. If the most ambitious gene-therapy trial attempted in humans was going to be measured in the frequency of runny noses and boxes of Kleenex, then it would be an embarrassment for the field. "It's a sham," Mulligan told a journalist when asked about the protocol. To test whether targeted genetic alterations could be introduced into human cells, and whether these genes would confer normal function safely and effectively, he proposed a careful, uncontaminated trial—"clean, chaste gene therapy," as he called it.

But by then, the ambitions of gene therapists had been frothed to such frenzy that "clean, chaste," careful experiments had become virtually impossible to perform. Following the reports of the T cell trials at the NIH, gene therapists envisaged novel cures for genetic diseases such as cystic fibrosis and Huntington's disease. Since genes could be delivered into virtually any cell, any cellular disease was a candidate for gene therapy: heart disease, mental illness, cancer. As the field readied itself to sprint forward, voices such as Mulligan's urged caution and restraint, but they were brushed aside. The enthusiasm would come at a steep price: it would bring the field of gene therapy, and human genetics, to the brink of disaster, and to the lowest, bleakest point in its scientific history.

Ⓘ

On September 9, 1999, almost exactly nine years after Ashi DeSilva had been treated with genetically modified white blood cells, a boy named

Jesse Gelsinger flew to Philadelphia to enroll in another gene-therapy trial. Gelsinger was eighteen years old. A motorcycling and wrestling enthusiast, with an easy, carefree manner, Gelsinger, like Ashi DeSilva and Cynthia Cutshall, was also born with a mutation in a single gene involved in metabolism. In Gelsinger's case, the gene was called ornithine transcarbamylase, or OTC, which encodes an enzyme synthesized in the liver. OTC, the enzyme, performs a critical step in the breakdown of proteins. In the absence of the enzyme, ammonia, a by-product of protein metabolism, accumulates in the body. Ammonia—the chemical found in cleaning fluid—damages blood vessels and cells, diffuses past the blood-brain barrier, and ultimately results in the slow poisoning of neurons in the brain. Most patients with mutations in OTC do not survive childhood. Even with strictly protein-free diets, they are poisoned by the breakdown of their own cells as they grow.

Among children born with an unfortunate disease, Gelsinger might have counted himself as especially fortunate, for his variant of OTC deficiency was mild. The mutation in his gene had not come from his father or his mother, but had occurred spontaneously in one of his cells *in utero*, probably when he was still a young embryo. Genetically, Gelsinger was a rare phenomenon—a human chimera, a patchwork cellular quilt, with some cells lacking functional OTC, and some with a working gene. Still, his ability to metabolize proteins was severely compromised. Gelsinger lived on a carefully calibrated diet—every calorie and portion weighed, measured, and accounted for—and took thirty-two pills a day to keep his ammonia level in check. Despite such extreme cautionary measures, Gelsinger had still suffered several life-threatening episodes. At four, he had joyfully eaten a peanut butter sandwich that had precipitated a coma.

In 1993, when Gelsinger was twelve years old, two pediatricians in Pennsylvania, Mark Batshaw and James Wilson, began to experiment with gene therapy to cure children with OTC deficiencies. A former college-level football player, Wilson was a risk taker fascinated by ambitious human experiments. He had formed a gene-therapy company, named Genova, and an Institute for Human Gene Therapy at the University of Pennsylvania. Both Wilson and Batshaw were intrigued by OTC deficiency. As with ADA deficiency, that OTC is caused by the dysfunction of a single gene made the illness an ideal test case for gene therapy. But the form of gene therapy that Wilson and Batshaw envisioned was vastly more radical: rather than extracting cells, genetically modifying them, and injecting them back into children (à la Anderson and Blaese), Batshaw and Wilson imagined inserting the corrected gene *directly* back into the body via a virus. This would not be gene-therapy lite: they would create a virus carrying the OTC gene and deliver the virus into the liver through the bloodstream, leaving the virus to infect cells *in situ*.

The virus-infected liver cells would start synthesizing the OTC enzyme, Batshaw and Wilson reasoned, and thus correct the enzyme deficiency. The telltale sign would be a reduction of ammonia in the blood. "It wasn't that subtle," Wilson recalls. To deliver the gene, Wilson and Batshaw chose adenovirus, a virus that typically causes a common cold but is not associated with any severe disease. It seemed like a safe, reasonable choice—the blandest of viruses used as the vehicle for one of the boldest human genetic experiments of the decade.

In the summer of 1993, Batshaw and Wilson began to inject the modified adenovirus into mice and monkeys. The mouse experiments worked as predicted: the virus reached the liver cells, disgorged the gene, and transformed the cells into microscopic factories for the functional OTC enzyme. But the monkey experiments were more complicated. At higher doses of the virus, an occasional monkey raised a brisk immune response to the virus, resulting in inflammation and liver failure. One monkey hemorrhaged to death. Wilson and Batshaw modified the virus, shaving off many of the viral genes that might elicit immunity, to make it a safer gene-delivery vehicle. They also reduced the potential human dose by seventeenfold to doubly ensure the safety of the virus. In 1997, they applied to the Recombinant DNA Advisory Committee, the gatekeeper for all gene-therapy experiments, for approval for a human trial. The RAC was resistant at first, but it too had changed: in the decade between the ADA trial and Wilson's, the once-fierce guardian of recombinant DNA had turned into an enthusiastic cheerleader of human gene therapy. The fizz of enthusiasm had even leached beyond the committee. Asked by the RAC to comment on Wilson's trial, bioethicists argued that treating children with full-blown OTC deficiency might result in "coercion": What parent *wouldn't* want to try a breakthrough therapy that might work on a dying child? Instead, ethicists recommended a trial on normal volunteers and patients with mild variants of OTC, such as Jesse Gelsinger.

①

In Arizona, Gelsinger, meanwhile, was chafing against the elaborate restrictions on his diet and medications ("All teenagers rebel," Gelsinger's father, Paul, told me, but teenage rebellion might feel particularly acute when it involves "a hamburger and a glass of milk"). In the summer of 1998, when he was seventeen, Gelsinger learned of the OTC trial at the University of Pennsylvania. Gelsinger was gripped by the thought of gene therapy. He wanted a respite from the grinding routine of his life. "But what got him even more excited," his father recalled, "was the idea that he was doing it for the babies. How do you say no to that?"

Gelsinger could hardly wait to sign on. In June 1999, he contacted the Pennsylvania team through his local doctors to enroll in the trial. That month, Paul and Jesse Gelsinger flew to Philadelphia to meet Wilson and Batshaw. Jesse and Paul were both impressed. The trial struck Paul Gelsinger as a "beautiful, beautiful thing." They visited the hospital and then wandered through the city in a haze of excitement and anticipation. Jesse stopped in front of the Rocky Balboa bronze outside the Spectrum Arena. Paul snapped a picture of his son, his arms raised in a boxer's victory stance.

On September 9, Jesse returned to Philadelphia with a duffel bag filled with clothes, books, and wrestling videos to start the trial at University Hospital. Jesse would stay with his uncle and cousins in the city and admit himself to the hospital on the appointed morning. The procedure was described as so quick and painless that Paul planned to pick up his son one week after the therapy had been completed to bring him back home on a commercial flight.

⓪

On the morning of September 13, the day chosen for the viral injection, Gelsinger's ammonia level was found to be hovering around seventy micromoles per liter—twice the normal level and at the upper edge of the cutoff value for the trial. The nurses brought news of the abnormal lab to Wilson and Batshaw. The protocol, meanwhile, was in full swing. The operating rooms were on standby. The viral liquid had been thawed and sat glistening in its plastic pouch. Wilson and Batshaw debated Gelsinger's eligibility, but decided that it was clinically safe to continue; the previous seventeen patients had, after all, tolerated the injection. At about 9:30 a.m., Gelsinger was wheeled down to the interventional radiology suite.

He was sedated, and two large catheters were snaked through his legs to reach an artery close to the liver. Around 11:00 a.m., a surgeon drew about thirty milliliters from a bag clouded with the concentrated adenovirus and injected the puff of virus into Gelsinger's artery. Hundreds of millions of invisible infectious particles carrying the OTC gene streamed into the liver. By noon, the procedure was done.

The afternoon was uneventful. That evening, back in his hospital room, Gelsinger spiked a fever to 104 degrees. His face was flushed. Wilson and Batshaw did not make much of the symptoms. The other patients had also experienced transient fevers. Jesse called Paul in Arizona on the phone and said, "I love you," before hanging up and drawing up his covers in bed. He slept fitfully through the night.

The next morning, a nurse noted that Jesse's eyeballs had turned the palest shade of yellow. A test confirmed that bilirubin, a product made in the liver and also stored in red blood cells, was spilling into his blood. The elevated bilirubin meant one of two things: either the liver was being injured or blood cells were being damaged. Both were ominous signs. In any other human, the small bump in cellular breakdown or liver failure might have been shrugged off. But in a patient with OTC deficiency, the combination of these two injuries might spark a perfect storm: the extra protein leaching out from the blood cells would not be metabolized, and the damaged liver, deficient in protein metabolism even in the best of times, would be even less capable of processing the excess protein load. The body would intoxicate itself on its own poisons. By noon, Gelsinger's ammonia level had climbed to a staggering 393 micromoles per liter—nearly ten times the normal level. Paul Gelsinger and Mark Batshaw were alerted. James Wilson heard the news from the surgeon who had inserted the catheter and injected the virus. Paul booked a red-eye to Pennsylvania, while a team of doctors swooped into the ICU to begin dialysis to avert a coma.

At eight o'clock the next morning, when Paul Gelsinger reached the hospital, Jesse was hyperventilating and confused. His kidneys were failing. The ICU team sedated him to try to use a mechanical ventilator to stabilize his breathing. Late that night, his lungs began to stiffen and collapse, filling up with fluids from the inflammatory response. The ventilator faltered, unable to push enough oxygen in, and so Jesse was hooked up to a device to force oxygen directly into the blood. His brain function was also deteriorating. A neurologist was called to examine him and noted Jesse's downcasting eyes—a sign of damage to the brain.

The next morning, Hurricane Floyd struck the East Coast, battering the shores of Pennsylvania and Maryland with shrieking winds and torrents of rain. Batshaw was stuck on a train getting to the hospital. He ran down the last minutes of his cell phone's battery talking to the nurses and doctors, then sat in the pitch darkness, stewing with anxiety. By the late afternoon, Jesse's condition worsened again. His kidneys shut down. His coma deepened. Stranded in his hotel room, with no taxi in sight, Paul Gelsinger walked a mile and a half through the whistling storm to the hospital to see Jesse in the ICU. He found his son unrecognizable—comatose, swollen, bruised, yellowed with jaundice, with dozens of lines and catheters crisscrossing his body. The ventilator puffed ineffectually against his inflamed lungs with the flat, dull sound of wind slapping water. The room buzzed and beeped with hundreds of instruments recording the slow decline of a boy in desperate physiological distress.

On the morning of Friday, September 17, on the fourth day after the gene delivery, Jesse was found to be brain-dead. Paul Gelsinger decided to withdraw life support. The chaplain came into the hospital room and put his hand on Jesse's head, anointing it with oil, and read the Lord's Prayer. The machines were shut off, one by one. The room fell into silence, except for Jesse's deep, agonal breaths. At 2:30 p.m., Jesse's heart stopped. He was officially pronounced dead.

"How could such a beautiful thing go so, so wrong?" When I spoke to Paul Gelsinger in the summer of 2014, he was still searching for an answer. A few weeks earlier, I had e-mailed Paul about my interest in Jesse's story. Gelsinger spoke to me on the phone, then agreed to meet me after my talk on the future of genetics and cancer at an open forum in Scottsdale, Arizona. While I stood in the lobby of the auditorium at the end of the talk, a man in a Hawaiian shirt with Jesse's round, open face—a face that I remembered vividly from pictures on the Web—pushed his way through the crowd and extended his hand.

In the aftermath of Jesse's death, Paul has become a one-man crusader against the overreach of clinical experimentation. He is not against medicine or innovation. He believes in the future of gene therapy. But he is suspicious of the hyperbaric atmosphere of enthusiasm and delusion that ultimately resulted in his son's death. The crowd thinned, and Paul turned around to leave. An acknowledgment passed between us: a doctor writing about the future of medicine and genetics, and a man whose story had been etched into its past. There had been an infinite horizon of grief in his voice. "They didn't have a handle on it yet," he said. "They tried it too quickly. They tried it without doing it right. They rushed this thing. They really rushed it."

①

The autopsy of an experiment gone "so, so wrong" began in earnest in October 1999 when the University of Pennsylvania launched an investigation into the OTC trial. By late October, an investigative journalist from the *Washington Post* had picked up the news of Gelsinger's death, and a tide of furor broke loose. In November, the US Senate, the House of Representatives, and the district attorney of Pennsylvania held independent hearings on Jesse Gelsinger's death. By December, the RAC and the FDA had launched an investigation of the University of Pennsylvania. Gelsinger's medical records, the pretrial animal experiments, consent forms, procedure notes, lab tests, and the records of all the other patients on the gene-therapy trial were pulled up from University Hospital's basement, and federal regulators plowed their way through the mounds of paper trying to exhume the cause of the boy's death.

The initial analysis revealed a damning pattern of incompetence, blunders, and neglect, compounded by fundamental gaps in knowledge. First, the animal experiments performed to establish the safety of the adenovirus had been performed hastily. One monkey inoculated with the highest doses of the virus had died, and while this death had been reported to the NIH and the dose reduced for human patients, no mention of the death was found in the forms given to the Gelsinger family. "Nothing about the consent forms," Paul Gelsinger recalled, "hinted clearly at the harm that the treatment could cause. It was portrayed like a perfect gamble—all upside and no downside." Second, even the human patients treated before Jesse had experienced side effects, some striking enough to have halted the trial or triggered reevaluations of the protocol. Fevers, inflammatory responses, and early signs of liver failure had been recorded, but these too had been underreported or ignored. That Wilson had a financial stake in the biotechnology company that stood to benefit from this gene-therapy experiment only further deepened the suspicion that the trial had been put together with inappropriate incentives.

The pattern of neglect was so damning that it nearly obscured the most important scientific lessons of the trial. Even if the doctors admitted that they had been negligent and impatient, Gelsinger's death was still a mystery: no one could explain why Jesse Gelsinger had suffered such a severe immune reaction to the virus, while the seventeen other patients had not. Clearly, the adenoviral vector—even the "third-generation" virus shorn of some of its immunogenic proteins—was capable of inciting a severe idiosyncratic response in some patients. The autopsy of Gelsinger's body showed that his physiology had become overwhelmed by this immune response. Notably, when his blood was analyzed, antibodies highly reactive to the virus were found, dating from even *before* the viral injection. Gelsinger's hyperactive immune response was likely related to prior exposure to a similar strain of adenovirus, possibly from a common cold. Exposures to pathogens are well-known to incite antibodies that remain in circulation for decades (this, after all, is how most vaccines work). In Jesse's case, this prior exposure had likely triggered a hyperactive immune response that had spiraled out of control for unknown reasons. Ironically, perhaps the choice of a "harmless," common virus as the initial vector for gene therapy had turned out to be the trial's key failing.

What, then, was the appropriate vector for gene therapy? What kind of virus could be used to deliver genes safely into humans? And which organs were appropriate targets? Just as the field of gene therapy was beginning to confront its most intriguing scientific problems, the entire discipline was placed under a strict moratorium. The litany of troubles uncovered in the OTC trial was not limited to that trial alone. In January 2000, when the FDA inspected twenty-eight other trials, nearly half of them required immediate remedial action. Justifiably alarmed, the FDA shut down nearly all the trials. "The entire field of gene therapy went into free fall," a journalist wrote. "Wilson was banned from working on FDA-regulated human clinical trials for five years. He stepped down from his position at the helm of the Institute for Human Gene Therapy, remaining as a professor at Penn. Soon afterward the institute itself was gone. In September 1999, gene therapy looked to be on the cusp of a breakthrough in medicine. By the end of 2000, it seemed like a cautionary tale of scientific overreach." Or, as Ruth Macklin, the bioethicist, put it bluntly, "Gene therapy is not yet therapy."

In science, there is a well-known aphorism that the most beautiful theory can be slayed by an ugly fact. In medicine, the same aphorism takes a somewhat different form: a beautiful therapy can be killed by an ugly trial. In retrospect, the OTC trial was nothing short of ugly—hurriedly designed, poorly planned, badly monitored, abysmally delivered. It was made twice as hideous by the financial conflicts involved; the prophets were in it for profits. But the basic concept behind the trial—delivering genes into human bodies or cells to correct genetic defects—was conceptually sound, as it had been for decades. In principle, the capacity to deliver genes into cells using viruses or other gene vectors should have led to powerful new medical technologies, had the scientific and financial ambitions of the early proponents of gene therapy not gotten in the way.

Gene therapy would eventually become therapy. It would rebound from the ugliness of the initial trials and learn the moral lessons implicit in the "cautionary tale of scientific overreach." But it would take yet another decade, and a lot more learning, for the science to cross the breach.

Genetic Diagnosis: "Previvors"

All that man is,
All mere complexities.
— W. B. Yeats, "Byzantium"

The anti-determinists want to say that DNA is a little
side-show, but every disease that's with us is caused by
DNA. And [every disease] can be fixed by DNA.
— George Church

While human gene therapy was exiled to wander its scientific tundra in the late 1990s, human genetic diagnosis experienced a remarkable renaissance. To understand this renaissance, we need to return to the "future's future" envisioned by Berg's students on the ramparts of the Sicilian castle. As the students had imagined it, the future of human genetics would be built on two fundamental elements. The first was "genetic diagnosis"— the idea that genes could be used to predict or determine illness, identity, choice, and destiny. The second was "genetic alteration"—that genes could be changed to change the future of diseases, choice, and destiny.

This second project—the intentional alteration of genes ("writing the genome")—had evidently faltered with the abrupt ban on gene-therapy trials. But the first—predicting future fate from genes ("reading the genome")—only gained more strength. In the decade following Jesse Gelsinger's death, geneticists uncovered scores of genes linked to some of the most complex and mysterious human diseases—illnesses for which genes had never been implicated as primary causes. These discoveries would enable the development of immensely powerful new technologies that would allow for the preemptive diagnosis of illness. But they would also force genetics and medicine to confront some of the deepest medical and moral conundrums in their history. "Genetic tests," as Eric Topol, the medical geneticist described it, "are also moral tests. When you decide to test for 'future risk,' you are also, inevitably, asking yourself, what kind of future am I willing to risk?"

①

Three case studies illustrate the power and the peril of using genes to predict "future risk." The first involves the breast cancer gene *BRCA1*. In the early 1970s, the geneticist Mary-Claire King began to study the inheritance of breast and ovarian cancer in large families. A mathematician by training, King had met Allan Wilson—the man who had dreamed up Mitochondrial Eve—at the University of California, Berkeley, and switched to the study of genes and the reconstruction of genetic lineages. (King's earlier studies, performed in Wilson's lab, had demonstrated that chimps and humans shared more than 90 percent genetic identity.)

After graduate school, King turned to a different sort of genetic history: reconstructing the lineages of human diseases. Breast cancer, in particular, intrigued her. Decades of careful studies on families had suggested that breast cancer came in two forms—sporadic and familial. In sporadic breast cancer, the illness appears in women without any family history. In familial breast cancer, the cancer courses through families across multiple generations. In a typical pedigree, a woman, her sister, her daughter, and her granddaughter might be affected—although the precise age of diagnosis, and the precise stage of cancer for each individual, might differ. The increased incidence of breast cancer in some of these families is often accompanied by a striking increase in the incidence of ovarian cancer, suggesting a mutation that is common to both forms of cancer.

In 1978, when the National Cancer Institute launched a survey on breast cancer patients, there was widespread disagreement about the cause of the disease. One camp of cancer experts argued that breast cancer was caused by a chronic viral infection, triggered by the overuse of oral contraceptives. Others blamed stress and diet. King asked to have two questions added to the survey: "Did the patient have a family history of breast cancer? Was there a family history of ovarian cancer?" By the end of the survey, the genetic connection vaulted out of the study: she had identified several families with deep histories of both breast and ovarian cancer. Between 1978 and 1988, King added hundreds of such families to her list and compiled enormous pedigrees of women with breast cancer. In one family with more than 150 members, she found 30 women affected by the illness.

A closer analysis of all the pedigrees suggested that a single gene was responsible for many of the familial cases—but identifying the gene was not easy. Although the culprit gene increased the cancer risk among carriers by more than tenfold, not everyone who inherited the gene had cancer. The breast cancer gene, King found, had "incomplete penetrance": even if the gene was mutated, its effect did not always fully "penetrate" into every individual to cause a symptom (i.e., breast or ovarian cancer).

Despite the confounding effect of penetrance, King's collection of cases was so large that she could use linkage analysis across multiple families, crossing multiple generations, to narrow the location of the gene to chromosome seventeen. By 1988, she had zoomed in farther on the gene: she had pinpointed it to a region on chromosome seventeen called 17q21. "The gene was still a hypothesis," she said, but at least it had a known physical presence on a human chromosome. "Being comfortable with uncertainty *for years* was the . . . lesson of the Wilson lab, and it is an essential part of what we do." She called the gene *BRCA1*, even though she had yet to isolate it.

The narrowing down of the chromosomal locus of *BRCA1* launched a furious race to identify the gene. In the early nineties, teams of geneticists across the globe, including King, set out to clone *BRCA1*. New technologies, such as the polymerase chain reaction (PCR), allowed researchers to make millions of copies of a gene in a test tube. These techniques, coupled with deft gene-cloning, gene-sequencing, and gene-mapping methods, made it possible to move rapidly from a chromosomal position to a gene. In 1994, a private company in Utah named Myriad Genetics announced the isolation of the *BRCA1* gene. In 1998, Myriad was granted a patent for the *BRCA1* sequence—one of the first-ever patents issued on a human gene sequence.

For Myriad, the real use of *BRCA1* in clinical medicine was genetic testing. In 1996, even before the patent on the gene had been granted, the company began marketing a genetic test for *BRCA1*. The test was simple: A woman at risk would be evaluated by a genetic counselor. If the family history was suggestive of breast cancer, a swab of cells from her mouth would be sent to a central lab. The lab would amplify parts of her *BRCA1* gene using the polymerase chain reaction, sequence the parts, and identify the mutant genes. It would report back "normal," "mutant," or "indeterminate" (some unusual mutations have not yet been fully categorized for breast cancer risk).

⊕

In the summer of 2008, I met a woman with a family history of breast cancer. Jane Sterling was a thirty-seven-year-old nurse from the North Shore of Massachusetts. The story of her family could have been plucked straight out of Mary-Claire King's case files: a great-grandmother with breast cancer at an early age; a grandmother who had had a radical mastectomy for cancer at forty-five; a mother who had had bilateral breast cancer at sixty. Sterling had two daughters. She had known about *BRCA1* testing for nearly a decade. When her first daughter was born, she had considered the test, but neglected to follow up. With the birth of the second daughter, and the diagnosis of breast cancer in a close friend, she came to terms with gene testing.

Sterling tested positive for a *BRCA1* mutation. Two weeks later, she returned to the clinic armed with sheaves of papers scribbled with questions. What would she do with the knowledge of her diagnosis? Women with *BRCA1* have an 80 percent lifetime risk of breast cancer. But the genetic test tells a woman nothing about when she might develop the cancer, nor the kind of cancer that she might have. Since the *BRCA1* mutation has incomplete penetrance, a woman with the mutation might develop inoperable, aggressive, therapy-resistant breast cancer at age thirty. She might develop a therapy-sensitive variant at age fifty, or a smoldering, indolent variant at age seventy-five. Or she might not develop cancer at all.

When should she tell her daughters about the diagnosis? "Some of these women [with *BRCA1* mutations] hate their mothers," one writer, who tested positive herself, wrote (the hatred of mothers, alone, illuminates the chronic misunderstanding of genetics, and its debilitating effects on the human psyche; the mutant *BRCA1* gene is as likely to be inherited from a mother as it is from a father). Would Sterling inform her sisters? Her aunts? Her second cousins?

The uncertainties about outcome were compounded by uncertainties about the choices of therapy. Sterling could choose to do nothing—to watch and wait. She could choose to have bilateral mastectomies and/or ovary removal to sharply diminish her risk of breast and ovarian cancer— "cutting off her breasts to spite her genes," as one woman with a *BRCA1* mutation described it. She could seek intensive screening with mammograms, self-examination, and MRIs to detect early breast cancer. Or she could choose to take a hormonal medicine, such as tamoxifen, which would decrease the risk of some, but not all, breast cancer.

Part of the reason for this vast variation in outcome reflects the fundamental biology of *BRCA1*. The gene encodes a protein that plays a critical role in the repair of damaged DNA. For a cell, a broken DNA strand is a catastrophe in the making. It signals the loss of information—a crisis. Soon after DNA damage, the *BRCA1* protein is recruited to the broken edges to repair the gap. In patients with the normal gene, the protein launches a chain reaction, recruiting dozens of proteins to the knife edge of the broken gene to swiftly plug the breach. In patients with the mutated gene, however, the mutant *BRCA1* is not appropriately recruited, and the breaks are not repaired. The mutation thus permits more mutations—like fire fueling fire—until the growth-regulatory and metabolic controls on the cell are snapped, ultimately leading to breast cancer. Breast cancer, even in *BRCA1*-mutated patients, requires multiple triggers. The environment clearly plays a role: add X-rays, or a DNA-damaging agent, and the mutation rate climbs even higher. Chance plays a role since the mutations that accumulate are random. And other genes accelerate or mitigate the effects of *BRCA1*—genes involved in repair of the DNA or the recruitment of the *BRCA1* protein to the broken strand.

The *BRCA1* mutation thus predicts a future, but not in the sense that a mutation in the cystic fibrosis gene or Huntington's disease gene predicts the future. The future of a woman carrying a *BRCA1* mutation is fundamentally changed by that knowledge—and yet it remains just as fundamentally uncertain. For some women, the genetic diagnosis is all-consuming; it is as if their lives and energies are spent anticipating cancer and imagining survivorship—from an illness that they have not yet developed. A disturbing new word, with a distinctly Orwellian ring, has been coined to describe these women: *previvors—pre*-survivors.

Φ

The second case study of genetic diagnosis concerns schizophrenia and bipolar disorder; it brings us full circle in our story. In 1908, the Swiss German psychiatrist Eugen Bleuler introduced the term *schizophrenia* to describe patients with a unique mental illness characterized by a terrifying form of cognitive disintegration—the collapse of thinking. Previously called *dementia praecox*, "precocious madness," schizophrenics were often young men who experienced a gradual but irreversible breakdown in their cognitive abilities. They heard spectral voices from within, commanding them to perform odd, out-of-place activities (recall Moni's hissing inner voice that kept repeating, "Piss here; piss here"). Phantasmic visions appeared and disappeared. The capacity to organize information or perform goal-oriented tasks collapsed, and new words, fears, and anxieties emerged, as if from the netherworlds of the mind. In the end, all organized thinking began to crumble, entrapping the schizophrenic in a maze of mental rubble. Bleuler argued that the principal characteristic of the illness was a splitting, or rather splintering, of the cognitive brain. This phenomenon inspired the word *schizo-phrenia*, from "split brain."

Like many other genetic diseases, schizophrenia also comes in two forms—familial and sporadic. In some families with schizophrenia, the disorder courses through multiple generations. Occasionally, some families with schizophrenia also have family members with bipolar disorder (Moni, Jagu, Rajesh). In sporadic or de novo schizophrenia, in contrast, the illness arises as a bolt from the blue: a young man from a family with no prior history might suddenly experience the cognitive collapse, often with little or no warning. Geneticists tried to make sense of these patterns, but could not draw a model of the disorder. How could the same illness have sporadic and familial forms? And what was the link between bipolar disorder and schizophrenia, two seemingly unrelated disorders of the mind?

The first clues about the etiology of schizophrenia came from twin studies. In the 1970s, studies demonstrated a striking degree of concordance among twins. Among identical twins, the chance of the second twin having schizophrenia was 30 to 50 percent, while among fraternal twins, the chance was 10 to 20 percent. If the definition of schizophrenia was broadened to include milder social and behavioral impairments, the concordance among identical twins rose to 80 percent.

Despite such tantalizing clues pointing to genetic causes, the idea that schizophrenia was a frustrated form of sexual anxiety gripped psychiatrists in the 1970s. Freud had famously attributed paranoid delusions to "unconscious homosexual impulses," apparently created by dominant mothers and weak fathers. In 1974, the psychiatrist Silvano Arieti attributed the illness to a "domineering, nagging and hostile mother who gives the child no chance to assert himself." Although the evidence from actual studies suggested nothing of the sort, Arieti's idea was so seductive—what headier mix than sexism, sexuality, and mental illness?—that it earned him scores of awards and distinctions, including the National Book Award for science.

It took the full force of human genetics to bring sanity to the study of madness. Throughout the 1980s, fleets of twin studies strengthened the case for a genetic cause of schizophrenia. In study upon study, the concordance among identical twins exceeded that of fraternal twins so strikingly that it was impossible to deny a genetic cause. Families with well-established histories of schizophrenia and bipolar disorder—such as mine—were documented across multiple generations, again demonstrating a genetic cause.

But what actual genes were involved? Since the late 1990s, a host of novel DNA-sequencing methods—called massively parallel DNA sequencing or next-generation sequencing—have allowed geneticists to sequence hundred of millions of base pairs from any human genome. Massively parallel sequencing is an enormous scale-up of the standard sequencing method: the human genome is fragmented into tens of thousands of shards, these DNA fragments are sequenced at the same time—i.e., in parallel—and the genome is "reassembled" using computers to find the overlaps between the sequences. The method can be applied to sequence the entire genome (termed *whole genome sequencing*) or to chosen parts of the genome, such as the protein-encoding exons (termed *exome sequencing*).

Massively parallel sequencing is especially effective for gene hunting when one closely related genome can be compared to another. If one member of a family has a disease, and all other members do not, then finding the gene becomes immeasurably simplified. Gene hunting devolves into a spot-the-odd-man-out game on a gigantic scale: by comparing the genetic sequences of all the closely related family members, a mutation that appears in the affected individual but not in the unaffected relatives can be found.

The sporadic variant of schizophrenia posed a perfect test case for the power of this approach. In 2013, an enormous study identified 623 young men and women with schizophrenia whose parents and siblings were unaffected. Gene sequencing was performed on these families. Since most parts of the genome are shared in any given family, only the putative culprit genes fell out as different.*

---

* Implicating a new mutation as the cause of a sporadic disease is not easy: an incidental mutation might be found by pure chance in a child and have nothing to do with the disease. Or specific environmental triggers might be required to release the disease: the so-called sporadic case may actually be a familial case that has been pushed over some tipping point by an environmental or genetic trigger.

In 617 such cases, a mutation was found in the child that was not present in either parent. On average, each child had only one mutation, although an occasional child had more. Nearly 80 percent of the mutations occurred in the chromosome derived from the father, and the father's age was a prominent risk factor, suggesting that the mutations may occur during spermiogenesis, particularly in older males. Many of these mutations, predictably, involved genes that affect the synapses between nerves, or the development of the nervous system. Although hundreds of mutations occurred in hundreds of genes across the 617 cases, occasionally the same mutant gene was found in several independent families, thereby vastly strengthening the likelihood of its links to the disorder.* By definition, these mutations are sporadic or de novo—i.e., they occurred during the conception of the child. Sporadic schizophrenia may be the consequence of alterations in neural development caused by the alterations of genes that specify the development of the nervous system. Strikingly, many of the genes found in this study have also been implicated in sporadic autism and bipolar disorder.†

①

But what about genes for *familial* schizophrenia? At first, you might imagine that finding genes for the familial variant would be easier. Schizophrenia that runs through families, like a saw blade cutting through generations, is more common to start with, and patients are easier to find and track. But counterintuitively, perhaps, identifying genes in complex familial diseases turns out to be much more difficult. Finding a gene that causes the sporadic or spontaneous variant of a disorder is like searching for a needle in haystack. You compare two genomes, trying to find small differences, and with enough data and computational power, such differ-

---

* An important class of mutations linked to schizophrenia is called copy number variation, or CNV—deletions of genes or duplications/triplications of the same gene. CNVs have also been found in cases of sporadic autism and other forms of mental illness.

† This method—of comparing the genome of a child with the sporadic or *de novo* variant of a disease versus the genome of his or her parents—was pioneered by autism researchers in the 2000s, and radically advanced the field of psychiatric genetics. The Simons Simplex Collection identified 2,800 families in which parents were not autistic, but only one child was born with an autism spectrum disorder. Comparison of the parental genome to the child's genome revealed several *de novo* mutations found in such children. Notably, several genes mutated in autism are also found to be mutated in schizophrenia, raising the possibility of deeper genetic links between the two diseases.

ences can generally be identified. But searching for multiple gene variants that cause a familial disease is like looking for a *haystack* in a haystack. Which parts of the "haystack"—i.e., which combinations of gene variants—increase the risk, and what parts are innocent bystanders? Parents and children naturally share parts of their genome, but which of those shared parts are relevant to the inherited disease? The first problem—"spot the outlier"—requires computational power. The second— "deconvolute the similarity"—demands conceptual subtlety.

Despite these hurdles, geneticists have begun systematic hunts for such genes, using combinations of genetic techniques, including linkage analysis to map the culprit genes to their physical locations on chromosomes, large association studies to identify genes that correlate with the disease, and next-generation sequencing to identify the genes and mutations. Based on the analysis of genomes, we know that there are at least 108 genes (or rather genetic regions) associated with schizophrenia—although we know the identity of only a handful of these culprits.* Notably, no single

---

* The strongest, and most intriguing, gene linked to schizophrenia is a gene associated with the immune system. The gene, called *C4*, comes in two closely related forms, called *C4A* and *C4B*, which sit, cheek by jowl, next to each other on the genome. Both forms encode proteins that may be used to recognize, eliminate, and destroy viruses, bacteria, cell debris, and dead cells—but the striking link between these genes and schizophrenia remained a tantalizing mystery.

In January 2016, a seminal study partly solved the puzzle. In the brain, nerve cells communicate with other nerve cells using specialized junctions or connections called *synapses*. These synapses are formed during the development of the brain, and their connectivity is the key to normal cognition—just as the connectivity of wires on a circuit-board is key to a computer's function.

During brain development, these synapses need to be pruned and reshaped, akin to the cutting and soldering of wires during the manufacture of a circuit-board. Astonishingly, the *C4* protein, the molecule thought to recognize and eliminate dead cells, debris, and pathogens, is "repurposed" and recruited to eliminate synapses—a process called *synaptic pruning*. In humans, synaptic pruning continues throughout childhood and into the third decade of adulthood—precisely the period of time that many symptoms of schizophrenia become manifest.

In patients with schizophrenia, variations in the *C4* genes increase the amount and activity of the *C4A* and *C4B* proteins, resulting in synapses that are "over-pruned" during development. Inhibitors of these molecules might restore the normal number of synapses in a susceptible child's or adolescent's brain.

Four decades of science—twin studies in the 1970s, linkage analysis in the 1980s, and neurobiology and cell biology in the 1990s and 2000s—converge on this discovery. For families such as mine, the discovery of *C4*'s link to schizophrenia opens remarkable prospects for the diagnosis and treatment of this illness—but also raises troubling questions about how and when such diagnostic tests or therapies may be deployed.

gene stands out as the sole driver of the risk. The contrast with breast cancer is revealing. There are certainly multiple genes implicated in hereditary breast cancer, but single genes, such as *BRCA1*, are powerful enough to drive the risk (even if we cannot predict *when* a woman with *BRCA1* will get breast cancer, she has a 70–80 percent lifetime risk of developing breast cancer). Schizophrenia generally does not seem to have such strong single drivers or predictors of disease. "There are lots of small, common genetic effects, scattered across the genome . . . ," one researcher said. "There are many different biological processes involved."

Familial schizophrenia (like normal human features such as intelligence and temperament) is thus highly *heritable* but only moderately *in*heritable. In other words, genes—hereditary determinants—are crucially important to the future development of the disorder. If you possess a particular combination of genes, the chance of developing the illness is extremely high: hence the striking concordance among identical twins. On the other hand, the inheritance of the disorder across generations is complex. Since genes are mixed and matched in every generation, the chance that you will inherit that exact permutation of variants from your father or mother is dramatically lower. In some families, perhaps, there are fewer gene variants, but with more potent effects—thereby explaining the recurrence of the disorder across generations. In other families, the genes may have weaker effects and require deeper modifiers and triggers—thereby explaining the infrequent inheritance. In yet other families, a single, highly penetrant gene is accidentally mutated in sperm or egg cells before conception, leading to the observed cases of sporadic schizophrenia.*

℗

Can we imagine a genetic test for schizophrenia? The first step would involve creating a compendium of all the genes involved—a gargantuan project for human genomics. But even such a compendium would be insufficient. Genetic studies clearly indicate that some mutations only act in concert with other mutations to cause the disease. We need to identify the combinations of genes that predict the actual risk.

---

* The distinction between "familial" and "sporadic" begins to tangle and collapse at a genetic level. Some genes mutated in familial diseases also turn out to be mutated in the sporadic disease. These genes are *most* likely to be powerful causes of the disease.

The next step is to contend with incomplete penetrance and variable expressivity. It is important to understand what "penetrance" and "expressivity" mean in these gene-sequencing studies. When you sequence the genome of a child with schizophrenia (or any genetic disease) and compare it to the genome of a normal sibling or parent, you are asking, "How are children diagnosed with schizophrenia genetically different from 'normal' children?" The question that you are not asking is the following: "If the mutated gene is present in a child, what are the chances that he or she will develop schizophrenia or bipolar disorder?"

The difference between the two questions is critical. Human genetics has become progressively adept at creating what one might describe as a "backward catalog"—a rearview mirror—of a genetic disorder: Knowing that a child has a syndrome, what are the genes that are mutated? But to estimate penetrance and expressivity, we also need to create a "forward catalog": If a child has a mutant gene, what are the chances that he or she will develop the syndrome? Is every gene fully predictive of risk? Does the same gene variant or gene combination produce highly variable phenotypes in individuals—schizophrenia in one, bipolar disorder in another, and a relatively mild variant of hypomania in a third? Do some combinations of variants require other mutations, or triggers, to push that risk over an edge?

①

There's a final twist to this puzzle of diagnosis—and to illustrate it, let me turn to a story. One night in 1946, a few months before his death, Rajesh came home from college with a riddle, a mathematical puzzle. The three younger brothers flung themselves at it, passing it back and forth like an arithmetic soccer ball. They were driven by the rivalry of siblings; the fragile pride of adolescence; the resilience of refugees; the terror of failure in an unforgiving city. I imagine the three of them—twenty-one, sixteen, thirteen—splayed on three corners of the pinched room, each spinning fantastical solutions, each attacking the problem with his distinctive strategy. My father: grim, purposeful, bullheaded, methodical, but lacking inspiration. Jagu: unconventional, oblique, out-of-the-box, but with no discipline to guide him. Rajesh: thorough, inspired, disciplined, often arrogant.

Night fell and the puzzle was still not solved. At about eleven at night, the brothers drifted off to sleep one by one. But Rajesh stayed up all night.

He paced the room, scribbling solutions and alternatives. By dawn, he had finally cracked it. The next morning, he wrote the solution on four sheets of paper and left it by the feet of one of his brothers.

This much of the story is imprinted in the myth and lore of my family. What happened next is not well-known. Years later, my father told me of the week of terror that followed that episode. Rajesh's first sleepless night turned into a second sleepless night, then a third. The all-nighter had tipped him into a burst of fulminant mania. Or perhaps it was the mania that came first and spurred the all-night marathon of problem solving and the solution. In either case, he disappeared for the next few days and could not be found. His brother Ratan was dispatched to find him, and Rajesh had to be forced back home. My grandmother, hoping to nip future breakdowns in the bud, banned puzzles and games from the house (she would remain permanently suspicious of games all her life. As children, we lived with a stifling moratorium on games at home). For Rajesh, this was a portent of the future—the first of many such breakdowns to come.

Abhed, my father had called heredity—"indivisible." There is an old trope in popular culture of the "crazy genius," a mind split between madness and brilliance, oscillating between the two states at the throw of a single switch. But Rajesh had no switch. There was no split or oscillation, no pendulum. The magic and the mania were perfectly contiguous—bordering kingdoms with no passports. They were part of the same whole, indivisible.

"We of the craft are all crazy," Lord Byron, the high priest of crazies, wrote. "Some are affected by gaiety, others by melancholy, but all are more or less touched." Versions of this story have been told, over and over, with bipolar disorder, with some variants of schizophrenia, and with rare cases of autism; all are "more or less touched." It is tempting to romanticize psychotic illness, so let me emphasize that the men and women with these mental disorders experience paralyzing cognitive, social, and psychological disturbances that send gashes of devastation through their lives. But also indubitably, some patients with these syndromes possess exceptional and unusual abilities. The effervescence of bipolar disorder has long been linked to extraordinary creativity; at times, the heightened creative impulse is manifest *during* the throes of mania.

In *Touched with Fire*, an authoritative study of the link between madness and creativity, the psychologist-writer Kay Redfield Jamison compiled a list of those "more or less touched" that reads like the Who's Who of cultural and artistic achievers: Byron (of course), van Gogh, Virginia

Woolf, Sylvia Plath, Anne Sexton, Robert Lowell, Jack Kerouac—and on and on. That list can be extended to include scientists (Isaac Newton, John Nash), musicians (Mozart, Beethoven), and an entertainer who built an entire genre out of mania before succumbing to depression and suicide (Robin Williams). Hans Asperger, one of the psychologists who first described children with autism, called them "little professors" for good reason. Withdrawn, socially awkward, or even language-impaired children, barely functional in one "normal" world, might produce the most ethereal version of Satie's *Gymnopédies* on the piano or calculate the factorial of eighteen in seven seconds.

The point is this: if you cannot separate the *phenotype* of mental illness from creative impulses, then you cannot separate the *genotype* of mental illness and creative impulse. The genes that "cause" one (bipolar disorder) will "cause" another (creative effervescence). This conundrum brings us to Victor McKusick's understanding of illness—not as absolute disability but as a relative incongruence between a genotype and an environment. A child with a high-functioning form of autism may be impaired in *this* world, but might be hyperfunctional in another—one in which, say, the performance of complex arithmetic calculations, or the sorting of objects by the subtlest gradations of color, is a requirement for survival or success.

What about that elusive genetic diagnosis for schizophrenia, then? Can we imagine a future in which we might eliminate schizophrenia from the human gene pool—by diagnosing fetuses using genetic tests and terminating such pregnancies? Not without acknowledging the aching uncertainties that remain unsolved. First, even though many variants of schizophrenia have been linked to mutations in single genes, hundreds of genes are involved—some known and some yet unknown. We do not know whether some combinations of genes are more pathogenic than others.

Second, even if we could create a comprehensive catalog of all genes involved, the vast universe of unknown factors might still alter the precise nature of the risk. We do not know what the penetrance of any individual gene is, or what modifies the risk in a particular genotype.

Finally, some of the genes identified in certain variants of schizophrenia or bipolar disorder actually *augment* certain abilities. If the most pathological variants of a mental illness can be sifted out or discriminated from the high-functioning variants by genes or gene combinations alone, then we can hope for such a test. But it is much more likely that such a test will have inherent limits: most of the genes that cause disease in one circumstance might be the very genes that cause hyperfunctional creativity in another. As Edvard Munch put it, "[My troubles] are part of me and my art. They are indistinguishable from me, and [treatment] would destroy my art. I want to keep those sufferings." These very "sufferings," we might remind ourselves, were responsible for one of the most iconic images of the twentieth century—of a man so immersed in a psychotic era that he could only scream a psychotic response to it.

The prospect of a genetic diagnosis for schizophrenia and bipolar disorder thus involves confronting fundamental questions about the nature of uncertainty, risk, and choice. We want to eliminate suffering, but we also want to "keep those sufferings." It is easy to understand Susan Sontag's formulation of illness as the "night-side of life." That conception works for many forms of illness—but not all. The difficulty lies in defining where twilight ends or where daybreak begins. It does not help that the very definition of illness in one circumstance becomes the definition of exceptional ability in another. Night on one side of the globe is often day, resplendent and glorious, on a different continent.

①

In the spring of 2013, I flew to San Diego to one of the most provocative meetings that I have ever attended. Entitled "The Future of Genomic Medicine," the meeting was at the Scripps Institute in La Jolla, at a conference center overlooking the ocean. The site was a monument to modernism—blond wood, angular concrete, mullions of steel. The light on the water was blindingly glorious. Joggers with post-human bodies ran lankily across the boardwalk. The population geneticist David Goldstein spoke about "Sequencing Undiagnosed Conditions of Childhood," an effort to extend massively parallel gene sequencing to undiagnosed childhood diseases. The physicist-turned-biologist Stephen Quake discussed the "Genomics of the Unborn," the prospect of diagnosing every mutation in a growing fetus by sampling the scraps of fetal DNA that spill naturally into maternal blood.

On the second morning of the conference, a fifteen-year-old girl—I'll call her Erika—was wheeled onstage by her mother. Erika wore a lacy, white dress and had a scarf slung across her shoulders. She had a story to tell—of genes, identity, fate, choices, and diagnosis. Erika has a genetic condition that has caused a severe, progressive degenerative disease. The symptoms began when she was one and half years old—small twitches in her muscles. By four, the tremors had progressed furiously; she could hardly keep her muscles still. She would wake up twenty or thirty times every night, drenched in sweat and racked by unstoppable tremors. Sleep seemed to worsen the symptoms, so her parents took shifts to stay awake with her, trying to console her into a few minutes of rest every night.

Clinicians suspected an unusual genetic syndrome, but all known genetic tests failed to diagnose the illness. Then in June 2011, Erika's father was listening to NPR when he heard about a pair of twins in California, Alexis and Noah Beery, who also had a long history of muscle problems. The twins had undergone gene sequencing and ultimately been diagnosed with a rare new syndrome. Based on that genetic diagnosis, the supplementation of a chemical, 5-hydroxytryptamine, or 5-HT, had dramatically reduced the twins' motor symptoms.

Erika hoped for a similar outcome. In 2012, she was the first patient to join a clinical trial that would attempt to diagnose her illness by sequencing her genome. By the summer of 2012, the sequence was back: Erika had not one but two mutations in her genome. One, in a gene called *ADCY5*, altered nerve cells' capacity to send signals to each other. The other was in a gene, *DOCK3*, that controls nerve signals that enable the coordinated movement of muscles. The combination of the two had precipitated the muscle-wasting and tremor-inducing syndrome. It was a genetic lunar eclipse—two rare syndromes superposed on each other, causing the rarest of rare illnesses.

After Erika's talk, as the audience spilled into the lobby outside the auditorium, I ran into Erika and her mother. Erika was utterly charming—modest, thoughtful, sober, mordantly funny. She seemed to carry the wisdom of a bone that has broken, repaired itself, and become stronger. She had written a book and was working on another. She maintained a blog, helped raise millions of dollars for research, and was, by far, among the most articulate, introspective teenagers that I have ever encountered. I asked her about her condition, and she spoke frankly of the anguish it had caused in her family. "Her biggest fear was that we wouldn't find anything. Not knowing would be the worst thing," her father once said.

But has "knowing" altered everything? Erika's fears have been alleviated, but there is very little that can be done about the mutant genes or their effects on her muscles. In 2012, she tried the medicine Diamox, known to alleviate muscle twitching in general, and had a brief reprieve.

There were eighteen nights of sleep—a lifetime's worth for a teenager who had hardly experienced a full night's sleep in her life—but the illness has relapsed. The tremors are back. The muscles are still wasting away. She is still in her wheelchair.

What if we could devise a prenatal test for this disease? Stephen Quake had just finished his talk on fetal genome sequencing—on "the genetics of the unborn." It will soon become feasible to scan every fetal genome for *all* potential mutations and rank many of them in order of severity and penetrance. We do not know all the details of the nature of Erika's genetic illness—perhaps, like some genetic forms of cancer, there are other, hidden "cooperative" mutations in her genome—but most geneticists suspect that she has only two mutations, both highly penetrant, causing her symptoms.

Should we consider allowing parents to fully sequence their children's genomes and potentially terminate pregnancies with such known devastating genetic mutations? We would certainly eliminate Erika's mutation from the human gene pool—but we would eliminate Erika as well. I will not minimize the enormity of Erika's suffering, or that of her family—but there is, indubitably, a deep loss in that. To fail to acknowledge the depth of Erika's anguish is to reveal a flaw in our empathy. But to refuse to acknowledge the price to be paid in this trade-off is to reveal, conversely, a flaw in our humanity.

A crowd was milling around Erika and her mother, and I walked down toward the beach, where sandwiches and drinks were being laid out. Erika's talk had sounded a resonantly sobering note through a conference otherwise tinged with optimism: you could sequence genomes hoping to find match-made medicines to alleviate specific mutations, but that would be a rare outcome. Prenatal diagnosis and the termination of pregnancies still remained the simplest choice for such rare devastating diseases—but also ethically the most difficult to confront. "The more technology evolves, the more we enter unknown territory. There's no doubt that we have to face incredibly tough choices," Eric Topol, the conference organizer, told me. "In the new genomics, there are very few free lunches."

Indeed, lunch had just ended. The bell chimed, and the geneticists returned to the auditorium to contemplate the future's future. Erika's mother wheeled her out of the conference center. I waved to her, but she did not notice me. As I entered the building, I saw her crossing the parking lot in her wheelchair, her scarf billowing in the wind behind her, like an epilogue.

①

I have chosen the three cases described here—Jane Sterling's breast cancer, Rajesh's bipolar disorder, and Erika's neuromuscular disease—because they span a broad spectrum of genetic diseases, and because they illuminate some of the most searing conundrums of genetic diagnosis. Sterling has an identifiable mutation in a single culprit gene (*BRCA1*) that leads to a common disease. The mutation has high penetrance—70 to 80 percent of carriers will eventually develop breast cancer—but the penetrance is incomplete (not 100 percent), and the precise form of the disease in the future, its timeline, and the extent of risk are unknown and perhaps unknowable. The prophylactic treatments—mastectomy, hormonal therapy—all entail physical and psychological anguish and carry risks in their own right.

Schizophrenia and bipolar disorder, in contrast, are illnesses caused by multiple genes, with much lower penetrance. No prophylactic treatments exist, and no cures. Both are chronic, relapsing diseases that shatter minds and splinter families. Yet the very genes that cause these illnesses can also, albeit in rare circumstances, potentiate a mystical form of creative urgency that is fundamentally linked to the illness itself.

And then there is Erika's neuromuscular disease—a rare genetic illness caused by one or two changes in the genome—that is highly penetrant, severely debilitating, and incurable. A medical therapy is not inconceivable, but is unlikely ever to be found. If gene sequencing of the fetal genome is coupled to the termination of pregnancies (or the selective implantation of embryos screened for these mutations), then such genetic diseases might be identifiable and could potentially be eliminated from the human gene pool. In a small number of cases, gene sequencing might identify a condition that is potentially responsive to medical therapy, or to gene therapy in the future (in the fall of 2015, a fifteen-month-old toddler with weakness, tremors, progressive blindness, and drooling—incorrectly diagnosed as having an "autoimmune disease"—was referred to a genetics clinic at Columbia University. Gene sequencing revealed a mutation in a gene linked to vitamin metabolism. Supplemented with vitamin B2, for which she was severely deficient, the girl recovered much of her neurological function).

Sterling, Rajesh, and Erika are all "previvors." Their future fates were latent in their genomes, yet the actual stories and choices of their previvorship could not be more varied. What do we do with this information? "My real résumé is in my cells," says Jerome, the young protagonist of the sci-fi film *GATTACA*. But how much of a person's genetic résumé can we read and understand? Can we decipher the kind of fate that is encoded within any genome in a usable manner? And under what circumstances can we—or should we—intervene?

⊕

Let's turn to the first question: How much of the human genome can we "read" in a usable or predictive sense? Until recently, the capacity to predict fate from the human genome was limited by two fundamental constraints. First, most genes, as Richard Dawkins describes them, are not "blueprints" but "recipes." They do not specify parts, but processes; they are formulas for forms. If you change a blueprint, the final product is changed in a perfectly predictable manner: eliminate a widget specified in the plan, and you get a machine with a missing widget. But the alteration of a recipe or formula does not change the product in a predictable manner: if you quadruple the amount of butter in a cake, the eventual effect is more complicated than just a quadruply buttered cake (try it; the whole thing collapses in an oily mess). By similar logic, you cannot examine most gene variants in isolation and decipher their influence on form and fate. That a mutation in the gene *MECP2*, whose normal function is to recognize chemical modifications to DNA, may cause a form of autism is far from self-evident (unless you understand how genes control the neurodevelopmental processes that make a brain).

The second constraint—possibly deeper in significance—is the intrinsically unpredictable nature of some genes. Most genes intersect with other triggers—environment, chance, behaviors, or even parental and prenatal exposures—to determine an organism's form and function, and its consequent effects on its future. Most of these interactions, we have already discovered, are not systematic: they happen as a result of chance, and there is no method to predict or model them with certainty. These interactions place powerful limits on genetic determinism: the eventual effects of these gene-environment intersections can *never* be reliably presaged by the genetics alone. Indeed, recent attempts to use illnesses in one twin to predict future illnesses in the other have come up with only modest successes.

But even with these uncertainties, several predictive determinants in the human genome will soon become knowable. As we investigate genes and genomes more deftly, more comprehensively, and with more computational power, we should be able to "read" the genome more thoroughly—at least in a probabilistic sense. Currently, only highly penetrant single-gene mutations (Tay-Sachs disease, cystic fibrosis, sickle-cell anemia), or alterations in entire chromosomes (Down syndrome), are used in genetic diagnosis in clinical settings. But there is no reason that the constraints on genetic diagnosis should be limited to diseases caused by mutations in single genes or chromosomes.* Nor, for that matter, is there any reason that "diagnosis" be restricted to disease. A powerful enough computer should be able to hack the understanding of a recipe: if you input an alteration, one should be able to compute its effect on the product.

By the end of this decade, permutations and combinations of genetic variants will be used to predict variations in human phenotype, illness, and destiny. Some diseases might never be amenable to such a genetic test, but perhaps the severest variants of schizophrenia or heart disease, or the most penetrant forms of familial cancer, say, will be predictable by the combined effect of a handful of mutations. And once an understanding of "process" has been built into predictive algorithms, the interactions between various gene variants could be used to compute ultimate effects on a whole host of physical and mental characteristics beyond disease alone. Computational algorithms could determine the probability of the development of heart disease or asthma or sexual orientation and assign a level of relative risk for various fates to each genome. The genome will thus be read not in absolutes, but in likelihoods—like a report card that does not contain grades but probabilities, or a résumé that does not list past experiences but future propensities. It will become a manual of previvorship.

⊕

In April 1990, as if to raise the stakes of human genetic diagnosis further, an article in *Nature* magazine announced the birth of a new technology

---

* The mutation or variation linked to the risk for a disease may not lie in the protein-coding region of a gene. The variation may lie in a regulatory region of a gene, or in a gene that does not code for proteins. Indeed, many of the genetic variations currently known to affect the risk for a particular disease or phenotype lie in regulatory, or non-coding regions of the genome.

that permitted genetic diagnosis to be performed on an embryo *before* implantation into a woman's body.

The technique relies on a peculiar idiosyncrasy of human embryology. When an embryo is produced by in vitro fertilization (IVF), it is typically grown for several days in an incubator before being implanted into a woman's womb. Bathed in a nutrient-rich broth in a moist incubator, the single-cell embryo divides to form a glistening ball of cells. At the end of three days, there are eight and then sixteen cells. Astonishingly, if you remove a few cells from that embryo, the remaining cells divide and fill in the gap of missing cells, and the embryo continues to grow normally as if nothing had happened. For a moment in our history, we are actually quite like salamanders or, rather, like salamanders' tails—capable of complete regeneration even after being cut by a fourth.

A human embryo can thus be biopsied at this early stage, the few cells extracted used for genetic tests. Once the tests have been completed, cherry-picked embryos possessing the correct genes can be implanted. With some modifications, even oocytes—a woman's eggs—can be genetically tested before fertilization. The technique is called "preimplantation genetic diagnosis," or PGD. From a moral standpoint, preimplantation genetic diagnosis achieves a seeming impossible sleight of hand. If you selectively implant the "correct" embryos and cryopreserve the others without killing them, you can select fetuses without aborting them. It is positive and negative eugenics in one go, without the concomitant death of a fetus.

Preimplantation genetic diagnosis was first used to select embryos by two English couples in the winter of 1989, one with a family history of a severe X-linked mental retardation, and another with a history of an X-linked immunological syndrome—both incurable genetic diseases only manifest in male children. The embryos were selected to be female. Female twins were born to both couples; as predicted, both sets of twins were disease-free.

The ethical vertigo induced by those two first cases was so acute that several countries moved immediately to place constraints on the technology. Perhaps understandably, among the first countries to put the most stringent limits on PGD were Germany and Austria—nations scarred by their legacies of racism, mass murder, and eugenics. In India, parts of which are home to some of the most blatantly sexist subcultures in the world, attempts to use PGD to "diagnose" the gender of a child were reported as early as 1995. Any form of sexual selection for male children was, and still is, prohibited by the Indian government, and PGD for gender selection was soon banned. Yet the government ban seems to have hardly staved the problem: readers from India and China might note, with some shame and sobriety, that the largest "negative eugenics" project in human history was not the systemic extermination of Jews in Nazi Germany or Austria in the 1930s. That ghastly distinction falls on India and China, where more than 10 million female children are missing from adulthood because of infanticide, abortion, and neglect of female children. Depraved dictators and predatory states are not an absolute requirement for eugenics. In the case of India, perfectly "free" citizens, left to their own devices, are capable of enacting grotesque eugenic programs— against females, in this case—without any state mandate.

Currently, PGD can be used to select against embryos carrying monogenic diseases, such as cystic fibrosis, Huntington's disease, and Tay-Sachs disease among many others. But in principle, nothing limits genetic diagnosis to monogenic diseases. It should not take a film such as *GATTACA* to remind us how deeply destabilizing that idea might be. We have no models or metaphors to apprehend a world in which a child's future is parsed into probabilities, or a fetus is diagnosed before birth, or becomes a "previvor" even before conception. The word *diagnosis* arises from the Greek "to know apart," but "knowing apart" has moral and philosophical consequences that lie far beyond medicine and science. Throughout our history, technologies of knowing apart have enabled us to identify, treat, and heal the sick. In their benevolent form, these technologies have allowed us to preempt illness through diagnostic tests and preventive measures, and to treat diseases appropriately (e.g., the use of the *BRCA1* gene to preemptively treat breast cancer). But they have also enabled stifling definitions of abnormalcy, partitioned the weak from the strong, or led, in their most gruesome incarnations, to the sinister excesses of eugenics. The history of human genetics has reminded us, again and again, that "knowing apart" often begins with an emphasis on "knowing," but often ends with an emphasis on "parting." It is not a coincidence that the vast anthropometric projects of Nazi scientists—the obsessive measurement of jaw sizes, head shapes, nose lengths, and heights—were also once legitimized as attempts to "know humans apart."

As the political theorist Desmond King puts it, "One way or another, we are all going to be dragged into the regime of 'gene management' that will, in essence, be eugenic. It will all be in the name of individual health rather than for the overall fitness of the population, and the managers will be you and me, and our doctors and the state. Genetic change will be managed by the invisible hand of individual choice, but the overall result will be the same: a coordinate attempt to 'improve' the genes of the next generation on the way."

℗

Until recently, three unspoken principles have guided the arena of genetic diagnosis and intervention. First, diagnostic tests have largely been restricted to gene variants that are singularly powerful determinants of illness—i.e., highly penetrant mutations, where the likelihood of developing the disease is close to 100 percent (Down syndrome, cystic fibrosis, Tay-Sachs disease). Second, the diseases caused by these mutations have generally involved extraordinary suffering or fundamental incompatibilities with "normal" life. Third, justifiable interventions—the decision to abort a child with Down syndrome, say, or intervene surgically on a woman with a *BRCA1* mutation—have been defined through social and medical consensus, and all interventions have been governed by complete freedom of choice.

The three sides of the triangle can be envisioned as moral lines that most cultures have been unwilling to transgress. The abortion of an embryo carrying a gene with, say, only a ten percent chance of developing cancer in the future violates the injunction against intervening on low-penetrance mutations. Similarly, a state-mandated medical procedure on a genetically ill person without the subject's consent (or parental consent in the case of a fetus) crosses the boundaries of freedom and noncoercion.

Yet it can hardly escape our attention that these parameters are inherently susceptible to the logic of self-reinforcement. *We* determine the definition of "extraordinary suffering." *We* demarcate the boundaries of "normalcy" versus "abnormalcy." *We* make the medical choices to intervene. *We* determine the nature of "justifiable interventions." Humans endowed with certain genomes are responsible for defining the criteria to define, intervene on, or even eliminate other humans endowed with other genomes. "Choice," in short, seems like an illusion devised by genes to propagate the selection of similar genes.

①

Even so, this triangle of limits—high-penetrance genes, extraordinary suffering, and noncoerced, justifiable interventions—has proved to be a useful guideline for acceptable forms of genetic interventions. But these boundaries are being breached. Take, for instance, a series of startlingly provocative studies that used a single gene variant to drive social-engineering choices. In the late 1990s, a gene called *5HTTLPR*, which encodes a molecule that modulates signaling between certain neurons in the brain, was found to be associated with the response to psychic stress. The gene comes in two forms or alleles—a long variant and a short variant. The short variant, called *5HTTLPR/short*, is carried by about 40 percent of the population and seems to produce significantly lower levels of the protein. The short variant has been repeatedly associated with anxious behavior, depression, trauma, alcoholism, and high-risk behaviors. The link is not strong, but it is broad: the short allele has been associated with increased suicidal risk among German alcoholics, increased depression in American college students, and a higher rate of PTSD among deployed soldiers.

In 2010, a team of researchers launched a research study, called the Strong African American Families project, or SAAF, in an impoverished rural belt in Georgia. It is a startlingly bleak place overrun by delinquency, alcoholism, violence, mental illness, and drug use. Abandoned clapboard houses with broken windows dot the landscape; crime abounds; vacant parking lots are strewn with hypodermic needles. Half the adults lack a high school education, and nearly half the families have single mothers.

Six hundred African-American families with early-adolescent children were recruited for the study. The families were randomly assigned to two groups. In one group, the children and their parents received seven weeks of intensive education, counseling, emotional support, and structured social interventions focused on preventing alcoholism, binge behaviors, violence, impulsiveness, and drug use. In the control group, the families received minimal interventions. Children in the intervention group and in the control group had the *5HTTLPR* gene sequenced.

The first result of this randomized trial was predictable from prior studies: in the control group, children with the short variant—i.e., the "high risk" form of the gene—were twice as likely to veer toward high-risk behaviors, including binge drinking, drug use, and sexual promiscuity as adolescents, confirming earlier studies that had suggested an increased risk within this genetic subgroup. The second result was more provocative: these very children were also the *most likely* to respond to the social interventions. In the intervention group, children with the high-risk allele were most strongly and rapidly "normalized"—i.e., the most drastically affected subjects were also the best responders. In a parallel study, orphaned infants with the short variant of *5HTTLRP* appeared more impulsive and socially disturbed than their long-variant counterparts at baseline—but were also the most likely to benefit from placement in a more nurturing foster-care environment.

In both cases, it seems, the short variant encodes a hyperactive "stress sensor" for psychic susceptibility, but also a sensor most likely to respond to an intervention that targets the susceptibility. The most brittle or fragile forms of psyche are the most likely to be distorted by trauma-inducing environments—but are also the most likely to be restored by targeted interventions. It is as if *resilience* itself has a genetic core: some humans are born resilient (but are less responsive to interventions), while others are born sensitive (but more likely to respond to changes in their environments).

The idea of a "resilience gene" has entranced social engineers. Writing in the *New York Times* in 2014, the behavioral psychologist Jay Belsky argued, "Should we seek to identify the most susceptible children and disproportionately target them when it comes to investing scarce intervention and service dollars? I believe the answer is yes." "Some children are—in one frequently used metaphor—like delicate orchids," Belsky wrote, "they quickly wither if exposed to stress and deprivation, but blossom if given a lot of care and support. Others are more like dandelions; they prove resilient to the negative effects of adversity, but at the same time do not particularly benefit from positive experiences." By identifying these "delicate orchid" versus "dandelion" children by gene profiling, Belsky proposes, societies might achieve vastly more efficient targeting with scarce resources. "One might even imagine a day when we could genotype all the children in an elementary school to ensure that those who could most benefit from help got the best teachers."

Genotyping all children in elementary school? Foster-care choices driven by genetic profiling? Dandelions and orchids? Evidently, the conversation around genes and predilections has already slipped past the original boundaries—from high-penetrance genes, extraordinary suffering, and justifiable interventions—to genotype-driven social engineering. What if genotyping identifies a child with a future risk for unipolar depression or bipolar disorder? What about gene profiling for violence, criminality, or impulsivity? What constitutes "extraordinary suffering," and which interventions are "justifiable"?

And what is normal? Are parents allowed to choose "normalcy" for their children? What if—obeying some sort of Heisenbergian principle of psychology—the very act of intervention reinforces the identity of abnormalcy?

Ⅾ

This book began as an intimate history—but it is the intimate future that concerns me. A child born to a parent with schizophrenia, we now know, has between a 13 to 30 percent chance of developing the disease by age sixty. If both parents are affected, the risk climbs to about 50 percent. With one uncle affected, a child runs a risk that is three- to fivefold higher than the general population. With two uncles and a cousin affected—Jagu, Rajesh, Moni—that number jumps to about tenfold the general risk. If my father, my sister, or my paternal cousins were to develop the disease (the symptoms can emerge later in life), the risk would again leap severalfold. It is a matter of waiting and watching, of spinning and respinning the teetotum of fate, of assessing and reassessing my genetic risk.

In the wake of the monumental studies on the genetics of familial schizophrenia, I have often wondered about sequencing my genome, and the genomes of selected members of my family. The technology exists: my own lab, as it turns out, is equipped to extract, sequence, and interpret genomes (I routinely use this technology to sequence the genes of my cancer patients). What is missing, still, is the identity of most of the gene variants, or combinations of variants, that increase the risk. But there is little doubt that many of these variants will be identified, and the nature of risk conferred by them quantified, by the end of the decade. For families such as mine, the prospect of genetic diagnosis will no longer remain an abstraction, but will transform into clinical and personal realities. The triangle of considerations—penetrance, extraordinary suffering, and justifiable choice—will be carved into our individual futures.

If the history of the last century taught us the dangers of empowering governments to determine genetic "fitness" (i.e., which person fits within the triangle, and who lives outside it), then the question that confronts our current era is what happens when this power devolves to the individual. It is a question that requires us to balance the desires of the individual—to carve out a life of happiness and achievement, without undue suffering— with the desires of a society that, in the short term, may be interested only in driving down the burden of disease and the expense of disability. And operating silently in the background is a third set of actors: our genes themselves, which reproduce and create new variants oblivious of our desires and compulsions—but, either directly or indirectly, acutely or obliquely, influence our desires and compulsions. Speaking at the Sorbonne in 1975, the cultural historian Michel Foucault once proposed that "a technology of abnormal individuals appears precisely when a regular network of knowledge and power has been established." Foucault was thinking about a "regular network" of humans. But it could just as easily be a network of genes.

Genetic Therapies: Post-Human

What do I fear? Myself? There's none else by.
—William Shakespeare,
Richard III, act 5, scene 3

There is in biology at the moment a sense of barely con-
tained expectations reminiscent of the physical sciences at
the beginning of the 20th century. It is a feeling of advancing
into the unknown and [a recognition] that where this
advance will lead is both exciting and mysterious. . . . The
analogy between 20th-century physics and 21st-century
biology will continue, for both good and ill.
—"Biology's Big Bang," 2007

In the summer of 1991, not long after the Human Genome Project had been launched, a journalist visited James Watson at the Cold Spring Harbor lab in New York. It was a sultry afternoon, and Watson was in his office, sitting by a window overlooking the gleaming bay. The interviewer asked Watson about the future of the Genome Project. What would happen once all the genes in our genome had been sequenced and scientists could manipulate human genetic information at will?

Watson chuckled and raised his eyebrows. "He ran a hand down his sparse strands of white hair . . . and a puckish gleam came into his eye. . . . 'A lot of people say they're worried about changing our genetic instructions. But those [genetic instructions] are just a product of evolution designed to adapt us for certain conditions that may not exist today. We all know how imperfect we are. Why not make ourselves a little better suited to survival?'"

"That's what we will do," he said. He looked at his interviewer and laughed suddenly, emitting that distinctive, high-pitched chortle that had become familiar to the scientific world as a prelude to a storm. "That's what we will do. We'll make ourselves a little better."

Watson's comment returns us to the second concern raised by the students at the Erice meeting: What if we learn to intentionally alter the human genome? Until the late 1980s, the only mechanism to reshape the human genome—to "make ourselves a little better" in a genetic sense—was to identify highly penetrant and seriously deleterious genetic mutations (such as those that cause Tay-Sachs disease or cystic fibrosis) *in utero* and terminate the pregnancy. In the 1990s, preimplantation genetic diagnosis (PGD) allowed parents to preemptively select and implant embryos without such mutations, substituting the moral dilemma of the termination of a life with the moral dilemma of choice. Still, human geneticists operated within the aforementioned triangle of boundaries: highly penetrant genetic lesions, extraordinary suffering, and justifiable, noncoerced interventions.

The advent of gene therapy in the late 1990s changed the terms of this discussion: genes could now be changed intentionally in human bodies. This was the rebirth of "positive eugenics." Rather than eliminating humans carrying deleterious genes, scientists could envision correcting defective human genes, thereby making the genome a "bit better."

Conceptually, gene therapy comes in two distinct flavors. The first involves modifying the genome of a *nonreproductive* cell—say a blood, brain, or muscle cell. The genetic modification of these cells affects their function, but it does not alter the human genome for more than one generation. If a genetic change is introduced into a muscle or blood cell, the change is not transmitted into a human embryo; the altered gene is lost when the cell dies. Ashi DeSilva, Jesse Gelsinger, and Cynthia Cutshall are all examples of humans treated with non-germ-line gene therapy: in all three cases, blood cells—but not germ-line cells (i.e., sperm and egg)—were altered by the introduction of foreign genes.

The second, more radical, form of gene therapy is to modify a human genome so that the change affects *reproductive* cells. Once a genomic change has been introduced into a sperm or egg—i.e., into the germ line of a human being—the change becomes self-propagating. The change is incorporated permanently into the human genome and transmitted from one generation to the next. The inserted gene becomes inextricably linked to the human genome.

Germ-line gene therapy in humans was not conceivable in the late 1990s: no reliable technique existed to transmit genetic changes into a human sperm or egg cell. But even *non*-germ-line therapy trials had been halted. Jesse Gelsinger's "biotech death," as the *New York Times Magazine* described it, had sent such tremors of anguish through the field that virtually all gene-therapy trials in the United States were frozen. Companies went bankrupt. Scientists left the field. The trial scorched the earth of all forms of gene therapies, leaving a permanent scar on the field.

But gene therapy has returned—step by cautious step. The seemingly stagnant decade between 1990 and 2000 was a decade of introspection and reconsideration. First, the litany of errors in the Gelsinger trial had to be meticulously dissected. Why had the introduction of a supposedly harmless virus carrying a gene into the liver caused such a devastating, fatal reaction? As physicians, scientists, and regulators sifted through the trial, the reasons for the failed experiment became evident. The vectors used to infect Gelsinger's cells had never been properly vetted in humans. But most important, Gelsinger's immune response to the virus should have been anticipated. Gelsinger had likely been naturally exposed to the strain of adenovirus that had been used in the gene-therapy experiment. His brisk immune response was not an aberration; it was the perfectly habitual response of a body fighting a pathogen that it had previously encountered, possibly during infection by a cold. In choosing a common human virus as their vehicle for gene delivery, gene therapists had made a crucial error of judgment: they had neglected to consider that genes were being delivered into a human body with a history, with scars, memories, and prior exposures. "How could such a beautiful thing go so, so wrong?" Paul Gelsinger had asked. We now know how: because—seeking only beauty—scientists were unprepared for catastrophe. The doctors pushing the frontiers of human medicine had forgotten to account for the common cold.

①

In the two decades that followed Gelsinger's death, the tools used in the original gene-therapy trials have largely been replaced by second- and third-generation technologies. New viruses are now used to deliver genes into human cells, and novel methods to monitor gene delivery have been developed. Many of these viruses have been purposefully selected because they are easy to manipulate in the lab and do not elicit the immune response that spiraled so devastatingly out of control in Gelsinger's body.

In 2014, a landmark study published in the *New England Journal of Medicine* announced the successful use of gene therapy to treat hemophilia. Hemophilia, the terrifying bleeding disease caused by a mutation in a blood-clotting factor, threads through the history of the gene in a continuous strand; it is the DNA in the story of DNA. It was the illness that had affected the Czarevitch Alexei from his birth in 1904 and thus inserted itself into the epicenter of political life in Russia in the early twentieth century. It was one of the first X-linked diseases to be identified in humans, thereby pointing to the physical presence of a gene on a chromosome. It was one of the first diseases to be definitively ascribed to a single gene. It was also one of the first genetic diseases for which an artificially engineered protein was created, by Genentech in 1984.

The idea of using gene therapy for hemophilia had first been broached in the mid-1980s. Since hemophilia is caused by the lack of a functional clotting protein, it was conceivable to use a virus to deliver the gene into cells so the body could produce the missing protein and thus restore the clotting of blood. In the early 2000s, after a nearly two-decade delay, gene therapists decided to try gene therapy for hemophilia again. Hemophilia comes in two major variants, classified by the particular clotting factor that is missing in blood. The variant of hemophilia chosen for the gene-therapy test was hemophilia B, in which the gene for clotting factor IX is mutated and fails to produce a normal protein.

The protocol for this test was simple: ten men with a severe variant of the disease were injected with a single dose of a virus carrying a gene for factor IX. The presence of the virus-encoded protein was monitored in the blood for several months. Notably, this trial tested not just safety, but efficacy: the ten virus-injected patients were monitored for bleeding episodes, and for their use of additional factor IX by injection. Although the injection of the virus-borne gene increased the factor IX concentration to just 5 percent of the normal value, the effect on bleeding episodes was startling. The patients experienced a *90 percent* reduction in bleeding incidents, and an equally dramatic reduction in their use of injected factor IX. The effect persisted over three years.

The potent therapeutic effect of a mere 5 percent replacement of a missing protein is a beacon for the aspirations of gene therapists. It reminds us of the power of degeneracy in human biology: if only 5 percent of a clotting factor is sufficient to restore virtually all clotting function in human blood, then 95 percent of the protein must be superfluous—a buffer, or reservoir, possibly maintained in the human body as a backup in the event of a truly catastrophic bleed. If the same principle holds for other genetic diseases caused by single genes—for cystic fibrosis, say—then gene therapy might be vastly more tractable than had previously been imagined. Even the inefficient delivery of a therapeutic gene to a small subset of cells might be sufficient to treat an otherwise fatal disease.

①

But what about that perennial fantasy of human genetics, the alteration of genes in reproductive cells to create permanently amended human genomes—"germ-line gene therapy"? What about the creation of the "post-humans" or "trans-humans"—i.e., human embryos with permanently modified genomes? By the early 1990s, the challenge of permanent human genome engineering had been reduced to three scientific hurdles. Each of these had once seemed like an impossible scientific challenge, but each is on the verge of being solved. The most remarkable fact about human genomic engineering today is not how far out of reach it is, but how perilously, tantalizingly near.

The first challenge was the establishment of a reliable human embryonic stem cell. ES cells are stem cells derived from the inner pith of early embryos. They live in transit between cells and organisms: they can be grown and manipulated like a cell line in the lab, but are also capable of forming all the tissue layers of a living embryo. The alteration of the genome of an ES cell is thus a convenient stepping-stone to the permanent alteration of an organism's genome: if the genome of an ES cell can be changed intentionally, then that genetic change can potentially be introduced into an embryo, into all organs formed within the embryo, and thus into an organism. The genetic modification of ES cells is the rather narrow pass through which every fantasy of germ-line genomic engineering has to travel.

In the late 1990s, James Thomson, an embryologist in Wisconsin, began to experiment with human embryos to derive stem cells from them. Although mouse ES cells had been known since the late 1970s, dozens of attempts to find human analogues had failed. Thomson traced these failures to two factors: bad seed and bad soil. The starting material for the establishment of human stem cells was often of poor quality, and the conditions for their growth were suboptimal. As a graduate student in the 1980s, Thomson had studied mouse ES cells intensively. Like a hothouse gardener capable of coaxing exotic plants to live and propagate outside their natural environments, Thomson had gradually learned the many eccentricities of ES cells. They were temperamental, volatile, and fussy. He knew of their propensity to fold up and die at the slightest provocation. He learned about their requirement for "nurse" cells to coddle them, their peculiar insistence on clumping together, and the translucent, refractive, hypnotic glow of the cells that transfixed him each time he saw them under a microscope.

In 1991, having moved to the Wisconsin Regional Primate Center, Thomson began to derive ES cells from monkeys. He plucked a six-day-old embryo from a pregnant rhesus monkey, then let the embryo grow in a petri dish. Six days later, he peeled away the outer layer of the embryo, as if skinning a cellular fruit, and extracted single cells from the pith of the inner cell mass. As with mouse cells, he learned to culture these cells in nests of nurse cells that could supply crucial growth factors; without these nurse cells, the ES cells died. In 1996, convinced that he could try his technique on humans, he asked the regulatory boards at the University of Wisconsin to allow him to create human ES cells.

But mouse and monkey embryos had been easy to find. Where might a scientist find freshly fertilized *human* embryos? Thomson stumbled on an obvious resource: IVF clinics. By the late 1990s, in vitro fertilization had become a common treatment for various forms of human infertility. To perform IVF, eggs are harvested from a woman after ovulation. A typical harvest yields multiple eggs—sometimes up to ten or twelve—and these eggs are fertilized by a man's sperm in a petri dish. The embryos are then grown briefly in an incubator before being implanted back into the uterus.

But not all IVF embryos are implanted. Implanting more than three embryos is unusual and unsafe, and the spare embryos are typically discarded (or rarely, implanted into other women's bodies, who carry such embryos as "surrogate" mothers). In 1996, having obtained permission from the University of Wisconsin, Thomson obtained thirty-six embryos from IVF clinics. Fourteen of them grew into glistening cellular spheres in the incubator. Using the technique that he had perfected on monkeys— the peeling away of the outer layers, the gentle coaxing of cell growth on "feeders" and nurse cells—Thomson isolated a few human embryonic stem cells. Implanted into mice, these cells were capable of generating all three layers of the human embryo—the primordial sources of all tissues, such as skin, bones, muscles, nerves, intestines, and blood.

The stem cells that Thomson had derived from IVF-discarded embryos recapitulated many features of human embryogenesis, but they still had a major limitation: although they were capable of making virtually all human tissues, they would not efficiently generate some tissues, such as sperm and egg cells. A genetic change introduced into these ES cells could thus be transmitted into all cells of the embryo—except to the most important ones: the cells capable of transmitting the gene to the next generation. In 1998, soon after Thomson's paper had been published in *Science*, groups of scientists around the world, including researchers from the United States, China, Japan, India, and Israel, began to derive dozens of embryonic stem cell lines from fetal embryonic tissues in the hopes of discovering a human ES cell capable of germ-line transmission of genes.

But then, with little warning, the field was frozen shut. In 2001, three years after Thomson's paper, President George W. Bush sharply restricted all federal ES cell research to the seventy-four cell lines that had already been created. No new lines could be derived, even from IVF-discarded embryonic tissues. Laboratories working on ES cells faced stringent oversight and cuts in funding. In 2006 and 2007, Bush repeatedly vetoed federal funding for the establishment of new cell lines. Advocates of stem cell research, including patients with degenerative diseases and neurological impairments, thronged the streets of Washington, threatening to sue federal agencies responsible for the ban. Bush countered these requests by holding press conferences flanked by children produced by the implantation of "discarded" IVF embryos that had been brought to life via surrogate mothers.

①

The ban on federal funding for new ES cells froze the ambitions of human genomic engineers, at least temporarily. But it could not stop the advance of the second step required to create permanent heritable changes in the human genome: a reliable, efficient method to introduce intentional changes in the genomes of ES cells that were already in existence.

At first, this too seemed like an insurmountable technological challenge. Virtually every technique to alter the human genome was crude and inefficient. Scientists could expose stem cells to radiation to mutate genes—but these mutations were sprinkled randomly throughout the genome, defying any attempt to directionally influence the mutation. Viruses carrying known genetic changes could insert their genes into the genome, but the site of insertion was typically random, and the inserted gene was often silenced. In the 1980s, another method to introduce a directional change in the genome—flooding cells with pieces of foreign DNA carrying a mutated gene—was invented. The foreign DNA was inserted directly into a cell's genetic material, or its message was copied into the genome. But although the process worked, it was notoriously inefficient and prone to errors. Reliable, efficient, *intentional* change—the deliberate alteration of specific genes in a specified manner—seemed impossible.

℈

In the spring of 2011, a researcher named Jennifer Doudna was approached by a bacteriologist, Emmanuelle Charpentier, about a conundrum that seemed, at first, to have little relevance to human genes or genomic engineering. Charpentier and Doudna were both attending a microbiology conference in Puerto Rico. As they walked through the alleyways of Old San Juan, past the fuchsia-and-ochre houses with arched doorways and painted façades, Charpentier told Doudna of her interest in bacterial immune systems—the mechanisms by which bacteria defend themselves against viruses. The war between viruses and bacteria has gone on for so long, and with such ferocity, that like ancient, conjoined enemies, each has become defined by the other: their mutual animosity has been imprinted in their genes. Viruses have evolved genetic mechanisms to invade and kill bacteria. And bacteria have counter-evolved genes to fight back. "A viral infection [is a] ticking time bomb," Doudna knew. "A bacterium has only a few minutes to defuse the bomb—before it gets destroyed itself."

In the mid-2000s, a pair of French scientists named Philippe Horvath and Rodolphe Barrangou stumbled on one such mechanism of bacterial self-defense. Horvath and Barrangou, both employees of the Danish food company Danisco, were working on cheese-producing and yogurt-making bacteria. Some of these bacterial species, they found, had evolved a system to deliver coordinated slashes in the genomes of invading viruses to paralyze them. The system—a molecular switchblade of sorts—recognized serial-offender viruses by their DNA sequence. The cuts were not delivered at random places, but at specific sites in viral DNA.

The bacterial defense system was soon found to involve at least two critical components. The first piece was the "seeker"—an RNA encoded in the bacterial genome that matched and recognized the DNA of the viruses. The principle for the recognition, yet again, was binding: the RNA "seeker" was able to find and recognize the DNA of an invading virus because it was a mirror image of that DNA—the yin to its yang. It was like carrying a permanent image of your enemy in your pocket—or, in the bacteria's case, an inverted photograph, etched indelibly into its genome.

The second element of the defense system was the "hitman." Once the viral DNA had been recognized and matched as foreign (by its reverse-image), a bacterial protein named Cas9 was deployed to deliver the lethal gash to the viral gene. The "seeker" and the "hitman" worked in concert: the Cas9 protein delivered its cuts to the genome only after the sequence had been matched by the recognition element. It was a classic combination of collaborators—spotter and executor, drone and rocket, Bonnie and Clyde.

Doudna, who had been immersed in the biology of RNA for most of her adult life, was intrigued by the system. At first, she thought of it as a curiosity—"the most obscure thing I ever worked on," as she later put it. But working with Charpentier, she began to meticulously break it down to its constituent components.

In 2012, Doudna and Charpentier realized that the system was "programmable." Bacteria, of course, only carry the images of viral genes so that they can seek and destroy viruses; they have no reason to recognize or cut other genomes. But Doudna and Charpentier learned enough about the self-defense system to trick it: by substituting a decoy recognition element, they could force the system to make intentional cuts in other genes and genomes. Switch the "seeker," they found, and a different gene might be sought and cut.

①

There is a phrase buried in that penultimate sentence that should send a restless ping of fantasy through the mind of any human geneticist. An "intentional cut" in a gene is the potential source of a mutation. Most mutations occur randomly in the genome; you cannot command an X-ray beam or direct a cosmic ray to selectively change only the cystic fibrosis gene or the Tay-Sachs gene. But in Doudna and Charpentier's case, the

471

mutation was not delivered randomly: the cut could be *programmed* to occur exactly at the site recognized by the self-defense system. By changing the recognition element, Doudna and Charpentier could redirect it to attack a selected gene, thereby mutating the gene at will.*

The system could be manipulated even further. When a gene is cut open, two ends of DNA are revealed, like severed string, and the ends are trimmed back. The cutting and trimming is meant to repair the broken gene—and the gene then tries to recover the lost information by seeking an intact copy. Matter has to conserve energy; the genome is designed to conserve information. Typically, a cut-open gene tries to recover the lost information from the other copy of the gene in the cell. But if a cell is flooded with foreign DNA, then the gene witlessly copies the information from this decoy DNA, rather than from its backup copy. The information written into the decoy DNA fragment is thus copied permanently into the genome—akin to erasing a word in a sentence and then forcibly writing a substitute in its place. A defined, predetermined genetic change can thus be written into a genome: the sequence ATGGGCCCG in a gene can be altered to ACCGCCGGG (or any desired sequence). A mutant cystic fibrosis gene can be corrected to the wild-type version; a gene to confer viral resistance can be introduced into an organism; the mutant *BRCA1* gene can be reverted to wild type; the mutated Huntington's gene, with its mirthless, singsong repeat, might be disrupted and deleted. The technique has been termed *genome editing*, or *genomic surgery*.

Doudna and Charpentier published their data on the microbial defense system, called CRISPR/Cas9, in *Science* magazine in 2012. The paper immediately ignited the imagination of biologists. In the three years since the publication of that landmark study, the use of this technique has exploded. The method still has some fundamental constraints: at times, the cuts are delivered to the wrong genes. Occasionally, the repair is not efficient, making it difficult to "rewrite" information into particular sites in the genome. But it works more easily, more powerfully, and more efficiently than virtually any other genome-altering method to date. Only a handful of such instances of scientific serendipity have occurred in the history of biology. An arcane microbial defense, devised

---

* Another system to deliver "programmable" cuts in specific genes using a DNA-cutting enzyme is also being developed. Termed a "TALEN," this enzyme can also be used for genome editing.

by microbes, discovered by yogurt engineers, and reprogrammed by RNA biologists, has created a trapdoor to the transformative technology that geneticists had sought so longingly for decades: *a method to achieve directed, efficient, and sequence-specific modification of the human genome.* Richard Mulligan, the pioneer of gene therapy, had once fantasized about "clean, chaste gene therapy." This system makes clean, chaste gene therapy feasible.

Ⓘ

One final step is necessary to achieve the intentional permanent modification of the genome in human organisms. Genetic changes that have been created in human ES cells have to become incorporated into human embryos. The direct transformation of a human ES cell into a viable human embryo is inconceivable, for both technical and ethical reasons. Even though human ES cells can generate all types of human tissues in a laboratory setting, it is impossible to envision implanting a human ES cell directly into a woman's womb in the hopes that the cell will autonomously organize itself into a viable human embryo. When human ES cells *have* been transplanted into animals, the most that the cells can achieve is a loose organization of the vital tissue layers of the human embryo—far from the anatomical and physiological coordination achieved by a fertilized egg during human embryogenesis.

One potential alternative is to attempt the genetic modification of an embryo *in toto* after it has achieved its basic anatomical form—i.e., a few days or weeks after conception. This strategy is also intractable: once organized, the human embryo becomes fundamentally obdurate to gene modification. Technical hurdles aside, the ethical qualms about such an experiment would vastly outweigh any other considerations: attempting genome modification in a living human embryo obviously raises a gamut of questions that reverberate far beyond biology and genetics. In most nations, such an experiment lies beyond the conceivable boundaries of permissibility.

But there's a third strategy that may be the most approachable. Suppose a genetic change is introduced into human ES cells using standard gene-modification technologies. And now imagine that the gene-modified ES cells can be converted into *reproductive* cells—sperm and eggs. If ES cells are truly pluripotent stem cells, then they should be able to give rise to human sperm and eggs (a real human embryo, after all, generates its own germ cells—sperm or egg).

Now consider a thought experiment: if a human embryo can be created by IVF using such gene-modified sperm or eggs, then the resultant embryo will necessarily carry these genetic changes in all its cells—including *its* sperm and egg cells. The preliminary steps of this process can be tested without changing or manipulating an actual human embryo—and can thus safely skirt the moral boundaries of human embryo manipulation.* Most critically, the process mimics the well-established protocols of IVF: a sperm and an egg are fertilized in vitro, and an early embryo is implanted into a woman's body—a procedure that raises few qualms. It is a shortcut to germ-line gene therapy, a back door to transhumanism: the introduction of a gene into the human germ *line* is facilitated by the conversion of ES cells into germ *cells*.

℗

This final challenge was largely on its way to being solved exactly as scientists were perfecting the systems to alter genomes. In the winter of 2014, a team of embryologists in Cambridge, England, and at the Weizmann Institute in Israel developed a system to make primordial germ cells—the precursors of sperm and eggs—out of human embryonic stem cells. Earlier experiments, using earlier versions of human ES cells, had failed to create such germ cells. In 2013, Israeli researchers modified these earlier studies to isolate new batches of ES cells that might be more capable of forming germ cells. A year later, collaborating with scientists at Cambridge, the team found that if they cultured these human ES cells under specific conditions, and shepherded their differentiation using specific coaxing agents, the cells would form clusters of sperm and egg precursors.

The technique is still cumbersome and inefficient. Obviously, due to stringent restrictions in the creation of artificial human embryos, whether these spermlike and egglike cells can give rise to human embryos capable of normal development is yet unknown. But the basic derivation of cells capable of transmitting heredity has been achieved. In principle,

---

* One important technical detail is that since individual ES cells can be cloned and expanded, cells with unintended mutations can be identified and discarded. Only *pre*screened ES cells, carrying the intended mutation, are transformed into sperm or egg.

if the parent ES cells can be modified using any genetic technique—including gene editing, genetic surgery, or the insertion of a gene using a virus—any genetic change can permanently and heritably be etched in the human genome.

⊕

It is one thing to manipulate genes. It is quite another thing to manipulate genomes. In the 1980s and 1990s, DNA-sequencing and gene-cloning technology allowed scientists to understand and manipulate genes and thereby control the biology of cells with extraordinary dexterity. But the manipulation of genomes in their native context, particularly in embryonic cells or germ cells, opens the door to a vastly more powerful technology. What is at stake is no longer a cell, but an organism—ourselves.

In the spring of 1939, Albert Einstein, mulling over recent advances in nuclear physics in his study at Princeton University, realized that every step required to achieve the creation of an unfathomably powerful weapon had been individually completed. The isolation of uranium, nuclear fission, the chain reaction, the buffering of the reaction, and its controlled release in a chamber had all fallen into place. All that was required was sequence: if you strung these reactions together in order, you obtained an atomic bomb. In 1972, at Stanford, Paul Berg stared at bands of DNA on a gel and found himself at a similar juncture. The cutting and pasting of genes, the creation of chimeras, and the introduction of these gene chimeras into bacterial and mammalian cells allowed scientists to engineer genetic hybrids between humans and viruses. All that was needed was the threading of these reactions into a sequence.

We are at a similar moment—a quickening—for human genome engineering. Consider the following steps in sequence: (a) the derivation of a true human embryonic stem cell (capable of forming sperm and eggs); (b) a method to create reliable, intentional genetic modifications in that cell line; (c) the directed conversion of that gene-modified stem cell into human sperm and eggs; (d) the production of human embryos from these modified sperm and eggs by IVF . . . and you arrive, rather effortlessly, at genetically modified humans.

There is no sleight of hand here; each of the steps lies within the reach of current technology. Of course, much remains unexplored: Can every gene be efficiently altered? What are the collateral effects of such alterations? Will the sperm and egg cells formed from ES cells truly generate functional human embryos? Many, many minor technical hurdles remain. But the pivotal pieces of the jigsaw puzzle have fallen into place.

Predictably, each of these steps is currently barricaded by strict regulations and bans. In 2009, after a prolonged ban on federally funded research on ES cells, the Obama administration lifted the injunction on the derivation of new ES cells in the United States. But even with the new regulations, the NIH categorically prohibits two kinds of research on human ES cells. First, scientists are not permitted to introduce these cells into humans or animals to enable their development into live embryos. And second, genome modifications on ES cells cannot be performed in circumstances that "might be transmitted into the germline"—i.e., into sperm or egg cells.

℗

In the spring of 2015, as I completed this book, a group of scientists, including Jennifer Doudna and David Baltimore, issued a joint statement seeking a moratorium on the use of gene-editing and gene-altering technologies in the clinical setting, and particularly in human ES cells. "The possibility of human germline engineering has long been a source of excitement and unease among the general public, especially in light of concerns about initiating a 'slippery slope' from disease-curing applications toward uses with less compelling or even troubling implications," the moratorium reads. "A key point of discussion is whether the treatment or cure of severe diseases in humans would be a responsible use of genome engineering, and if so, under what circumstances. For example, would it be appropriate to use the technology to change a disease-causing genetic mutation to a sequence more typical among healthy people? Even this seemingly straightforward scenario raises serious concerns . . . because there are limits to our knowledge of human genetics, gene-environment interactions, and the pathways of disease."

Many scientists find the call for a moratorium understandable, even necessary. "Gene editing," the stem cell biologist George Daley noted, "raises the most fundamental of issues about how we are going to view our humanity in the future and whether we are going to take the dramatic step of modifying our own germ line and in a sense take control of our genetic destiny, which raises enormous peril for humanity."

In many ways, the proposed scheme of restrictions is reminiscent of the Asilomar moratorium. It seeks to limit the use of technology until the ethical, political, social, and legal implications of the technology can be ascertained. It calls for a public appraisal of the science and its future. It is also a frank acknowledgment of how tantalizingly close we are to making embryos with permanently altered human genomes. "It is very clear that people will try to do gene editing in humans," Rudolf Jaenisch, the MIT biologist who created the first mouse embryos from ES cells, said. "We need some principled agreement that we want to enhance humans in this way or we don't."

The word to watch in that last sentence is *enhance*, for it signals a radical departure from the conventional limits of genomic engineering. Prior to the invention of genome-editing technologies, techniques such as embryo selection allowed us to cull information away from the human genome: by selecting embryos via preimplantation genetic diagnosis (PGD), the Huntington's disease mutation, or the cystic fibrosis mutation, could be eliminated from a particular family's lineage.

CRISPR/Cas9-based genomic engineering, in contrast, allows us to *add* information to the genome: a gene can be changed in an intentional manner, and new genetic code can be written into the human genome. "This reality means that germline manipulation would largely be justified by attempts to 'improve ourselves,'" Francis Collins wrote to me. "That means that someone is empowered to decide what an 'improvement' is. Anyone contemplating such action should be aware of their hubris."

The crux, then, is not genetic emancipation (freedom from the bounds of hereditary illnesses), but genetic enhancement (freedom from the current boundaries of form and fate encoded by the human genome). The distinction between the two is the fragile pivot on which the future of genome editing whirls. If one man's illness is another man's normalcy, as this history teaches us, then one person's understanding of enhancement may be another's conception of emancipation ("why not make ourselves a little better?" as Watson asks).

But can humans responsibly "enhance" our own genomes? What are the consequences of augmenting the natural information encoded by our genes? Can we make our genomes a "little better" without risking the possibility of making ourselves substantially worse?

⓪

In the spring of 2015, a laboratory in China announced that it had casually crossed the barricade. At the Sun Yat-sen University in Guangzhou, a team led by Junjiu Huang obtained eighty-six human embryos from an IVF clinic and tried to use the CRISPR/Cas9 system to correct a gene responsible for a common blood disorder (only embryos that were nonviable in the long term were chosen). Seventy-one embryos survived. Of the fifty-four embryos tested, only four were found to have the corrected gene inserted. More portentously, the system was found to have inaccuracies: in one-third of all the embryos tested, unintentional mutations in other genes were also introduced, including mutations in genes essential for normal development and survival. The experiment was stopped.

It was a daring, if slapdash, experiment, meant to provoke a response—and it did. Around the world, scientists reacted to the attempted modification of a human embryo with extreme anguish and concern. The highest-ranking scientific journals, including *Nature*, *Cell*, and *Science*, refused to publish the results, citing broad violations of safety and ethical concerns (the results were eventually published in a scarcely read online journal, *Protein + Cell*). Yet, even as they read the study with apprehension and horror, biologists already knew that this was just the first step past the breach point. The Chinese researchers had taken the shortest route to permanent human genome engineering, and predictably, the embryos had been littered with unforeseen mutations. But the technique could be modified with multiple variations to make it potentially more efficient and accurate. If embryonic stem cells, and stem-cell-derived sperm and eggs, had been used, for instance, these cells could have been screened up front to cull away any deleterious mutations, and the efficiency of gene targeting might have been greatly increased.

Junjiu Huang told a journalist that he was "planning to decrease the number of off-target mutations [using] different strategies—tweaking the enzymes to guide them more precisely to the desired spot, introducing the enzymes in a different format that could help to regulate their lifespans and thus allow them to be shut down before mutations accumulate." In a few months, he hoped to attempt another variation of the experiment—this time, he expected, with much higher efficiency and fidelity. He was not exaggerating: the technology to modify the genome of a human embryo may be complex, inefficient, and inaccurate—but it does not lie out of scientific reach.

While scientists in the West continue to watch Junjiu Huang's experiments on human embryos with justified apprehension, Chinese scientists are far more sanguine about such experiments. "I don't think China wants to take a moratorium," one scientist reported in the *New York Times* in late June 2015. A Chinese bioethicist clarified, "Confucian thinking says someone becomes a person after they are born. That is different from the United States and other countries with a Christian influence, where because of religion they may feel research on embryos is not okay. Our 'red line' here is that you can only experiment on embryos that are younger than fourteen days old."

Another scientist wrote of the Chinese approach, "Do first, think later." Several public commentators seemed to agree with this strategy; in the comments section of the *New York Times*, readers advocated lifting the bans on human genomic engineering and urged a ramp-up in experimentation in the West, in part to remain competitive with the efforts in Asia. The Chinese experiments had evidently raised the stakes throughout the world. As one writer put it, "If we don't do this work, China will." The drive to change the genome of a human embryo has turned into an intercontinental arms race.

As of this writing, four other groups in China are reportedly working on introducing permanent mutations in human embryos. By the time this book is published, I would not be surprised if the first successful targeted genome modification of a human embryo had been achieved in a laboratory. The first "post-genomic" human might be on his or her way to being born.

Ⓞ

We need a manifesto—or at least a hitchhiker's guide—for a post-genomic world. The historian Tony Judt once told me that Albert Camus's novel *The Plague* was about the plague in the same sense that *King Lear* is about a king named Lear. In *The Plague*, a biological cataclysm becomes the testing ground for our fallibilities, desires, and ambitions. You cannot read *The Plague* except as a thinly disguised allegory of human nature. The genome is also a testing ground for our fallibilities and desires, although reading it does not require understanding allegories or metaphors. What we read and write into our genome *is* our fallibilities, desires, and ambitions. It *is* human nature.

The task of writing that complete manifesto belongs to another generation, but perhaps we can scribe its opening salvos by recalling the scientific, philosophical, and moral lessons of this history:

1. *A gene is the basic unit of hereditary information.* It carries the information needed to build, maintain, and repair organisms. Genes collaborate with other genes, with inputs from the environment, with triggers, and with random chance to produce the ultimate form and function of an organism.

2. *The genetic code is universal.* A gene from a blue whale can be inserted into a microscopic bacterium and it will be deciphered accurately and with nearly perfect fidelity. A corollary: there is nothing particularly special about human genes.

3. *Genes influence form, function, and fate, but these influences typically do not occur in a one-to-one manner.* Most human attributes are the consequence of more than one gene; many are the result of collaborations between genes, environments, and chance. Most of these interactions are nonsystematic—i.e., they occur through the intersection between a genome and fundamentally unpredictable events. And some genes tend to influence only propensities and tendencies. We can thus reliably predict the ultimate effect of a mutation or variation on an organism for only a minor subset of genes.

4. *Variations in genes contribute to variations in features, forms, and behaviors.* When we use the colloquial terms *gene for blue eyes* or *gene for height*, we are really referring to a variation (or allele) that specifies an eye color or height. These variations constitute an extremely minor portion of the genome. They are magnified in our imagination because of cultural, and possibly biological, tendencies that tend to amplify differences. A six-foot man from Denmark and a four-foot man from Demba share the same anatomy, physiology, and biochemistry. Even the two most extreme human variants—male and female—share 99.688 percent of their genes.

5. *When we claim to find "genes for" certain human features or functions, it is by virtue of defining that feature narrowly.* It makes

sense to define "genes for" blood type or "genes for" height since these biological attributes have intrinsically narrow definitions. But it is an old sin of biology to confuse the definition of a feature with the feature itself. If we define "beauty" as having blue eyes (and only blue eyes), then we will, indeed, find a "gene for beauty." If we define "intelligence" as the performance on only one kind of problem in only one kind of test, then we will, indeed, find a "gene for intelligence." The genome is only a mirror for the breadth or narrowness of human imagination. It is Narcissus, reflected.

6. *It is nonsense to speak about "nature" or "nurture" in absolutes or abstracts.* Whether nature—i.e., the gene—or nurture—i.e., the environment—dominates in the development of a feature or function depends, acutely, on the individual feature and the context. The *SRY* gene determines sexual anatomy and physiology in a strikingly autonomous manner; it is all nature. Gender identity, sexual preference, and the choice of sexual roles are determined by intersections of genes and environments—i.e., nature plus nurture. The manner in which "masculinity" versus "femininity" is enacted or perceived in a society, in contrast, is largely determined by an environment, social memory, history, and culture; this is all nurture.

7. *Every generation of humans will produce variants and mutants; it is an inextricable part of our biology.* A mutation is only "abnormal" in a statistical sense: it is the less common variant. The desire to homogenize and "normalize" humans must be counterbalanced against biological imperatives to maintain diversity and abnormalcy. Normalcy is the antithesis of evolution.

8. *Many human diseases—including several illnesses previously thought to be related to diet, exposure, environment, and chance— are powerfully influenced or caused by genes.* Most of these diseases are polygenic—i.e., influenced by multiple genes. These illnesses are heritable—i.e., caused by the intersection of a particular permutation of genes—but not easily *in*heritable—i.e., likely to be transmitted intact to the next generation, since the

permutations of genes are remixed in each generation. Instances of each single-gene—monogenic—disease are rare, but, in sum, they turn out to be surprisingly common. More than ten thousand such diseases have been defined thus far. Between one in a hundred and one in two hundred children will be born with a monogenic disease.

9. *Every genetic "illness" is a mismatch between an organism's genome and its environment.* In some cases, the appropriate medical intervention to mitigate a disease might be to alter the environment to make it "fit" an organismal form (building alternative architectural realms for those with dwarfism; imagining alternative educational landscapes for children with autism). In other cases, conversely, it might mean changing genes to fit environments. In yet other cases, the match may be impossible to achieve: the severest forms of genetic illnesses, such as those caused by nonfunction of essential genes, are incompatible with all environments. It is a peculiar modern fallacy to imagine that the definitive solution to illness is to change nature—i.e., genes—when the environment is often more malleable.

10. *In exceptional cases, the genetic incompatibility may be so deep that only extraordinary measures, such as genetic selection, or directed genetic interventions, are justified.* Until we understand the many unintended consequences of selecting genes and modifying genomes, it is safer to categorize such cases as exceptions rather than rules.

11. *There is nothing about genes or genomes that makes them inherently resistant to chemical and biological manipulation.* The standard notion that "most human features are the result of complex gene-environment interactions and most are the result of multiple genes" is absolutely true. But while these complexities constrain the ability to manipulate genes, they leave plenty of opportunity for potent forms of gene modification. Master regulators that affect dozens of genes are common in human biology. An epigenetic modifier may be designed to change the state

of hundreds of genes with a single switch. The genome is replete with such nodes of intervention.

12. *A triangle of considerations—extraordinary suffering, highly penetrant genotypes, and justifiable interventions—has, thus far, constrained our attempts to intervene on humans.* As we loosen the boundaries of this triangle (by changing the standards for "extraordinary suffering" or "justifiable interventions"), we need new biological, cultural, and social precepts to determine which genetic interventions may be permitted or constrained, and the circumstances in which these interventions become safe or permissible.

13. *History repeats itself, in part because the genome repeats itself. And the genome repeats itself, in part because history does.* The impulses, ambitions, fantasies, and desires that drive human history are, at least in part, encoded in the human genome. And human history has, in turn, selected genomes that carry these impulses, ambitions, fantasies, and desires. This self-fulfilling circle of logic is responsible for some of the most magnificent and evocative qualities in our species, but also some of the most reprehensible. It is far too much to ask ourselves to escape the orbit of this logic, but recognizing its inherent circularity, and being skeptical of its overreach, might protect the weak from the will of the strong, and the "mutant" from being annihilated by the "normal."

Perhaps even that skepticism exists somewhere in our twenty-one thousand genes. Perhaps the compassion that such skepticism enables is also encoded indelibly in the human genome.

Perhaps it is part of what makes us human.

Epilogue: *Bheda, Abheda*

Sura-na Bheda Pramaana Sunaavo;
Bheda, Abheda, Pratham kara Jaano.

Show me that you can divide the notes of a song;
But first, show me that you can discern
Between what can be divided
And what cannot.

> —An anonymous musical composition
> inspired by a classical Sanskrit poem

Abhed, my father had called genes—"indivisible." *Bhed*, the opposite, is its own kaleidoscope of a word: "to discriminate" (in its verb form), "to excise, to determine, to discern, to divide, to cure." It shares linguistic roots with *vidya*, "knowledge," and with *ved*, "medicine." The Hindu scriptures, the Vedas, acquired their name from the same root. It arises from the ancient Indo-European word *uied*, "to know" or "to discern meaning."

Scientists divide. We discriminate. It is the inevitable occupational hazard of our profession that we must break the world into its constituent parts—genes, atoms, bytes—before making it whole again. We know of no other mechanism to understand the world: to create the sum of the parts, we must begin by dividing it into the parts of the sum.

But there is a hazard implicit in this method. Once we perceive organisms—humans—as assemblages built from genes, environments, and gene-environment interactions, our view of humans is fundamentally changed. "No sane biologist believes that we are entirely the product of their genes," Berg told me, "but once you bring genes into the picture, then our perception of ourselves can no longer be the same." A whole assembled from the sum of the parts is different from the whole before it was broken into the parts.

As the poem in Sanskrit goes:

> *Show me that you can divide the notes of a song;*
> *But first, show me that you can discern*
> *Between what can be divided*
> *And what cannot.*

①

Three enormous projects lie ahead for human genetics. All three concern discrimination, division, and eventual reconstruction. The first is to discern the exact nature of information coded in the human genome. The Human Genome Project provided the starting point for this inquiry, but it raised a series of intriguing questions about what, precisely, is "encoded" by the 3 billion nucleotides of human DNA. What are the functional elements in the genome? There are protein-coding genes, of course—about twenty-one to twenty-four thousand in all—but also regulatory sequences of genes, and stretches of DNA (introns) that split genes into modules. There is information to build tens of thousands of RNA molecules that do not get translated into proteins but still seem to perform diverse roles in cellular physiology. There are long highways of "junk" DNA that are unlikely to be junk after all and may encode hundreds of yet-unknown functions. There are kinks and folds that allow one part of the chromosome to associate with another part in three-dimensional space.

To understand the role of each of these elements, a vast international project, launched in 2013, hopes to create a compendium of every functional element in the human genome—i.e., any part of any sequence in any chromosome that has a coding or instructional function. Ingeniously termed the Encyclopedia of DNA Elements (ENC-O-DE), this project will cross-annotate the sequence of the human genome against all the information contained within it.

Once these functional "elements" have been identified, biologists can move to the second challenge: understanding how the elements can be combined in time and space to enable human embryology and physiology, the specification of anatomical parts, and the development of an organism's unique features and characteristics.* One humbling fact about our understanding of the human genome is how little we know of the *human* genome: much of our knowledge of our genes and their functions is inferred from similar-looking genes in yeast, worms, flies, and mice. As David Botstein writes: "Very few human genes have been studied directly." Part of the task of the new genomics is to close the gap between mice and men—to determine how human genes function in the context of the human organism.

For medical genetics, this project promises several particularly important payoffs. The functional annotation of the human genome will enable biologists to discover novel mechanisms of illness. New genomic elements will be linked to complex medical diseases, and these links will allow us to determine the ultimate causes of diseases. We still do not know, for instance, how the intersection between genetic information, behavioral exposures, and random chance causes hypertension, schizophrenia, depression, obesity, cancer, or heart disease. Finding the correct functional elements in the genome that are linked to these diseases is the first step to solving the mechanisms by which they arise.

Understanding these links will also reveal the predictive power of the human genome. In an influential review published in 2011, the psychologist Eric Turkheimer wrote, "A century of familial studies of twins, siblings, parents and children, adoptees, and whole pedigrees has established, beyond a shadow of a doubt, that genes play a crucial role in the explanation of *all* human differences, from the medical to the normal, the biological to the behavioral." Yet, despite the strengths of these links, the "genetic world," as Turkheimer calls it, has proved much more difficult to map and deconvolute than expected. Until recently, the only genetic changes that were powerfully predictive of future illnesses were those with high penetrance that caused the severest of phenotypes. Combinations of gene variants were particularly difficult to decipher. It was impossible to determine how a particular permutation of genes (i.e., a genotype) would determine a particular outcome in the future (i.e., phenotype), especially if that outcome was governed by a multitude of genes.

---

* To understand how genes become actualized into organisms, it is necessary to understand not just genes, but also RNA, proteins, and epigenetic marks. Future studies will need to reveal how the genome, all the variants of proteins (the proteome), and all the epigenetic marks (the epigenome) are coordinated to build and maintain humans.

But this barrier might soon collapse. Imagine a thought experiment that might seem far-fetched at first glance. Suppose we could comprehensively sequence the genomes of one hundred thousand children *prospectively*—i.e., before anything is known about the future of any child—and create a database of all the variations and combinations of the functional elements of each child's genome (one hundred thousand is an arbitrary number; the experiment can be extended to any number of children). Imagine, now, creating a "fate map" of this cohort of children: every illness or physiological aberrancy identified and recorded on a parallel database. We might describe this map as a human "phenome"—the full set of all phenotypes (attributes, features, behaviors) of an individual. And now imagine a computational engine that mines the data from these gene map/ fate map pairs to determine how one might predict the other. Despite remnant uncertainties—even deep ones—the prospective mapping of one hundred thousand human genomes to one hundred thousand human phenomes would provide an extraordinary data set. It would begin to describe the nature of fate that is encoded by a genome.

The extraordinary feature of this fate map is that it need not be restricted to illness; it can be as wide and deep and detailed as we would like it to be. It could include the low birth weight of a child, a learning disability in preschool, the transient tumult of a particular adolescence, a teenage infatuation, an impulsive marriage, coming out, infertility, a midlife crisis, a propensity for addiction, a cataract in the left eye, premature baldness, depression, a heart attack, an early death from ovarian cancer or breast cancer. Such an experiment would have been inconceivable in the past. But the combined power of computing technology, data storage, and gene sequencing has made it conceivable in the future. It is a gargantuan twin study—except without twins: millions of virtual genetic "twins" are created computationally by matching genomes across space and time and these permutations are then annotated against life events.

It's important to recognize the inherent limitations of such projects or, more generally, of trying to predict diseases and destinies from genomes. "Perhaps," as one observer complained, "the fate of genetic explanations will [culminate in] de-contextualising etiological processes, underrepresenting the role of environments, producing some stunning medical interventions, [but] revealing little about the fate of populations." But the power of such studies is precisely that they "decontextualize" illness; *genes* provide the context to understand development and fate. Situations that are context dependent or environment dependent get diluted and filtered out—and only those powerfully affected by genes remain. With enough subjects, and enough computational power, nearly all of the predictive capacity of the genome can, in principle, be determined and computed.

①

The final project is perhaps the most far-reaching. As much as the ability to predict human phenomes from human genomes was limited by the lack of computational technologies, the ability to intentionally *change* human genomes was bounded by the paucity of biological technologies. Gene delivery methods such as viruses were inefficient and unreliable at best, and lethal at worst—and intentional gene delivery into the human embryo was virtually impossible.

These barriers too have started to collapse. Novel "gene editing" technologies now allow geneticists to make remarkably precise alterations in the human genome with equally remarkable specificity. In principle, a single letter of DNA can be mutated to another letter in a directional manner, leaving the 3 billion other bases of the genome largely untouched (this technology might be likened to a copyediting device that scans sixty-six volumes of the *Encyclopaedia Britannica* and finds, erases, and changes one word, leaving all other words untouched). Between 2010 and 2014, a postdoctoral researcher in my laboratory tried to introduce a defined genetic change into a cell line using the standard gene-delivery viruses, but with little success. In 2015, having switched to the new CRISPR-based technology, she engineered fourteen alterations of genes in fourteen human genomes, including the genomes of human embryonic stem cells, in six months—a feat unimaginable in the past. Geneticists and gene therapists around the world are currently exploring the possibility of changing the human genome with renewed verve and urgency—in part, because the current technologies have brought us to a precipice. A combination of stem cell technologies, nuclear transfer and epigenetic modulation, and gene-editing methods has made it conceivable that the human genome can be broadly manipulated, and that transgenic humans can be created.

We have no knowledge of the fidelity or efficiency of these techniques in practice. Does making an intentional change in a gene run the risk of

489

creating an unintended change in another part of the genome? Are some genes more easily "edited" than others—and what governs the pliability of a gene? Nor do we know whether making a directed change in one gene might cause the entire genome to become dysregulated. If some genes are indeed "recipes," as in Dawkins's formulation, then altering one gene may cause far-reaching consequences for gene regulation—potentially unleashing a volley of downstream consequences, akin to the proverbial butterfly effect. If such butterfly-effect genes are common in the genome, then they will represent fundamental limitations for gene-editing technologies. The discontinuity of genes—the discreteness and autonomy of each individual unit of heredity—will turn out to be an illusion: genes may yet be more interconnected than we think.

> *But first, show me that you can discern*
> *between what can be divided*
> *and what cannot.*

℗

Imagine, then, a world in which these technologies can be routinely deployed. When a child is conceived, every parent is given the choice of testing the fetus using comprehensive genome sequencing *in utero*. Mutations that cause the most severe impairments are identified, and parents are given the option of aborting such fetuses at the earliest stages of pregnancy, or selectively implanting only the "normal" fetuses after comprehensive genetic screening (we might call this comprehensive preimplantation genetic diagnosis, or c-PGD).*

More complex combinations of genes that might cause *tendencies* toward disease are also identified by genome sequencing. When children with such predicted tendencies are born, they are offered selective

---

* Comprehensive testing of fetal genomes has already entered clinical practice, under the name of NIPT, or Non-Invasive Prenatal Testing. In 2014, a Chinese company reported that it had tested 150,000 fetuses for chromosomal disorders, and was extending the test to capture single-gene mutations. Although these tests seem to detect chromosomal abnormalities, such as Down syndrome, with just as much fidelity as amniocentesis, a major issue of the test is "false positives"—i.e., the fetal DNA is thought to carry a chromosomal abnormality, but it is actually normal. These false positive rates will decrease dramatically as technologies advance.

interventions throughout childhood. A child with a predilection toward a genetic form of obesity, for instance, might be monitored for changes in body mass, treated with an alternative diet, or metabolically "reprogrammed" using hormones, drugs, or genetic therapies in childhood. A child with a tendency for an attention-deficit or hyperactivity syndrome might undergo behavioral therapy or be placed in an enriched classroom.

If and when the illnesses do emerge or advance, gene-based therapies are deployed to treat or cure them. Corrected genes are delivered directly into the affected tissues: the functioning cystic fibrosis gene, for instance, is aerosolized and injected into the lungs of patients, where it partially restores the normal function of the lung. A girl born with ADA deficiency is transplanted with bone marrow stem cells carrying the correct gene. For more complex genetic diseases, genetic diagnostics are combined with genetic therapies, with drugs and with "environment therapies." Cancers are comprehensively analyzed by documenting the mutations responsible for driving malignant growth of one particular cancer. These mutations are used to identify culprit pathways that fuel the growth of cells and to devise exquisitely targeted therapies to kill malignant cells and spare normal cells.

"Imagine you are a soldier returning from war with PTSD," the psychiatrist Richard Friedman wrote in the *New York Times* in 2015. "With a simple blood test looking at gene variants, we could discover whether you were biologically adept at fear extinction. . . . If you had a mutation that reduced your ability to extinguish fear, your therapist would know you might just need more exposure—more treatment sessions—to recover. Or, perhaps a different therapy altogether that doesn't rely on exposure, like interpersonal therapy, or medication." Perhaps drugs that can erase epigenetic marks are prescribed in combination with talk therapy. Perhaps the erasure of cellular memories can ease the erasure of historical memories.

Genetic diagnoses and genetic interventions are also used to screen and correct mutations in human embryos. When "intervenable" mutations in certain genes are identified in the germ line, parents are given the choice of genetic surgery to alter their sperm and eggs before conception, or prenatal screening of embryos to avoid implanting mutant embryos in the first place. Genes that cause the most detrimental variants of illness are thus excised from the human genome up front by positive or negative selection, or by genome modification.

①

If you read that scenario carefully, it inspires both wonder and a certain moral queasiness. The individual interventions may not push the boundaries of transgression—indeed some, such as the targeted treatment of cancer, schizophrenia, and cystic fibrosis, represent landmark goals for medicine—but aspects of this world seem distinctly and even repulsively alien. It is a world inhabited by "previvors" and "post-humans": men and women who have been screened for genetic vulnerabilities or created with altered genetic propensities. Illness might progressively vanish, but so might identity. Grief might be diminished, but so might tenderness. Traumas might be erased but so might history. Mutants would be eliminated but so would human variation. Infirmities might disappear, but so might vulnerability. Chance would become mitigated, but so, inevitably, would choice.*

In 1990, writing about the Human Genome Project, the worm geneticist John Sulston wondered about the philosophical quandary raised by an intelligent organism that has "learned to read its own instructions." But an infinitely deeper quandary is raised when an intelligent organism learns to write its own instructions. If genes determine the nature and fate of an organism, and if organisms now begin to determine the nature and fate of their genes, then a circle of logic closes on itself. Once we start thinking of genes as destiny, manifest, then it is inevitable to begin imagining the human genome as manifest destiny.

⌀

On the way back from Moni's institution in Calcutta, my father wanted to stop again outside the house where he had grown up—the place where they brought Rajesh back in the throes of his mania, thrashing like a wild bird. We drove in silence. His memories had formed the walls of a room

* Even seemingly simple scenarios of genetic screening force us to enter arenas of unnerving moral hazard. Take Friedman's example of using a blood test to screen soldiers for genes that predispose to PTSD. At first glance, such a strategy would seem to mitigate the trauma of war: soldiers incapable of "fear extinction" might be screened and treated with intensive psychiatric therapies or medical therapies to return them to normalcy. But what if, extending the logic, we screen soldiers for PTSD risk *before* deployment? Would that really be desirable? Do we truly want to select soldiers incapable of registering trauma, or genetically "augmented" with the capacity to extinguish the psychic anguish of violence? Such a form of screening would seem to me to be precisely undesirable: a mind incapable of "fear extinction" is exactly the dangerous sort of mind to be avoided in war.

492

around him. We left the car at the narrow inlet of Hayat Khan Lane, and walked into the cul-de-sac on foot. It was about six in the evening. The houses were lit by an oblique, smoky light, and the air threatened rain.

"Bengalis have only one event in their history: Partition," my father said. He was looking up at the balconies jutting above us and trying to recall the names of his former neighbors: Ghosh, Talukdar, Mukherjee, Chatterjee, Sen. A gentle drizzle began to descend on us—or perhaps it was just the drippings from the laundry, hanging thickly on clotheslines tacked across the houses. "Partition was the defining event for every man and woman in this city," he said. "Either you lost your own home, or your home became a shelter for someone else." He pointed to the colonnades of windows above our heads. "Every family here had another family living within it." There were households within households, rooms inside rooms, microcosms lodged inside microcosms.

"When we arrived here from Barisal, with our four steel trunks and our few salvaged possessions, we thought we were beginning a new life. We had experienced a catastrophe, but it was also a fresh start." Every house on that street, I knew, had its own story of steel trunks and salvaged possessions. It was as if all the inhabitants had been equalized, like a garden cut back to the roots in the winter.

For a cohort of men, including my father, the journey from East to West Bengal involved a radical resetting of all clocks. Thus began Year Zero. Time was splintered into two halves: the era before the cataclysm, and the era after. BP and AP. This vivisection of history—the partition of Partition—produced a strangely dissonant experience: the men and women of my father's generation perceived themselves as unwitting participants in a natural experiment. Once the clocks had been reset to zero, it was as if you could watch the lives, fates, and choices of human beings played out from some starting gate, or from the beginning of time. My father had experienced this experiment all too acutely. One brother had devolved into mania and depression. Another's sense of reality had been shattered. My grandmother had acquired her lifelong suspicion of all forms of change. My father had acquired his taste for adventure. It seemed as if distinct futures—like homunculi—had been folded into each person, waiting to be unfurled.

What force, or mechanism, might explain such widely divergent fates and choices of individual human beings? In the eighteenth century, an individual's destiny was commonly described as a series of events ordained by God. Hindus had long believed that a person's fate was derived, with near-arithmetic precision, by some calculus of the good and evil acts that he had performed in a previous life. (God, in this scheme, was a glorified moral tax-accountant, tallying and divvying out portions of good and bad fate based on past investments and losses.) The Christian God, capable of inexplicable compassion and equally inexplicable wrath, was a more mercurial bookkeeper—but He too was the ultimate, if more inscrutable, arbiter of destiny.

Nineteenth- and twentieth-century medicine offered more secular conceptions of fate and choice. Illness—perhaps the most concrete and universal of all acts of fate—could now be described in mechanistic terms, not as the arbitrary visitation of divine vengeance, but as the consequence of risks, exposures, predispositions, conditions, and behaviors. Choice was understood as an expression of an individual's psychology, experiences, memories, traumas, and personal history. By the mid-twentieth century, identity, affinity, temperament, and preference (straightness versus gayness or impulsivity versus caution) were increasingly described as phenomena caused by the intersections of psychological impulses, personal histories, and random chance. An *epidemiology* of destiny and choice was born.

In the early decades of the twenty-first century, we are learning to speak yet another language of cause and effect, and constructing a new epidemiology of self: we are beginning to describe illness, identity, affinity, temperament, preferences—and, ultimately, fate and choice—in terms of genes and genomes. This is not to make the absurd claim that genes are the only lenses through which fundamental aspects of our nature and destiny can be viewed. But it is to propose and to give serious consideration to one of the most provocative ideas about our history and future: that the influence of genes on our lives and beings is richer, deeper, and more unnerving than we had imagined. This idea becomes even more provocative and destabilizing as we learn to interpret, alter, and manipulate the genome intentionally, thereby acquiring the ability to alter future fates and choices. "[Nature] may, after all, be entirely approachable," Thomas Morgan wrote in 1919. "Her much-advertised inscrutability has once more been found to be an illusion." We are now trying to extend Morgan's conclusions—not just to nature, but to human nature.

I have often thought about the possible trajectories of Jagu's and Rajesh's lives if they had been born in the future, say fifty or a hundred years from now. Would our knowledge of their heritable vulnerabilities be used to find cures for the illnesses that had devastated their lives? Would that knowledge be used to "normalize" them—and if so, what moral, social, and biological hazards would that entail? Would such forms of knowledge enable new kinds of empathy and understanding? Or would they nucleate novel forms of discrimination? Would the knowledge be used to redefine what is "natural"?

But what *is* "natural"? I wonder. On one hand: variation, mutation, change, inconstancy, divisibility, flux. And on the other: constancy, permanence, indivisibility, fidelity. *Bhed. Abhed.* It should hardly surprise us that DNA, the molecule of contradictions, encodes an organism of contradictions. We seek constancy in heredity—and find its opposite: variation. Mutants are necessary to maintain the essence of our selves. Our genome has negotiated a fragile balance between counterpoised forces, pairing strand with opposing strand, mixing past and future, pitting memory against desire. It is the most human of all things that we possess. Its stewardship may be the ultimate test of knowledge and discernment for our species.

Acknowledgments

When I completed the final draft of the six-hundred-page *Emperor of All Maladies* in May 2010, I never thought I would lift a pen to write another book. The physical exhaustion of writing *Emperor* was easy to fathom and overcome, but the exhaustion of imagination was unexpected. When the book won the Guardian *First Book Prize* that year, one reviewer complained that it should have been nominated for the *Only Book Prize*. The critique cut to the bone of my fears. *Emperor* had sapped all my stories, confiscated my passports, and placed a lien on my future as a writer; I had nothing more to tell.

But there *was* another story: of normalcy before it tips into malignancy. If cancer, to twist the description of the monster from *Beowulf*, is the "distorted version of our normal selves," then what generates the undistorted variants of our normal selves? *Gene* is that story—of the search for normalcy, identity, variation, and heredity. It is a prequel to *Emperor's* sequel.

There are innumerable people to thank. Books about family and heredity are not so much written as lived. Sarah Sze, my wife, my most passionate interlocutor and reader, and my daughters, Leela and Aria, were daily reminders of my stake in genetics and the future. My father, Sibeswar, and mother, Chandana, are an inextricable part of this story. My sister, Ranu, and her husband, Sanjay, provided moral interceptions when needed. Judy and Chia-Ming Sze, and David Sze and Kathleen Donohue, sustained discussions about the family and the future.

Extraordinarily generous readers ensured the factual accuracy of this book and provided comments on content, including Paul Berg (genetics and cloning), David Botstein (gene mapping), Eric Lander and Robert Waterston (Human Genome Project), Robert Horvitz and David Hirsh (worm biology), Tom Maniatis (molecular biology), Sean

Carroll (evolution and gene regulation), Harold Varmus (cancer), Nancy Segal (twin studies), Inder Verma (gene therapy), Nancy Wexler (human gene mapping), Marcus Feldman (human evolution), Gerald Fishbach (schizophrenia and autism), David Allis and Timothy Bestor (epigenetics), Francis Collins (gene mapping and the Human Genome Project), Eric Topol (human genetics), and Hugh Jackman (Wolverine; mutants).

Ashok Rai, Nell Breyer, Bill Helman, Gaurav Majumdar, Suman Shirodkar, Meru Gokhale, Chiki Sarkar, David Blistein, Azra Raza, Chetna Chopra, and Sujoy Bhattacharyya read early manuscripts and supplied immensely valuable comments. Conversations with Lisa Yuskavage, Matvey Levenstein, Rachel Feinstein, and John Currin were indispensable. A passage from this book appeared in an essay on Yuskavage's work ("Twins"), and another passage in my essay *The Laws of Medicine, 2015*. Brittany Rush patiently (and brilliantly) compiled all eight hundred–odd references and worked on mind-numbing aspects of production; Daniel Loedel read and edited the manuscript over a weekend to prove that it could be achieved. Mia Crowley-Hald and Anna-Sophia Watts provided invaluable copy editing, and Kate Lloyd was a supremely capable publicist.

Nan Graham: Did you read all sixty-eight drafts? You did—and with Stuart Williams and indomitable Sarah Chalfant, who first saw this book through the pinhole of a two-paragraph proposal, you gave *Gene* shape, form, clarity, gravity, and urgency. Thank you.

Glossary

Allele: A variant or alternative form of a gene. Alleles are usually created by mutations, and can be responsible for phenotypic variations. A gene can have multiple alleles.

Central dogma, or Central theory: The theory that biological information in most organisms moves from genes in DNA to messenger RNA to proteins. This theory has been modified several times. Retroviruses contain enzymes that can be used to build DNA from an RNA template.

Chromatin: The material from which chromosomes are composed. Chromatin takes its name from *chroma* ("color"), since it was initially found by staining cells with dyes. Chromatin may consist of DNA, RNA, and proteins.

Chromosome: A structure within a cell comprised of DNA and proteins that stores genetic information.

DNA: Deoxyribonucleic acid, a chemical that carries genetic information in all cellular organisms. It is usually present in the cell as two paired, complementary strands. Each strand is a chemical chain made up of four chemical units—abbreviated A, C, T, and G. Genes are carried in the form of a genetic "code" in the strand and the sequence is converted (transcribed) into RNA and then translated into proteins.

Enzyme: A protein that accelerates a biochemical reaction.

Epigenetics: The study of phenotypic variations that are not caused by changes in the primary DNA sequence (i.e., A, C, T, G) but by chemical alterations of DNA (e.g., methylation) or changes in packaging of DNA via DNA-binding proteins (e.g., histones). Some of these alterations are heritable.

Gene: A unit of inheritance, normally comprised of a stretch of DNA that codes for a protein or for an RNA chain (in special cases, genes might be carried in RNA form).

Genome: The full complement of all genetic information within the organism. A genome includes protein-encoding genes, genes that do not encode proteins, the regulatory regions of genes, and sequences of DNA with yet-unknown functions.

GLOSSARY

Genotype: An organism's collection of genetic information that determines its physical, chemical, biological, and intellectual characteristics (see "phenotype").

Mutation: An alteration in the chemical structure of DNA. Mutations can be silent—i.e., the change might not affect any function of the organism—or can result in a change in the function or structure of an organism.

Nucleus: A membrane-enclosed cellular structure or organelle that is found in animal and plant cells, but not in bacterial cells. Chromosomes (and genes) in animal cells are contained in the nucleus. In animal cells, most genes are nuclear, although some genes are also found in mitochondria.

Organelle: A specialized subunit within a cell that is typically dedicated to a specific function. Individual organelles are usually separately enclosed within their own membranes. Mitochondria are organelles dedicated to the production of energy.

Penetrance: The proportion of organisms that carry a particular variant of a gene that also expresses the associated trait or phenotype. In medical genetics, penetrance refers to the proportion of individuals carrying a genotype that manifest the symptoms of an illness.

Phenotype: The set of an individual's biological, physical, and intellectual traits, such as skin color or eye color. Phenotypes can also include complex traits, such as temperament or personality. Phenotypes are determined by genes, epigenetic alterations, environments, and random chance.

Protein: A chemical comprised, at its core, of a chain of amino acids that is created when a gene is translated. Proteins carry out the bulk of cellular functions, including relaying signals, providing structural support, and accelerating biochemical reactions. Genes usually "work" by providing the blueprint for proteins. Proteins can be modified chemically by the addition of small chemicals such as phosphates or sugars or lipids.

Reverse transcription: The process by which an enzyme (reverse transcriptase) uses a chain of RNA as a template to build a chain of DNA. Reverse transcriptase is found in retroviruses.

Ribosome: A cellular structure composed of protein and RNA that is responsible for the decoding of messenger RNA into proteins.

RNA: Ribonucleic acid, a chemical that performs several functions in the cells, including acting as an "intermediate" message for a gene to be translated into a protein. RNA is composed of a chain of bases—A, C, G, and U—strung together along a sugar-phosphate backbone. Typically, RNA is found as a single strand in a cell (unlike DNA, which is always double-stranded), although double-stranded RNA can be formed under special conditions. Some organisms, such as retroviruses, use RNA to carry their genetic information.

Traits, dominant and recessive: A physical or biological feature of an organism. Traits are typically encoded by genes. Many genes might encode a single trait, and a single gene might also encode many traits. A dominant trait is one that usually asserts itself when both dominant and recessive alleles are present, while a recessive trait is one that asserts itself only when both alleles are recessive. Genes can also be co-dominant: in this case, an intermediate trait is manifested when both dominant and recessive alleles are present.

Transcription: The process of generating an RNA copy of a gene. In transcription, the genetic code in DNA (ATG-CAC-GGG) is used to build an RNA "copy" (AUG-CAC-GGG).

Transformation: The horizontal transfer of genetic material from one organism to another. Typically, bacteria can exchange genetic information without reproduction by the transfer of genetic material between organisms.

Translation (of genes): The process by which genetic information is converted from the RNA message into a protein by the ribosome. During translation, a codon consisting of a triplet of bases in RNA (e.g., AUG) is used to add amino acids to a protein (e.g., Methionine). A chain of RNA can thus encode a chain of amino acids.

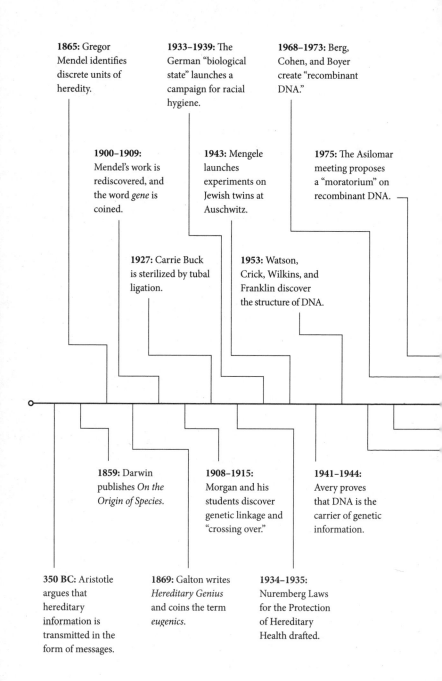

1865: Gregor Mendel identifies discrete units of heredity.

1933–1939: The German "biological state" launches a campaign for racial hygiene.

1968–1973: Berg, Cohen, and Boyer create "recombinant DNA."

1900–1909: Mendel's work is rediscovered, and the word *gene* is coined.

1943: Mengele launches experiments on Jewish twins at Auschwitz.

1975: The Asilomar meeting proposes a "moratorium" on recombinant DNA.

1927: Carrie Buck is sterilized by tubal ligation.

1953: Watson, Crick, Wilkins, and Franklin discover the structure of DNA.

1859: Darwin publishes *On the Origin of Species*.

1908–1915: Morgan and his students discover genetic linkage and "crossing over."

1941–1944: Avery proves that DNA is the carrier of genetic information.

350 BC: Aristotle argues that hereditary information is transmitted in the form of messages.

1869: Galton writes *Hereditary Genius* and coins the term *eugenics*.

1934–1935: Nuremberg Laws for the Protection of Hereditary Health drafted.

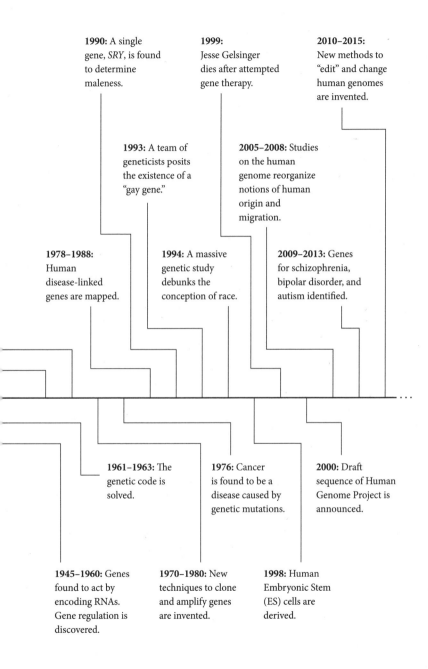

1990: A single gene, *SRY*, is found to determine maleness.

1999: Jesse Gelsinger dies after attempted gene therapy.

2010–2015: New methods to "edit" and change human genomes are invented.

1993: A team of geneticists posits the existence of a "gay gene."

2005–2008: Studies on the human genome reorganize notions of human origin and migration.

1978–1988: Human disease-linked genes are mapped.

1994: A massive genetic study debunks the conception of race.

2009–2013: Genes for schizophrenia, bipolar disorder, and autism identified.

1961–1963: The genetic code is solved.

1976: Cancer is found to be a disease caused by genetic mutations.

2000: Draft sequence of Human Genome Project is announced.

1945–1960: Genes found to act by encoding RNAs. Gene regulation is discovered.

1970–1980: New techniques to clone and amplify genes are invented.

1998: Human Embryonic Stem (ES) cells are derived.

Notes

Epigraph

ix *An exact determination of the laws of heredity:* W. Bateson, "Problems of Heredity as a Subject for Horticultural Investigation," in *A Century of Mendelism in Human Genetics*, ed. Milo Keynes, A.W.F. Edwards, and Robert Peel (Boca Raton, FL: CRC Press, 2004), 153.

ix *Human beings are ultimately nothing but carriers:* Haruki Murakami, *1Q84* (London: Vintage, 2012), 231.

Prologue: Families

1 *The blood of your parents is not lost in you:* Charles W. Eliot, *The Harvard Classics: The Odyssey of Homer*, ed. Charles W. Eliot (Danbury, CT: Grolier Enterprises, 1982), 49.

1 *They fuck you up, your mum and dad:* Philip Larkin, *High Windows* (New York: Farrar, Straus and Giroux, 1974).

8 *In 2012, several further studies:* Maartje F. Aukes et al., "Familial clustering of schizophrenia, bipolar disorder, and major depressive disorder," *Genetics in Medicine* 14, no. 3 (2012): 338–41; and Paul Lichtenstein et al., "Common genetic determinants of schizophrenia and bipolar disorder in Swedish families: A population-based study," *Lancet* 373, no. 9659 (2009): 234–39.

9 *Three profoundly destabilizing: Atoms, Bytes and Genes: Public Resistance and Techno-Scientific Responses* by Martin W. Bauer, Routledge Advances in Sociology (New York: Routledge, 2015).

10 *"In the sum of the parts, there are only the parts":* Helen Vendler, *Wallace Stevens: Words Chosen out of Desire* (Cambridge, MA: Harvard University Press, 1984), 21.

10 *"The whole organic world":* Hugo de Vries, *Intracellular Pangenesis: Including a Paper on Fertilization and Hybridization* (Chicago: Open Court, 1910), 13.

10 *"Alchemy could not become chemistry until":* Arthur W. Gilbert, "The Science of Genetics," *Journal of Heredity* 5, no. 6 (1914): 239.

11 *"That the fundamental aspects of heredity":* Thomas Hunt Morgan, *The Physical Basis of Heredity* (Philadelphia: J. B. Lippincott, 1919), 14.

12 *"the quest for eternal youth":* Jeff Lyon and Peter Gorner, *Altered Fates: Gene Therapy and the Retooling of Human Life* (New York: W. W. Norton, 1996), 9–10.

PART ONE: THE "MISSING SCIENCE OF HEREDITY"

16 *This missing science of heredity:* Herbert G. Wells, *Mankind in the Making* (Leipzig: Tauchnitz, 1903), 33.

16 *JACK: Yes, but you said yourself:* Oscar Wilde, *The Importance of Being Earnest* (New York: Dover Publications, 1990), 117.

The Walled Garden

17 *The students of heredity:* G. K. Chesterton, *Eugenics and Other Evils* (London: Cassell, 1922), 66.

18 *the Augustinians, fortunately, saw no conflict:* Gareth B. Matthews, *The Augustinian Tradition* (Berkeley: University of California Press, 1999).

18 *In October 1843, a young man from Silesia:* Details of Mendel's life and the Augustinian monastery are from several sources, including Gregor Mendel, Alain F. Corcos, and Floyd V. Monaghan, *Gregor Mendel's Experiments on Plant Hybrids: A Guided Study* (New Brunswick, NJ: Rutgers University Press, 1993); Edward Edelson, *Gregor Mendel: And the Roots of Genetics* (New York: Oxford University Press, 1999); and Robin Marantz Henig, *The Monk in the Garden: The Lost and Found Genius of Gregor Mendel, the Father of Genetics* (Boston: Houghton Mifflin, 2000).

18 *The tumult of 1848:* Edward Berenson, *Populist Religion and Left-Wing Politics in France, 1830–1852* (Princeton, NJ: Princeton University Press, 1984).

18 *"Seized by an unconquerable timidity":* Henig, *Monk in the Garden*, 37.

18 *he applied for a job to teach mathematics:* Ibid., 38.

19 *In the late spring of 1850, an eager Mendel:* Harry Sootin, *Gregor Mendel: Father of the Science of Genetics* (New York: Random House Books for Young Readers, 1959).

19 *On July 20, in the midst of an enervating heat wave:* Henig, *Monk in the Garden*, 62.

19 *On August 16, he appeared before his examiners:* Ibid., 47.

20 *In 1842, Doppler, a gaunt, acerbic:* Jagdish Mehra and Helmut Rechenberg, *The Historical Development of Quantum Theory* (New York: Springer-Verlag, 1982).

20 *But in 1845, Doppler had loaded a train:* Kendall F. Haven, *100 Greatest Science Discoveries of All Time* (Westport, CT: Libraries Unlimited, 2007), 75–76.

20 *But these categories, originally devised by the Swedish botanist:* Margaret J. Anderson, *Carl Linnaeus: Father of Classification* (Springfield, NJ: Enslow Publishers, 1997).

21 *"Not the true parent is the woman's":* Aeschylus, *The Greek Classics: Aeschylus—Seven Plays* (n.p.: Special Edition Books, 2006), 240.

21 *"She doth but nurse the seed":* Ibid.

22 *from Indian or Babylonian geometers:* Maor Eli, *The Pythagorean Theorem: A 4,000-Year History* (Princeton, NJ: Princeton University Press, 2007).

22 *A century after Pythagoras's death:* Plato, *The Republic*, ed. and trans. Allan Bloom (New York: Basic Books, 1968).

22 *In one of the most intriguing passages:* Plato, *The Republic* (Edinburgh: Black & White Classics, 2014), 150.

22 *"For when your guardians are ignorant":* Ibid.

23 *The result, a compact treatise:* Aristotle, *Generation of Animals* (Leiden: Brill Archive, 1943).

23 *"And from deformed":* Aristotle, *History of Animals, Book VII*, ed. and trans. D. M. Balme (Cambridge, MA: Harvard University Press, 1991).

23 *"just as lame come to be from lame"*: Ibid., 585b28–586a4.

23 *"Men generate before they yet have certain characters"*: Aristotle, *The Complete Works of Aristotle: The Revised Oxford Translation*, ed. Jonathan Barnes (Princeton, NJ: Princeton University Press, 1984), bk. 1, 1121.

24 *Aristotle offered an alternative theory*: Aristotle, *The Works of Aristotle*, ed. and trans. W. D. Ross (Chicago: Encyclopædia Britannica, 1952), "Aristotle: Logic and Metaphysics."

24 *"[Just as] no material part comes from the carpenter"*: Aristotle, *Complete Works of Aristotle*, 1134.

24 *biologist Max Delbrück would joke that Aristotle*: Daniel Novotny and Lukás Novák, *Neo-Aristotelian Perspectives in Metaphysics* (New York: Routledge, 2014), 94.

25 *In the 1520s, the Swiss-German alchemist Paracelsus*: Paracelsus, *Paracelsus: Essential Readings*, ed. and trans. Nicholas Godrick-Clarke (Wellingborough, Northamptonshire, England: Crucible, 1990).

25 *"floating . . . in our First Parent's loins"*: Peter Hanns Reill, *Vitalizing Nature in the Enlightenment* (Berkeley: University of California Press, 2005), 160.

26 *In 1694, Nicolaas Hartsoeker, the Dutch physicist*: Nicolaas Hartsoeker, *Essay de dioptrique* (Paris: Jean Anisson, 1694).

26 *"In nature there is no generation"*: Matthew Cobb, "Reading and writing the book of nature: Jan Swammerdam (1637–1680)," *Endeavour* 24, no. 3 (2000): 122–28.

26 *In 1768, the Berlin embryologist Caspar Wolff*: Caspar Friedrich Wolff, "De formatione intestinorum praecipue," *Novi commentarii Academiae Scientiarum Imperialis Petropolitanae* 12 (1768): 43–47. Wolff also wrote about *essentialis corporis* in 1759: Richard P. Aulie, "Caspar Friedrich Wolff and his 'Theoria Generationis,' 1759," *Journal of the History of Medicine and Allied Sciences* 16, no. 2 (1961): 124–44.

27 *"The opposing views of today were in existence centuries ago"*: Oscar Hertwig, *The Biological Problem of To-day: Preformation or Epigenesis? The Basis of a Theory of Organic Development* (London: Heinneman's Scientific Handbook, 1896), 1.

"The Mystery of Mysteries"

28 *They mean to tell us all was rolling blind*: Robert Frost, *The Robert Frost Reader: Poetry and Prose*, ed. Edward Connery Lathem and Lawrance Thompson (New York: Henry Holt, 2002).

28 *Charles Darwin, boarded a ten-gun brig-sloop*: Charles Darwin, *The Autobiography of Charles Darwin*, ed. Francis Darwin (Amherst, NY: Prometheus Books, 2000), 11.

28 *He had tried, unsuccessfully, to study medicine*: Jacob Goldstein, "Charles Darwin, Medical School Dropout," *Wall Street Journal*, February 12, 2009, http://blogs.wsj .com/health/2009/02/12/charles-darwin-medical-school-dropout/.

28 *Christ's College in Cambridge*: Darwin, *Autobiography of Charles Darwin*, 37.

28 *Holed up in a room*: Adrian J. Desmond and James R. Moore, *Darwin* (New York: Warner Books, 1991), 52.

28 *John Henslow, the botanist and geologist*: Duane Isely, *One Hundred and One Botanists* (Ames: Iowa State University, 1994), "John Stevens Henslow (1796–1861)."

29 *The first*, Natural Theology, *published in 1802*: William Paley, *The Works of William Paley . . . Containing His Life, Moral and Political Philosophy, Evidences of Christianity, Natural Theology, Tracts, Horae Paulinae, Clergyman's Companion, and Sermons, Printed Verbatim from the Original Editions. Complete in One Volume* (Philadelphia: J. J. Woodward, 1836).

29 *The second book,* A Preliminary Discourse: John F. W. Herschel, *A Preliminary Discourse on the Study of Natural Philosophy. A Facsim. of the 1830 Ed.* (New York: Johnson Reprint, 1966).

29 *"To ascend to the origin of things":* Ibid., 38.

30 *"Battered relics of past ages":* Martin Gorst, *Measuring Eternity: The Search for the Beginning of Time* (New York: Broadway Books, 2002), 158.

30 *"mystery of mysteries":* Charles Darwin, *On the Origin of Species by Means of Natural Selection* (London: Murray, 1859), 7.

30 *dominated by so-called parson-naturalists:* Patrick Armstrong, *The English Parson-Naturalist: A Companionship between Science and Religion* (Leominster, MA: Gracewing, 2000), "Introducing the English Parson-Naturalist."

31 *In August 1831, two months after his graduation:* John Henslow, "Darwin Correspondence Project," Letter 105, https://www.darwinproject.ac.uk/letter/entry-105.

31 *The Beagle lifted anchor on December 27, 1831:* Darwin, *Autobiography of Charles Darwin,* "Voyage of the 'Beagle.' "

32 *Charles Lyell's Principles of Geology:* Charles Lyell, *Principles of Geology: Or, The Modern Changes of the Earth and Its Inhabitants Considered as Illustrative of Geology* (New York: D. Appleton, 1872).

32 *Lyell had argued (radically, for his time):* Ibid., "Chapter 8: Difference in Texture of the Older and Newer Rocks."

32 *In September 1832, exploring the gray cliffs:* Charles Darwin, *Geological Observations on the Volcanic Islands and Parts of South America Visited during the Voyage of H.M.S. "Beagle"* (New York: D. Appleton, 1896), 76–107.

32 *The skull belonged to a megatherium:* David Quammen, "Darwin's first clues," *National Geographic* 215, no. 2 (2009): 34–53.

33 *In 1835, the ship left Lima:* Charles Darwin, *Charles Darwin's Letters: A Selection, 1825–1859,* ed. Frederick Burkhardt (Cambridge: University of Cambridge, 1996), "To J. S. Henslow 12 [August] 1835," 46–47.

33 *On October 20, Darwin returned to sea:* G. T. Bettany and John Parker Anderson, *Life of Charles Darwin* (London: W. Scott, 1887), 47.

35 *rather than all species radiating out:* Duncan M. Porter and Peter W. Graham, *Darwin's Sciences* (Hoboken, NJ: Wiley-Blackwell, 2015), 62–63.

35 *As an afterthought, he added, "I think":* Ibid., 62.

36 *In the spring of 1838, as Darwin tore into a new journal:* Timothy Shanahan, *The Evolution of Darwinism: Selection, Adaptation, and Progress in Evolutionary Biology* (Cambridge: Cambridge University Press, 2004), 296.

36 *But the answer that came to him in October 1838:* Barry G. Gale, "After Malthus: Darwin Working on His Species Theory, 1838–1859" (PhD diss., University of Chicago, 1980).

37 *In 1798, writing under a pseudonym, Malthus:* Thomas Robert Malthus, *An Essay on the Principle of Population* (Chicago: Courier Corporation, 2007).

37 *"sickly seasons, epidemics, pestilence and plague":* Arno Karlen, *Man and Microbes: Disease and Plagues in History and Modern Times* (New York: Putnam, 1995), 67.

37 *"It at once struck me":* Charles Darwin, *On the Origin of Species by Means of Natural Selection,* ed. Joseph Carroll (Peterborough, Canada: Broadview Press, 2003), 438.

37 *the phrase survival of the fittest was borrowed:* Gregory Claeys, "The 'Survival of the Fittest' and the Origins of Social Darwinism," *Journal of the History of Ideas* 61, no. 2 (2000): 223–40.

38 *In 1844, he distilled the crucial parts:* Charles Darwin, *The Foundations of the Origin of Species, Two Essays Written in 1842 and 1844,* ed. Francis Darwin (Cambridge: Cambridge University Press, 1909), "Essay of 1844."

38 *Alfred Russel Wallace, published a paper:* Alfred R. Wallace, "XVIII.—On the law which has regulated the introduction of new species," *Annals and Magazine of Natural History* 16, no. 93 (1855): 184–96.

38 *Wallace had been born to a middle-class family:* Charles H. Smith and George Beccaloni, *Natural Selection and Beyond: The Intellectual Legacy of Alfred Russel Wallace* (Oxford: Oxford University Press, 2008), 10.

38 *but on the hard-back benches of the free library:* Ibid., 69.

39 *Like Darwin, Wallace had also embarked:* Ibid., 12.

39 *Wallace moved from the Amazon basin:* Ibid., ix.

39 *"The answer was clearly":* Benjamin Orange Flowers, "Alfred Russel Wallace," *Arena* 36 (1906): 209.

39 *In June 1858, Wallace sent Darwin a tentative draft:* Alfred Russel Wallace, *Alfred Russel Wallace: Letters and Reminiscences,* ed. James Marchant (New York: Arno Press, 1975), 118.

39 *On July 1, 1858, Darwin's and Wallace's papers were read:* Charles Darwin, *The Correspondence of Charles Darwin,* vol. 13, ed. Frederick Burkhardt, Duncan M. Porter, and Sheila Ann Dean, et al. (Cambridge: Cambridge University Press, 2003), 468.

39 *The next May, the president of the society remarked:* E. J. Browne, *Charles Darwin: The Power of Place* (New York: Alfred A. Knopf, 2002), 42.

39 *"I heartily hope that my Book":* Charles Darwin, *The Correspondence of Charles Darwin,* vol. 7, ed. Frederick Burkhardt and Sydney Smith (Cambridge: Cambridge University Press, 1992), 357.

40 *"All copies were sold [on the] first day":* Charles Darwin, *The Life and Letters of Charles Darwin* (London: John Murray, 1887), 70.

40 *"The conclusions announced by Mr. Darwin are such":* "Reviews: Darwin's Origins of Species," *Saturday Review of Politics, Literature, Science and Art* 8 (December 24, 1859): 775–76.

40 *"We imply that his work [is] one of the most important that":* Ibid.

40 *"light will be thrown on the origin of man":* Charles Darwin, *On the Origin of Species,* ed. David Quammen (New York: Sterling, 2008), 51.

40 *"intellectual husks":* Richard Owen, "Darwin on the Origin of Species," *Edinburgh Review* 3 (1860): 487–532.

40 *"One's imagination must fill up very wide blanks":* Ibid.

The "Very Wide Blank"

41 *The "Very Wide Blank":* Darwin, *Correspondence of Charles Darwin,* Darwin's letter to Asa Gray, September 5, 1857, https://www.darwinproject.ac.uk/letter/entry-2136.

41 *Now, I wonder if:* Alexander Wilford Hall, *The Problem of Human Life: Embracing the "Evolution of Sound" and "Evolution Evolved,"* with a Review of the Six Great Modern Scientists, Darwin, Huxley, Tyndall, Haeckel, Helmholtz, and Mayer (London: Hall & Company, 1880), 441.

42 *In Lamarck's view:* Monroe W. Strickberger, *Evolution* (Boston: Jones & Bartlett, 1990), "The Lamarckian Heritage."

42 *"with a power proportional to the length of time":* Ibid., 24.

43 *driving himself to the brink:* James Schwartz, *In Pursuit of the Gene: From Darwin to DNA* (Cambridge, MA: Harvard University Press, 2008), 2.

43 *minute particles containing hereditary information*—gemmules: Ibid., 2–3.

43 *blending inheritance—was already familiar:* Brian Charlesworth and Deborah Charlesworth, "Darwin and genetics," *Genetics* 183, no. 3 (2009): 757–66.

44 *Darwin dubbed his theory pangenesis:* Ibid., 759–60.

44 *a new manuscript,* The Variation of Animals: Charles Darwin, *The Variation of Animals and Plants under Domestication*, vol. 2 (London: O. Judd, 1868).

44 *"It is a rash and crude hypothesis":* Darwin, *Correspondence of Charles Darwin*, vol. 13, "Letter to T. H. Huxley," 151.

44 *"Pangenesis will be called a mad dream":* Charles Darwin, *The Life and Letters of Charles Darwin: Including Autobiographical Chapter*, vol. 2., ed. Francis Darwin (New York: Appleton, 1896), "C. Darwin to Asa Gray," October 16, 1867, 256.

44 *"The [variant] will be swamped":* Fleeming Jenkin, "The Origin of Species," *North British Review* 47 (1867): 158.

46 *There was no denying:* In fairness to Darwin, he had sensed the problem in "blending inheritance" even without Jenkin's interjection. "If varieties be allowed freely to cross, such varieties will be constantly demolished . . . any small tendency in them to vary will be constantly counteracted," he wrote in his notes.

46 *"Experiments in Plant Hybridization":* G. Mendel, "Versuche über Pflanzen-Hybriden," *Verhandlungen des naturforschenden Vereins Brno* 4 (1866): 3–47 (*Journal of the Royal Horticultural Society* 26 [1901]: 1–32).

46 *he made extensive handwritten notes on pages 50, 51, 53, and 54:* David Galton, "Did Darwin read Mendel?" *Quarterly Journal of Medicine* 102, no. 8 (2009): 588, doi:10.1093/qjmed/hcp024.

"Flowers He Loved"

47 *"Flowers He Loved":* Edward Edelson, *Gregor Mendel and the Roots of Genetics* (New York: Oxford University Press, 1999), "Clemens Janetchek's Poem Describing Mendel after His Death," 75.

47 *"We want only to disclose the [nature of] matter and its force":* Jiri Sekerak, "Gregor Mendel and the scientific milieu of his discovery," ed. M. Kokowski (The Global and the Local: The History of Science and the Cultural Integration of Europe, Proceedings of the 2nd ICESHS, Cracow, Poland, September 6–9, 2006).

47 *"The whole organic world is the result":* Hugo de Vries, *Intracellular Pangenesis; Including a Paper on Fertilization and Hybridization* (Chicago: Open Court, 1910), "Mutual Independence of Hereditary Characters."

47 *Gregor Mendel decided to return to Vienna:* Henig, *Monk in the Garden*, 60.

48 *"remained constant without exception":* Eric C. R. Reeve, *Encyclopedia of Genetics* (London: Fitzroy Dearborn, 2001), 62.

49 *Contrary to later belief:* Mendel had several predecessors who had studied plant hybrids just as intensively, except, perhaps, without Mendel's immersion in numbers and quantification. In the 1820s, English botanists, such as T. A. Knight, John Goss, Alexander Seton, and William Herbert—attempting to breed more vigorous agricultural plants—had performed experiments with plant hybrids that were strikingly similar to Mendel's. In France, Augustine Sageret's work on melon hybrids was also similar to Mendel's work. The most intensive work on plant hybrids immediately pre-

ceding Mendel was performed by the German botanist Josef Kölreuter, who had bred *Nicotania* hybrids. Kölreuter's work was followed by the work of Karl von Gaertner and Charles Naudin in Paris. Darwin had actually read Sageret's and Naudin's studies, both of which suggested the particulate quality of hereditary information, but Darwin had failed to appreciate their importance.

49 *"the history of the evolution of organic forms":* Gregor Mendel, *Experiments in Plant Hybridisation* (New York: Cosimo, 2008), 8.

49 *By the late summer of 1857, the first hybrid peas:* Henig, *Monk in the Garden*, 81. More details in "Chapter 7: First Harvest."

50 *"How small a thought it takes to fill":* Ludwig Wittgenstein, *Culture and Value*, trans. Peter Winch (Chicago: University of Chicago Press, 1984), 50e.

51 *Mendel termed these overriding traits:* Henig, *Monk in the Garden*, 86.

51 *In some of these third-generation crosses:* Ibid., 130.

52 *"It requires indeed some courage":* Mendel, *Experiments in Plant Hybridization*, 8.

53 *Mendel presented his paper:* Henig, *Monk in the Garden*, "Chapter 11: Full Moon in February," 133–47. A second portion of Mendel's paper was read on March 8, 1865.

53 *Mendel's paper was published in:* Mendel, "Experiments in Plant Hybridization," www.mendelweb.org/Mendel.html.

53 *It is likely that he sent one to Darwin:* Galton, "Did Darwin Read Mendel?" 587.

54 *"one of the strangest silences in the history of biology":* Leslie Clarence Dunn, *A Short History of Genetics: The Development of Some of the Main Lines of Thought, 1864–1939* (Ames: Iowa State University Press, 1991), 15.

54 *"only empirical . . . cannot be proved rational":* Gregor Mendel, "Gregor Mendel's letters to Carl Nägeli, 1866–1873," *Genetics* 35, no. 5, pt. 2 (1950): 1.

54 *"I knew that the results I obtained":* Allan Franklin et al., *Ending the Mendel-Fisher Controversy* (Pittsburgh, PA: University of Pittsburgh Press, 2008), 182.

54 *"an isolated experiment might be doubly dangerous":* Mendel, "Letters to Carl Nägeli," April 18, 1867, 4.

55 *In November 1873, Mendel wrote his last letter to Nägeli:* Ibid., November 18, 1867, 30–34.

55 *"I feel truly unhappy that I have to neglect":* Gian A. Nogler, "The lesser-known Mendel: His experiments on *Hieracium*," *Genetics* 172, no. 1 (2006): 1–6.

55 *On January 6, 1884, Mendel died:* Henig, *Monk in the Garden*, 170.

55 *"Gentle, free-handed, and kindly . . . Flowers he loved":* Edelson, *Gregor Mendel*, "Clemens Janetchek's Poem Describing Mendel after His Death," 75.

"A Certain Mendel"

56 *The origin of species is a natural phenomenon:* Lucius Moody Bristol, *Social Adaptation: a Study in the Development of the Doctrine of Adaptation as a Theory of Social Progress* (Cambridge, MA: Harvard University Press, 1915), 70.

56 *The origin of species is an object of inquiry:* Ibid.

56 *The origin of species is an object of experimental investigation:* Ibid.

56 *In the summer of 1878:* Peter W. van der Pas, "The correspondence of Hugo de Vries and Charles Darwin," *Janus* 57: 173–213.

57 *"margin was too small":* Mathias Engan, *Multiple Precision Integer Arithmetic and Public Key Encryption* (M. Engan, 2009), 16–17.

57 *"In another work I shall discuss"*: Charles Darwin, *The Variation of Animals & Plants under Domestication*, ed. Francis Darwin (London: John Murray, 1905), 5.

57 *Darwin died in 1882*: "Charles Darwin," Famous Scientists, http://www.famousscientists .org/charles-darwin/.

57 *In 1883, with rather grim determination*: James Schwartz, *In Pursuit of the Gene: From Darwin to DNA* (Cambridge, MA: Harvard University Press, 2008), "Pangenes."

57 *Weismann called this hereditary material* germplasm: August Weismann, William Newton Parker, and Harriet Rönnfeldt, *The Germ-Plasm; a Theory of Heredity* (New York: Scribner's, 1893).

58 *In a landmark paper written in 1897*: Schwartz, *In Pursuit of the Gene*, 83.

58 *He called these particles "pangenes"*: Ida H. Stamhuis, Onno G. Meijer, and Erik J. A. Zevenhuizen, "Hugo de Vries on heredity, 1889–1903: Statistics, Mendelian laws, pangenes, mutations," *Isis* (1999): 238–67.

59 *"I know that you are studying hybrids"*: Iris Sandler and Laurence Sandler, "A conceptual ambiguity that contributed to the neglect of Mendel's paper," *History and Philosophy of the Life Sciences* 7, no. 1 (1985): 9.

59 *"Modesty is a virtue"*: Edward J. Larson, *Evolution: The Remarkable History of a Scientific Theory* (New York: Modern Library, 2004).

59 *That same year de Vries published his monumental study*: Hans-Jörg Rheinberger, "Mendelian inheritance in Germany between 1900 and 1910. The case of Carl Correns (1864–1933)," *Comptes Rendus de l'Académie des Sciences—Series III—Sciences de la Vie* 323, no. 12 (2000): 1089–96, doi:10.1016/s0764-4469(00)01267-1.

60 *"I too still believed that I had found something new"*: Url Lanham, *Origins of Modern Biology* (New York: Columbia University Press, 1968), 207.

60 *"by a strange coincidence"*: Carl Correns, "G. Mendel's law concerning the behavior of progeny of varietal hybrids," *Genetics* 35, no. 5 (1950): 33–41.

60 *de Vries stumbled on an enormous, invasive*: Schwartz, *In Pursuit of the Gene*, 111.

61 *He called them* mutants: Hugo de Vries, *The Mutation Theory*, vol. 1 (Chicago: Open Court, 1909).

61 *For William Bateson, the English biologist*: John Williams Malone, *It Doesn't Take a Rocket Scientist: Great Amateurs of Science* (Hoboken, NJ: Wiley, 2002), 23.

62 *"We are in the presence of a new principle"*: Schwartz, *In Pursuit of the Gene*, 112.

62 *"I am writing to ask you"*: Nicholas W. Gillham, "Sir Francis Galton and the birth of eugenics," *Annual Review of Genetics* 35, no. 1 (2001): 83–101.

62 *First, he independently confirmed Mendel's work*: Other scientists, including Reginald Punnett and Lucien Cuenot, provided crucial experimental support for Mendel's laws. In 1905, Punnett authored *Mendelism*, considered the first textbook of modern genetics.

62 *"His linen is foul. I daresay"*: Alan Cock and Donald R. Forsdyke, *Treasure Your Exceptions: The Science and Life of William Bateson* (Dordrecht: Springer Science & Business Media, 2008), 186.

62 *Nicknamed "Mendel's bulldog"*: Ibid., "Mendel's Bulldog (1902–1906)," 221–64.

62 *"man's outlook on the world"*: William Bateson, "Problems of heredity as a subject for horticultural investigation," *Journal of the Royal Horticultural Society* 25 (1900–1901): 54.

62 *"No single word in common use"*: William Bateson and Beatrice (Durham) Bateson, *William Bateson, F.R.S., Naturalist; His Essays & Addresses, Together with a Short Account of His Life* (Cambridge: Cambridge University Press, 1928), 93.

62 *In 1905, still struggling for an alternative*: Schwartz, *In Pursuit of the Gene*, 221.

63   *"What will happen when . . . enlightenment actually comes to pass"*: Bateson and Bateson, *William Bateson, F.R.S.*, 456.

Eugenics

64   *Improved environment and education:* Herbert Eugene Walter, *Genetics: An Introduction to the Study of Heredity* (New York: Macmillan, 1938), 4.

64   *Most Eugenists are Euphemists:* G. K. Chesterton, *Eugenics and Other Evils* (London: Cassell, 1922), 12–13.

64   *In 1883, one year after Charles Darwin's death:* Francis Galton, *Inquiries into Human Faculty and Its Development* (London: Macmillan, 1883).

64   *"We greatly want a brief word to express":* Roswell H. Johnson, "Eugenics and So-Called Eugenics," *American Journal of Sociology* 20, no. 1 (July 1914): 98–103, http://www.jstor.org/stable/2762976.

65   *"at least a neater word . . . than* viriculture*":* Ibid., 99.

65   *"Believing, as I do, that human eugenics":* Galton, *Inquiries into Human Faculty*, 44.

65   *A child prodigy, Galton:* Dean Keith Simonton, *Origins of Genius: Darwinian Perspectives on Creativity* (New York: Oxford University Press, 1999), 110.

65   *He tried studying medicine, but then switched:* Nicholas W. Gillham, *A Life of Sir Francis Galton: From African Exploration to the Birth of Eugenics* (New York: Oxford University Press, 2001), 32–33.

65   *"I saw enough of savage races":* Niall Ferguson, *Civilization: The West and the Rest* (Duisburg: Haniel-Stiftung, 2012), 176.

65   *"initiated into an entirely new province of knowledge":* Francis Galton to C. R. Darwin, December 9, 1859, https://www.darwinproject.ac.uk/letter/entry-2573.

66   *Galton tried transfusing rabbits:* Daniel J. Fairbanks, *Relics of Eden: The Powerful Evidence of Evolution in Human DNA* (Amherst, NY: Prometheus Books, 2007), 219.

66   *"Man is born, grows up and dies":* Adolphe Quetelet, *A Treatise on Man and the Development of His Faculties: Now First Translated into English*, trans. T. Smibert (New York: Cambridge University Press, 2013), 5.

66   *He tabulated the chest breadth and height:* Jerald Wallulis, *The New Insecurity: The End of the Standard Job and Family* (Albany: State University of New York Press, 1998), 41.

67   *"Whenever you can":* Karl Pearson, *The Life, Letters and Labours of Francis Galton* (Cambridge: Cambridge University Press, 1914), 340.

67   *"Keenness of Sight and Hearing":* Sam Goldstein, Jack A. Naglieri, and Dana Princiotta, *Handbook of Intelligence: Evolutionary Theory, Historical Perspective, and Current Concepts* (New York: Springer, 2015), 100.

67   *To marshal further evidence, Galton began:* Gillham, *Life of Sir Francis Galton*, 156.

67   *Galton published much of this data:* Francis Galton, *Hereditary Genius* (London: Macmillan, 1892).

68   *"You have made a convert":* Charles Darwin, *More Letters of Charles Darwin: A Record of His Work in a Series of Hitherto Unpublished Letters*, vol. 2 (New York: D. Appleton, 1903), 41.

69   *Galton called this the Ancestral Law of Heredity:* John Simmons, *The Scientific 100: A Ranking of the Most Influential Scientists, Past and Present* (Secaucus, NJ: Carol Publishing Group, 1996), "Francis Dalton," 441.

69   Basset Hound Club Rules, *a compendium:* Schwartz, *In Pursuit of the Gene*, 61.

69 *Two prominent biologists:* Ibid., 131.

70 *But as Darbishire analyzed his own first-generation:* Gillham, *Life of Sir Francis Galton*, "The Mendelians Trump the Biometricians," 303–23.

70 *In the spring of 1905:* Karl Pearson, *Walter Frank Raphael Weldon, 1860–1906* (Cambridge: Cambridge University Press, 1906), 48–49.

70 *trying . . . to rework the data to fit Galtonian theory:* Ibid., 49.

70 *"To Weldon I owe the chief awakening of my life":* Schwartz, *In Pursuit of the Gene*, 143.

71 *"Each of us who now looks at his own patch":* William Bateson, *Mendel's Principles of Heredity: A Defence*, ed. Gregor Mendel (Cambridge: Cambridge University Press, 1902), v.

71 *"We have only touched the edge":* Ibid., 208.

71 *"is second to no branch of science":* Ibid., ix.

71 *Johannsen shortened the word to* gene: Johan Henrik Wanscher, "The history of Wilhelm Johannsen's genetical terms and concepts from the period 1903 to 1926," *Centaurus* 19, no. 2 (1975): 125–47.

71 *"Language is not only our servant":* Wilhelm Johannsen, "The genotype conception of heredity," *International Journal of Epidemiology* 43, no. 4 (2014): 989–1000.

72 *"The science of genetics is so new":* Arthur W. Gilbert, "The science of genetics," *Journal of Heredity* 5, no. 6 (1914): 235–44, http://archive.org/stream/journalofheredit 05amer/journalofheredit05amer_djvu.txt.

72 *"the technology of the industrial revolution confirmed":* Daniel J. Kevles, *In the Name of Eugenics: Genetics and the Uses of Human Heredity* (New York: Alfred A. Knopf, 1985), 3.

73 *"forces which bring greatness to the social group":* *Problems in Eugenics: First International Eugenics Congress, 1912* (New York: Garland, 1984), 483.

73 *In the spring of 1904, Galton presented his argument:* Paul B. Rich, *Race and Empire in British Politics* (Cambridge: Cambridge University Press, 1986), 234.

73 *"introduced into national consciousness, like a new religion":* *Papers and Proceedings—First Annual Meeting—American Sociological Society*, vol. 1 (Chicago: University of Chicago Press, 1906), 128.

73 *"All creatures would agree that it was better":* Francis Galton, "Eugenics: Its definition, scope, and aims," *American Journal of Sociology* 10, no. 1 (1904): 1–25.

73 *"if unsuitable marriages from the eugenic point of view":* Andrew Norman, *Charles Darwin: Destroyer of Myths* (Barnsley, South Yorkshire: Pen and Sword, 2013), 242.

73 *Henry Maudsley, the psychiatrist:* Galton, "Eugenics," comments by Maudsley, doi:10.1017/s0364009400001161.

74 *"He had five brothers," Maudsley noted:* Ibid., 7.

74 *"It is in the sterilization of failure":* Ibid., comments by H. G. Wells; and H. G. Wells and Patrick Parrinder, *The War of the Worlds* (London: Penguin Books, 2005).

75 *"A pleasant sort o' soft woman":* George Eliot, *The Mill on the Floss* (New York: Dodd, Mead, 1960), 12.

75 *In 1911, Havelock Ellis, Galton's colleague:* Lucy Bland and Laura L. Doan, *Sexology Uncensored: The Documents of Sexual Science* (Chicago: University of Chicago Press, 1998), "The Problem of Race-Regeneration: Havelock Ellis (1911)."

76 *On July 24, 1912:* R. Pearl, "The First International Eugenics Congress," *Science* 36, no. 926 (1912): 395–96, doi:10.1126/science.36.926.395.

77 *Davenport's 1911 book:* Charles Benedict Davenport, *Heredity in Relation to Eugenics* (New York: Holt, 1911).

77 *Van Wagenen suggested, and "they are totally"*: First International Eugenics Congress, *Problems in Eugenics* (1912; repr., London: Forgotten Books, 2013), 464–65.

77 *"We endeavor to keep track"*: Ibid., 469.

"Three Generations of Imbeciles Is Enough"

78 *If we enable the weak and the deformed*: Theodosius G. Dobzhansky, *Heredity and the Nature of Man* (New York: New American Library, 1966), 158.

78 *And from deformed [parents] deformed [offspring]*: Aristotle, *History of Animals, Book VII*, 6, 585b28–586a4.

78 *In the spring of 1920, Emmett Adaline Buck*: Many of the details of the Buck family story are from J. David Smith, *The Sterilization of Carrie Buck* (Liberty Corner, NJ: New Horizon Press, 1989).

78 *Her husband, Frank Buck*: Much of the information in this chapter is from Paul Lombardo, *Three Generations, No Imbeciles: Eugenics, the Supreme Court, and* Buck v. Bell (Baltimore: Johns Hopkins University Press, 2008).

78 *A cursory mental examination*: "Buck v. Bell," Law Library, American Law and Legal Information, http://law.jrank.org/pages/2888/Buck-v-Bell-1927.html.

79 *Of these, an idiot was the easiest to classify*: *Mental Defectives and Epileptics in State Institutions: Admissions, Discharges, and Patient Population for State Institutions for Mental Defectives and Epileptics*, vol. 3 (Washington, DC: US Government Printing Office, 1937).

80 *On January 23, 1924*: "Carrie Buck Committed (January 23, 1924)," *Encyclopedia Virginia*, http://www.encyclopediavirginia.org/Carrie_Buck_Committed_January _23_1924.

80 *On March 28, 1924*: Ibid.

80 *"Moron, Middle Grade"*: Stephen Murdoch, *IQ: A Smart History of a Failed Idea* (Hoboken, NJ: John Wiley & Sons, 2007), 107.

80 *Carrie Buck was asked to appear*: Ibid., "Chapter 8: From Segregation to Sterilization."

81 *On March 29, 1924, with Priddy's help*: "Period during which sterilization occurred," Virginia Eugenics, doi:www.uvm.edu/~lkaelber/eugenics/VA/VA.html.

81 *"Do you care to say anything"*: Lombardo, *Three Generations*, 107.

83 *"A cross between"*: Madison Grant, *The Passing of the Great Race* (New York: Scribner's, 1916).

83 *"the menace of race deterioration"*: Carl Campbell Brigham and Robert M. Yerkes, *A Study of American Intelligence* (Princeton, NJ: Princeton University Press, 1923), "Foreword."

83 *"The Eugenic ravens are croaking"*: A. G. Cock and D. R. Forsdyke, *Treasure Your Exceptions: The Science and Life of William Bateson* (New York: Springer, 2008), 437–38n3.

83 *"It is better for all the world"*: Jerry Menikoff, *Law and Bioethics: An Introduction* (Washington, DC: Georgetown University Press, 2001), 41.

84 *"Three generations of imbeciles is enough*: Ibid.

84 *In 1927, the state of Indiana passed*: *Public Welfare in Indiana* 68–75 (1907): 50. In 1907, a new law passed by the state legislature and signed by the governor of Indiana provided for the involuntary sterilization of "confirmed criminals, idiots, imbeciles and rapists." Although it was eventually found to be unconstitutional, this law

is widely regarded as the first eugenics sterilization legislation passed in the world. In 1927, a revised law was implemented and before it was repealed in 1974, over 2,300 of the state's most vulnerable citizens were involuntarily sterilized. In addition, Indiana established a state-funded Committee on Mental Defectives that carried out eugenic family studies in over twenty counties and was home to an active "better babies" movement that encouraged scientific motherhood and infant hygiene as routes to human improvement. http://www.iupui.edu/~eugenics/.

85 *Better Babies Contests:* Laura L. Lovett, "Fitter Families for Future Firesides: Florence Sherbon and Popular Eugenics," *Public Historian* 29, no. 3 (2007): 68–85.

85 *"You should score 50% for heredity":* Charles Davenport to Mary T. Watts, June 17, 1922, Charles Davenport Papers, American Philosophical Society Archives, Philadelphia, PA. Also see Mary Watts, "Fitter Families for Future Firesides," *Billboard* 35, no. 50 (December 15, 1923): 230–31.

85 *In 1927, a film called* Are You Fit to Marry?: Martin S. Pernick and Diane B. Paul, *The Black Stork: Eugenics and the Death of "Defective" Babies in American Medicine and Motion Pictures since 1915* (New York: Oxford University Press, 1996).

PART TWO: "IN THE SUM OF THE PARTS,
THERE ARE ONLY THE PARTS"

87 *"In the Sum of the Parts":* Wallace Stevens, *The Collected Poems of Wallace Stevens* (New York: Alfred A. Knopf, 2011), "On the Road Home," 203–4.

88 *It was when I said:* Ibid.

"Abhed"

89 *I am the family face:* Thomas Hardy, *The Collected Poems of Thomas Hardy* (Ware, Hertfordshire, England: Wordsworth Poetry Library, 2002), "Heredity," 204–5.

92 *In 1907, when William Bateson visited:* William Bateson, "Facts limiting the theory of heredity," in *Proceedings of the Seventh International Congress of Zoology*, vol. 7 (Cambridge: Cambridge University Press Warehouse, 1912).

92 *"Morgan is a blockhead":* Schwartz, *In Pursuit of the Gene*, 174.

92 *"Cell biologists look; geneticists count; biochemists clean":* Arthur Kornberg, author interview, 1993.

92 *"We are interested in heredity not primarily":* "Review: Mendelism up to date," *Journal of Heredity* 7, no 1 (1916): 17–23.

92 *Walter Sutton, a grasshopper-collecting farm boy:* David Ellyard, *Who Discovered What When* (Frenchs Forest, New South Wales, Australia: New Holland, 2005), "Walter Sutton and Theodore Boveri: Where Are the Genes?"

93 *In 1905, using cells from the common mealworm:* Stephen G. Brush, "Nettie M. Stevens and the Discovery of Sex Determination by Chromosome," *Isis* 69, no. 2 (1978): 162–72.

93 *The students called his laboratory the Fly Room:* Ronald William Clark, *The Survival of Charles Darwin: A Biography of a Man and an Idea* (New York: Random House, 1984), 279.

94 *He had visited Hugo de Vries's:* Russ Hodge, *Genetic Engineering: Manipulating the Mechanisms of Life* (New York: Facts On File, 2009), 42.

95 *For Morgan, this genetic linkage:* Thomas Hunt Morgan, *The Mechanism of Mendelian Heredity* (New York: Holt, 1915), "Chapter 3: Linkage."

95 *genes had to be* physically *linked to each other:* Morgan was exceptionally lucky in choosing fruit flies for his experiments, since flies have an unusually low number of chromosomes—just four. If flies had multiple chromosomes, linkage might have been much harder to prove.

95 *It was a material* thing: Thomas Hunt Morgan, "The Relation of Genetics to Physiology and Medicine," Nobel Lecture (June 4, 1934), in *Nobel Lectures, Physiology and Medicine, 1922–1941* (Amsterdam: Elsevier, 1965), 315.

98 *The czarina of Russia, Alexandra:* Daniel L. Hartl and Elizabeth W. Jones, *Essential Genetics: A Genomics Perspective* (Boston: Jones and Bartlett, 2002), 96–97.

99 *Grigory Rasputin:* Helen Rappaport, *Queen Victoria: A Biographical Companion* (Santa Barbara, CA: ABC-CLIO, 2003), "Hemophilia."

99 *Rasputin was poisoned:* Andrew Cook, *To Kill Rasputin: The Life and Death of Grigori Rasputin* (Stroud, Gloucestershire: Tempus, 2005), "The End of the Road."

100 *On the evening of July 17, 1918:* "Alexei Romanov," *History of Russia*, http://historyofrussia.org/alexei-romanov/.

100 *In 2007, an archaeologist:* "DNA Testing Ends Mystery Surrounding Czar Nicholas II Children," *Los Angeles Times*, March 11, 2009.

Truths and Reconciliations

101 *All changed, changed utterly:* William Butler Yeats, *Easter, 1916* (London: Privately printed by Clement Shorter, 1916).

103 *In 1909, a young mathematician:* Eric C. R. Reeve and Isobel Black, *Encyclopedia of Genetics* (London: Fitzroy Dearborn, 2001), "Darwin and Mendel United: The Contributions of Fisher, Haldane and Wright up to 1932."

104 *In 1918, Fisher published:* Ronald Fisher, "The Correlation between Relatives on the Supposition of Mendelian Inheritance," *Transactions of the Royal Society of Edinburgh* 52 (1918): 399–433.

105 *Hugo de Vries had proposed that* mutations: Hugo de Vries, *The Mutation Theory; Experiments and Observations on the Origin of Species in the Vegetable Kingdom*, trans. J. B. Farmer and A. D. Darbishire (Chicago: Open Court, 1909).

105 *In the 1930s, Theodosius Dobzhansky:* Robert E. Kohler, *Lords of the Fly:* Drosophila *Genetics and the Experimental Life* (Chicago: University of Chicago Press, 1994), "From Laboratory to Field: Evolutionary Genetics."

106 *In September 1943, Dobzhansky:* Th. Dobzhansky, "Genetics of natural populations IX. Temporal changes in the composition of populations of *Drosophila pseudoobscura,*" *Genetics* 28, no. 2 (1943): 162.

109 *Dobzhansky could demonstrate it experimentally:* Details of Dobzhansky's experiments are sourced from Theodosius Dobzhansky, "Genetics of natural populations XIV. A response of certain gene arrangements in the third chromosome of *Drosophila pseudoobscura* to natural selection," *Genetics* 32, no. 2 (1947): 142; and S. Wright and T. Dobzhansky, "Genetics of natural populations; experimental reproduction of some of the changes caused by natural selection in certain populations of *Drosophila pseudoobscura,*" *Genetics* 31 (March 1946): 125–56. Also see T. Dobzhansky, Studies on Hybrid Sterility. II. Localization of Sterility Factors in *Drosophila Pseudoobscura* Hybrids. *Genetics* (March 1, 1936) vol 21, 113–135.

Transformation

111 *If you prefer an "academic life":* H. J. Muller, "The call of biology," *AIBS Bulletin* 3, no. 4 (1953). Copy with handwritten notes, http://libgallery.cshl.edu/archive/files /c73e9703aa1b65ca3f4881b9a2465797.jpg.

111 *We do deny that:* Peter Pringle, *The Murder of Nikolai Vavilov: The Story of Stalin's Persecution of One of the Great Scientists of the Twentieth Century* (Simon & Schuster, 2008), 209.

111 *Grand Synthesis:* Ernst Mayr and William B. Provine, *The Evolutionary Synthesis: Perspectives on the Unification of Biology* (Cambridge, MA: Harvard University Press, 1980).

112 *Transformation was discovered:* William K. Purves, *Life, the Science of Biology* (Sunderland, MA: Sinauer Associates, 2001), 214–15.

113 *Griffith performed an experiment:* Werner Karl Maas, *Gene Action: A Historical Account* (Oxford: Oxford University Press, 2001), 59–60.

114 *"this tiny man who . . . barely spoke above a whisper":* Alvin Coburn to Joshua Lederberg, November 19, 1965, Rockefeller Archives, Sleepy Hollow, NY, http://www .rockarch.org/.

114 *Griffith published his data:* Fred Griffith, "The significance of pneumococcal types," *Journal of Hygiene* 27, no. 2 (1928): 113–59.

114 *In 1920, Hermann Muller:* "Hermann J. Muller—biographical," http://www.nobel prize.org/nobel_prizes/medicine/laureates/1946/muller-bio.html.

115 *accumulated mutations—dozens of them:* H. J. Muller, "Artificial transmutation of the gene," *Science* 22 (July 1927): 84–87.

115 *In Darwin's scheme:* James F. Crow and Seymour Abrahamson, "Seventy years ago: Mutation becomes experimental," *Genetics* 147, no. 4 (1997): 1491.

116 *"There is no permanent* status quo *in nature":* Jack B. Bresler, *Genetics and Society* (Reading, MA: Addison-Wesley, 1973), 15.

116 *struck him as frankly sinister:* Kevles, *In the Name of Eugenics,* "A New Eugenics," 251–68.

117 *befriended the novelist and social activist Theodore Dreiser:* Sam Kean, *The Violinist's Thumb: And Other Lost Tales of Love, War, and Genius, as Written by Our Genetic Code* (Boston: Little, Brown, 2012), 33.

117 *The FBI launched:* William DeJong-Lambert, *The Cold War Politics of Genetic Research: An Introduction to the Lysenko Affair* (Dordrecht: Springer, 2012), 30.

Lebensunwertes Leben (Lives Unworthy of Living)

119 *He wanted to be God:* Robert Jay Lifton, *The Nazi Doctors: Medical Killing and the Psychology of Genocide* (New York: Basic Books, 2000), 359.

119 *A hereditarily ill person costs 50,000 reichsmarks:* Susan Bachrach, "In the name of public health—Nazi racial hygiene," *New England Journal of Medicine* 351 (2004): 417–19.

119 *Nazism, the biologist Fritz Lenz once said:* Erwin Baur, Eugen Fischer, and Fritz Lenz, *Human Heredity* (London: G. Allen & Unwin, 1931), 417. Also used by Hess, Hitler's deputy, the phrase was originally coined by Fritz Lenz as part of a review of *Mein Kampf.*

120 *had coined the phrase as early as 1895:* Alfred Ploetz. *Grundlinien Einer Rassen-Hygiene* (Berlin: S. Fischer, 1895); and Sheila Faith Weiss, "The race hygiene movement in Germany," *Osiris* 3 (1987): 193–236.

120 *In 1914, Ploetz's colleague Heinrich Poll:* Heinrich Poll, "Über Vererbung beim Menschen," *Die Grenzboten* 73 (1914): 308.

120 *Kaiser Wilhelm Institute for Anthropology:* Robert Wald Sussman, *The Myth of Race: The Troubling Persistence of an Unscientific Idea* (Cambridge, MA: Harvard University Press, 2014), "Funding of the Nazis by American Institutes and Businesses," 138.

120 *Hitler, imprisoned for leading the Beer Hall Putsch:* Harold Koenig, Dana King, and Verna B. Carson, *Handbook of Religion and Health* (Oxford: Oxford University Press, 2012), 294.

121 *Sterilization Law:* US Chief Counsel for the Prosecution of Axis Criminality, *Nazi Conspiracy and Aggression*, vol. 5 (Washington, DC: US Government Printing Office, 1946), document 3067-PS, 880–83 (English translation accredited to Nuremberg staff; edited by GHI staff).

121 *Films such as* Das Erbe: "Nazi Propaganda: Racial Science," USHMM Collections Search, http://collections.ushmm.org/search/catalog/fv3857.

121 *and* Erbkrank: "1936—Rassenpolitisches Amt der NSDAP—*Erbkrank*," Internet Archive, https://archive.org/details/1936-Rassenpolitisches-Amt-der-NSDAP-Erbkrank.

121 *in Leni Riefenstahl's* Olympia: *Olympia*, directed by Leni Riefenstahl, 1936.

121 *In November 1933:* "Holocaust timeline," History Place, http://www.historyplace.com/worldwar2/holocaust/timeline.html.

121 *In October 1935, the Nuremberg Laws:* "Key dates: Nazi racial policy, 1935," US Holocaust Memorial Museum, http://www.ushmm.org/outreach/en/article.php?ModuleId=10007696.

122 *By 1934, nearly five thousand adults:* "Forced sterilization," US Holocaust Memorial Museum, http://www.ushmm.org/learn/students/learning-materials-and-resources/mentally-and-physically-handicapped-victims-of-the-nazi-era/forced-sterilization.

122 *to euthanize their child, Gerhard:* Christopher R. Browning and Jürgen Matthäus, *The Origins of the Final Solution: The Evolution of Nazi Jewish Policy, September 1939–March 1942* (Lincoln: University of Nebraska, 2004), "Killing the Handicapped."

122 *Working with Karl Brandt:* Ulf Schmidt, *Karl Brandt: The Nazi Doctor, Medicine, and Power in the Third Reich* (London: Hambledon Continuum, 2007).

123 *No. 4 Tiergartenstrasse in Berlin:* Götz Aly, Peter Chroust, and Christian Pross, *Cleansing the Fatherland*, trans. Belinda Cooper (Baltimore: Johns Hopkins University Press, 1994), "Chapter 2: Medicine against the Useless."

124 *The Sterilization Law had achieved:* Roderick Stackelberg, *The Routledge Companion to Nazi Germany* (New York: Routledge, 2007), 303.

124 *"banality of evil":* Hannah Arendt, *Eichmann in Jerusalem: A Report on the Banality of Evil* (New York: Viking, 1963).

124 *In a rambling treatise entitled:* Otmar Verschuer and Charles E. Weber, *Racial Biology of the Jews* (Reedy, WV: Liberty Bell Publishing, 1983).

125 *First they came for the Socialists:* J. Simkins, "Martin Niemoeller," Spartacus Educational Publishers, 2012, www. spartacus.schoolnet.co.uk/GERniemoller.htm.

126 *Trofim Lysenko:* Jacob Darwin Hamblin, *Science in the Early Twentieth Century: An Encyclopedia* (Santa Barbara, CA: ABC-CLIO, 2005), "Trofim Lysenko," 188–89.

126 *"gives one the feeling of a toothache":* David Joravsky, *The Lysenko Affair* (Chicago: University of Chicago Press, 2010), 59. Also see Zhores A. Medvedev, *The Rise and Fall of T. D. Lysenko*, trans. I. Michael Lerner (New York: Columbia University Press, 1969), 11–16.

126 *The gene, he argued:* T. Lysenko, *Agrobiologia*, 6th ed. (Moscow: Selkhozgiz, 1952), 602–6.

127 *In 1940, Lysenko:* "Trofim Denisovich Lysenko," *Encyclopaedia Britannica Online,* http://www.britannica.com/biography/Trofim-Denisovich-Lysenko.

127 *"I am nothing but dung now":* Pringle, *Murder of Nikolai Vavilov,* 278.

127 *died a few weeks later:* A number of Vavilov's colleagues, including Karpechenko, Govorov, Levitsky, Kovalev, and Flayksberger, were also arrested. Lysenko's influence virtually emptied the Soviet academy of all geneticists. Biology in the Soviet Union would be hobbled for decades.

128 *Having coined the phrase:* James Tabery, *Beyond Versus: The Struggle to Understand the Interaction of Nature and Nurture* (Cambridge, MA: MIT Press, 2014), 2.

128 *In 1924, Hermann Werner Siemens:* Hans-Walter Schmuhl, *The Kaiser Wilhelm Institute for Anthropology, Human Heredity, and Eugenics, 1927–1945: Crossing Boundaries* (Dordrecht: Springer, 2008), "Twin Research."

129 *Between 1943 and 1945:* Gerald L. Posner and John Ware, *Mengele: The Complete Story* (New York: McGraw-Hill, 1986).

130 *"We were always sitting together—always nude":* Lifton, *Nazi Doctors,* 349.

131 *In April 1933:* Wolfgang Benz and Thomas Dunlap, *A Concise History of the Third Reich* (Berkeley: University of California Press, 2006), 142.

131 *"Hitler may have ruined":* George Orwell, *In Front of Your Nose, 1946–1950,* ed. Sonia Orwell and Ian Angus (Boston: D. R. Godine, 2000), 11.

132 *a lecture later published as:* Erwin Schrödinger, *What Is Life?: The Physical Aspect of the Living Cell* (Cambridge: Cambridge University Press, 1945).

"That Stupid Molecule"

133 *Never underestimate the power of . . . stupidity:* Walter W. Moore Jr., *Wise Sayings: For Your Thoughtful Consideration* (Bloomington, IN: AuthorHouse, 2012), 89.

133 *"The Fess":* "The Oswald T. Avery Collection: Biographical information," National Institutes of Health, http://profiles.nlm.nih.gov/ps/retrieve/Narrative/CC/p-nid/35.

134 *No one knew or understood the chemical structure:* Robert C. Olby, *The Path to the Double Helix: The Discovery of DNA* (New York: Dover Publications, 1994), 107.

134 *Swiss biochemist, Friedrich Miescher:* George P. Sakalosky, *Notio Nova: A New Idea* (Pittsburgh, PA: Dorrance, 2014), 58.

135 *extremely "unsophisticated" structure:* Olby, *Path to the Double Helix,* 89.

135 *"stupid molecule":* Garland Allen and Roy M. MacLeod, eds., *Science, History and Social Activism: A Tribute to Everett Mendelsohn,* vol. 228 (Dordrecht: Springer Science & Business Media, 2013), 92.

135 *"structure-determining, supporting substance":* Olby, *Path to the Double Helix,* 107.

137 *"primordial sea":* Richard Preston, *Panic in Level 4: Cannibals, Killer Viruses, and Other Journeys to the Edge of Science* (New York: Random House, 2009), 96.

137 *"Who could have guessed it?":* Letter from Oswald T. Avery to Roy Avery, May 26, 1943, Oswald T. Avery Papers, Tennessee State Library and Archives.

137 *Avery wanted to be doubly sure:* Maclyn McCarty, *The Transforming Principle: Discovering That Genes Are Made of DNA* (New York: W. W. Norton, 1985), 159.

137 *"cloth from which genes were cut":* Lyon and Gorner, *Altered Fates,* 42.

137 *Oswald Avery's paper on DNA was published:* O. T. Avery, Colin M. MacLeod, and Maclyn McCarty, "Studies on the chemical nature of the substance inducing transformation of pneumococcal types: Induction of transformation by a deoxyribonucleic

acid fraction isolated from pneumococcus type III," *Journal of Experimental Medicine* 79, no. 2 (1944): 137–58.

137 *That year, an estimated 450,000 were gassed:* US Holocaust Memorial Museum, "Introduction to the Holocaust," *Holocaust Encyclopedia*, http://www.ushmm.org/wlc /en/article.php?ModuleId=10005143.

137 *In the early spring of 1945:* Ibid.

138 *The Eugenics Record Office:* Steven A. Farber, "U.S. scientists' role in the eugenics movement (1907–1939): A contemporary biologist's perspective," *Zebrafish* 5, no. 4 (2008): 243–45.

"Important Biological Objects Come in Pairs"

139 *One could not be a successful scientist:* James D. Watson, *The Double Helix: A Personal Account of the Discovery of the Structure of DNA* (London: Weidenfeld & Nicolson, 1981), 13.

139 *It is the molecule that has the glamour:* Francis Crick, *What Mad Pursuit: A Personal View of Scientific Discovery* (New York: Basic Books, 1988), 67.

139 *Science [would be] ruined:* Donald W. Braben, *Pioneering Research: A Risk Worth Taking* (Hoboken, NJ: John Wiley & Sons, 2004), 85.

140 *Among the early converts:* Maurice Wilkins, *Maurice Wilkins: The Third Man of the Double Helix: An Autobiography* (Oxford: Oxford University Press, 2003).

140 *Ernest Rutherford:* Richard Reeves, *A Force of Nature: The Frontier Genius of Ernest Rutherford* (New York: W. W. Norton, 2008).

140 *"Life . . . is a chemical incident":* Arthur M. Silverstein, *Paul Ehrlich's Receptor Immunology: The Magnificent Obsession* (San Diego, CA: Academic, 2002), 2.

143 *Wilkins found an X-ray diffraction machine:* Maurice Wilkins, correspondence with Raymond Gosling on the early days of DNA research at King's College, 1976, Maurice Wilkins Papers, King's College London Archives.

144 *It was, as one friend of Franklin's:* Letter of June 12, 1985, notes on Rosalind Franklin, Maurice Wilkins Papers, no. ad92d68f-4071-4415-8df2-dcfe041171fd.

144 *the relationship soon froze into frank, glacial hostility:* Daniel M. Fox, Marcia Meldrum, and Ira Rezak, *Nobel Laureates in Medicine or Physiology: A Biographical Dictionary* (New York: Garland, 1990), 575.

144 *She "barks often, doesn't succeed in biting me":* James D. Watson, *The Annotated and Illustrated Double Helix*, ed. Alexander Gann and J. A. Witkowski (New York: Simon & Schuster, 2012), letter to Crick, 151.

144 *"Now she's trying to drown me":* Brenda Maddox, *Rosalind Franklin: The Dark Lady of DNA* (New York: HarperCollins, 2002), 164.

144 *Franklin found most of her male colleagues "positively repulsive":* Watson, *Annotated and Illustrated Double Helix*, letter from Rosalind Franklin to Anne Sayre, March 1, 1952, 67.

144 *It was not just sexism:* Crick never believed that Franklin was affected by sexism. Unlike Watson, who eventually wrote a generous recapitulation of Franklin's work highlighting the adversities that she had faced as a scientist, Crick maintained that Franklin was unaffected by the atmosphere at King's. Franklin and Crick would eventually become close friends in the late 1950s; Crick and his wife were especially helpful to Franklin during her prolonged illness and in the months preceding her untimely death. Crick's fondness for Franklin can be found in Crick, *What Mad Pursuit*, 82–85.

145 *passionate Marie Curie, with her chapped palms:* "100 years ago: Marie Curie wins 2nd Nobel Prize," *Scientific American*, October 28, 2011, http://www.scientific american.com/article/curie-marie-sklodowska-greatest-woman-scientist/.

145 *ethereal Dorothy Hodgkin at Oxford:* "Dorothy Crowfoot Hodgkin—biographical," Nobelprize.org, http://www.nobelprize.org/nobel_prizes/chemistry/laureates/1964 /hodgkin-bio.html.

145 *an "affable looking housewife":* Athene Donald, "Dorothy Hodgkin and the year of crystallography," *Guardian*, January 14, 2014.

145 *ingenious apparatus that bubbled hydrogen:* "The DNA riddle: King's College, London, 1951–1953," Rosalind Franklin Papers, http://profiles.nlm.nih.gov/ps/retrieve /Narrative/KR/p-nid/187.

145 *J. D. Bernal, the crystallographer:* J. D. Bernal, "Dr. Rosalind E. Franklin," *Nature* 182 (1958): 154.

145 *"shirttails flying, knees in the air":* Max F. Perutz, *I Wish I'd Made You Angry Earlier: Essays on Science, Scientists, and Humanity* (Cold Spring Harbor, NY: Cold Spring Harbor Laboratory Press, 1998), 70.

146 *Wilkins showed little, if any, excitement:* Watson Fuller, "For and against the helix," Maurice Wilkins Papers, no. 00c0a9ed-e951-4761-955c-7490e0474575.

146 *"Before Maurice's talk":* Watson, *Double Helix*, 23.

146 *"Maurice was English":* http://profiles.nlm.nih.gov/ps/access/SCBBKH.pdf.

146 *"nothing about the X-ray diffraction":* Watson, *Double Helix*, 22.

146 *"a complete flop":* Ibid., 18.

146 *"The fact that I was unable":* Ibid., 24.

146 *Watson had moved to Cambridge:* Officially, Watson had moved to Cambridge to help Perutz and another scientist, John Kendrew, with their work on a protein called myoglobin. Watson then switched to the study of the structure of a virus called tobacco mosaic virus, or TMV. But he was vastly more interested in DNA and soon abandoned all other projects to focus on DNA. Watson, *Annotated and Illustrated Double Helix*, 127.

147 *"A youthful arrogance":* Crick, *What Mad Pursuit*, 64.

147 *"The trouble is, you see, that there is":* Watson, *Annotated and Illustrated Double Helix*, 107.

148 *Pauling's seminal paper:* L. Pauling, R. B. Corey, and H. R. Branson, "The structure of proteins: Two hydrogen-bonded helical configurations of the polypeptide chain," *Proceedings of the National Academy of Sciences* 37, no. 4 (1951): 205–11.

148 *"product of common sense":* Watson, *Annotated and Illustrated Double Helix*, 44.

148 *"like trying to determine the structure of a piano":* http://www.diracdelta.co.uk/science /source/c/r/crick%20francis/source.html#.Vh8XlaJeGKI.

149 *The experimental data would generate the models:* Crick, *What Mad Pursuit*, 100–103. Crick always maintained that Franklin fully understood the importance of model building.

149 *"How dare you interpret my data for me?":* Victor K. McElheny, *Watson and DNA: Making a Scientific Revolution* (Cambridge, MA: Perseus, 2003), 38.

150 *"Big helix with several chains":* Alistair Moffat, *The British: A Genetic Journey* (Edinburgh: Birlinn, 2014); and from Rosalind Franklin's laboratory notebooks, dated 1951.

150 *"Superficially, the X-ray data":* Watson, *Annotated and Illustrated Double Helix,* 73.

151 *"check it with":* Ibid.

151 *Wilkins, Franklin, and her student, Ray Gosling:* Bill Seeds and Bruce Fraser accompanied them on this visit.

151 *As Gosling recalled, "Rosalind let rip"*: Watson, *Annotated and Illustrated Double Helix*, 91.

152 *"His mood"*: Ibid., 92.

153 *In the first weeks of January 1953*: Linus Pauling and Robert B. Corey, "A proposed structure for the nucleic acids," *Proceedings of the National Academy of Sciences* 39, no. 2 (1953): 84–97.

153 *"V.Good. Wet Photo"*: http://profiles.nlm.nih.gov/ps/access/KRBBJF.pdf.

154 *"important biological objects come in pairs"*: Watson, *Double Helix*, 184.

155 *he would later write defensively*: Anne Sayre, *Rosalind Franklin & DNA* (New York: W. W. Norton, 1975), 152.

155 *"Suddenly I became aware"*: Watson, *Annotated and Illustrated Double Helix*, 207.

156 *"Upon his arrival"*: Ibid., 208.

156 *"winged into the Eagle"*: Ibid., 209.

157 *"We see it as a rather stubby double helix"*: John Sulston and Georgina Ferry, *The Common Thread: A Story of Science, Politics, Ethics, and the Human Genome* (Washington, DC: Joseph Henry Press, 2002), 3.

158 *Maurice Wilkins came to take a look*: Most likely on March 11 or 12, 1953. Crick informed Delbrück of the model on Thursday, March 12. Also see Watson Fuller, "Who said helix?" with related papers, Maurice Wilkins Papers, no. c065700f-b6d9-46cf -902a-b4f8e078338a.

158 *"The model was standing high"*: June 13, 1996, Maurice Wilkins Papers.

158 *"I think you're a couple of old rogues"*: Letter from Maurice Wilkins to Francis Crick, March 18, 1953, Wellcome Library, Letter Reference no. 62b87535-040a-448c-9b73 -ff3a3767db91. http://wellcomelibrary.org/player/b20047198#?asi=0&ai=0&z=0.12 15%2C0.2046%2C0.5569%2C0.3498.

158 *"I like the idea"*: Fuller, "Who said helix?" with related papers.

158 *"The positioning of the backbone"*: Watson, *Annotated and Illustrated Double Helix*, 222.

158 *On April 25, 1953*: J. D. Watson and F. H. C. Crick, "Molecular structure of nucleic acids: A structure for deoxyribose nucleic acid," *Nature* 171 (1953): 737–38.

159 *"the enigma of how the vast amount"*: Fuller, "Who said helix?" with related papers.

"That Damned, Elusive Pimpernel"

161 *In the protein molecule*: "1957: Francis H. C. Crick (1916–2004) sets out the agenda of molecular biology," *Genome News Network*, http://www.genomenewsnetwork.org /resources/timeline/1957_Crick.php.

161 *In 1941*: "1941: George W. Beadle (1903–1989) and Edward L. Tatum (1909–1975) show how genes direct the synthesis of enzymes that control metabolic processes," *Genome News Network*, http://www.genomenewsnetwork.org/resources/timeline /1941_Beadle_Tatum.php.

161 *a student of Thomas Morgan's*: Edward B. Lewis, "Thomas Hunt Morgan and his legacy," Nobelprize.org, http://www.nobelprize.org/nobel_prizes/medicine/laureates /1933/morgan-article.html.

162 *the "action" of a gene*: Frank Moore Colby et al., *The New International Year Book: A Compendium of the World's Progress, 1907–1965* (New York: Dodd, Mead, 1908), 786.

163 *"A gene," Beadle wrote in 1945*: George Beadle, "Genetics and metabolism in *Neurospora*," *Physiological reviews* 25, no. 4 (1945): 643–63.

164 *"For over a year"*: James D. Watson, *Genes, Girls, and Gamow: After the Double Helix* (New York: Alfred A. Knopf, 2002), 31.

164 *"I am playing with complex organic"*: http://scarc.library.oregonstate.edu/coll/pauling /dna/corr/sci9.001.43-gamow-lp-19531022-transcript.html.

164 *Gamow called it the RNA Tie Club*: Ted Everson, *The Gene: A Historical Perspective* (Westport, CT: Greenwood, 2007), 89–91.

164 *"It always had a rather ethereal existence"*: "Francis Crick, George Gamow, and the RNA Tie Club," Web of Stories. http://www.webofstories.com/play/francis.crick/84.

164 *"Do or die, or don't try"*: Sam Kean, *The Violinist's Thumb: And Other Lost Tales of Love, War, and Genius, as Written by Our Genetic Code* (New York: Little, Brown, 2012).

164 *was required for the translation of DNA into proteins*: Arthur Pardee and Monica Riley had also proposed a variant of this idea.

165 *Is he in heaven, is he in hell?*: Cynthia Brantley Johnson, *The Scarlet Pimpernel* (Simon & Schuster, 2004), 124.

166 "It's the magnesium": "Albert Lasker Award for Special Achievement in Medical Science: Sydney Brenner," Lasker Foundation, http://www.laskerfoundation.org /awards/2000special.htm.

166 *Like DNA, these RNA molecules were built*: Two other scientists, Elliot Volkin and Lazarus Astrachan, had proposed an RNA intermediate for genes in 1956. The two seminal papers published by the Brenner/Jacob group and the Watson/Gilbert group in 1961 are: F. Gros et al., "Unstable ribonucleic acid revealed by pulse labeling of Escherichia coli," *Nature* 190 (May 13, 1960): 581–85; and S. Brenner, F. Jacob, and M. Meselson, "An unstable intermediate carrying information from genes to ribosomes for protein synthesis," *Nature* 190 (May 13, 1960): 576–81.

167 *"It seems likely . . . that the precise sequence"*: J. D. Watson and F. H. C. Crick, "Genetical implications of the structure of deoxyribonucleic acid," *Nature* 171, no. 4361 (1953): 965.

170 *In 1904, a single image*: David P. Steensma, Robert A. Kyle, and Marc A. Shampo, "Walter Clement Noel—first patient described with sickle cell disease," *Mayo Clinic Proceedings* 85, no. 10 (2010).

170 *In 1951, working with Harvey Itano*: "Key participants: Harvey A. Itano," *It's in the Blood! A Documentary History of Linus Pauling, Hemoglobin, and Sickle Cell Anemia*, http://scarc.library.oregonstate.edu/coll/pauling/blood/people/itano.html.

Regulation, Replication, Recombination

172 *It is absolutely necessary to find the origin*: Quoted in Sean Carrol, *Brave Genius: A Scientist, a Philosopher, and Their Daring Adventures from the French Resistance to the Nobel Prize* (New York: Crown, 2013), 133.

173 *"the properties implicit in genes"*: Thomas Hunt Morgan, "The relation of genetics to physiology and medicine," *Scientific Monthly* 41, no. 1 (1935): 315.

173 *Jacques Monod, the French biologist*: Agnes Ullmann, "Jacques Monod, 1910–1976: His life, his work and his commitments," *Research in Microbiology* 161, no. 2 (2010): 68–73.

177 *Pardee, Jacob, and Monod published*: Arthur B. Pardee, François Jacob, and Jacques Monod, "The genetic control and cytoplasmic expression of 'inducibility' in the synthesis of β=galactosidase by *E. coli*," *Journal of Molecular Biology* 1, no. 2 (1959): 165–78.

178 *"The genome contains"*: François Jacob and Jacques Monod, "Genetic regulatory mechanisms in the synthesis of proteins," *Journal of Molecular Biology* 3, no. 3 (1961): 318–56.

179 *1953 paper:* Watson and Crick, "Molecular structure of nucleic acids," 738.

180 *"He called it DNA polymerase":* Arthur Kornberg, "Biologic synthesis of deoxyribo-nucleic acid," *Science* 131, no. 3412 (1960): 1503–8.

180 *"Five years ago":* Ibid.

From Genes to Genesis

185 *In the beginning:* Richard Dawkins, *The Selfish Gene* (Oxford: Oxford University Press, 1989), 12.

185 *Am not I:* Nicholas Marsh, *William Blake: The Poems* (Houndmills, Basingstoke, England: Palgrave, 2001), 56.

186 *Lewis studied mutants:* Many of these mutants had initially been created by Alfred Sturtevant and Calvin Bridges. Details of the mutants and the relevant genes can be found in Ed Lewis's Nobel lecture, December 8, 1995.

190 *"Is it sin":* Friedrich Max Müller, *Memories: A Story of German Love* (Chicago: A. C. McClurg, 1902), 20.

190 *In Leo Lionni's classic children's book:* Leo Lionni, *Inch by Inch* (New York: I. Obolen-sky, 1960).

191 *"We propose to identify every cell in the worm":* James F. Crow and W. F. Dove, *Perspectives on Genetics: Anecdotal, Historical, and Critical Commentaries, 1987–1998* (Madison: University of Wisconsin Press, 2000), 176.

191 *"like watching a bowl of hundreds of grapes":* Robert Horvitz, author interview, 2012.

192 *"There is no history":* Ralph Waldo Emerson, *The Journals and Miscellaneous Notebooks of Ralph Waldo Emerson*, vol. 7, ed. William H. Gilman (Cambridge, MA: Belknap Press of Harvard University Press, 1960), 202.

193 *131 extra cells had somehow disappeared:* Ning Yang and Ing Swie Goping, *Apoptosis* (San Rafael, CA: Morgan & Claypool Life Sciences, 2013), "*C. elegans* and Discovery of the Caspases."

193 *he called it* apoptosis: John F. R. Kerr, Andrew H. Wyllie, and Alastair R. Currie, "Apoptosis: A basic biological phenomenon with wide-ranging implications in tissue kinetics," *British Journal of Cancer* 26, no. 4 (1972): 239.

193 *In another mutant, dead cells:* This mutant was initially identified by Ed Hedgecock. Robert Horvitz, author interview, 2013.

194 *Horvitz and Sulston discovered:* J. E. Sulston and H. R. Horvitz, "Post-embryonic cell lineages of the nematode, *Caenorhabditis elegans*," *Developmental Biology* 56. no. 1 (March 1977): 110–56. Also see Judith Kimble and David Hirsh, "The postembryonic cell lineages of the hermaphrodite and male gonads in *Caenorhabditis elegans*," *Developmental Biology* 70, no. 2 (1979): 396–417.

194 *But even natural ambiguity:* Judith Kimble, "Alterations in cell lineage following laser ablation of cells in the somatic gonad of *Caenorhabditis elegans*," *Developmental Biology* 87, no. 2 (1981): 286–300.

195 *The British way, Brenner wrote:* W. J. Gehring, *Master Control Genes in Development and Evolution: The Homeobox Story* (New Haven, CT: Yale University Press, 1998), 56.

195 *began to study the effects of sharp perturbations on cell fates:* The method had been pioneered by John White and John Sulston. Robert Horvitz, author interview, 2013.

196 *As one scientist described it:* Gary F. Marcus, *The Birth of the Mind: How a Tiny Number of Genes Creates the Complexities of Human Thought* (New York: Basic Books, 2004), "Chapter 4: Aristotle's Impetus."

196 *The geneticist Antoine Danchin:* Antoine Danchin, *The Delphic Boat: What Genomes Tell Us* (Cambridge, MA: Harvard University Press, 2002).

197 *Some genes, Dawkins suggests:* Richard Dawkins, *A Devil's Chaplain: Reflections on Hope, Lies, Science, and Love* (Boston: Houghton Mifflin, 2003), 105.

PART THREE: "THE DREAMS OF GENETICISTS"

202 *Progress in science depends on new techniques:* Sydney Brenner, "Life sentences: Detective Rummage investigates," *Scientist—the Newspaper for the Science Professional* 16, no. 16 (2002): 15.

202 *If we are right . . . it is possible to induce:* "DNA as the 'stuff of genes': The discovery of the transforming principle, 1940–1944," Oswald T. Avery Collection, National Institutes of Health, http://profiles.nlm.nih.gov/ps/retrieve/Narrative/CC/p-nid/157.

"Crossing Over"

203 *A biochemist by training:* Details of Paul Berg's education and sabbatical are from the author's interview with Paul Berg, 2013; and "The Paul Berg Papers," Profiles in Science, National Library of Medicine, http://profiles.nlm.nih.gov/CD/.

204 *a "piece of bad news wrapped in a protein coat":* M. B. Oldstone, "Rous-Whipple Award Lecture. Viruses and diseases of the twenty-first century," *American Journal of Pathology* 143, no. 5 (1993): 1241.

204 *Unlike many viruses, Berg learned:* David A. Jackson, Robert H. Symons, and Paul Berg, "Biochemical method for inserting new genetic information into DNA of simian virus 40: circular SV40 DNA molecules containing lambda phage genes and the galactose operon of Escherichia coli," *Proceedings of the National Academy of Sciences* 69, no. 10 (1972): 2904–09.

205 *Peter Lobban, had written a thesis:* P. E. Lobban, "The generation of transducing phage in vitro," (essay for third PhD examination, Stanford University, November 6, 1969).

205 *Avery, after all, had boiled it:* Oswald T. Avery, Colin M. MacLeod, and Maclyn McCarty. "Studies on the chemical nature of the substance inducing transformation of pneumococcal types: Induction of transformation by a desoxyribonucleic acid fraction isolated from pneumococcus type III," *Journal of Experimental Medicine* 79, no. 2 (1944): 137–58.

207 *"none of the individual procedures, manipulations, and reagents:* P. Berg and J. E. Mertz, "Personal reflections on the origins and emergence of recombinant DNA technology," *Genetics* 184, no. 1 (2010): 9–17, doi:10.1534/genetics.109.112144.

207 *In the winter of 1970, Berg and David Jackson:* Jackson, Symons, and Berg, "Biochemical method for inserting new genetic information into DNA of simian virus 40," *Proceedings of the National Academy of Sciences* 69, no. 10 (1972): 2904–09.

209 *In June 1971, Mertz traveled from Stanford:* Kathi E. Hanna, ed., *Biomedical politics* (Washington, DC: National Academies Press, 1991), 266.

210 *"You can stop splitting the atom":* Erwin Chargaff, "On the dangers of genetic meddling," *Science* 192, no. 4243 (1976): 938.

210 *"My first reaction was: this was absurd":* "Reaction to Outrage over Recombinant DNA, Paul Berg." DNA Learning Center, doi:https://www.dnalc.org/view/15017-Reaction -to-outrage-over-recombinant-DNA-Paul-Berg.html.

210 *Dulbecco had even offered to* drink *SV40*: Shane Crotty, *Ahead of the Curve: David Baltimore's Life in Science* (Berkeley: University of California Press, 2001), 95.

210 *"In truth, I knew the risk was little"*: Paul Berg, author interview, 2013.

211 *"Janet really made the process vastly more efficient"*: Ibid.

211 *Boyer had arrived in San Francisco in the summer of '66*: Details of the story of Boyer and Cohen come from the following resources: John Archibald, *One Plus One Equals One: Symbiosis and the Evolution of Complex Life* (Oxford: Oxford University Press, 2014). Also see Stanley N. Cohen et al., "Construction of biologically functional bacterial plasmids in vitro," *Proceedings of the National Academy of Sciences* 70, no. 11 (1973): 3240–44.

212 *Late that evening, Boyer:* Details of this episode are from several sources including Stanley Falkow, "I'll Have the Chopped Liver Please, Or How I Learned to Love the Clone," *ASM News* 67, no. 11 (2001); Paul Berg, author interview, 2015; Jane Gitschier, "Wonderful life: An interview with Herb Boyer," *PLOS Genetics* (September 25, 2009).

The New Music

215 *Each generation needs a new music:* Crick, *What Mad Pursuit*, 74.

215 *People now made music from everything:* Richard Powers, *Orfeo: A Novel* (New York: W. W. Norton, 2014), 330.

216 *In the early 1950s, Sanger had solved:* Frederick Sanger, "The arrangement of amino acids in proteins," *Advances in Protein Chemistry* 7 (1951): 1–67.

216 *Frederick Banting, and his medical student:* Frederick Banting et al., "The effects of insulin on experimental hyperglycemia in rabbits," *American Journal of Physiology* 62, no. 3 (1922).

217 *In 1958, Sanger won the Nobel Prize:* "The Nobel Prize in Chemistry 1958," Nobel prize.org, http://www.nobelprize.org/nobel_prizes/chemistry/laureates/1958/.

217 *his "lean years":* Frederick Sanger, *Selected Papers of Frederick Sanger: With Commentaries*, vol. 1, ed. Margaret Dowding (Singapore: World Scientific, 1996), 11–12.

217 *In the summer of 1962, Sanger moved:* George G. Brownlee, *Fred Sanger—Double Nobel Laureate: A Biography* (Cambridge: Cambridge University Press, 2014), 20.

218 *On February 24, 1977, Sanger used:* F. Sanger et al., "Nucleotide sequence of bacteriophage ΦI74 DNA," *Nature* 265, no. 5596 (1977): 687–95, doi:10.1038/265687a0.

218 *"The sequence identifies many of the features":* Ibid.

219 *In 1977, two scientists working independently:* Sayeeda Zain et al., "Nucleotide sequence analysis of the leader segments in a cloned copy of adenovirus 2 fiber mRNA," *Cell* 16, no. 4 (1979): 851–61. Also see "Physiology or Medicine 1993—press release," Nobelprize.org, http://www.nobelprize.org/nobel_prizes/medicine/laureates/1993 /press.html.

222 *The "arsenal of chemical manipulations":* Walter Sullivan, "Genetic decoders plumbing the deepest secrets of life processes," *New York Times*, June 20, 1977.

222 *"Genetic engineering . . . implies deliberate":* Jean S. Medawar, *Aristotle to Zoos: A Philosophical Dictionary of Biology* (Cambridge, MA: Harvard University Press, 1985), 37–38.

222 *"By learning to manipulate genes experimentally":* Paul Berg, author interview, September 2015.

222 *T cells sense the presence of invading cells*: J. P Allison, B. W. McIntyre, and D. Bloch, "Tumor-specific antigen of murine T-lymphoma defined with monoclonal antibody," *Journal of Immunology* 129 (1982): 2293–2300; K. Haskins et al, "The major his-

tocompatibility complex-restricted antigen receptor on T cells: I. Isolation with a monoclonal antibody," *Journal of Experimental Medicine* 157 (1983): 1149–69.

223   *In 1970, David Baltimore and Howard Temin*: "Physiology or Medicine 1975—Press Release," Nobelprize.org. Nobel Media AB 2014. Web. 5 Aug 2015. http://www.nobel prize.org/nobel_prizes/medicine/laureates/1975/press.html.

224   *In 1984, this technique was deployed*: S. M. Hedrick et al., "Isolation of cDNA clones encoding T cell-specific membrane-associated proteins," *Nature* 308 (1984): 149–53; Y. Yanagi et al., "A human T cell-specific cDNA clone encodes a protein having extensive homology to immunoglobulin chains," *Nature* 308 (1984): 145–49.

224   *"liberated by cloning"*: Steve McKnight, "Pure genes, pure genius," *Cell* 150, no. 6 (September 14, 2012): 1100–1102.

Einsteins on the Beach

225   *I believe in the inalienable right*: Sydney Brenner, "The influence of the press at the Asilomar Conference, 1975," Web of Stories, http://www.webofstories.com/play/sydney .brenner/182;jsessionid=2c147f1c4222a58715e708eabd868e58.

225   *In the summer of 1972*: Crotty, *Ahead of the Curve*, 93.

226   *"the beginning of a new era"*: Herbert Gottweis, *Governing Molecules: The Discursive Politics of Genetic Engineering in Europe and the United States* (Cambridge, MA: MIT Press, 1998).

227   *"Asilomar I," as Berg would later call*: Details of Berg's account of Asilomar come from conversations and interviews with Paul Berg, 1993 and 2013; and Donald S. Fredrickson, "Asilomar and recombinant DNA: The end of the beginning," in *Biomedical Politics*, ed. Hanna, 258–92.

227   *The Asilomar conference produced an important book*: Alfred Hellman, Michael Neil Oxman, and Robert Pollack, *Biohazards in Biological Research* (Cold Spring Harbor, NY: Cold Spring Harbor Laboratory Press, 1973).

227   *summer of 1973 when Boyer and Cohen*: Cohen et al., "Construction of biologically functional bacterial plasmids," 3240–44.

227   *"'safe' viruses, plasmids and bacteria"*: Crotty, *Ahead of the Curve*, 99.

228   *"Well, if we had any guts at all"*: Ibid.

228   *"Don't put toxin genes into E. coli"*: "The moratorium letter regarding risky experiments, Paul Berg," DNA Learning Center, https://www.dnalc.org/view/15021-The -moratorium-letter-regarding-risky-experiments-Paul-Berg.html.

228   *In 1974, the "Berg letter" ran*: P. Berg et al., "Potential biohazards of recombinant DNA molecules," *Science* 185 (1974): 3034. See also *Proceedings of the National Academy of Sciences* 71 (July 1974): 2593–94.

229   *"are specious"*: Herb Boyer interview, 1994, by Sally Smith Hughes, UCSF Oral History Program, Bancroft Library, University of California, Berkeley, http://content .cdlib.org/view?docId=kt5d5nb0zs&brand=calisphere&doc.view=entire_text.

229   *On New Year's Day 1974*: John F. Morrow et al., "Replication and transcription of eukaryotic DNA in *Escherichia coli*," *Proceedings of the National Academy of Sciences* 71, no. 5 (1974): 1743–47.

229   *Asilomar II—one of the most unusual*: Paul Berg et al., "Summary statement of the Asilomar Conference on recombinant DNA molecules," *Proceedings of the National Academy of Sciences* 72, no. 6 (1975): 1981–84.

230   *"You fucked the plasmid group"*: Crotty, *Ahead of the Curve*, 107.

230 *He was promptly accused of:* Brenner, "The influence of the press."

230 *"Some people got sick of it all":* Crotty, *Ahead of the Curve*, 108.

231 *"The new techniques, which permit":* Gottweis, *Governing Molecules*, 88.

231 *To mitigate the risks, the document:* Berg et al., "Summary statement of the Asilomar Conference," 1981–84.

232 *two-page letter written in August 1939:* Albert Einstein, "Letter to Roosevelt, August 2, 1939," Albert Einstein's Letters to Franklin Delano Roosevelt, http://hypertext book.com/eworld/einstein.shtml#first.

232 *As Alan Waterman, the head:* Attributed to Alan T. Waterman, in Lewis Branscomb, "Foreword," *Science, Technology, and Society, a Prospective Look: Summary and Conclusions of the Bellagio Conference* (Washington, DC: National Academy of Sciences, 1976).

232 *Nixon, fed up with his scientific advisers:* F. A. Long, "President Nixon's 1973 Reorganization Plan No. 1," *Science and Public Affairs* 29, no. 5 (1973): 5.

233 *"was to demonstrate that scientists were capable":* Paul Berg, author interview, 2013.

233 *"The public's trust was undeniably increased":* Paul Berg, "Asilomar and recombinant DNA," Nobelprize.org, http://www.nobelprize.org/nobel_prizes/chemistry/laureates /1980/berg-article.html.

233 *"Did the organizers and participants":* Ibid.

"Clone or Die"

236 *If you know the question:* Herbert W. Boyer, "Recombinant DNA research at UCSF and commercial application at Genentech: Oral history transcript, 2001," Online Archive of California, 124, http://www.oac.cdlib.org/search?style=oac4;titlesAZ=r ;idT=UCb11453293x.

236 *Any sufficiently advanced technology:* Arthur Charles Clark, *Profiles of the Future: An Inquiry Into the Limits of the Possible* (New York: Harper & Row, 1973).

236 *"may completely change the pharmaceutical industry's":* Doogab Yi, *The Recombinant University: Genetic Engineering and the Emergence of Stanford Biotechnology* (Chicago: University of Chicago Press, 2015), 2.

236 *In May, the* San Francisco Chronicle *ran:* "Getting Bacteria to Manufacture Genes," *San Francisco Chronicle*, May 21, 1974.

237 *Cohen also received:* Roger Lewin, "A View of a Science Journalist," in *Recombinant DNA and Genetic Experimentation*, ed. J. Morgan and W. J. Whelan (London: Elsevier, 2013), 273.

237 *Cohen and Boyer filed a patent:* "1972: First recombinant DNA," Genome.gov, http:// www.genome.gov/25520302.

237 *"to commercial ownership of the techniques for cloning all possible DNAs":* P. Berg and J. E. Mertz, "Personal reflections on the origins and emergence of recombinant DNA technology," *Genetics* 184, no. 1 (2010): 9–17, doi:10.1534/genetics.109.112144.

238 *Swanson came to see Boyer in January 1976:* Sally Smith Hughes, *Genentech: The Beginnings of Biotech* (Chicago: University of Chicago Press, 2011), "Prologue."

239 *Boyer rejected Swanson's suggestion of HerBob:* Felda Hardymon and Tom Nicholas, "Kleiner-Perkins and Genentech: When venture capital met science," Harvard Business School Case 813-102, October 2012, http://www.hbs.edu/faculty/Pages/item.aspx?num=43569.

239 *In 1869, a Berlin medical student:* A. Sakula, "Paul Langerhans (1847–1888): A centenary tribute," *Journal of the Royal Society of Medicine* 81, no. 7 (1988): 414.

239 *Two decades later, two surgeons:* J. v. Mering and Oskar Minkowski, "Diabetes mel-

litus nach Pankreasexstirpation," *Naunyn-Schmiedeberg's Archives of Pharmacology* 26, no. 5 (1890): 371–87.

240 *Ultimately, in 1921, Banting and Best:* F. G. Banting et al., "Pancreatic extracts in the treatment of diabetes mellitus," *Canadian Medical Association Journal* 12, no. 3 (1922): 141.

240 *In 1953, after three more decades:* Frederick Sanger and E. O. P. Thompson, "The amino-acid sequence in the glycyl chain of insulin. 1. The identification of lower peptides from partial hydrolysates," *Biochemical Journal* 53, no. 3 (1953): 353.

241 *To synthesize the somatostatin gene:* Hughes, *Genentech*, 59–65.

241 *"I thought about it all the time":* "Fierce Competition to Synthesize Insulin, David Goeddel," DNA Learning Center, https://www.dnalc.org/view/15085-Fierce-competition -to-synthesize-insulin-David-Goeddel.html.

243 *"Gilbert was, as he had for many days past":* Hughes, *Genentech*, 93.

243 *460 Point San Bruno Boulevard:* Ibid., 78.

244 *"You'd go through the back of Genentech's door":* "Introductory materials," First Chief Financial Officer at Genentech, 1978–1984, http://content.cdlib.org/view?docId=kt 8k40159r&brand=calisphere&doc.view=entire_text.

244 *Gilbert recalled. The UCSF team:* Hughes, *Genentech*, 93.

244 *In the summer of 1978, Boyer learned:* Payne Templeton, "Harvard group produces insulin from bacteria," *Harvard Crimson*, July 18, 1978.

244 *August 21, 1978, Goeddel joined:* Hughes, *Genentech*, 91.

245 *On October 26, 1982, the US Patent:* "A history of firsts," Genentech: Chronology, http://www.gene.com/media/company-information/chronology.

245 *"effectively, the patent claimed":* Luigi Palombi, *Gene Cartels: Biotech Patents in the Age of Free Trade* (London: Edward Elgar Publishing, 2009), 264.

246 *Many newspapers accusingly termed it:* "History of AIDS up to 1986," http://www .avert.org/history-aids-1986.htm.

248 *In April, exactly two years:* Gilbert C. White, "Hemophilia: An amazing 35-year journey from the depths of HIV to the threshold of cure," *Transactions of the American Clinical and Climatological Association* 121 (2010): 61.

249 *90 percent would acquire HIV:* "HIV/AIDS," National Hemophilia Foundation, https://www.hemophilia.org/Bleeding-Disorders/Blood-Safety/HIV/AIDS.

250 *Of the several million variants:* John Overington, Bissan Al-Lazikani, and Andrew Hopkins, "How many drug targets are there?" *Nature Reviews Drug Discovery* 5 (December 2006): 993–96, "Table 1 | Molecular targets of FDA-approved drugs," http:// www.nature.com/nrd/journal/v5/n12/fig_tab/nrd2199_T1.html.

251 *On October 14, 1980, Genentech sold:* "Genentech: Historical stock info," Gene.com, http://www.gene.com/about-us/investors/historical-stock-info.

251 *In the summer of 2001, Genentech launched:* Harold Evans, Gail Buckland, and David Lefer, *They Made America: From the Steam Engine to the Search Engine—Two Centuries of Innovators* (London: Hachette UK, 2009), "Hebert Boyer and Robert Swanson: The biotech industry," 420–31.

PART FOUR: "THE PROPER STUDY OF MANKIND IS MAN"

254 *Know then thyself:* Alexander Pope, *Essay on Man* (Oxford: Clarendon Press, 1869).

254 *Albany: How have you known:* William Shakespeare and Jay L. Halio, *The Tragedy of King Lear* (Cambridge: Cambridge University Press, 1992), act 5, sc. 3.

The Birth of a Clinic

259 *I start with the premise that:* Lyon and Gorner, *Altered Fates.*

259 *the* New York Times *published:* John A. Osmundsen, "Biologist hopeful in solving secrets of heredity this year," *New York Times*, February 2, 1962.

259 *"The most important contribution to medicine":* Thomas Morgan, "The relation of genetics to physiology and medicine," Nobel Lecture, June 4, 1934, Nobelprize.org, http://www.nobelprize.org/nobel_prizes/medicine/laureates/1933/morgan-lecture.html.

260 *In 1947, Victor McKusick:* "From 'musical murmurs' to medical genetics, 1945–1960," Victor A. McKusick Papers, NIH, http://profiles.nlm.nih.gov/ps/retrieve/narrative/jq/p-nid/305.

260 *McKusick described the case:* Harold Jeghers, Victor A. McKusick, and Kermit H. Katz, "Generalized intestinal polyposis and melanin spots of the oral mucosa, lips and digits," *New England Journal of Medicine* 241, no. 25 (1949): 993–1005, doi:10.1056/nejm194912222412501.

260 *In 1899, Archibald Garrod:* Archibald E. Garrod, "A contribution to the study of alkaptonuria," *Medico-chirurgical Transactions* 82 (1899): 367.

261 *"The phenomena of obesity":* Archibald E. Garrod, "The incidence of alkaptonuria: A study in chemical individuality," *Lancet* 160, no. 4137 (1902): 1616–20, doi:10.1016/s0140-6736(01)41972-6.

261 *for decades, some medical historians:* Harold Schwartz, *Abraham Lincoln and the Marfan Syndrome* (Chicago: American Medical Association, 1964).

261 *By the mid-1980s, McKusick and his students:* J. Amberger et al., "McKusick's Online Mendelian Inheritance in Man," *Nucleic Acids Research* 37 (2009): (database issue) D793–D796, fig. 1 and 2, doi:10.1093/nar/gkn665.

262 *By the twelfth edition of his book:* "Beyond the clinic: Genetic studies of the Amish and little people, 1960–1980s," Victor A. McKusick Papers, NIH, http://profiles.nlm.nih.gov/ps/retrieve/narrative/jq/p-nid/307.

265 *"The imperfect is our paradise":* Wallace Stevens, *The Collected Poems of Wallace Stevens* (New York: Alfred A. Knopf, 1954), "The Poems of Our Climate," 193–94.

266 *In November 1961: Fantastic Four #1* (New York: Marvel Comics, 1961), http://marvel.com/comics/issue/12894/fantastic_four_1961_1.

266 *"a fantastic amount of radioactivity":* Stan Lee et al., *Marvel Masterworks: The Amazing Spider-Man* (New York: Marvel Publishing, 2009), "The Secrets of Spider-Man."

266 *the* X-Men, *launched in September 1963: Uncanny X-Men #1* (New York: Marvel Comics, 1963), http://marvel.com/comics/issue/12413/uncanny_x-men_1963_1.

266 *in the spring of 1966:* Alexandra Stern, *Telling Genes: The Story of Genetic Counseling in America* (Baltimore: Johns Hopkins University Press, 2012), 146.

267 *Fetal cells from the amnion:* Leo Sachs, David M. Serr, and Mathilde Danon, "Analysis of amniotic fluid cells for diagnosis of foetal sex," *British Medical Journal* 2, no. 4996 (1956): 795.

267 *On May 31, 1968:* Carlo Valenti, "Cytogenetic diagnosis of down's syndrome in utero," *Journal of the American Medical Association* 207, no. 8 (1969): 1513, doi:10.1001/jama.1969.03150210097018.

268 *In September 1969:* Details of McCorvey's life are from Norma McCorvey with Andy Meisler, *I Am Roe: My Life, Roe v. Wade, and Freedom of Choice* (New York: HarperCollins, 1994).

268 *"with dirty instruments scattered around the room"*: Ibid.

268 *Blackmun wrote*: Roe v. Wade, Legal Information Institute, https://www.law.cornell.edu/supremecourt/text/410/113.

268 *"The individual's [i.e., mother's]"*: Alexander M. Bickel, *The Morality of Consent* (New Haven: Yale University Press, 1975), 28.

269 *control of the fetal genome to medicine*: Jeffrey Toobin, "The people's choice," *New Yorker*, January 28, 2013, 19–20.

269 *In some states*: H. Hansen, "Brief reports decline of Down's syndrome after abortion reform in New York State," *American Journal of Mental Deficiency* 83, no. 2 (1978): 185–88.

269 *By the mid-1970s*: Daniel J. Kevles, *In the Name of Eugenics: Genetics and the Uses of Human Heredity* (New York: Alfred A. Knopf, 1985), 257.

269 *"Tiny fault after tiny fault"*: M. Susan Lindee, *Moments of Truth in Genetic Medicine* (Baltimore: Johns Hopkins University Press, 2005), 24.

269 *McKusick published a new edition*: V. A. McKusick and R. Claiborne, eds., *Medical Genetics* (New York: HP Publishing, 1973).

269 *Joseph Dancis, the pediatrician, wrote*: Ibid., Joseph Dancis, "The prenatal detection of hereditary defects," 247.

270 *In June 1969, a woman named Hetty Park*: Mark Zhang, "Park v. Chessin (1977)," *The Embryo Project Encyclopedia*, January 31, 2014, https://embryo.asu.edu/pages/park-v-chessin-1977.

271 *One commentator noted, "The court asserted"*: Ibid.

"Interfere, Interfere, Interfere"

272 *After millennia in which most people*: Gerald Leach, "Breeding Better People," *Observer*, April 12, 1970.

272 *No newborn should be declared human*: Michelle Morgante, "DNA scientist Francis Crick dies at 88," *Miami Herald*, July 29, 2004.

274 *"The old eugenics was limited"*: Lily E. Kay, *The Molecular Vision of Life: Caltech, the Rockefeller Foundation, and the Rise of the New Biology* (New York: Oxford University Press, 1993), 276.

274 *In 1980, Robert Graham*: David Plotz, "Darwin's Engineer," *Los Angeles Times*, June 5, 2005, http://www.latimes.com/la-tm-spermbank23jun05-story.html#page=1.

274 *The physicist William Shockley*: Joel N. Shurkin, *Broken Genius: The Rise and Fall of William Shockley, Creator of the Electronic Age* (London: Macmillan, 2006), 256.

275 *"cruel, blundering and inefficient"*: Kevles, *In the Name of Eugenics*, 263.

275 *"moral obligation of the medical profession"*: Departments of Labor and Health, Education, and Welfare Appropriations for 1967 (Washington, DC: Government Printing Office, 1966), 249.

276 *"Near the end of his terms of office"*: Victor McKusick, in *Legal and Ethical Issues Raised by the Human Genome Project: Proceedings of the Conference in Houston, Texas, March 7–9, 1991*, ed. Mark A. Rothstein (Houston: University of Houston, Health Law and Policy Institute, 1991).

277 *"needle in a haystack"*: Matthew R. Walker and Ralph Rapley, *Route Maps in Gene Technology* (Oxford: Blackwell Science, 1997), 144.

A Village of Dancers, an Atlas of Moles

278 *Glory be to God for dappled things:* W. H. Gardner, *Gerard Manley Hopkins: Poems and Prose* (Taipei: Shu lin, 1968), "Pied Beauty."

278 *We suddenly came upon two women:* George Huntington, "Recollections of Huntington's chorea as I saw it at East Hampton, Long Island, during my boyhood," *Journal of Nervous and Mental Disease* 37 (1910): 255–57.

278 *In 1978, two geneticists:* Robert M. Cook-Deegan, *The Gene Wars: Science, Politics, and the Human Genome* (New York: W. W. Norton, 1994), 38.

279 *By studying Mormons in Utah:* K. Kravitz et al., "Genetic linkage between hereditary hemochromatosis and HLA," *American Journal of Human Genetics* 31, no. 5 (1979): 601.

280 *When Botstein and Davis had first discovered:* David Botstein et al., "Construction of a genetic linkage map in man using restriction fragment length polymorphisms," *American Journal of Human Genetics* 32, no. 3 (1980): 314.

280 *The poet Louis MacNeice once wrote:* Louis MacNeice, "Snow," in *The New Cambridge Bibliography of English Literature*, vol. 3, ed. George Watson (Cambridge: Cambridge University Press, 1971).

280 *In 1978, two other researchers:* Y. Wai Kan and Andree M. Dozy, "Polymorphism of DNA sequence adjacent to human beta-globin structural gene: Relationship to sickle mutation," *Proceedings of the National Academy of Sciences* 75, no. 11 (1978): 5631–35.

281 *"We can give you markers":* Victor K. McElheny, *Drawing the Map of Life: Inside the Human Genome Project* (New York: Basic Books, 2010), 29.

281 *"We describe a new basis":* Botstein et al., "Construction of a genetic linkage map," 314.

282 *"like watching a giant puppet show":* N. Wexler, "Huntington's Disease: Advocacy Driving Science," *Annual Review of Medicine*, no. 63 (2012): 1–22.

283 *life devolves into a grim roulette:* Wexler NS. "Genetic 'Russian Roulette': The Experience of Being At Risk for Huntington's Disease," *Genetic Counseling: Psychological Dimensions*, ed. S. Kessler (New York, Academic Press, 1979).

283 *"waiting game for the onset of symptoms":* "New discovery in fight against Huntington's disease," NUI Galway, February 22, 2012, http://www.nuigalway.ie/about-us/news-and-events/news-archive/2012/february2012/new-discovery-in-fight-against-huntingtons-disease-1.html.

283 *"I don't know the point where":* Gene Veritas, "At risk for Huntington's disease," September 21, 2011, http://curehd.blogspot.com/2011_09_01_archive.html.

283 *Milton Wexler, Nancy's father, a clinical psychologist:* Details of the Wexler family story came from Alice Wexler, *Mapping Fate: A Memoir of Family, Risk, and Genetic Research* (Berkeley: University of California Press, 1995); Lyon and Gorner, *Altered Fates*; and "Makers profile: Nancy Wexler, neuropsychologist & president, Hereditary Disease Foundation," MAKERS: The Largest Video Collection of Women's Stories, http://www.makers.com/nancy-wexler.

283 *"Each one of you has a one-in-two":* Ibid.

283 *That year, Milton Wexler launched:* "History of the HDF," Hereditary Disease Foundation, http://hdfoundation.org/history-of-the-hdf/.

283 *In one nursing home:* Wexler, Nancy, "Life In The Lab" *Los Angeles Times Magazine*, February 10, 1991.

283 *Leonore died on May 14, 1978:* Associated Press, "Milton Wexler; Promoted Huntington's Research," *Washington Post*, March 23, 2007, http://www.washingtonpost.com/wp-dyn/content/article/2007/03/22/AR2007032202068.html.

283 *In October 1979:* Wexler, *Mapping Fate,* 177.

284 *"There have been a few times in my life":* Ibid., 178.

284 *At first glance, a visitor to Barranquitas:* Description of Barranquitas from "Nancy Wexler in Venezuela Huntington's disease," BBC, 2010, YouTube, https://www.you tube.com/watch?v=D6LbkTW8fDU.

285 *When the Venezuelan neurologist Américo Negrette:* M. S. Okun and N. Thommi, "Américo Negrette (1924 to 2003): Diagnosing Huntington disease in Venezuela," *Neurology* 63, no. 2 (2004): 340–43, doi:10.1212/01.wnl.0000129827.16522.78.

285 *In some parts:* for data on prevalence, see http://www.cmmt.ubc.ca/research/diseases/ huntingtons/HD_Prevalence.

285 *two copies of the mutated Huntington's disease gene—i.e., "homozygotes":* see "What Is a Homozygote?", Nancy Wexler, *Gene Hunter: The Story of Neuropsychologist Nancy Wexler,* (Women's Adventures in Science, Joseph Henry Press), October 30, 2006: 51.

286 *"It was a clash of total bizarreness":* Jerry E. Bishop and Michael Waldholz, *Genome: The Story of the Most Astonishing Scientific Adventure of Our Time* (New York: Simon & Schuster, 1990), 82–86.

286 *They assiduously collected:* This pedigree would eventually grow to contain more than 18,000 individuals over 10 generations. All have descended from a common ancestor, a woman named Maria Concepión—a strangely apt name—who conceived the first family that carried the abnormal gene to these villages in the nineteenth century.

286 *Here too the illness:* The American family was not big enough to prove linkage, but the Venezuelan family was. By adding the two together, the scientists could prove the existence of a DNA marker traveling with HD. See Gusella JF, Wexler NS, Conneally PM, Naylor SL, Anderson MA, Tanzi RE, Watkins PC, Ottina K, Wallace MR, Sakaguchi AY, Young AB, Shoulson I, Bonilla E, and Martin JB. "A Polymorphic DNA Marker Genetically Linked to Huntington's Disease." *Nature,* 1983 Nov 17–23; 306 (5940): 234–8.

286 *In August 1983, Wexler, Gusella, and Conneally:* James F. Gusella et al., "A polymorphic DNA marker genetically linked to Huntington's disease," *Nature* 306, no. 5940 (1983): 234–38, doi:10.1038/306234a0.

287 *The candidate gene had been found:* Karl Kieburtz et al., "Trinucleotide repeat length and progression of illness in Huntington's disease," *Journal of Medical Genetics* 31, no. 11 (1994): 872–74.

288 *"We've got it, we've got it":* Lyon and Gorner, *Altered Fates,* 424.

288 *A remarkable feature of the inheritance:* Nancy S. Wexler, "Venezuelan kindreds reveal that genetic and environmental factors modulate Huntington's disease age of onset," *Proceedings of the National Academy of Sciences* 101, no. 10 (2004): 3498–503.

289 *In 1857, a Swiss almanac: The Almanac of Children's Songs and Games from Switzerland* (Leipzig: J. J. Weber, 1857).

289 *"Inside the pericardium":* "The History of Cystic Fibrosis," cysticfibrosismedicine .com, http://www.cfmedicine.com/history/earlyyears.htm.

289 *In 1985, Lap-Chee Tsui:* Lap-Chee Tsui et al., "Cystic fibrosis locus defined by a genetically linked polymorphic DNA marker," *Science* 230, no. 4729 (1985): 1054–57.

290 *By the spring of 1989, Collins:* Wanda K. Lemna et al., "Mutation analysis for heterozygote detection and the prenatal diagnosis of cystic fibrosis," *New England Journal of Medicine* 322, no. 5 (1990): 291–96.

291 *Over the last decade:* V. Scotet et al., "Impact of public health strategies on the birth

prevalence of cystic fibrosis in Brittany, France," *Human Genetics* 113, no. 3 (2003): 280–85.

291 *In 1993, a New York hospital:* D. Kronn, V. Jansen, and H. Ostrer, "Carrier screening for cystic fibrosis, Gaucher disease, and Tay-Sachs disease in the Ashkenazi Jewish population: The first 1,000 cases at New York University Medical Center, New York, NY," *Archives of Internal Medicine* 158, no. 7 (1998): 777–81.

292 *As the physicist and historian Evelyn Fox Keller:* Elinor S. Shaffer, ed., *The Third Culture: Literature and Science*, vol. 9 (Berlin: Walter de Gruyter, 1998), 21.

292 *"a new horizon in the history of man":* Robert L. Sinsheimer, "The prospect for designed genetic change," *American Scientist* 57, no. 1 (1969): 134–42.

292 *"Some may smile and may feel":* Jay Katz, Alexander Morgan Capron, and Eleanor Swift Glass, *Experimentation with Human Beings: The Authority of the Investigator, Subject, Professions, and State in the Human Experimentation Process* (New York: Russell Sage Foundation, 1972), 488.

292 *"no beliefs, no values, no institutions":* John Burdon Sanderson Haldane, *Daedalus or Science and the Future* (New York: E. P. Dutton, 1924), 48.

"To Get the Genome"

293 *Our ability to read out this sequence:* Sulston and Ferry, *Common Thread*, 264.

294 *In 1977, when Fred Sanger had sequenced:* Cook-Deegan, *The Gene Wars*, 62.

294 *The human genome contains 3,095,677,412 base pairs:* "OrganismView: Search organisms and genomes," CoGe: OrganismView, https://genomevolution.org/coge //organismview.pl?gid=7029.

294 BRCA1, *was only identified in 1994:* Yoshio Miki et al., "A strong candidate for the breast and ovarian cancer susceptibility gene *BRCA1*," *Science* 266, no. 5182 (1994): 66–71.

294 *such as chromosome jumping:* F. Collins et al., "Construction of a general human chromosome jumping library, with application to cystic fibrosis," *Science* 235, no. 4792 (1987): 1046–49, doi:10.1126/science.2950591.

294 *"There was no shortage of exceptionally clever":* Mark Henderson, "Sir John Sulston and the Human Genome Project," Wellcome Trust, May 3, 2011, http://genome.well come.ac.uk/doc_wtvm051500.html.

294 *"But even with the immense power":* *Departments of Labor, Health and Human Services, Education, and Related Agencies Appropriations for 1996: Hearings before a Subcommittee of the Committee on Appropriations, House of Representatives, One Hundred Fourth Congress, First Session* (Washington, DC: Government Printing Office, 1995), http://catalog.hathitrust.org/Record/003483817.

295 *in 1872, Hilário de Gouvêa, a Brazilian ophthalmologist:* Alvaro N. A. Monteiro and Ricardo Waizbort, "The accidental cancer geneticist: Hilário de Gouvêa and hereditary retinoblastoma," *Cancer Biology & Therapy* 6, no. 5 (2007): 811–13, doi:10.4161 /cbt.6.5.4420.

297 *Vogelstein had already discovered that cancers:* Bert Vogelstein and Kenneth W. Kinzler, "The multistep nature of cancer," *Trends in Genetics* 9, no. 4 (1993): 138–41.

298 *Schizophrenia, in particular, sparked a furor:* Valrie Plaza, *American Mass Murderers* (Raleigh, NC: Lulu Press, 2015), "Chapter 57: James Oliver Huberty."

298 *NAS study found that identical twins possessed:* "Schizophrenia in the National Academy of Sciences–National Research Council Twin Registry: A 16-year up-

date," *American Journal of Psychiatry* 140, no. 12 (1983): 1551–63, doi:10.1176/ajp.140.12.1551.

298 *An earlier study, published by:* D. H. O'Rourke et al., "Refutation of the general single-locus model for the etiology of schizophrenia," *American Journal of Human Genetics* 34, no. 4 (1982): 630.

298 *For identical twins with the severest form:* Peter McGuffin et al., "Twin concordance for operationally defined schizophrenia: Confirmation of familiality and heritability," *Archives of General Psychiatry* 41, no. 6 (1984): 541–45.

300 *Populist anxieties about genes, mental illness:* James Q. Wilson and Richard J. Herrnstein, *Crime and Human Nature: The Definitive Study of the Causes of Crime* (New York: Simon & Schuster, 1985).

300 *"bad friends, bad neighborhoods, bad labels":* Matt DeLisi, "James Q. Wilson," in *Fifty Key Thinkers in Criminology,* ed. Keith Hayward, Jayne Mooney, and Shadd Maruna (London: Routledge, 2010), 192–96.

301 *another meeting of scientists was called to evaluate whether:* Doug Struck, "The Sun (1837–1988)," *Baltimore Sun,* February 2, 1986, 79.

302 *The most important technical breakthrough:* Kary Mullis, "Nobel Lecture: The polymerase chain reaction," December 8, 1993, Nobelprize.org, http://www.nobelprize.org/nobel_prizes/chemistry/laureates/1993/mullis-lecture.html.

302 *To sequence all 3 billion base pairs:* Sharyl J. Nass and Bruce Stillman, *Large-Scale Biomedical Science: Exploring Strategies for Future Research* (Washington, DC: National Academies Press, 2003), 33.

303 *"The only way to give Rufus a life":* McElheny, *Drawing the Map of Life,* 65.

303 *By 1989 after several:* "About NHGRI: A Brief History and Timeline," Genome.gov, http://www.genome.gov/10001763.

304 *In January 1989, a twelve-member council:* McElheny, *Drawing the Map of Life,* 89.

304 *"We are initiating an unending study":* Ibid.

304 *On January 28, 1983:* J. David Smith, "Carrie Elizabeth Buck (1906–1983)," *Encyclopedia Virginia,* http://www.encyclopediavirginia.org/Buck_Carrie_Elizabeth_1906-1983.

304 *Vivian Dobbs—the child who:* Ibid.

The Geographers

306 *So Geographers in Afric-maps:* Jonathan Swift and Thomas Roscoe, *The Works of Jonathan Swift, DD: With Copious Notes and Additions and a Memoir of the Author,* vol. 1 (New York: Derby, 1859), 247–48.

306 *More and more, the Human Genome Project:* Justin Gillis, "Gene-mapping controversy escalates; Rockville firm says government officials seek to undercut its effort," *Washington Post,* March 7, 2000.

306 *Craig Venter, proposed a shortcut:* L. Roberts, "Gambling on a Shortcut to Genome Sequencing," *Science* 252, no. 5013 (1991): 1618–19.

306 *In 1986, he had heard of:* Lisa Yount, *A to Z of Biologists* (New York: Facts On File, 2003), 312.

306 *"my future in a crate":* J. Craig Venter, *A Life Decoded: My Genome, My Life* (New York: Viking, 2007), 97.

308 *the NIH technology transfer office contacted:* R. Cook-Deegan and C. Heaney, "Patents in genomics and human genetics," *Annual Review of Genomics and Human Genetics* 11 (2010): 383–425, doi:10.1146/annurev-genom-082509-141811.

308 *In 1984, Amgen had filed a patent:* Edmund L. Andrews, "Patents; Unaddressed Question in Amgen Case," *New York Times,* March 9, 1991.

308 *"Patents (or so I had believed) are designed":* Sulston and Ferry, *Common Thread,* 87.

308 *"It's a quick and dirty land grab":* Pamela R. Winnick, *A Jealous God: Science's Crusade against Religion* (Nashville, TN: Nelson Current, 2005), 225.

309 *"Could you patent an elephant":* Eric Lander, author interview, 2015.

309 *Walter Bodmer, the English geneticist, warned:* L. Roberts, "Genome Patent Fight Erupts," *Science* 254, no. 5029 (1991): 184–86.

309 The *Institute for Genomic Research:* Venter, *Life Decoded,* 153.

309 *Working with a new ally, Hamilton Smith:* Hamilton O. Smith et al., "Frequency and distribution of DNA uptake signal sequences in the *Haemophilus influenzae* Rd genome," *Science* 269, no. 5223 (1995): 538–40.

310 *"The final [paper] took forty drafts":* Venter, *Life Decoded,* 212.

310 *"thrilled by the first glimpse":* Ibid., 219.

311 *"What if you took a word":* Eric Lander, author interview, October 2015.

311 *"The real challenge of the Human Genome Project":* Ibid.

312 *TIGR had been set up:* HGS was launched by William Haseltine, a former Harvard professor, who hoped to use genomics to discover novel drugs.

312 *In December 1998:* "1998: Genome of roundworm *C. elegans* sequenced," Genome .gov, http://www.genome.gov/25520394.

313 *A gene called* ceh-13, *for instance:* Borbála Tihanyi et al., "The *C. elegans Hox* gene *ceh-13* regulates cell migration and fusion in a non-colinear way. Implications for the early evolution of *Hox* clusters," *BMC Developmental Biology* 10, no. 78 (2010), doi:10.1186/1471-213X-10-78.

315 *The* C. elegans *genome—published to universal: Science* 282, no. 5396 (1998): 1945–2140.

315 *125 semiautomated sequencing machines:* Mike Hunkapiller was partly responsible for a crucial technological development in genome sequencing: semiautomated sequencing machines that could rapidly sequence thousands of bases of DNA.

315 *its one-billionth human base pair:* David Dickson and Colin Macilwain, "'It's a G': The one-billionth nucleotide," *Nature* 402, no. 6760 (1999): 331.

315 *it had sequenced the genome of the fruit fly:* Declan Butler, "Venter's *Drosophila* 'success' set to boost human genome efforts," *Nature* 401, no. 6755 (1999): 729–30.

316 *In March 2000,* Science *published:* "The *Drosophila* genome," *Science* 287, no. 5461 (2000): 2105–364.

316 *Of the 289 human genes known to be:* David N. Cooper, *Human Gene Evolution* (Oxford: BIOS Scientific Publishers, 1999), 21.

316 *177 genes:* William K. Purves, *Life: The Science of Biology* (Sunderland, MA: Sinauer Associates, 2001), 262.

316 *"a man like me":* Marsh, *William Blake,* 56.

316 *"The lesson is that the complexity":* Quote from the director of the Berkeley *Drosophila* Genome Project, Gerry Rubin, in Robert Sanders, "UC Berkeley collaboration with Celera Genomics concludes with publication of nearly complete sequence of the genome of the fruit fly," press release, UC Berkeley, March 24, 2000, http://www .berkeley.edu/news/media/releases/2000/03/03-24-2000.html.

317 *"between a human and a nematode worm":* The Age of the Genome, BBC Radio 4, http://www.bbc.co.uk/programmes/b00ss2rk.

317 *"Fix this!":* James Shreeve, *The Genome War: How Craig Venter Tried to Capture the Code of Life and Save the World* (New York: Alfred A. Knopf, 2004), 350.

318 *That initial meeting in Ari Patrinos's basement:* For details of this story see ibid. Also see Venter, *Life Decoded*, 97.

318 *At 10:19 a.m. on the morning of June 26:* "June 2000 White House Event," Genome .gov, https://www.genome.gov/10001356.

318 *Clinton spoke first, comparing the map:* "President Clinton, British Prime Minister Tony Blair deliver remarks on human genome milestone," CNN.com Transcripts, June 26, 2000.

319 *We have sequenced the genome:* the sequence described by Venter's group had representations from males and females of each group, but the complete sequence of any of these individuals had not been completed.

319 *"My greatest success":* Shreeve, *Genome War*, 360.

320 *Lander recruited yet another team of scientists:* McElheny, *Drawing the Map of Life*, 163.

320 *"In the history of scientific writing since the 1600s":* Eric Lander, author interview, October 2015.

320 *"genome tossed salad":* Shreeve, *Genome War*, 364.

The Book of Man (in Twenty-Three Volumes)

322 *It encodes about 20,687 genes in total:* Details of the Human Genome Project come from "Human genome far more active than thought," Wellcome Trust, Sanger Institute, September 5, 2012, http://www.sanger.ac.uk/about/press/2012/120905.html; Venter, *Life Decoded*; and Committee on Mapping and Sequencing the Human Genome, *Mapping and Sequencing the Human Genome* (Washington, DC: National Academy Press, 1988), http://www.nap.edu/read/1097/chapter/1.

PART FIVE: THROUGH THE LOOKING GLASS

328 *How nice it would be:* Lewis Carroll, *Alice in Wonderland* (New York: W. W. Norton, 2013).

"So, We's the Same"

329 *"So, We's the Same":* Kathryn Stockett, *The Help* (New York: Amy Einhorn Books/ Putnam, 2009), 235.

329 *We got to have a re-vote:* "Who is blacker Charles Barkley or Snoop Dogg," YouTube, January 19, 2010, https://www.youtube.com/watch?v=yHfX-11ZHXM.

329 *What have I in common with Jews?:* Franz Kafka, *The Basic Kafka* (New York: Pocket Books, 1979), 259.

329 *This mirror writing can result:* Everett Hughes, "The making of a physician: General statement of ideas and problems," *Human Organization* 14, no. 4 (1955): 21–25.

330 *"as absurd as defining the organs":* Allen Verhey, *Nature and Altering It* (Grand Rapids, MI: William B. Eerdmans, 2010), 19. Also see Matt Ridley, *Genome: The Autobiography of a Species In 23 Chapters* (New York: Harper Collins, 1999), 54.

330 *"Encoded in the DNA sequence are fundamental":* Committee on Mapping and Sequencing, *Mapping and Sequencing*, 11.

332 *"Had Mr. Darwin or his followers furnished":* Louis Agassiz, "On the origins of species," *American Journal of Science and Arts* 30 (1860): 142–54.

332 *In 1848, stone diggers in a limestone quarry:* Douglas Palmer, Paul Pettitt, and Paul G. Bahn, *Unearthing the Past: The Great Archaeological Discoveries That Have Changed History* (Guilford, CT: Globe Pequot, 2005), 20.

332 *"an early time in the evolution of man":* Popular Science Monthly 100 (1922).

333 *Allan Wilson began to use genetic tools:* Rebecca L. Cann, Mork Stoneking, and Allan C. Wilson, "Mitochondrial DNA and human evolution," *Nature* 325 (1987): 31–36.

335 *The genes lodged within mitochondria:* See Chuan Ku et al., "Endosymbiotic origin and differential loss of eukaryotic genes," *Nature* 524 (2015): 427–32.

335 *First, when Wilson measured the overall diversity:* Thomas D. Kocher et al., "Dynamics of mitochondrial DNA evolution in animals: Amplification and sequencing with conserved primers," *Proceedings of the National Academy of Sciences* 86, no. 16 (1989): 6196–200.

336 *By 1991, Wilson could use his method:* David M. Irwin, Thomas D. Kocher, and Allan C. Wilson, "Evolution of the cytochrome-b gene of mammals," *Journal of Molecular Evolution* 32, no. 2 (1991): 128–44; Linda Vigilant et al., "African populations and the evolution of human mitochondrial DNA," *Science* 253, no. 5027 (1991): 1503–7; and Anna Di Rienzo and Allan C. Wilson, "Branching pattern in the evolutionary tree for human mitochondrial DNA," *Proceedings of the National Academy of Sciences* 88, no. 5 (1991): 1597–601.

336 *In November 2008, a seminal study:* Jun Z. Li et al., "Worldwide human relationships inferred from genome-wide patterns of variation," *Science* 319, no. 5866 (2008): 1100–104.

336 *"You get less and less variation":* John Roach, "Massive genetic study supports 'out of Africa' theory," *National Geographic News*, February 21, 2008.

336 *The oldest human populations:* Lev A. Zhivotovsky, Noah A. Rosenberg, and Marcus W. Feldman, "Features of evolution and expansion of modern humans, inferred from genomewide microsatellite markers," *American Journal of Human Genetics* 72, no. 5 (2003): 1171–86.

336 *The "youngest" humans:* Noah Rosenberg et al., "Genetic structure of human populations," *Science* 298, no. 5602 (2002): 2381–85. A map of human migrations can be found in L. L. Cavalli-Sforza and Marcus W. Feldman, "The application of molecular genetic approaches to the study of human evolution," *Nature Genetics* 33 (2003): 266–75.

336 *It is called the Out of Africa theory:* For the origin of humans in Southern Africa, see Brenna M. Henn et al., "Hunter-gatherer genomic diversity suggests a southern African origin for modern humans," *Proceedings of the National Academy of Sciences* 108, no. 13 (2011): 5154–62. Also see Brenna M. Henn, L. L. Cavalli-Sforza, and Marcus W. Feldman, "The great human expansion," *Proceedings of the National Academy of Sciences* 109, no. 44 (2012): 17758–64.

339 *"Sexual intercourse began":* Philip Larkin, "Annus Mirabilis," *High Windows*.

339 *"In terms of modern humans":* Christopher Stringer, "Rethinking 'out of Africa,'" editorial, *Edge*, November 12, 2011, http://edge.org/conversation/rethinking-out-of-africa.

340 *Others have proposed:* H. C. Harpending et al., "Genetic traces of ancient demography," *Proceedings of the National Academy of Sciences* 95 (1998): 1961–67; R. Gonser et al., "Microsatellite mutations and inferences about human demography," *Genetics* 154 (2000): 1793–1807; A. M. Bowcock et al., "High resolution of human evolutionary trees with polymorphic microsatellites," *Nature* 368 (1994): 455–57; and C. Dib et al., "A comprehensive genetic map of the human genome based on 5,264 microsatellites," *Nature* 380 (1996): 152–54.

341 *The most recent estimates suggest that:* Anthony P. Polednak, *Racial and Ethnic Differences in Disease* (Oxford: Oxford University Press, 1989), 32–33.

342 *As Marcus Feldman and Richard Lewontin put it:* M. W. Feldman and R. C. Lewontin, "Race, ancestry, and medicine," in *Revisiting Race in a Genomic Age*, ed. B. A. Koenig, S. S. Lee, and S. S. Richardson (New Brunswick, NJ: Rutgers University Press, 2008). Also see Li et al., "Worldwide human relationships inferred from genome-wide patterns of variation," 1100–104.

342 *In his monumental study on human genetics:* L. Cavalli-Sforza, Paola Menozzi, and Alberto Piazza, *The History and Geography of Human Genes* (Princeton, NJ: Princeton University Press, 1994), 19.

343 *"So, we's the same":* Stockett, *Help.*

343 *In 1994, the very year:* Cavalli-Sforza, Menozzi, and Piazza, *The History and Geography.*

343 *a very different kind of book about:* Richard Herrnstein and Charles Murray, *The Bell Curve* (New York: Simon & Schuster, 1994).

343 *"a flame-throwing treatise on class":* "The 'Bell Curve' agenda," *New York Times*, October 24, 1994.

343 *his 1985 book,* Crime and Human Nature: Wilson and Herrnstein. *Crime and Human Nature.*

344 *In 1904, Charles Spearman, a British statistician:* Charles Spearman, " 'General Intelligence,' objectively determined and measured," *American Journal of Psychology* 15, no. 2 (1904): 201–92.

344 *Recognizing that this measurement varied with age:* The concept of IQ was initially developed by William Stern, the German psychologist.

345 *Developmental psychologists such as Louis Thurstone:* Louis Leon Thurstone, "The absolute zero in intelligence measurement," *Psychological Review* 35, no. 3 (1928): 175; and L. Thurstone, "Some primary abilities in visual thinking," *Proceedings of the American Philosophical Society* (1950): 517–21. Also see Howard Gardner and Thomas Hatch, "Educational implications of the theory of multiple intelligences," *Educational Researcher* 18, no. 8 (1989): 4–10.

345 *Drawing heavily from an earlier article:* Herrnstein and Murray, *Bell Curve*, 284.

345 *In the 1950s, a series of reports:* George A. Jervis, "The mental deficiencies," *Annals of the American Academy of Political and Social Science* (1953): 25–33. Also see Otis Dudley Duncan, "Is the intelligence of the general population declining?" *American Sociological Review* 17, no. 4 (1952): 401–7.

346 *They limited the tests to only those administered after 1960:* The particular variables assessed by Murray and Herrnstein deserve mention. They wondered whether a deep disenchantment with tests and scores might pervade African-Americans, making them reluctant to engage with IQ tests. But subtle experiments to measure and excise any such "test disengagement" could not erase the 15-point difference. They considered the possibility that the tests were culturally biased (perhaps the most notorious example, borrowed from an SAT examination, asks students to consider the analogy "oarsmen:regatta." It hardly takes an expert on language and culture to know that most inner-city children, black or white, might have little knowledge of what a regatta is, let alone what an oarsman does in one). Yet even after removing such culture-specific and class-specific items from the tests, Murray and Herrnstein wrote, a difference of 15-odd points remained.

347 *In the 1990s, the psychologist Eric Turkheimer:* Eric Turkheimer, "Consensus and controversy about IQ," *Contemporary Psychology* 35, no. 5 (1990): 428–30. Also see Eric Turkheimer et al., "Socioeconomic status modifies heritability of IQ in young children," *Psychological Science* 14, no. 6 (2003): 623–28.

348 *In a blistering article written:* Stephen Jay Gould, "Curve ball," *New Yorker*, November 28, 1994, 139–40.

348 *The Harvard historian Orlando Patterson:* Orlando Patterson, "For Whom the Bell Curves," in *The Bell Curve Wars: Race, Intelligence, and the Future of America*, ed. Steven Fraser (New York: Basic Books, 1995).

348 *black children do worse at tests:* William Wright, *Born That Way: Genes, Behavior, Personality* (London: Routledge, 2013), 195.

348 *a fact buried so inconspicuously:* Herrnstein and Murray, *Bell Curve*, 300–305.

348 *Sandra Scarr and Richard Weinberg in 1976:* Sandra Scarr and Richard A. Weinberg, "Intellectual similarities within families of both adopted and biological children," *Intelligence* 1, no. 2 (1977): 170–91.

350 *"When nobody read":* Alison Gopnik, "To drug or not to drug," *Slate*, February 22, 2010, http://www.slate.com/articles/arts/books/2010/02/to_drug_or_not_to _drug.2.html.

The First Derivative of Identity

352 *For several decades, anthropology has participated:* Paul Brodwin, "Genetics, identity, and the anthropology of essentialism," *Anthropological Quarterly* 75, no. 2 (2002): 323–30.

357 *"Sex is not inherited":* Frederick Augustus Rhodes, *The Next Generation* (Boston: R. G. Badger, 1915), 74.

357 *"The egg, as far as sex is concerned":* Editorials, *Journal of the American Medical Association* 41 (1903): 1579.

358 *She termed it the sex chromosome:* Nettie Maria Stevens, *Studies in Spermatogenesis: A Comparative Study of the Heterochromosomes in Certain Species of Coleoptera, Hemiptera and Lepidoptera, with Especial Reference to Sex Determination* (Baltimore: Carnegie Institution of Washington, 1906).

360 *"punk meets new romantic":* Kathleen M. Weston, *Blue Skies and Bench Space: Adventures in Cancer Research* (Cold Spring Harbor, NY: Cold Spring Harbor Laboratory Press, 2012), "Chapter 8: Walk This Way."

361 *In 1955, Gerald Swyer, an English endocrinologist:* G. I. M. Swyer, "Male pseudohermaphroditism: A hitherto undescribed form," *British Medical Journal* 2, no. 4941 (1955): 709.

361 *Page called the gene ZFY:* Ansbert Schneider-Gädicke et al., "*ZFX* has a gene structure similar to *ZFY*, the putative human sex determinant, and escapes X inactivation," *Cell* 57, no. 7 (1989): 1247–58.

362 *intronless gene called SRY:* Philippe Berta et al., "Genetic evidence equating *SRY* and the testis-determining factor," *Nature* 348, no. 6300 (1990): 448–50.

362 *the mice developed as anatomically male:* Ibid.; John Gubbay et al., "A gene mapping to the sex-determining region of the mouse Y chromosome is a member of a novel family of embryonically expressed genes," *Nature* 346 (1990): 245–50; Ralf J. Jäger et al., "A human XY female with a frame shift mutation in the candidate testis-determining gene *SRY* gene," *Nature* 348 (1990): 452–54; Peter Koopman et al., "Expression of a candidate sex-determining gene during mouse testis differentiation," *Nature* 348 (1990): 450–52; Peter Koopman et al., "Male development of chromosomally female mice transgenic for *SRY* gene," *Nature* 351 (1991): 117–21; and Andrew H. Sinclair et al., "A gene from the human sex-determining region encodes a protein with homology to a conserved DNA-binding motif," *Nature* 346 (1990): 240–44.

363 *"I didn't fit in well"*: "IAmA young woman with Swyer syndrome (also called XY gonadal dysgenesis)," Reddit, 2011, https://www.reddit.com/r/IAmA/comments /e792p/iama_young_woman_with_swyer_syndrome_also_called/.

363 *On the morning of May 5, 2004*: Details of the story of David Reimer are from John Colapinto, *As Nature Made Him: The Boy Who Was Raised as a Girl* (New York: HarperCollins, 2000).

364 *Based on Money's advice, "Brenda"*: John Money, *A First Person History of Pediatric Psychoendocrinology* (Dordrecht: Springer Science & Business Media, 2002), "Chapter 6: David and Goliath."

364 *"Gender identity is sufficiently incompletely"*: Gerald N. Callahan, *Between XX and XY* (Chicago: Chicago Review Press, 2009), 129.

365 *"my leather-and-lace look"*: J. Michael Bostwick and Kari A. Martin, "A man's brain in an ambiguous body: A case of mistaken gender identity," *American Journal of Psychiatry* 164, no. 10 (2007): 1499–505.

365 *"I feel like I have the brain of a man"*: Ibid.

366 *In 2005, a team of researchers at Columbia University*: Heino F. L. Meyer-Bahlburg, "Gender identity outcome in female-raised 46,XY persons with penile agenesis, cloacal exstrophy of the bladder, or penile ablation," *Archives of Sexual Behavior* 34, no. 4 (2005): 423–38.

367 *"Is it really the case that all"*: Otto Weininger, *Sex and Character: An Investigation of Fundamental Principles* (Bloomington: Indiana University Press, 2005), 2.

368 *these animals might be anatomically female*: Carey Reed, "Brain 'gender' more flexible than once believed, study finds," *PBS NewsHour*, April 5, 2015, http://www.pbs.org /newshour/rundown/brain-gender-flexible-believed-study-finds/. Also see Bridget M. Nugent et al., "Brain feminization requires active repression of masculinization via DNA methylation," *Nature Neuroscience* 18 (2015): 690–97.

The Last Mile

370 *Like sleeping dogs, unknown twins*: Wright, *Born That Way*, 27.

370 *"It is the consensus of many contemporary"*: Sándor Lorand and Michael Balint, ed., *Perversions: Psychodynamics and Therapy* (New York: Random House, 1956; repr., London: Ortolan Press, 1965), 75.

370 *"The homosexual's real enemy"*: Bernard J. Oliver Jr., *Sexual Deviation in American Society* (New Haven, CT: New College and University Press, 1967), 146.

370 *"close-binding and [sexually] intimate"*: Irving Bieber, *Homosexuality: A Psychoanalytic Study* (Lanham, MD: Jason Aronson, 1962), 52.

370 *"a homosexual is a person"*: Jack Drescher, Ariel Shidlo, and Michael Schroeder, *Sexual Conversion Therapy: Ethical, Clinical and Research Perspectives* (Boca Raton, FL: CRC Press, 2002), 33.

371 *"homosexuality is more of a choice"*: "The 1992 campaign: The vice president; Quayle contends homosexuality is a matter of choice, not biology," *New York Times*, September 14, 1992, http://www.nytimes.com/1992/09/14/us/1992-campaign-vice-president -quayle-contends-homosexuality-matter-choice-not.html.

371 *In July 1993, the discovery of the*: David Miller, "Introducing the 'gay gene': Media and scientific representations," *Public Understanding of Science* 4, no. 3 (1995): 269–84, http://www.academia.edu/3172354/Introducing_the_Gay_Gene_Media_and _Scientific_Representations.

371 *"What do we say of the woman"*: C. Sarler, "Moral majority gets its genes all in a twist," *People*, July 1993, 27.

372 *The second book, Richard Lewontin's:* Richard C. Lewontin, Steven P. R. Rose, and Leon J. Kamin, *Not in Our Genes: Biology, Ideology, and Human Nature* (New York: Pantheon Books, 1984).

372 *"There is no acceptable evidence that"*: Ibid., 261.

373 *In the 1980s, a professor of psychology:* J. Michael Bailey and Richard C. Pillard, "A genetic study of male sexual orientation," *Archives of General Psychiatry* 48, no. 12 (1991): 1089–96.

374 *The brothers, who looked virtually identical:* Frederick L. Whitam, Milton Diamond, and James Martin, "Homosexual orientation in twins: A report on 61 pairs and three triplet sets," *Archives of Sexual Behavior* 22, no. 3 (1993): 187–206.

375 *Protocol #92-C-0078 was launched:* Dean Hamer, *Science of Desire: The Gay Gene and the Biology of Behavior* (New York: Simon & Schuster, 2011), 40.

376 *"gay Roots project"*: Ibid., 91–104.

377 *"There were TV cameramen lined up"*: "The 'gay gene' debate," *Frontline*, PBS, http://www.pbs.org/wgbh/pages/frontline/shows/assault/genetics/.

377 *"science could be used to eradicate it"*: Richard Horton, "Is homosexuality inherited?" *Frontline*, PBS, http://www.pbs.org/wgbh/pages/frontline/shows/assault/genetics/nyreview.html.

377 *"does identify a chromosomal region"*: Timothy F. Murphy, *Gay Science: The Ethics of Sexual Orientation Research* (New York: Columbia University Press, 1997), 144.

377 *Hamer was attacked left and right:* M. Philip, "A review of Xq28 and the effect on homosexuality," *Interdisciplinary Journal of Health Science* 1 (2010): 44–48.

378 *Since Hamer's 1993 paper in* Science: Dean H. Hamer et al., "A linkage between DNA markers on the X chromosome and male sexual orientation," *Science* 261, no. 5119 (1993): 321–27.

378 *In 2005, in perhaps the largest study:* Brian S. Mustanski et al., "A genomewide scan of male sexual orientation," *Human Genetics* 116, no. 4 (2005): 272–78.

378 *In 2015, in yet another detailed analysis of 409:* A. R. Sanders et al., "Genome-wide scan demonstrates significant linkage for male sexual orientation," *Psychological Medicine* 45, no. 7 (2015): 1379–88.

378 *One gene that sits:* Elizabeth M. Wilson, "Androgen receptor molecular biology and potential targets in prostate cancer," *Therapeutic Advances in Urology* 2, no. 3 (2010): 105–17.

379 *In 1971, in a book titled:* Macfarlane Burnet, *Genes, Dreams and Realities* (Dordrecht: Springer Science & Business Media, 1971), 170.

379 *"An environmentalist view"*: Nancy L. Segal, *Born Together—Reared Apart: The Landmark Minnesota Twin Study* (Cambridge: Harvard University Press, 2012), 4.

380 *"random access memory onto which"*: Wright, *Born That Way*, viii.

380 *"Whatever back-porch wisdom"*: Ibid., vii.

381 *Minnesota Study of Twins:* Thomas J. Bouchard et al., "Sources of human psychological differences: The Minnesota study of twins reared apart," *Science* 250, no. 4978 (1990): 223–28.

382 *"Empathy, altruism, sense of equity"*: Richard P. Ebstein et al., "Genetics of human social behavior," *Neuron* 65, no. 6 (2010): 831–44.

382 *"A surprisingly high genetic component"*: Wright, *Born That Way*, 52.

382 *Daphne Goodship and Barbara Herbert:* Ibid., 63–67.

383 *"Both drove Chevrolets"*: Ibid., 28.

383 *Two other women, also separated at birth:* Ibid., 74.

383 *oxford shirts with epaulets:* Ibid., 70.

384 *to describe the odd habit: squidging:* Ibid., 65.

384 *"door-knobs, needles and fishhooks":* Ibid., 80.

385 *The most extreme novelty seekers, he discovered:* Richard P. Ebstein et al., "Dopamine D4 receptor (*D4DR*) exon III polymorphism associated with the human personality trait of novelty seeking," *Nature Genetics* 12, no. 1 (1996): 78–80.

386 *Perhaps the subtle drive caused by:* Luke J. Matthews and Paul M. Butler, "Novelty-seeking DRD4 polymorphisms are associated with human migration distance out-of-Africa after controlling for neutral population gene structure," *American Journal of Physical Anthropology* 145, no. 3 (2011): 382–89.

387 *"How nice it would be":* Lewis Carroll, *Alice in Wonderland* (New York: W. W. Norton, 2013).

389 *Forty-three studies, performed:* Eric Turkheimer, "Three laws of behavior genetics and what they mean," *Current Directions in Psychological Science* 9, no. 5 (2000): 160–64; and E. Turkheimer and M. C. Waldron, "Nonshared environment: A theoretical, methodological, and quantitative review," *Psychological Bulletin* 126 (2000): 78–108.

389 *"unsystematic, idiosyncratic, serendipitous events":* Robert Plomin and Denise Daniels, "Why are children in the same family so different from one another?" *Behavioral and Brain Sciences* 10, no. 1 (1987): 1–16.

390 *"a devil, a born devil":* William Shakespeare, *The Tempest*, act 4, scene 1.

The Hunger Winter

391 *Identical twins have exactly the same:* Nessa Carey, *The Epigenetics Revolution: How Modern Biology Is Rewriting Our Understanding of Genetics, Disease, and Inheritance* (New York: Columbia University Press, 2012), 5.

391 *Genes have had a glorious run in the 20th century:* Evelyn Fox Keller, quoted in Margaret Lock and Vinh-Kim Nguyen, *An Anthropology of Biomedicine* (Hoboken, NJ: John Wiley & Sons, 2010).

392 *When a songbird encounters a new:* Erich D. Jarvis et al., "For whom the bird sings: Context-dependent gene expression," *Neuron* 21, no. 4 (1998): 775–88.

392 *In the 1950s, Conrad Waddington:* Conrad Hal Waddington, *The Strategy of the Genes: A Discussion of Some Aspects of Theoretical Biology* (London: Allen & Unwin, 1957), ix, 262.

393 *"only [consists of] a stomach":* Max Hastings, *Armageddon: The Battle for Germany, 1944–1945* (New York: Alfred A. Knopf, 2004), 414.

394 *In the 1980s, however:* Bastiaan T. Heijmans et al., "Persistent epigenetic differences associated with prenatal exposure to famine in humans," *Proceedings of the National Academy of Sciences* 105, no. 44 (2008): 17046–49.

396 *"aptitude for doing things on a small scale":* John Gurdon, "Nuclear reprogramming in eggs," *Nature Medicine* 15, no. 10 (2009): 1141–44.

397 *In 1961, Gurdon began to test:* J. B. Gurdon and H. R. Woodland, "The cytoplasmic control of nuclear activity in animal development," *Biological Reviews* 43, no. 2 (1968): 233–67.

397 *It would lead, famously, to the cloning of Dolly:* "Sir John B. Gurdon—facts," Nobelprize.org, http://www.nobelprize.org/nobel_prizes/medicine/laureates/2012/gurdon-facts.html.

398 *the only other "observed case"*: John Maynard Smith, interview in the *Web of Stories*. www.webofstories.com/play/john.maynard.smith/78.

399 *Lyon found: in one cell:* The Japanese scientist Susumu Ohno had hypothesized about X inactivation before the phenomenon was discovered.

401 *simple organisms, such as yeast:* K. Raghunathan et al., "Epigenetic inheritance uncoupled from sequence-specific recruitment," *Science* 348 (April 3, 2015): 6230.

403 *In his remarkable story "Funes the Memorious"*: Jorge Luis Borges, *Labyrinths*, trans. James E. Irby (New York: New Directions, 1962), 59–66.

405 *One of the four genes used by Yamanaka:* K. Takahashi and S. Yamanaka, "Induction of pluripotent stem cells from mouse embryonic and adult fibroblast cultures by defined factors," *Cell* 126, no. 4 (2006): 663–76. Also see M. Nakagawa et al., "Generation of induced pluripotent stem cells without *Myc* from mouse and human fibroblasts," *Nature Biotechnology* 26, no. 1 (2008): 101–6.

409 *"It sometimes seems as if curbing entropy"*: James Gleick, *The Information: A History, a Theory, a Flood* (New York: Pantheon Books, 2011).

411 *At Harvard, a soft-spoken biochemist:* Itay Budin and Jack W. Szostak, "Expanding roles for diverse physical phenomena during the origin of life," *Annual Review of Biophysics* 39 (2010): 245–63; and Alonso Ricardo and Jack W. Szostak, "Origin of life on Earth," *Scientific American* 301, no. 3 (2009): 54–61.

411 *followed the work of Stanley Miller:* The original experiments were performed by Miller in conjunction with Harold Urey at the University of Chicago; John Sutherland, in Manchester, also performed key experiments.

411 *Subsequent variations of the Miller experiment:* Ricardo and Szostak, "Origin of life on Earth," 54–61.

411 *Szostak has demonstrated that such micelles:* Jack W. Szostak, David P. Bartel, and P. Luigi Luisi, "Synthesizing life," *Nature* 409, no. 6818 (2001): 387–90. Also see Martin M. Hanczyc, Shelly M. Fujikawa, and Jack W. Szostak, "Experimental models of primitive cellular compartments: Encapsulation, growth, and division," *Science* 302, no. 5645 (2003): 618–22.

412 *"It is relatively easy to see how"*: Ricardo and Szostak, "Origin of life on Earth," 54–61.

PART SIX: POST-GENOME

416 *Those who promise us paradise on earth:* Elias G. Carayannis and Ali Pirzadeh, *The Knowledge of Culture and the Culture of Knowledge: Implications for Theory, Policy and Practice* (London: Palgrave Macmillan, 2013), 90.

416 *It's only we humans:* Tom Stoppard, *The Coast of Utopia* (New York: Grove Press, 2007), "Act Two, August 1852."

The Future of the Future

417 *Probably no DNA science is at once:* Gina Smith, *The Genomics Age: How DNA Technology Is Transforming the Way We Live and Who We Are* (New York: AMACOM, 2004).

417 *Clear the air!:* Thomas Stearns Eliot, *Murder in the Cathedral* (Boston: Houghton Mifflin Harcourt, 2014).

418 *In 1974, barely three years after:* Rudolf Jaenisch and Beatrice Mintz, "Simian virus

40 DNA sequences in DNA of healthy adult mice derived from preimplantation blastocysts injected with viral DNA," *Proceedings of the National Academy of Sciences* 71, no. 4 (1974): 1250–54.

418 *biologists stumbled on a critical discovery:* M. J. Evans and M. H. Kaufman, "Establishment in culture of pluripotential cells from mouse embryos," *Nature* 292 (1981): 154–56.

419 *"Nobody seems to be interested in my cells":* M. Capecchi, "The first transgenic mice: An interview with Mario Capecchi. Interview by Kristin Kain," *Disease Models & Mechanisms* 1, no. 4–5 (2008): 197.

420 *With ES cells, however, scientists:* See for instance M. R. Capecchi, "High efficiency transformation by direct microinjection of DNA into cultured mammalian cells," *Cell* 22 (1980): 479–88; and K. R. Thomas and M. R. Capecchi, "Site-directed mutagenesis by gene targeting in mouse embryo–derived stem cells," *Cell* 51 (1987): 503–12.

420 *You could choose to change the insulin gene:* O. Smithies et al., "Insertion of DNA sequences into the human chromosomal-globin locus by homologous re-combination," *Nature* 317 (1985): 230–34.

421 *The "watchmaker" of evolution, as Richard Dawkins:* Richard Dawkins, *The Blind Watchmaker: Why the Evidence of Evolution Reveals a Universe without Design* (W. W. Norton, 1986).

421 *They are the savants of the rodent world:* Kiyohito Murai et al., "Nuclear receptor TLX stimulates hippocampal neurogenesis and enhances learning and memory in a transgenic mouse model," *Proceedings of the National Academy of Sciences* 111, no. 25 (2014): 9115–20.

422 *"It may be the field's dirty little secret":* Karen Hopkin, "Ready, reset, go," *The Scientist*, March 11, 2011, http://www.the-scientist.com/?articles.view/articleno/29550/title /ready—reset—go/.

422 *In 1988, a two-year-old girl:* Details of the story of Ashanti DeSilva are from W. French Anderson, "The best of times, the worst of times," *Science* 288, no. 5466 (2000): 627; Lyon and Gorner, *Altered Fates*; and Nelson A. Wivel and W. French Anderson, "24: Human gene therapy: Public policy and regulatory issues," *Cold Spring Harbor Monograph Archive* 36 (1999): 671–89.

423 *"Mommy, you shouldn't have had":* Lyon and Gorner, *Altered Fates*, 107.

423 *The Bubble Boy, as David was called:* "David Phillip Vetter (1971–1984)," *American Experience*, PBS, http://www.pbs.org/wgbh/amex/bubble/peopleevents/p_vetter.html.

423 *Richard Mulligan, a virologist and geneticist:* Luigi Naldini et al., "In vivo gene delivery and stable transduction of nondividing cells by a lentiviral vector," *Science* 272, no. 5259 (1996): 263–67.

424 *led by William French Anderson and Michael Blaese:* "Hope for gene therapy," *Scientific American Frontiers*, PBS, http://www.pbs.org/saf/1202/features/genetherapy.htm.

424 *In the early 1980s, Anderson and Blaese:* W. French Anderson et al., "Gene transfer and expression in nonhuman primates using retroviral vectors," *Cold Spring Harbor Symposia on Quantitative Biology* 51 (1986): 1073–81.

424 *"Nobody knows what may happen":* Lyon and Gorner, *Altered Fates*, 124.

425 *Perhaps predictably, the RAC rejected the protocol outright:* Lisa Yount, *Modern Genetics: Engineering Life* (New York: Infobase Publishing, 2006), 70.

427 *"A cosmic moment has come and gone":* Lyon and Gorner, *Altered Fates*, 239.

427 *"Jesus Christ himself could walk by":* Ibid., 240.

428 *"It's not a big improvement":* Ibid., 268.

429 *At four, he had joyfully eaten:* Barbara Sibbald, "Death but one unintended conse-

quence of gene-therapy trial," *Canadian Medical Association Journal* 164, no. 11 (2001): 1612.

429 *In 1993, when Gelsinger was:* For details of the Jesse Gelsinger story see Evelyn B. Kelly, *Gene Therapy* (Westport, CT: Greenwood Press, 2007); Lyon and Gorner, *Altered Fates*; and Sally Lehrman, "Virus treatment questioned after gene therapy death," *Nature* 401, no. 6753 (1999): 517–18.

432 *By noon, the procedure was done:* James M. Wilson, "Lessons learned from the gene therapy trial for ornithine transcarbamylase deficiency," *Molecular Genetics and Metabolism* 96, no. 4 (2009): 151–57.

433 *"How could such a beautiful thing":* Paul Gelsinger, author interview, November 2014 and April 2015.

434 *That Wilson had a financial stake in:* Robin Fretwell Wilson, "Death of Jesse Gelsinger: New evidence of the influence of money and prestige in human research," *American Journal of Law and Medicine* 36 (2010): 295.

435 *In January 2000, when the FDA inspected:* Sibbald, "Death but one unintended consequence," 1612.

435 *"The entire field of gene therapy":* Carl Zimmer, "Gene therapy emerges from disgrace to be the next big thing, again," *Wired*, August 13, 2013.

435 *"Gene therapy is not yet therapy":* Sheryl Gay Stolberg, "The biotech death of Jesse Gelsinger," *New York Times*, November 27, 1999, http://www.nytimes.com/1999/11/28 /magazine/the-biotech-death-of-jesse-gelsinger.html.

436 *"cautionary tale of scientific overreach":* Zimmer, "Gene therapy emerges."

Genetic Diagnosis: "Previvors"

437 *All that man is:* W. B. Yeats, *The Collected Poems of W. B. Yeats*, ed. Richard Finneran (New York: Simon & Schuster, 1996), "Byzantium," 248.

437 *The anti-determinists want to say:* Jim Kozubek, "The birth of 'transhumans,'" *Providence (RI) Journal*, September 29, 2013.

438 *"Genetic tests," as Eric Topol:* Eric Topol, author interview, 2013.

438 *Between 1978 and 1988, King added:* Mary-Claire King, "Using pedigrees in the hunt for BRCA1," DNA Learning Center, https://www.dnalc.org/view/15126-Using-pedigress -in-the-hunt-for-BRCA1-Mary-Claire-King.html.

439 *she had pinpointed it to a region:* Jeff M. Hall et al., "Linkage of early-onset familial breast cancer to chromosome 17q21," *Science* 250, no. 4988 (1990): 1684–89.

439 *"Being comfortable with uncertainty":* Jane Gitschier, "Evidence is evidence: An interview with Mary-Claire King," *PLOS*, September 26, 2013.

439 *In 1998, Myriad was granted:* E. Richard Gold and Julia Carbone, "Myriad Genetics: In the eye of the policy storm," *Genetics in Medicine* 12 (2010): S39–S70.

440 *"Some of these women [with BRCA1 mutations]":* Masha Gessen, *Blood Matters: From BRCA1 to Designer Babies, How the World and I Found Ourselves in the Future of the Gene* (Boston: Houghton Mifflin Harcourt, 2009), 8.

441 *In 1908, the Swiss German psychiatrist:* Eugen Bleuler and Carl Gustav Jung, "Komplexe und Krankheitsursachen bei Dementia praecox," *Zentralblatt für Nervenheilkunde und Psychiatrie* 31 (1908): 220–27.

442 *In the 1970s, studies demonstrated:* Susan Folstein and Michael Rutte, "Infantile autism: A genetic study of 21 twin pairs," *Journal of Child Psychology and Psychiatry* 18, no. 4 (1977): 297–321.

442 *"domineering, nagging and hostile mother"*: Silvano Arieti and Eugene B. Brody, *Adult Clinical Psychiatry* (New York: Basic Books, 1974), 553.

443 *National Book Award for science:* "1975: Interpretation of Schizophrenia by Silvano Arieti," National Book Award Winners: 1950–2014, National Book Foundation, http://www.nationalbook.org/nbawinners_category.html#.vcnit7fxhom.

443 *In 2013, an enormous study identified:* Menachem Fromer et al., "De novo mutations in schizophrenia implicate synaptic networks," *Nature* 506, no. 7487 (2014): 179–84.

445 *108 genes (or rather genetic regions):* Schizophrenia Working Group of the Psychiatric Genomics, *Nature* 511 (2014): 421–27.

445 *The strongest, and most:* "Schizophrenia risk from complex variation of complement component 4," Sekar et al. *Nature* 530, 177–183.

446 *"There are lots of"*: Benjamin Neale, quoted in Simon Makin, "Massive study reveals schizophrenia's genetic roots: The largest-ever genetic study of mental illness reveals a complex set of factors," *Scientific American*, November 1, 2014.

448 *"We of the craft are all crazy"*: *Carey's Library of Choice Literature*, vol. 2 (Philadelphia: E. L. Carey & A. Hart, 1836), 458.

448 *In* Touched with Fire, *an authoritative:* Kay Redfield Jamison, *Touched with Fire* (New York: Simon & Schuster, 1996).

449 *Hans Asperger, the psychologist who first:* Tony Attwood, *The Complete Guide to Asperger's Syndrome* (London: Jessica Kingsley, 2006).

450 *As Edvard Munch put it:* Adrienne Sussman, "Mental illness and creativity: A neurological view of the 'tortured artist,'" *Stanford Journal of Neuroscience* 1, no. 1 (2007): 21–24.

450 *illness as the "night-side of life"*: Susan Sontag, *Illness as Metaphor and AIDS and Its Metaphors* (New York: Macmillan, 2001).

450 *Entitled "The Future of Genomic Medicine"*: Details of the conference can be found in "The future of genomic medicine VI," Scripps Translational Science Institute, http://www.slideshare.net/mdconferencefinder/the-future-of-genomic-medicine -vi-23895019; Eryne Brown, "Gene mutation didn't slow down high school senior," *Los Angeles Times*, July 5, 2015, http://www.latimes.com/local/california/la-me-lilly -grossman-update-20150702-story.html; and Konrad J. Karczewski, "The future of genomic medicine is here," *Genome Biology* 14, no. 3 (2013): 304.

451 *Alexis and Noah Beery:* "Genome maps solve medical mystery for California twins," National Public Radio broadcast, June 16, 2011.

451 *Based on that genetic diagnosis:* Matthew N. Bainbridge et al., "Whole-genome sequencing for optimized patient management," *Science Translational Medicine* 3, no. 87 (2011): 87re3.

454 *That a mutation in the gene MECP2:* Antonio M. Persico and Valerio Napolioni, "Autism genetics," *Behavioural Brain Research* 251 (2013): 95–112; and Guillaume Huguet, Elodie Ey, and Thomas Bourgeron, "The genetic landscapes of autism spectrum disorders," *Annual Review of Genomics and Human Genetics* 14 (2013): 191–213.

454 *the eventual effects of these gene-environment:* Albert H. C. Wong, Irving I. Gottesman, and Arturas Petronis, "Phenotypic differences in genetically identical organisms: The epigenetic perspective," *Human Molecular Genetics* 14, suppl. 1 (2005): R11–R18. Also see Nicholas J. Roberts et al., "The predictive capacity of personal genome sequencing," *Science Translational Medicine* 4, no. 133 (2012): 133ra58.

455 *an article in* Nature *magazine announced:* Alan H. Handyside et al., "Pregnancies from biopsied human preimplantation embryos sexed by Y-specific DNA amplification," *Nature* 344, no. 6268 (1990): 768–70.

457 *As the political theorist Desmond King puts it:* D. King, "The state of eugenics," *New Statesman & Society* 25 (1995): 25–26.

459 *Take, for instance, a series of startlingly provocative:* K. P. Lesch et al., "Association of anxiety-related traits with a polymorphism in the serotonergic transporter gene regulatory region," *Science* 274 (1996): 1527–31.

459 *the short allele has been associated with:* Douglas F. Levinson, "The genetics of depression: A review," *Biological Psychiatry* 60, no. 2 (2006): 84–92.

459 *In 2010, a team of researchers launched:* "Strong African American Families Program," Blueprints for Healthy Youth Development, http://www.blueprintsprograms.com/evaluationAbstracts.php?pid=f76b2ea6b45eff3bc8e4399145cc17a0601f5c8d.

459 *Six hundred African-American families with early-adolescent:* Gene H. Brody et al., "Prevention effects moderate the association of 5-HTTLPR and youth risk behavior initiation: Gene × environment hypotheses tested via a randomized prevention design," *Child Development* 80, no. 3 (2009): 645–61; and Gene H. Brody, Yi-fu Chen, and Steven R. H. Beach, "Differential susceptibility to prevention: GABAergic, dopaminergic, and multilocus effects," *Journal of Child Psychology and Psychiatry* 54, no. 8 (2013): 863–71.

460 *Writing in the* New York Times *in 2014:* Jay Belsky, "The downside of resilience," *New York Times,* November 28, 2014.

462 *"a technology of abnormal individuals":* Michel Foucault, *Abnormal: Lectures at the Collège de France, 1974–1975,* vol. 2 (New York: Macmillan, 2007).

Genetic Therapies: Post-Human

463 *There is in biology at the moment:* "Biology's Big Bang," *Economist,* June 14, 2007.

463 *a journalist visited James Watson at:* Lyon and Gorner, *Altered Fates,* 537.

465 *Jesse Gelsinger's "biotech death":* Stolberg, "Biotech death of Jesse Gelsinger," 136–40.

466 *In 2014, a landmark study:* Amit C. Nathwani et al., "Long-term safety and efficacy of factor IX gene therapy in hemophilia B," *New England Journal of Medicine* 371, no. 21 (2014): 1994–2004.

469 *In 1998, soon after Thomson's paper:* James A. Thomson et al., "Embryonic stem cell lines derived from human blastocysts," *Science* 282, no. 5391 (1998): 1145–47.

469 *President George W. Bush sharply restricted:* Dorothy C. Wertz, "Embryo and stem cell research in the United States: History and politics," *Gene Therapy* 9, no. 11 (2002): 674–78.

472 *Doudna and Charpentier published their data:* Martin Jinek et al., "A programmable dual-RNA-guided DNA endonuclease in adaptive bacterial immunity," *Science* 337, no. 6096 (2012): 816–21.

472 *this technique has exploded:* Key contributors to the use of CRISPR/Cas9 in human cells include Feng Zhang (MIT) and George Church (Harvard). See, for instance, L. Cong et al., "Multiplex genome engineering using CRISPR/Cas systems," *Science* 339, no. 6121 (2013): 819–23; and F. A. Ran, "Genome engineering using the CRISPR-Cas9 system," *Nature Protocols* 11 (2013): 2281–308. Also see P. Mali et al., "RNA-Guided Human Genome Engineering via Cas9," *Science* 339, no. 6121 (2013): 823–26.

474 *In the winter of 2014, a team:* Walfred W. C. Tang et al., "A unique gene regulatory network resets the human germline epigenome for development," *Cell* 161, no. 6 (2015): 1453–67; and "In a first, Weizmann Institute and Cambridge University

scientists create human primordial germ cells," Weizmann Institute of Science, December 24, 2014, http://www.newswise.com/articles/in-a-first-weizmann-institute -and-cambridge-university-scientists-create-human-primordial-germ-cells.

476 *Jennifer Doudna and David Baltimore:* B. D. Baltimore et al., "A prudent path forward for genomic engineering and germline gene modification," *Science* 348, no. 6230 (2015): 36–38; and Cormac Sheridan, "CRISPR germline editing reverberates through biotech industry," *Nature Biotechnology* 33, no. 5 (2015): 431–32.

477 *"It is very clear that people will try":* Nicholas Wade, "Scientists seek ban on method of editing the human genome," *New York Times*, March 19, 2015.

477 *"This reality means":* Francis Collins, Letter to the author, October 2015.

478 *In the spring of 2015, a laboratory:* David Cyranoski and Sara Reardon, "Chinese scientists genetically modify human embryos," *Nature* (April 22, 2015).

478 *The highest-ranking scientific journals:* Chris Gyngell and Julian Savulescu, "The moral imperative to research editing embryos: The need to modify nature and science," Oxford University, April 23, 2015, Blog.Practicalethics.Ox.Ac.Uk/2015/04/the-Moral -Imperative-to-Research-Editing-Embryos-the-Need-to-Modify-Nature-and-Science/.

478 *the results were eventually published in:* Puping Liang et al., "CRISPR/Cas9-mediated gene editing in human tripronuclear zygotes," *Protein & Cell* 6, no. 5 (2015): 1–10.

478 *"planning to decrease the number of off-target":* Cyranoski and Reardon, "Chinese scientists genetically modify human embryos."

479 *"I don't think China wants":* Didi Kristen Tatlow, "A scientific ethical divide between China and West," *New York Times*, June 29, 2015.

Epilogue: *Bheda, Abheda*

485 *"No sane biologist believes":* Paul Berg, author interview, 1993.

487 *"very few human genes":* David Botstein, letter to the author, October 2015.

487 *In an influential review published in 2011:* Eric Turkheimer, "Still missing," *Research in Human Development* 8, nos. 3–4 (2011): 227–41.

488 *"Perhaps," as one observer complained:* Peter Conrad, "A mirage of genes," *Sociology of Health & Illness* 21, no. 2 (1999): 228–41.

491 *"Imagine you are a soldier returning from war":* Richard A. Friedman, "The feel-good gene," *New York Times*, March 6, 2015.

494 *"[Nature] may, after all, be entirely approachable":* Morgan, *Physical Basis of Heredity*, 15.

Acknowledgments

497 *"distorted version of our normal selves":* H.Varmus, Nobel lecture, 1989. http://www .nobelprize.org/nobel_prizes/medicine/laureates/1989/varmus-lecture.html. For the paper describing the existence of endogenous proto-oncogenes in cells see D. Stehelin et al., "DNA related to the transforming genes of avian sarcoma viruses is present in normal DNA," *Nature* 260, no. 5547 (1976). 170–73. Also see Harold Varmus to Dominique Stehelin, February 3, 1976, Harold Varmus Papers, National Library of Medicine Archives.

Selected Bibliography

Arendt, Hannah. *Eichmann in Jerusalem: A Report on the Banality of Evil.* New York: Viking, 1963.

Aristotle. *Generation of Animals.* Leiden: Brill Archive, 1943.

Aristotle, and D. M. Balme, ed. *History of Animals.* Cambridge: Harvard University Press, 1991.

Aristotle, and Jonathan Barnes, ed. *The Complete Works of Aristotle.* Revised Oxford Translation. Princeton, NJ: Princeton University Press, 1984.

Berg, Paul, and Maxine Singer. *Dealing with Genes: The Language of Heredity.* Mill Valley, CA: University Science Books, 1992.

———. *George Beadle, An Uncommon Farmer: The Emergence of Genetics in the 20th Century.* Cold Spring Harbor, NY: Cold Spring Harbor Laboratory Press, 2003.

Bliss, Catherine. *Race Decoded: The Genomic Fight for Social Justice.* Palo Alto, CA: Stanford University Press, 2012.

Browne, E. J. *Charles Darwin: A Biography.* New York: Alfred A. Knopf, 1995.

Cairns, John, Gunther Siegmund Stent, and James D. Watson, eds. *Phage and the Origins of Molecular Biology.* Cold Spring Harbor, NY: Cold Spring Harbor Laboratory Press, 1968.

Carey, Nessa. *The Epigenetics Revolution: How Modern Biology Is Rewriting Our Understanding of Genetics, Disease, and Inheritance.* New York: Columbia University Press, 2012.

Chesterton, G. K. *Eugenics and Other Evils.* London: Cassell, 1922.

Cobb, Matthew. *Generation: The Seventeenth-Century Scientists Who Unraveled the Secrets of Sex, Life, and Growth.* New York: Bloomsbury Publishing, 2006.

Cook-Deegan, Robert M. *The Gene Wars: Science, Politics, and the Human Genome.* New York: W. W. Norton, 1994.

Crick, Francis. *What Mad Pursuit: A Personal View of Scientific Discovery.* New York: Basic Books, 1988.

Crotty, Shane. *Ahead of the Curve: David Baltimore's Life in Science.* Berkeley: University of California Press, 2001.

Darwin, Charles. *On the Origin of Species by Means of Natural Selection.* London: Murray, 1859.

Darwin, Charles, and Francis Darwin, ed. *The Autobiography of Charles Darwin.* Amherst, NY: Prometheus Books, 2000.

Dawkins, Richard. *The Blind Watchmaker: Why the Evidence of Evolution Reveals a Universe without Design*. New York: W. W. Norton, 1986.

———. *The Selfish Gene*. Oxford: Oxford University Press, 1989.

Desmond, Adrian, and James Moore. *Darwin*. New York: Warner Books, 1991.

De Vries, Hugo. *The Mutation Theory*. Vol. 1. Chicago: Open Court, 1909.

Dobzhansky, Theodosius. *Genetics and the Origin of Species*. New York: Columbia University Press, 1937.

———. *Heredity and the Nature of Man*. New York: New American Library, 1966.

Edelson, Edward. *Gregor Mendel, and the Roots of Genetics*. New York: Oxford University Press, 1999.

Feinstein, Adam. *A History of Autism: Conversations with the Pioneers*. West Sussex: Wiley-Blackwell, 2010.

Flynn, James. *Intelligence and Human Progress: The Story of What Was Hidden in Our Genes*. Oxford: Elsevier, 2013.

Fox Keller, Evelyn. *The Century of the Gene*. Cambridge: Harvard University Press, 2009.

Fredrickson, Donald S. *The Recombinant DNA Controversy: A Memoir: Science, Politics, and the Public Interest 1974–1981*. Washington, DC: American Society for Microbiology Press, 2001.

Friedberg, Errol C. *A Biography of Paul Berg: The Recombinant DNA Controversy Revisited*. Singapore: World Scientific Publishing, 2014.

Gardner, Howard E. *Frames of Mind: The Theory of Multiple Intelligences*. New York: Basic Books, 2011.

———. *Intelligence Reframed: Multiple Intelligences for the 21st Century*. New York: Perseus Books Group, 2000.

Glimm, Adele. *Gene Hunter: The Story of Neuropsychologist Nancy Wexler*. New York: Franklin Watts, 2005.

Hamer, Dean. *Science of Desire: The Gay Gene and the Biology of Behavior*. New York: Simon & Schuster, 2011

Happe, Kelly E. *The Material Gene: Gender, Race, and Heredity after the Human Genome Project*. New York: NYU Press, 2013.

Harper, Peter S. *A Short History of Medical Genetics*. Oxford: Oxford University Press, 2008.

Hausmann, Rudolf. *To Grasp the Essence of Life: A History of Molecular Biology*. Berlin: Springer Science & Business Media, 2013.

Henig, Robin Marantz. *The Monk in the Garden: The Lost and Found Genius of Gregor Mendel, the Father of Genetics*. Boston: Houghton Mifflin, 2000.

Herring, Mark Youngblood. *Genetic Engineering*. Westport, CT: Greenwood, 2006.

Herrnstein, Richard, and Charles Murray. *The Bell Curve*. New York: Simon & Schuster, 1994.

Herschel, John F. W. *A Preliminary Discourse on the Study of Natural Philosophy. A Facsim. of the 1830 Ed*. New York: Johnson Reprint, 1966.

Hodge, Russ. *The Future of Genetics: Beyond the Human Genome Project*. New York: Facts on File, 2010.

Hughes, Sally Smith. *Genentech: The Beginnings of Biotech*. Chicago: University of Chicago Press, 2011.

Jamison, Kay Redfield. *Touched with Fire*. New York: Simon & Schuster, 1996.

Judson, Horace Freeland. *The Eighth Day of Creation*. New York: Simon & Schuster, 1979.

———. *The Search for Solutions*. New York: Holt, Rinehart, and Winston, 1980.

Kevles, Daniel J. *In the Name of Eugenics: Genetics and the Uses of Human Heredity*. New York: Alfred A. Knopf, 1985.

Kornberg, Arthur. *For the Love of Enzymes: The Odyssey of a Biochemist*. Cambridge: Harvard University Press, 1991.

———. *The Golden Helix: Inside Biotech Ventures*. Sausalito, CA: University Science Books, 2002.

Kornberg, Arthur, Adam Alaniz, and Roberto Kolter. *Germ Stories*. Sausalito, CA: University Science Books, 2007.

Kornberg, Arthur, and Tania A. Baker. *DNA Replication*. San Francisco: W. H. Freeman, 1980.

Krimsky, Sheldon. *Genetic Alchemy: The Social History of the Recombinant DNA Controversy*. Cambridge: MIT Press, 1982.

———. *Race and the Genetic Revolution: Science, Myth, and Culture*. New York: Columbia University Press, 2011.

Kush, Joseph C., ed. *Intelligence Quotient: Testing, Role of Genetics and the Environment and Social Outcomes*. New York: Nova Science, 2013.

Larson, Edward John. *Evolution: The Remarkable History of a Scientific Theory*. Vol. 17. New York: Random House Digital, 2004.

Lombardo, Paul A. *Three Generations, No Imbeciles: Eugenics, the Supreme Court, and Buck v. Bell*. Baltimore: Johns Hopkins University Press, 2008.

Lyell, Charles. *Principles of Geology: Or, The Modern Changes of the Earth and Its Inhabitants Considered as Illustrative of Geology*. New York: D. Appleton & Company, 1872.

Lyon, Jeff, and Peter Gorner. *Altered Fates: Gene Therapy and the Retooling of Human Life*. New York: W. W. Norton, 1996.

Maddox, Brenda. *Rosalind Franklin: The Dark Lady of DNA*. UK: HarperCollins, 2002.

McCabe, Linda L., and Edward R. B. McCabe. *DNA: Promise and Peril*. Berkeley: University of California Press, 2008.

McElheny, Victor K. *Drawing the Map of Life: Inside the Human Genome Project*. New York: Basic Books, 2012.

———. *Watson and DNA: Making a Scientific Revolution*. Cambridge: Perseus, 2003.

Mendel, Gregor, Alain F. Corcos, and Floyd V. Monaghan, eds. *Gregor Mendel's Experiments on Plant Hybrids: A Guided Study*. New Brunswick, NJ: Rutgers University Press, 1993.

Morange, Michel. *A History of Molecular Biology*. Trans. Matthew Cobb. Cambridge: Harvard University Press, 1998.

Morgan, Thomas Hunt. *The Mechanism of Mendelian Heredity*. New York: Holt, 1915.

———. *The Physical Basis of Heredity*. Philadelphia: J. B. Lippincott, 1919.

Müller-Wille, Staffan, and Hans-Jörg Rheinberger. *A Cultural History of Heredity*. Chicago: University of Chicago Press, 2012.

Olby, Robert C. *The Path to the Double Helix: The Discovery of DNA*. New York: Dover Publications, 1994.

Paley, William. *The Works of William Paley*. Philadelphia: J. J. Woodward, 1836.

Patterson, Paul H. *The Origins of Schizophrenia*. New York: Columbia University Press, 2013.

Portugal, Franklin H., and Jack S. Cohen. *A Century of DNA: A History of the Discovery of the Structure and Function of the Genetic Substance*. Cambridge: MIT Press, 1977.

Posner, Gerald L., and John Ware. *Mengele: The Complete Story*. New York: McGraw-Hill, 1986.

Ridley, Matt. *Genome: The Autobiography of a Species in 23 Chapters*. New York: HarperCollins, 1999.

Sambrook, Joseph, Edward F. Fritsch, and Tom Maniatis. *Molecular Cloning*. Vol. 2. Cold Spring Harbor, NY: Cold Spring Harbor Laboratory Press, 1989.

Sayre, Anne. *Rosalind Franklin and DNA*. New York: W. W. Norton, 2000.

Schrödinger, Erwin. *What Is Life?: The Physical Aspect of the Living Cell*. Cambridge: Cambridge University Press, 1945.

Schwartz, James. *In Pursuit of the Gene: From Darwin to DNA*. Cambridge: Harvard University Press, 2008.

Seedhouse, Erik. *Beyond Human: Engineering Our Future Evolution*. New York: Springer, 2014.

Shapshay, Sandra. *Bioethics at the Movies*. Baltimore: Johns Hopkins University Press, 2009.

Shreeve, James. *The Genome War: How Craig Venter Tried to Capture the Code of Life and Save the World*. New York: Alfred A. Knopf, 2004.

Singer, Maxine, and Paul Berg. *Genes & Genomes: a Changing Perspective*. Sausalito, CA: University Science Books, 1991.

Stacey, Jackie. *The Cinematic Life of the Gene*. Durham, NC: Duke University Press, 2010.

Sturtevant, A. H. *A History of Genetics*. New York: Harper & Row, 1965.

Sulston, John, and Georgina Ferry. *The Common Thread: A Story of Science, Politics, Ethics, and the Human Genome*. Washington, DC: Joseph Henry Press, 2002.

Thurstone, Louis L. *Learning Curve Equation*. Princeton, NJ: Psychological Review Company, 1919.

———. *Multiple-Factor Analysis: A Development & Expansion of the Vectors of Mind*. Chicago: University of Chicago Press, 1947.

———. *The Nature of Intelligence*. London: Routledge, Trench, Trubner, 1924.

Venter, J. Craig. *A Life Decoded: My Genome, My Life*. New York: Viking, 2007.

Wade, Nicholas. *Before the Dawn: Recovering the Lost History of Our Ancestors*. New York: Penguin, 2006.

Wailoo, Keith, Alondra Nelson, and Catherine Lee, eds. *Genetics and the Unsettled Past: The Collision of DNA, Race, and History*. New Brunswick, NJ: Rutgers University Press, 2012.

Watson, James D. *The Double Helix: A Personal Account of the Discovery of the Structure of DNA*. London: Weidenfeld & Nicolson, 1981.

———. *Recombinant DNA: Genes and Genomes: A Short Course*. New York: W. H. Freeman, 2007.

Watson, James D., and John Tooze. *The DNA Story: A Documentary History of Gene Cloning*. San Francisco: W. H. Freeman, 1981.

Wells, Herbert G. *Mankind in the Making*. Leipzig: Tauchnitz, 1903.

Wells, Spencer, and Mark Read. *The Journey of Man: A Genetic Odyssey*. Princeton, NJ: Princeton University Press, 2002.

Wexler, Alice. *Mapping Fate: A Memoir of Family, Risk, and Genetic Research*. Berkeley: University of California Press, 1995.

Wilkins, Maurice. *Maurice Wilkins: The Third Man of the Double Helix: An Autobiography*. Oxford: Oxford University Press, 2003.

Wright, William. *Born That Way: Genes, Behavior, Personality*. London: Routledge, 2013.

Yi, Doogab. *The Recombinant University: Genetic Engineering and the Emergence of Stanford Biotechnology*. Chicago: University of Chicago Press, 2015.

Index

Photo Credits

Picture research by Alexandra Truitt & Jerry Marshall, www.pictureresearching.com.

Page 1: *homunculus:* © Science Source; *trees of lineage:* © HIP/Art Resource, NY; *Charles Darwin and his "tree of life":* © Huntington Library/SuperStock.com

Page 2: *Gregor Mendel:* © James King-Holmes/Science Source; *William Bateson and Wilhelm Johannsen:* © 2013 The American Philosophical Society; *Francis Galton:* © Paul D. Stewart/Science Source

Page 3: *twin studies:* Archives of the Max Planck Society, Berlin; *family history charts:* © ullstein bild/The Image Works; *Better Babies contests:* Library of Congress Prints & Photographs Division; *"eugenics tree":* © 2013 The American Philosophical Society

Page 4: *Carrie and Emma Buck:* Arthur Estabrook Papers. M. E. Grenander Department of Special Collections and Archives. University at Albany Libraries; *Morgan in his Caltech Fly Room:* Courtesy of the Archives, California Institute of Technology; *Rosalind Franklin looks down a microscope:* Museum of London/The Art Archive at Art Resource, NY; *Franklin's photograph of a DNA crystal:* King's College London Archives

Page 5: *James Watson and Francis Crick:* © A. Barrington Brown/Science Source; *Victor McCusick:* Alan Mason Chesney Medical Archives. The John Hopkins Medical Institutions. Victor Almon McKusick Collection. Reproduced with permission of Betty Malashuk; *Nancy Wexler:* Photo by Acey Harper/The LIFE Images Collection/Getty Images

Page 6: *"perfect race":* Courtesy of the National Institutes of Health; *Herb Boyer and Robert Swanson:* Genentech archives; *Asilomar meeting:* Courtesy of the National Library of Medicine; *Frederick Sanger:* Courtesy of MRC Laboratory of Molecular Biology

Page 7: *Jesse Gelsinger:* © Mickie Gelsinger via MBR/KRT/Newscom; Science *magazine cover:* © Photography by Ann Elliott Cutting. From *Science,* 16 Feb 2001. Vol. 291 No. 5507. Reprinted with permission from AAAS; *Craig Venter, President Bill Clinton, and Francis Collins:* © AP Photo/Ron Edmonds

Page 8: *babies:* © Stringer/Reuters/Corbis; *gene-sequencing machines:* © David Parker/ Science Source; *Jennifer Doudna and coworker:* UC-Berkeley Public Affairs

penguin.co.uk/vintage